Structural Biology of Membrane Proteins

RSC Biomolecular Sciences

Editorial Board

This Series is devoted to coverage of the interface between the chemical and biological sciences, especially structural biology, chemical biology, bio- and chemo-informatics, drug discovery and development, chemical enzymology and biophysical chemistry.

Ideal as reference and state-of-the-art guides at the graduate and post-graduate level.

Visit our website on www.rsc.org/biomolecularsciences

For further information please contact:
Sales and Customer Services
Royal Society of Chemistry
Thomas Graham House
Science Park, Milton Road
Cambridge CB4 0WF, UK
Telephone +44 (0)1223 432360, Fax +44 (0)1223 426017, Email sales@rsc.org

Structural Biology of Membrane Proteins

Edited by

Reinhard Grisshammer and Susan K. Buchanan
Laboratory of Molecular Biology, National Institutes of Health, Bethesda, Maryland, USA

RSCPublishing

The cover illustration shows the structure of the outer membrane porin MspA from the soil bacterium *Mycobacter smegmatis*. Mycobacteria, which have an unusual outer membrane, are considered a third group equidistant from the Gram-positive and Gram-negative bacteria. The MspA protein consists of eight subunits, forming a goblet-like structure around a single central channel. The upper part of the goblet consists of eight rim domains. The lower part contains two consecutive β-barrels with 8 × 2 = 16 strands forming the stem and the base regions of the goblet. For details about crystallization and structure, see the chapter by G.E. Schulz.

ISBN-10: 0-85404-361-6
ISBN-13: 978-0-85404-361-3

A catalogue record for this book is available from the British Library

Published by The Royal Society of Chemistry,
Thomas Graham House, Science Park, Milton Road,
Cambridge CB4 0WF, UK

Registered Charity Number 207890

For further information see our web site at www.rsc.org

Typeset by Macmillan India Ltd, Bangalore, India
Printed by Henry Ling Ltd, Dorchester, Dorset, UK

Preface

The past few years have seen exciting advances in the field of membrane protein structural biology. Although membrane proteins still constitute a small fraction of the total number of solved protein structures (see Hartmut Michel's and Stephen White's summaries at http://www.mpibp-frankfurt.mpg.de/michel/public/memprotstruct.html, http://blanco.biomol.uci.edu/Membrane_Proteins_xtal.html), an exponential increase of membrane protein structures has been observed[1] and is anticipated to continue. This book addresses a number of issues pertaining to membrane protein structural biology. The first section describes approaches for expression and purification of membrane proteins, and a general introduction to detergents. The major challenge in this field is with eukaryotic membrane proteins, which is reflected by the contributions in this section. The second section addresses selected methods for structure determination of membrane proteins, such as solution and solid-state nuclear magnetic resonance, atomic force microscopy, electron microscopy including single particle analysis and lipidic cubic phase crystallization. The final section highlights some of the recently solved membrane protein structures.

The content of the individual chapters varies from basic introduction to very detailed description of a particular theme. Therefore, the experienced membrane protein biochemist, as well as the novice, will find useful information. While we have endeavored to provide a balanced overview, not all topics could be covered. Clearly, this book is not an exhaustive reference for all concepts relating to membrane protein structural biology. However, the following paragraphs give some further information with key references on topics not addressed by this book.

Heterologous expression and purification of membrane proteins have been covered by a number of reviews.[2–5] Here, the reader will find information on both prokaryotic and eukaryotic membrane proteins. Until recently, all structures of eukaryotic membrane proteins, for example the visual pigment rhodopsin,[6,7] had been solved with material from natural sources, reflecting the difficulty of producing functional eukaryotic membrane proteins in heterologous expression systems. The new structure of a recombinant mammalian voltage-gated potassium channel, produced in the yeast *Pichia pastoris*, is therefore exciting.[8]

Structure determination of any membrane protein necessitates that the purified protein retain its native fold and full functionality. Progress on refolding of membrane proteins, deposited as aggregates in the cytoplasm of *Escherichia coli*, has been demonstrated for seven-helix G-protein-coupled receptors[9,10] and bacterial outer membrane proteins (for reviews, see Refs. 11 and 12). Likewise, cell-free *in vitro* expression has been reported for some membrane proteins,[13–15] and the potential usefulness of this method for generating functional membrane protein for structural studies is under investigation.

Overproduction of correctly folded membrane proteins is intimately linked to efficient membrane insertion. Great advances have been made in understanding the mechanism of membrane protein insertion into membranes, but a rational approach to predicting good 'overexpressors' from their amino acid sequences is still lacking. It is simply fascinating to see how the expression levels of closely related proteins, such as seven-helix G-protein-coupled receptors, vary in a particular host and under particular growth conditions (see for example Ref. 16). Two recent topics warrant highlighting: (a) Elegant work combining theory and biochemical experiments provides the basic features of how transmembrane helices are recognized by the endoplasmic reticulum translocon ('biological' hydrophobicity scale[17,18]). (b) Successful recombinant overproduction of membrane proteins appears to be linked to avoidance of stress responses in the host cell[19] and can be related to the differential expression of genes involved in membrane protein secretion and cellular physiology.[20] This implies that simple approaches for increasing recombinant membrane protein yields may be unsuccessful, emphasizing the benefit of considering a 'whole-cell' approach for developing general strategies to obtain high levels of functional membrane proteins, especially of eukaryotic targets.

The process of membrane protein crystallization usually starts with the extraction of the target from membranes by using detergents, followed by purification. However, the yields of purified membrane proteins can be low, which has limited the number of crystallization parameters to be explored. An obvious (but expensive) solution to this problem is to conserve protein by using nanoliter-pipetting robots (see for example Ref. 21). Likewise, smaller crystals can now be analyzed due to improvements in synchrotron beam lines. Another avenue for improving the chance of successful crystallization is to employ Fv or Fab antibody fragments for co-crystallization. This approach relies on expanding the hydrophilic surface of a membrane protein, facilitating formation of crystal contacts – the reader is referred to[22–26] for more information. Along the same lines, complexes between an integral membrane protein with little hydrophilic surface, and its soluble partner, may be more amenable to successful crystallization than the membrane protein alone. The recent structure of the mammalian voltage-gated potassium channel, Kv1.2-β2, illustrates this concept nicely.[8] For the general principles of protein crystallization, the reader is referred to Refs.27 and 28.

Mass spectrometry has found increasing use for the characterization and analysis of integral membrane proteins, despite the unfavorable spectrometer response caused by detergent and lipid molecules associated with membrane proteins. The reader is referred to the following publications for recent advances in this field.[29–33]

Another spectroscopic method, utilizing spin labeling for electron paramagnetic resonance experiments, has provided dynamic structural information on membrane proteins, in contrast to static views obtained from X-ray crystal structures. This technique has been applied, for example, to channels,[34,35] seven-helix G-protein-coupled receptors[36] and bacterial outer membrane transporters.[37]

This book presents a selected number of recent membrane protein structures, but many other exciting structures could not be included here. Among these are ATP-binding-cassette transporters,[38,39] major-facilitator superfamily transporters,[40,41] the sodium/proton antiporter from *E. coli*,[42] a bacterial homolog of a neurotransmitter transporter,[43] bacterial and mammalian potassium channels,[8,26,44–46] chloride channels/transporters,[25,47] plant light harvesting complexes[48,49] and cytochrome b6f.[50,51] A complete list of all membrane protein structures can be found at the web sites listed in the first paragraph.

Many integral membrane proteins, including eukaryotic members, can already be produced in quantities sufficient to consider structural and functional analyses. Further technological advances will continue to promote the exponential growth of membrane protein structures. Likewise, protocols to assess the dynamics and functions of membrane proteins continue to be developed, especially in the fields of solution and solid-state nuclear magnetic resonance. Most importantly, the dedication and enthusiasm of researchers, combined with good biochemistry, will propel membrane protein structural biology to new levels.

Reinhard Grisshammer and Susan Buchanan
October 2005

Acknowledgment

The research of the editors is supported by the Intramural Research Program of the NIH, National Institute of Diabetes and Digestive and Kidney Diseases.

References

1. S.H. White, *Protein Sci.*, 2004, **13**, 1948–1949.
2. R. Grisshammer and C.G. Tate, *Q. Rev. Biophys.*, 1995, **28**, 315–422.
3. C.G. Tate, *FEBS Lett.*, 2001, **504**, 94–98.
4. V. Sarramegna, F. Talmont, P. Demange and A. Milon, *Cell Mol. Life Sci.*, 2003, **60**, 1529–1546.
5. Special Issue on Overexpression of Integral Membrane Proteins, *Biochim. Biophys. Acta*, 2003, **1610**, 1–153.
6. K. Palczewski, T. Kumasaka, T. Hori, C.A. Behnke, H. Motoshima, B.A. Fox, I. Le Trong, D.C. Teller, T. Okada, R.E. Stenkamp, M. Yamamoto, M. Miyano, *Science*, 2000, **289**, 739–745.
7. J. Li, P.C. Edwards, M. Burghammer, C. Villa and G.F.X. Schertler, *J. Mol. Biol.*, 2004, **343**, 1409–1438.
8. S.B. Long, E.B. Campbell and R. MacKinnon, *Science*, 2005, **309**, 897–903.
9. H. Kiefer, *Biochim. Biophys. Acta*, 2003, **1610**, 57–62.
10. J.L. Baneres, A. Martin, P. Hullot, J.P. Girard, J.C. Rossi and J. Parello, *J. Mol. Biol.*, 2003, **329**, 801–814.
11. S.K. Buchanan, *Curr. Opin. Struct. Biol.*, 1999, **9**, 455–461.
12. M. Bannwarth and G.E. Schulz, *Biochim. Biophys. Acta*, 2003, **1610**, 37–45.
13. Y. Elbaz, S. Steiner-Mordoch, T. Danieli and S. Schuldiner, *Proc. Natl. Acad. Sci. USA*, 2004, **101**, 1519–1524.
14. G. Ishihara, M. Goto, M. Saeki, K. Ito, T. Hori, T. Kigawa, M. Shirouzu and S. Yokoyama, *Protein Expr. Purif.*, 2005, **41**, 27–37.
15. C. Klammt, F. Löhr, B. Schäfer, W. Haase, V. Dötsch, H. Rüterjans, C. Glaubitz and F. Bernhard, *Eur. J. Biochem.*, 2004, **271**, 568–580.

16. M. Akermoun, M. Koglin, D. Zvalova-Iooss, N. Folschweiller, S.J. Dowell and K.L. Gearing, *Protein Expr. Purif.*, 2005, **44**, 65–74.

17. T. Hessa, H. Kim, K. Bihlmaier, C. Lundin, J. Boekel, H. Andersson, I. Nilsson, S.H. White, G. von Heijne, *Nature*, 2005, **433**, 377–381.

18. T. Hessa, S.H. White and G. von Heijne, *Science*, 2005, **307**, 1427.

19. D.A. Griffith, C. Delipala, J. Leadsham, S.M. Jarvis and D. Oesterhelt, *FEBS Lett.*, 2003, **553**, 45–50.

20. N. Bonander, K. Hedfalk, C. Larsson, P. Mostad, C. Chang, L. Gustafsson and R.M. Bill, *Protein Sci.*, 2005, **14**, 1729–1740.

21. D. Stock, O. Perisic and J. Lowe, *Prog. Biophys. Mol. Biol.*, 2005, **88**, 311–327.

22. Y. Zhou, J.H. Morais-Cabral, A. Kaufman and R. MacKinnon, *Nature*, 2001, **414**, 43–48.

23. S. Iwata, C. Ostermeier, B. Ludwig and H. Michel, *Nature*, 1995, **376**, 660–669.

24. C. Hunte, J. Koepke, C. Lange, T. Rossmanith and H. Michel, *Structure Fold. Des.*, 2000, **8**, 669–684.

25. R. Dutzler, E.B. Campbell and R. MacKinnon, *Science*, 2003, **300**, 108–112.

26. Y. Jiang, A. Lee, J. Chen, V. Ruta, M. Cadene, B.T. Chait and R. MacKinnon, *Nature*, 2003, **423**, 33–41.

27. A. McPherson, *Crystallization of Biological Macromolecules*, Cold Spring Harbor Laboratory Press, Cold Spring Harbor, NY, 1999.

28. S. Iwata, *Methods & Results in Crystallization of Membrane Proteins*, International University Line, La Jolla, California, USA, 2003.

29. M. Cadene and B.T. Chait, *Anal. Chem.*, 2000, **72**, 5655–5658.

30. A.B. Weinglass, J.P. Whitelegge, Y. Hu, G.E. Verner, K.F. Faull and H.R. Kaback, *EMBO J.*, 2003, **22**, 1467–1477.

31. A.B. Weinglass, M. Soskine, J.L. Vazquez-Ibar, J.P. Whitelegge, K.F. Faull, H.R. Kaback and S. Schuldiner, *J. Biol. Chem.*, 2005, **280**, 7487–7492.

32. M. Trester-Zedlitz, A. Burlingame, B. Kobilka and M. von Zastrow, *Biochemistry*, 2005, **44**, 6133–6143.

33. L.L. Ilag, I. Ubarretxena-Belandia, C.G. Tate and C.V. Robinson, *J. Am. Chem. Soc.*, 2004, **126**, 14362–14363.

34. E. Perozo, A. Kloda, D.M. Cortes and B. Martinac, *J. Gen. Physiol.*, 2001, **118**, 193–206.

35. L.G. Cuello, D.M. Cortes and E. Perozo, *Science*, 2004, **306**, 491–495.

36. W.L. Hubbell, C. Altenbach, C.M. Hubbell and H.G. Khorana, *Adv. Protein Chem.*, 2003, **63**, 243–290.

37. N. Cadieux, P.G. Phan, D.S. Cafiso and R.J. Kadner, *Proc. Natl. Acad. Sci. USA*, 2003, **100**, 10688–10693.

38. K.P. Locher, A.T. Lee and D.C. Rees, *Science*, 2002, **296**, 1091–1098.

39. C.L. Reyes and G. Chang, *Science*, 2005, **308**, 1028–1031.

40. J. Abramson, I. Smirnova, V. Kasho, G. Verner, H.R. Kaback and S. Iwata, *Science*, 2003, **301**, 610–615.

41. Y. Huang, M.J. Lemieux, J. Song, M. Auer and D.N. Wang, *Science*, 2003, **301**, 616–620.

42. C. Hunte, E. Screpanti, M. Venturi, A. Rimon, E. Padan and H. Michel, *Nature*, 2005, **435**, 1197–1202.

43. A. Yamashita, S.K. Singh, T. Kawate, Y. Jin and E. Gouaux, *Nature*, 2005, **437**, 215–223.

44. D.A. Doyle, J. Morais-Cabral, R.A. Pfuetzner, A. Kuo, J.M. Gulbis, S.L. Cohen, B.T. Chait and R. MacKinnon, *Science*, 1998, **280**, 69–77.

45. Y. Jiang, A. Lee, J. Chen, M. Cadene, B.T. Chait and R. MacKinnon, *Nature*, 2002, **417**, 515–522.

46. A. Kuo, J.M. Gulbis, J.F. Antcliff, T. Rahman, E.D. Lowe, J. Zimmer, J. Cuthbertson, F.M. Ashcroft, T. Ezaki, D.A. Doyle. *Science*, 2003, **300**, 1922–1926.

47. J.A. Mindell, M. Maduke, C. Miller and N. Grigorieff, *Nature*, 2001, **409**, 219–223.

48. Z. Liu, H. Yan, K. Wang, T. Kuang, J. Zhang, L. Gui, X. An and W. Chang, *Nature*, 2004, **428**, 287–292.

49. R. Standfuss, A.C.T. van Scheltinga, M. Lamborghini and W. Kühlbrandt, *EMBO J.*, 2005, **24**, 919–928.

50. G. Kurisu, H. Zhang, J.L. Smith, and W.A. Cramer, *Science*, 2003, **302**, 1009–1014.

51. D. Stroebel, Y. Choquet, J.L. Popot and D. Picot, *Nature*, 2003, **426**, 413–418.

Contents

**Chapter 3   Expression of Recombinant G-Protein Coupled
               Receptors for Structural Biology   29**
               Filippo Mancia and Wayne A. Hendrickson

Section 2 Methods for Structural Characterization of Membrane Proteins

Section 3   New Membrane Protein Structures

Chapter 11   Aquaporins: Integral Membrane
Channel Proteins   195

*Robert M. Stroud, William E.C. Harries, John Lee,
Shahram Khademi and David Savage*

Chapter 12   Gas Channels for Ammonia   212

Shahram Khademi and Robert M. Stroud

Chapter 19   Glutamate Receptor Ion Channels: Structural Insights into Molecular Mechanisms   349

Avinash Gill and Dean R. Madden

Section 1

Expression and Purification of
Membrane Proteins

CHAPTER 1

Refolding of G-Protein-Coupled Receptors

JEAN-LOUIS BANÈRES

UMR 5074 CNRS, "Chimie Biomoléculaire et Interactions Biologiques", Faculté de Pharmacie, 15 av. Ch. Flahault, BP14491, 34093 Montpellier Cedex 05, France

1 Introduction

G-protein-coupled receptors (GPCRs) are transmembrane receptors that are involved in the recognition and transduction of messages as diverse as light, Ca^{2+} ions, odorants, small molecules (including amino-acids, nucleotides, peptides) and proteins.[1,2] Although different classes of receptors have been described,[3] they all share a common structural motif composed of seven α-helices spanning the plasma membrane.

Although significant progress has been made within the last few years in dissecting GPCR-mediated signal transduction pathways, understanding the mechanisms underlying ligand recognition and signal transduction across the membrane has been hampered by the lack of information at the molecular level. This is largely due to the low abundance of most GPCRs in cellular membranes. Furthermore, few expression systems have proven satisfactory for producing these receptors in a functional state and sufficient yields.[4–6] Structural information on the GPCR family is therefore very sparse, with the exception of rhodopsin for which X-ray[7] and electron[8] diffraction data have been obtained.

Recombinant expression has been one of the major bottlenecks in structural biology of GPCRs.[9] One of the most widely used expression systems for structural biology is *Escherichia coli*. However, in general, bacterial expression has been hampered by the relatively low yields of GPCRs owing to the toxic effects caused by these 7TM receptors when inserted into the bacterial membrane.[6] To circumvent this toxicity problem, GPCRs can be directed to bacterial inclusion bodies. This leads to high expression levels of the receptor (in the range of 10–50 mg of protein per liter of bacterial culture). However, the highly expressed recombinant receptors are inactive and require refolding into a functional form. This explains why intensive work has been developed during the last years in analyzing the refolding of GPCRs and in devising efficient strategies for refolding receptors solubilized under denaturing conditions.

Understanding the basic principles of membrane protein folding is also of great interest in a fundamental perspective. Many studies have been dedicated in the past years to understand the trafficking of GPCRs.[10] The quality control process in the endoplasmic reticulum involves a variety of mechanisms. These mechanisms ensure that only correctly folded proteins are directed to the plasma membrane. Despite this stringent quality control mechanism, gain- or loss-of-function mutations affecting protein folding in the endoplasmic reticulum, which have been described, can have profound effects on the health of an organism. Understanding the molecular mechanisms of protein folding could therefore help in correcting the structural abnormalities associated with misfolded receptors.

There are essentially two main types of membrane-spanning structures: transmembrane α-helices and β-barrels. The latter appear to be limited to outer membrane proteins. The present work will focus on the refolding of α-helical membrane proteins. This is the most general case and the most interesting from a pharmacological point of view. It applies, in particular, to the GPCR family.

Investigating the *in vitro* refolding of membrane proteins is a difficult task, in particular, due to the hydrophobic nature of integral membrane proteins. Indeed, to work with isolated membrane proteins, one has to manipulate refolding solvents, generally composed of detergents or detergent/lipid mixtures, which poorly mimic the natural membranes. Nevertheless, biophysical studies on model systems have begun to provide a sound physical basis for membrane protein folding.[11]

2 Refolding of Membrane Proteins

As stated above, one of the greatest problems in setting up conditions for studying a membrane protein *in vitro* arises from the relative instability of these proteins in detergent solutions. This problem, associated with the low abundance of most GPCRs in cellular membranes, explains the limited number of examples of refolding studies with GPCRs. Most of the biophysical studies on α-helical membrane protein folding have been carried out with bacteriorhodopsin as a model system, since it is a membrane protein that can be purified in high yields and is relatively stable in solution.[12] Moreover, high-resolution structures of bacteriorhodopsin are available[13,14] that help to understand the folding of this protein on a molecular basis. Bacteriorhodopsin functions as a light-driven proton pump in the purple membrane of the archaebacterium *Halobacterium salinarium*.[15] As is the case with GPCRs, it possesses seven transmembrane α-helical segments. Although it may not be fully representative of GPCRs at the folding level,[16] it is nevertheless a good model system to gain a better understanding of receptor folding. Some general rules for α-helical membrane protein folding have been inferred from folding/unfolding studies with bacteriorhodopsin. In particular, a two-stage model has been proposed for the folding of these proteins that decomposes this process.[17,18] First, individually stable transmembrane helices form and then these helices pack to form a functional protein (Figure 1).

The first step in the two-stage model thus involves the formation of the helical segments, and transmembrane helices display some characteristic features.[19] In general, they are largely hydrophobic sequences that include a limited number of polar or

unfolded

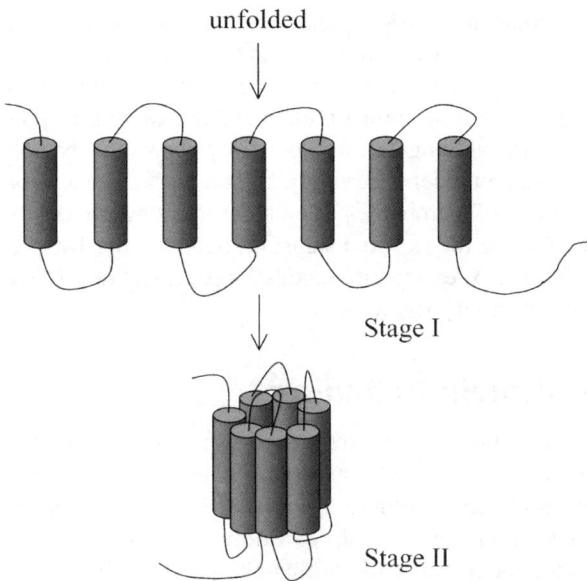

Stage I

Stage II

Figure 1 *Schematic representation of the two-stage model for membrane protein folding (18). The transmembrane helices are considered as individually folding units. In the first stage of the folding, these domains form separately due to specific hydrophobic effects and hydrogen bonding. These helical domains then establish intramolecular contacts that lead to the native fold*

potentially charged groups. In contrast to what is observed in globular proteins, prolines and glycines are also often found in membrane helices. The bending of a helical segment induced by a proline residue could be an important feature for membrane protein function. This is, for example, the case for the GPCR rhodopsin, where kink-inducing proline residues are found at key positions of the three-dimensional structure.[7] Individual transmembrane helical structures are stable essentially because of hydrophobic effects and hydrogen bonds that are strong in the low dielectric environment of the membrane. Many helices found in membrane proteins are likely to be considered as stable folding units.[20] In agreement with this view, individual helical segments or fragments of bacteriorhodopsin or GPCRs, containing only a reduced number of helices, can reach a stable fold in a membrane mimicking environment (see below).

If individual helices are formed in response to main-chain hydrogen bonding and hydrophobic effects, other interactions are likely to be involved in their assembly in the second stage of refolding. An important factor is certainly the way that the helices fit together, guided by Van der Waals interactions and side-chain rotamers.[18,19] Another factor that is likely to influence the assembly of the transmembrane helices is the lipid environment. Besides its general role as a solvent, the lipid may also stabilize membrane proteins through specific interactions. Indeed, there are many examples of specific associations of individual lipids or of classes of lipids with membrane proteins.[21–23] The interhelical loops can also have a role in the formation of helix/helix contacts, although the fact that proteins can in some cases assemble from fragments to form functional species suggests that the constraint induced by the loops may not be essential for folding. The loops could nevertheless promote folding events that bring transmembrane helices together. In agreement with this view, in the case of bacteriorhodopsin, the specific conformation of all the protein loops, except the DE loop, contributes to protein stability and is required for the correct folding and function of the protein.[24]

In closing this part, although it may not fully represent the folding process of GPCRs,[16] the two-stage model is certainly a good working basis for a better understanding of the folding of membrane proteins. A three stage model has also been proposed that, in addition to the two first steps, introduces a third step corresponding to ligand binding, folding of extramembranous loops, insertion of peripheral domains and formation of quaternary structures.[25]

3   *In Vitro* Protein Refolding

Inclusion body production is a recurrent theme in recombinant protein technology. Refolding of inclusion bodies consists first in solubilizing the protein under denaturing conditions and then refolding is initiated by the removal of the denaturing agent. This can be done by dialysis or dilution. Protein folding has also been achieved by binding the protein to a chromatographic resin in the unfolded state and subsequent washing with an appropriate buffer that contains no denaturant. Affinity resins, such as Ni-NTA Sepharose[26,27] or heparin Sepharose[28] have been used for poly-histidine-tagged or poly-arginine-tagged proteins, respectively. Compared to folding by dialysis or rapid dilution, this method has the main advantage of preventing aggregation due to intermolecular interaction of partly folded protein species. However, an interference of the chromatographic support with the folding protein molecule may be detrimental, especially in the case of the highly hydrophobic membrane proteins, possibly causing precipitation of the protein on the matrix.

The efficiency of refolding depends on the competition between protein refolding and aggregation. One of the main difficulties in refolding membrane proteins is thus to find conditions that favor refolding over aggregation. A delicate balance must be reached between too harsh or too mild environments. Protein folding screens for identification of optimal folding conditions have been developed during the past years to screen different factors that may influence globular protein refolding.[27–32] In a general manner, this consists in screening multiple conditions, in which different parameters (additives, pH, salt and protein concentration) are altered. Such folding screens can be adapted to integral membrane proteins, keeping in mind that one of the most crucial parameters to test will be the nature of the detergent and/or detergent/lipid mixture. This is indeed likely to be the factor with the most dramatic effect on membrane protein refolding efficiency.

4   GPCR *In Vitro* Refolding

4.1   Resolubilization from Inclusion Bodies

In most cases, refolding of GPCRs has been carried out with material recovered from bacterial inclusion bodies.[27,33–35] As far as expression in *E. coli* is considered, it seems that there is no general strategy to be used for the efficient accumulation of

GPCRs in inclusion bodies, even if some rules have been inferred from studies with different receptors.[36] In some cases, the receptor simply fused to a T7 tag was efficiently expressed in *E. coli*. This is the case, for example, for the leukotriene B_4 receptor BLT1[27] or, more recently, for the V2 vasopressin receptor.[37] However, it must be noted that in the case of BLT1, protein expression levels dramatically vary from one clone to another. In other cases, fusion of the receptor to a protein partner was absolutely required for its expression. Different partners, such as glutathione S-transferase (GST)[33,36] or ketosteroid isomerase (KSI)[34] have been used. It must be emphasized again that no system allowing the expression of "all" GPCRs in inclusion bodies has been described so far. For example, no expression was observed with the 5-HT4a receptor fused to GST, whereas this receptor was efficiently produced when fused to KSI.[34] Similarly, in our hands, among all the receptors we tested, KSI fusions gave inclusion bodies only with the 5-HT4a receptor. The best approach may therefore be to test different fusion partners and then quantify the expression levels.

Before starting the refolding, the inclusion body material has to be solubilized. Globular proteins in inclusion bodies can be solubilized in the presence of high concentrations of chaotropic agents, such as guanidinium hydrochloride or urea. In contrast, aggregated membrane proteins require detergents (or organic solvents) for efficient solubilization, due to the predominance of hydrophobic effects in the aggregated material. Usually, SDS is used as a strong denaturing detergent. However, it must be kept in mind that, in SDS, helical membrane proteins, such as bacteriorhodopsin or GPCRs are not totally unfolded, but they can retain a significant amount of secondary structure. Indeed, sequence regions that, in the folded structure, form transmembrane helices tend to locally adopt an α-helical structure even in SDS. Bacteriorhodopsin, for example, is about 40–45% helical in SDS.[38–40] The 5-HT4a receptor is also 30–35% helical in SDS-containing buffers (Banères, unpublished data). The SDS-solubilized starting point for refolding is thus not to be considered as a fully unfolded state but rather as a partially folded state, as far as the secondary structure is concerned.

4.2 Refolding

As stated above, "unfolded" membrane proteins are first solubilized in harsh detergents, such as SDS or lauroylsarcosine. Refolding will then consist in replacing this denaturing detergent by a detergent that will stabilize the three-dimensional fold of the membrane protein. Under these conditions, based on the two-stage model, regions that have a propensity to fold will do so and then interactions between protein segments will appear. Those can be intramolecular, leading to refolding or intermolecular, leading to aggregation. One of the main challenges in refolding membrane proteins is therefore to find conditions, in particular, the nature of the detergent environment, which will favor the intramolecular over intermolecular contacts, and therefore refolding over aggregation. This implies finding the right balance between harsh and mild environments. For GPCRs, however, in the absence of extensive examples of successful receptor refolding, it is difficult to infer a general rule for the lipid and detergent requirements for maximal refolding efficiency.

Two different refolding studies have been reported so far for GPCRs. In the first case, the refolding involved, as described for bacteriorhodopsin, peptides encompassing one or several transmembrane domains obtained either by peptide synthesis, restricted protease digestion of receptors or bacterial expression in inclusion bodies.[41,42] In the second case, the intact receptor was produced in bacterial inclusion bodies and then refolded *in vitro*.[27,34,43]

4.3   Refolding of GPCR Fragments

We will not consider here the refolding of the extracellular ectodomains of GPCRs that has been described for several receptors,[29,44,45] since these domains behave as typical globular proteins. We will focus here only on the transmembrane regions of GPCRs. Several papers reporting the refolding of fragments containing some of the transmembrane helices of GPCRs have been published so far.[41,42] These fragments range from a single transmembrane domain to several transmembrane helices. As predicted by the two-stage model for membrane protein refolding, these fragments form folded domains when transferred from denaturing conditions to a milder environment. This suggests that GPCR transmembrane helices can also be considered as individual folding units. Besides their interest for a better understanding of the molecular processes involved in receptor folding, these studies indicate that the study of receptor fragments could be an alternative to the analysis of the structural properties of GPCRs. We will provide here two recent examples to illustrate this aspect of GPCR refolding.

The first is that of the μ-opioid receptor for which the refolding of an 80-residue fragment has recently been described.[41] The fragment, produced in *E. coli* inclusion bodies as a fusion with GST, encompassed the second and third transmembrane segments as well as the first extracellular loop of the receptor. In this case, simply by exchanging the harsh detergent lauroylsarcosine, used for inclusion body solubilization, to milder lysophosphatidylcholine micelles, a significant amount of secondary structure was recovered. Under these conditions, the receptor fragment adopted approximately 50% α-helical structure, consistent with the assumption of an α-helical structure in the two membrane-spanning regions and a non-helical structure in the loop region connecting the second and the third transmembrane domains.

The production of the seven transmembrane domains of the adenosine receptor as individual synthetic peptides has also been recently reported.[42] In this case, circular dichroism (CD) spectra indicated that each of the seven peptides form stable, independent α-helical structures in both detergent micelles and lipid vesicles. In particular, the peptides corresponding to the third, fourth, sixth and seventh transmembrane domains exhibit high-helical structure content, close to the predicted maximum for their transmembrane segments. The peptide corresponding to the first transmembrane domain also adopts a relatively high content of α-helical structure. Interestingly, the measured helical content of some transmembrane domains does not directly correlate with the predicted helicity based on amino acid sequence. This points out that, while hydrophobic interactions can be a major determinant for folding of transmembrane peptides, other factors, such as helix–helix interactions may play significant roles for specific transmembrane domains. Such an observation suggests that, although the transmembrane peptides may essentially be considered independent folding units, interactions between the transmembrane domains could be required to some degree for proper insertion and folding of some helical domains.

4.4 Refolding of Intact GPCRs

In vitro refolding of some intact receptors has also been reported during the last 10 years. The first bacterially expressed GPCR to be successfully refolded *in vitro* has been the OR5 olfactory receptor.[43] Subsequently, the refolding of two other GPCRs, namely the leukotriene B_4 receptor BLT1[27] and the serotonin receptor 5-HT4a,[34] has been described. For all three receptors, the same general approach was used (see Table 1), and the reader is referred to Ref. 27, 34 and 43 for a step-by-step procedure. They all were produced in *E. coli* IB, thus allowing large protein quantities to be recovered (in the 10 mg range). The receptors were then solubilized with a harsh detergent. In the case of OR5, the overexpressed protein was solubilized in the strong, negatively charged detergent, lauroylsarcosine, whereas BLT1 and the 5-HT4a receptors were solubilized in the presence of both urea and SDS. The receptors were then refolded by solvent exchange. In all cases, solvent exchange was carried out with the receptor immobilized on a solid Ni-NTA matrix. This could indicate that for integral membrane proteins, preventing non-specific protein/protein interactions that favor aggregation, could be a crucial point to get acceptable refolding yields. At least in our hands, the matrix-assisted procedure systematically yielded significantly higher refolding yields compared to dialysis or dilution (Banères, unpublished data).

Lacking a detailed theoretical understanding of the various factors affecting GPCR refolding yields and tendency to aggregate, the strategy used in all the reported cases of receptor refolding was to vary different parameters in a systematic way and quantify the refolding yield. One of the most crucial parameters was, as expected, the composition of detergent/lipid micelles. The OR5 was first reconstituted in the non-denaturing detergent digitonin. In this detergent, the receptor was able to bind its ligand (fluorescence-monitored ligand-binding assays) indicating

Table 1 *Summary of the expression, solubilization and refolding conditions for three efficiently refolded GPCRs, the OR5 receptor,[43] the BLT1 receptor[27] and the 5-HT4a receptor.[34] CHS: cholesteryl hemisuccinate, CHAPS: 3-[(3-Cholamidopropyl) dimethylammonio]-1-propanesulfonate, DDM: dodecyl maltoside, DMPC: dimyristoyl-phosphatidylcholine, GST: glutathione S-transferase, KSI: ketosteroid isomerase, IB: E. coli inclusion bodies, LDAO: lauryl-N,N-dimethylamine-N-oxide, POPC: 1-palmitoyl-2-oleoyl-sn-glycerol-3-phosphocholine and POPG: 1-palmitoyl-2-oleoyl-sn-glycero-3-[phospho-rac-(1-glycerol)]. The lipid and/or lipid/detergent weight ratios used for refolding are indicated*

Receptor	Expression	Solubilization	Refolding method	Detergent and/or lipid	References
OR5	as IB GST fusion	sarcosyl	matrix-assisted (Ni-NTA)	POPC/POPG (4:1)	43
BLT1	as IB T7-tag fusion	urea/SDS	matrix-assisted (Ni-NTA)	LDAO	27
5-HT4a	as IB KSI fusion	urea/SDS	matrix-assisted (Ni-NTA)	CHAPS/DMPC/CHS (2:1:0.02)	34

that, under these conditions, the receptor was properly folded. Subsequently, mixed micelles, composed of dodecyl maltoside (DDM) as the detergent, and of 1-palmitoyl-2-oleoyl-*sn*-glycerol-3-phosphocholine (POPC) and 1-palmitoyl-2-oleoyl-*sn*-glycero-3-[phospho-*rac*-(1-glycerol)] (POPG) as the lipids, were added and the detergent was removed by treatment with hydrophobic beads. Under these conditions, the protein was stabilized into a fully active state as assessed by photoaffinity labeling. In the case of the BLT1 receptor, the best refolding yields were achieved with lauryl-*N,N*-dimethylamine-*N*-oxide (LDAO) as a detergent. For BLT1, different detergents were tested and the best yields were systematically achieved from those with a long alkyl chain, *i.e.* above C12. The observation of a prominent role of the alkyl chain length on BLT1 refolding is reminiscent of what had been previously observed with rhodopsin, where detergents with long alkyl chains (above C10) stabilized the protein better than detergents with shorter chains.[46] It is to be noted that BLT1 and OR5 (see above), were first refolded in solutions containing only detergents. However, in this case, subsequent reconstitution of the receptor in a detergent/lipid mixed micelle or in a lipid vesicle dramatically increases the stability of the isolated BLT1 receptor in solution (Banères, unpublished data). This suggests that, even if the detergent can mimic the membrane environment for receptor refolding, the presence of lipids and maybe specific lipid/protein contacts, are essential to ensure a full stabilization of the protein in solution. Finally, the 5-HT4a receptor was stabilized in a functional conformation only in the presence of mixed detergent/lipid micelles. The micelles in this case were composed of 3-[(3-cholamidopropyl)dimethylammonio]-1-propanesulfonate (CHAPS), dimyristoyl-phosphatidylcholine (DMPC) and cholesterol (CH) (see Table 1). In contrast to what had been observed with BLT1, the 5-HT4a receptor could not be refolded in the presence of a detergent only, emphasizing the importance of lipids for the stabilization of the receptor conformation. Also, the presence of CH was required not only for increasing the refolding yields of the 5-HT4a receptor, but also to increase the long-term stability and optimize the ligand-binding properties of the receptor in solution (see Figure 2). The differences in the detergent requirements for stabilizing the functional conformations of the OR5, BLT1 and 5-HT4a receptors indicate that each protein may be a specific case in terms of detergent and lipid requirements for refolding. In the absence of more published data, it is difficult to infer a general rule as to which are the most appropriate detergent and lipids for GPCR refolding. The most straightforward strategy to find suitable folding conditions seems therefore to test different detergents and detergent/lipid mixtures in a systematic way.

Besides the step involving the search for a detergent that could promote receptor folding, the data obtained for the receptors cited above emphasize the fact that one of the main problems, when screening for the best conditions for protein folding, is to determine the yield of functional protein. This is easier to achieve for membrane proteins that display well-defined ligand-binding properties. In the case of OR5, BLT1 and 5-HT4a, the main criterion used to assess the correct refolding was the ligand-binding properties of the reconstituted receptor. It is to be noted in this context that one must be cautious while using an agonist to determine if the receptor has been successfully refolded. In the case of an isolated receptor, the agonist affinity may be significantly lower in the absence of G-proteins. Receptors isolated from

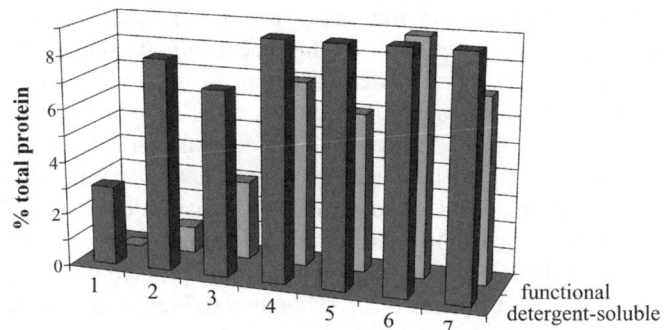

Figure 2 *Percentage of detergent-soluble and functional (i.e. competent to bind antagonist) 5-HT4a receptor after refolding in different detergent and/or detergent/lipid mixtures. (1) octyl glucoside, (2) LDAO, (3) CHAPS, (4) CHAPS/DMPC (detergent/ lipid ratio 2:1), (5) CHAPS/DOPC (2:1), (6) CHAPS/DMPC/CHS (2:1:0.02), and (7) CHAPS/DOPC/CHS (2:1:0.02) (see legend to Table 1 for the abbreviations of the detergents and lipids)*

membrane fractions may still be bound to endogenous G-proteins, and hence display higher agonist affinities. Other parameters can also be used to assess correct folding of isolated GPCRs. These include the ability of the refolded receptor to interact with and/or to activate intracellular partners, such as G-proteins or arrestins.[47,48] Determining the amount of correctly refolded protein is more difficult when no biological assay is available. Centrifugation or filtration can remove precipitates resulting from protein aggregation, but solubility cannot be used as a stringent criterion simply because partially folded non-active intermediates can be soluble in the presence of detergents. Another method that has been largely applied to the refolding of globular proteins is limited proteolysis. Partially folded intermediates are assumed to be more susceptible to limited proteolysis than the fully folded protein. However, despite this method having successfully been applied to globular proteins,[31] one must be cautious when working with membrane proteins, since inaccessibility to the protease could simply be due to the masking of the cleavage sites by the detergent and not to the correct folding of the protein. This method has nevertheless been successfully applied to membrane proteins, but rather to test, for example, for correct insertion in a lipid membrane.[49,50] Spectroscopic methods, such as CD or fluorescence, that give access to the structural characteristics of the protein, can also be used to monitor refolding. CD will provide the secondary structure content of the protein,[51] whereas tryptophan fluorescence spectroscopy can be used to detect folded protein, since unfolded conformations, folding intermediates and fully folded proteins may be distinguishable in their respective spectra.[52] However, one must again be cautious while using these criteria to assess for correct membrane protein folding. For example, a good CD profile does not mean that the protein is well folded. If one considers the two-stage model for protein refolding, a possibility is to get a partially folded protein where the secondary structure elements are formed so that its far-UV CD properties are close to those of the fully folded protein, but where the intramolecular helical contacts are not properly established so that the receptor is not able to bind its ligands. One must also keep in mind that even in harsh detergents, such as SDS, helical membrane proteins, bacteriorhodopsin or GPCRs usually can retain a certain amount of secondary structure. A good alternative in this case could be the use of CD in the near-UV regions, rather than in the far-UV regions. Indeed, the near-UV region is sensitive to the three-dimensional folding of the protein and is therefore likely to be affected by the packing of the transmembrane helices.[27,34] As for fluorescence, it can be difficult to assess correct refolding in the absence of a reference spectrum of the functional protein. In a general manner, for most of the methods described above, one of the main problems indeed arises from the absence of well-folded protein to be taken as a reference due to the low abundance of most GPCRs in cellular membranes. Finally, as a consequence of the possible occurrence of detergent-soluble misfolded proteins, a proper way to purify the functional receptor is also crucial. In this context, the best method seems to be, whenever possible, affinity chromatography with an immobilized ligand column, since the main goal of this purification step is to discriminate between active and inactive receptors.

To give an example emphasizing the importance of a functional assay for assessing receptor refolding, we can consider the case of the 5-HT4a receptor. When refolding was carried out with only detergents, two different protein fractions were recovered after the refolding step. The first one, highly aggregated, was simply removed by centrifugation on a sucrose gradient. The second one was totally soluble in the detergent-containing solution and its secondary structure properly recovered as assessed by far-UV CD. However, only a small amount of receptor in this fraction was able to bind a 5-HT4a antagonist ligand, as assessed by direct ligand-binding experiments (Figure 2). It is only after adding lipids in the refolding buffers that a fully functional protein fraction was recovered.

5 Conclusion

Membrane protein refolding, in particular GPCR refolding, has been the focus of intensive work during the past years. This is due to the possible implications of receptor misfolding in some diseases as well as to the interest of GPCR refolding in the context of high-yield protein production for structural studies. Indeed, the increasing number of reports of production of GPCRs in *E. coli* inclusion bodies makes refolding a central step for producing functional receptors. Although further development of the techniques and refolding systems are still required, an understanding of the process is being gained. The possibility to study the refolding of model proteins, such as bacteriorhodopsin, and now a few GPCRs, will certainly provide us with a more detailed model of the refolding mechanism. It is likely that future success in refolding more membrane proteins and/or reaching higher refolding yields *in vitro* will depend on how we understand the structural properties of the membrane proteins, as well as the way they interact with their lipid environment. In particular, some work will certainly have to focus on the factors that influence the interactions between helical domains in the membrane protein since this is, at least in our hands, the limiting step for reaching efficient refolding of all the GPCRs we have studied so far. Such fundamental work on receptor refolding will without doubt help in finding the factors that currently limit high-level *in vitro* refolding of GPCRs, especially since high-level expression of "unfolded proteins" seems now somehow to be ensured, at least in the bacterial system.

References

1. J. Bockaert and J.-P. Pin, *EMBO J.*, 1999, **18**, 1723.
2. J. Bockaert, S. Claeysen, C. Becamel, S. Pinloche and A. Dumuis, *Int Rev. Cytol.*, 2002, **212**, 63.
3. S.M. Foord, T.I. Bonner, R.R. Neubig, E.M. Rosser, J.P. Pin, A.P. Davenport, M. Spedding and A.J. Harmar, *Pharmacol. Rev.*, 2005, **57**, 279–288.
4. C.G. Tate and R. Grisshammer, *Trends Biotechnol.*, 1996, **14**, 426.
5. C.G. Tate, *FEBS Lett.*, 2001, **504**, 94.
6. V. Sarramegna, F. Talmont, P. Demange and A. Milon, *Cell. Mol. Life Sci.*, 2003, **60**, 1529.
7. K. Palczewski, T. Kumusaka, T. Hori, C.A. Behnke, H. Motoshima, B.A. Fox, I. Le Trong, D.C. Teller, T. Okada, R.E. Stenkamp, M. Yamamoto and M. Miyamo, *Science*, 2000, **289**, 739.
8. J.J. Ruprecht, T. Mielke, R. Vogel, C. Villa and G.F. Schertler, *EMBO J.*, 2004, **23**, 3609.
9. K. Lundstrom, *Trends Biotechnol.*, 2005, **23**, 103.
10. C.M. Tan, A.E. Brady, H.H. Nickols, Q. Wang and L.E. Limbird, *Annu. Rev. Pharmacol. Toxicol.*, 2004, **44**, 559.
11. P.J. Booth and A.R. Curran, *Curr. Opin. Struct. Biol.*, 1999, **9**, 115.
12. J.K. Lanyi and H. Luecke, *Curr. Opin. Struct. Biol.*, 2001, **11**, 415.
13. K. Edman, P. Nollert, A. Royant, H. Belrhali, E. Pebay-Peyroula, J. Hajdu, R. Neutze and E.M. Landau, 1999, *Nature*, **401**, 822.
14. H. Luecke, B. Schobert, H.T. Richter, J.P. Cartailler and J.K. Lanyi, *J. Mol. Biol.*, 1999, **291**, 899.
15. J.K. Lanyi, *Annu. Rev. Physiol.*, 2004, **66**, 665.
16. J. Klein-Seetharaman, *Trends Pharmacol. Sci.*, 2005, **26**, 183.
17. J.-L. Popot, S.E. Gerchman and D.M. Engelman, *J. Mol. Biol.*, 1987, **198**, 655.
18. J.-L. Popot and D.M. Engelman, *Biochemistry*, 1990, **29**, 4031.
19. J.-L. Popot and D.M. Engelman, *Annu. Rev. Biochem.*, 2000, **69**, 881.
20. J.-L. Popot, *Curr. Opin. Struct. Biol.*, 1993, **3**, 532.
22. A.D. Ferguson, E. Hofmann, J.W. Coulton, K. Diederichs and W. Welte, *Science*, 1998, **282**, 2215.
23. H. Belrhali, P. Nollert, A. Royant, C. Menzel, J.P. Rosenbuch, E.M. Landau and E. Pebay-Peyroula, *Struct. Fold Des.*, 1999, **7**, 909.
24. J.-M. Kim, P.J. Booth, S.J. Allen and H.G. Khorana, *J. Mol. Biol.*, 2001, **308**, 409.
25. D.M. Engelman, Y. Chen, C.N. Chin, A.R. Curran, A.M. Dixon, A.D. Dupuy, A.S. Lee, U. Lehnert, E.E. Matthews, Y.K. Reshetnyak, A. Senes and J.-L. Popot, *FEBS Lett.*, 2003, **555**, 122–125.
25. M. Weik, G. Zaccaï, N.A. Dencher, D. Oesterhelt and T. Hauss, *J. Mol. Biol.*, 1998, **275**, 625.

26. J.-L. Banères, F. Roquet, M. Green, H. LeCalvez and J. Parello, *J. Biol. Chem.*, 1998, **273**, 24744.

27. J.-L. Banères, A. Martin, P. Hullot, J.-P. Girard, J.-C. Rossi and J. Parello, *J. Mol. Biol.*, 2003, **329**, 801.

28. G. Stempfer, B. Hall-Neugebauer and R. Rudolph, *Nat. Biotechnol.*, 1996, **14**, 329.

29. G.Q. Chen and E. Gouaux, *Proc. Natl. Acad. Sci. USA*, 1997, **94**, 13431.

30. N. Armstrong, A. de Lencastre and E. Gouaux, *Protein Sci.*, 1999, **8**, 1475.

31. C. Heiring and Y.A. Muller, *Protein Eng.*, 2001, **14**, 183.

32. D.A. Tobbell, B.J. Middleton, S. Raines, M.R. Needham, I.W. Taylor, J.Y. Beveridge and W.M. Abbott, *Protein Exp. Purif.*, 2002, **24**, 242.

33. H. Kiefer, K. Maier and R. Vogel, *Biochem. Soc. Trans.*, **27**, 908.

34. J.-L. Banères, D. Mesnier, A. Martin, L. Joubert, A. Dumuis and J. Bockaert, *J. Biol. Chem.*, 2005, **280**, 20253.

35. H. Kiefer, *Biochim. Biophys. Acta*, 2003, **1610**, 57.

36. H. Kiefer, R. Vogel and K. Maier, *Receptor. Channel.*, 2000, **7**, 109.

37. C. Tian, R.M. Breyer, H.J. Kim, M.D. Karra, D.B. Friedman, A. Karpay and C.R. Sanders, *J. Am. Chem. Soc.*, 2005, **127**, 8010.

38. K.S. Huang, H. Bayley, M.J. Liao, E. London and H.G. Khorana, *J. Biol. Chem.*, 1981, **256**, 3802.

39. E. London and H.G. Khorana, *J. Biol. Chem.*, 1982, **257**, 7003.

40. P.J. Booth, *Fold. Des.*, 1997, **2**, R85.

41. A. Kerman and V.S. Ananthanarayanan, *Biochim. Biophys. Acta*, 2005, **1747**, 133.

42. T. Lazarova, K.A. Brewin, K. Stoeber and C.R. Robinson, *Biochemistry*, 2004, **43**, 12945.

43. H. Kiefer, J. Krieger, J.D. Olszewski, G. Von Heijne, G.D. Prestwich and H. Breer, *Biochemistry*, 1996, **35**, 16077.

44. U. Grauschopf, H. Lilie, K. Honold, M. Wozny, D. Reusch, A. Esswein, W. Schafer, K.P. Rucknagel and R. Rudolph, *Biochemistry*, 2000, **39**, 8878.

45. H. Vlase, N. Matsuoka, P.N. Graves, R.P. Magnusson and T.F. Davies, *Endocrinology*, 1997, **8**, 1658.

46. P. Knudsen and W.L. Hubbel, *Membr. Biochem.*, 1978, **1**, 297.

47. B. Bertin, M. Freissmuth, R.M. Breyer, W. Schutz, A.D. Strosberg and S. Marullo, *J. Biol. Chem.*, 1992, **267**, 8200.

48. T.A. Key, T.A. Bennett, T.D. Foutz, V.V. Gurevich, L.A. Sklar and E.R. Prossnitz, *J. Biol. Chem.*, 2001, **276**, 49204.

49. J.-L. Rigaud, A. Bluzat and S. Büschlen, *Biochem. Biophys. Res. Comm.*, 1983, **111**, 373.

50. J. Cladera, J.-L. Rigaud and M. Dunach, *Eur. J. Biochem.*, 1997, **243**, 798.

51. G.D. Fasman, *Circular Dichroism and the Conformational Analysis of Biomolecules*, Plenum Press, New York, 1996.

52. C.A. Royer, *Methods Mol. Biol.*, 1995, **40**, 65.

CHAPTER 2

Expression of Genes Encoding Eukaryotic Membrane Proteins in Mammalian Cells

PHILIP J. REEVES

Department of Biological Sciences, University of Essex, Wivenhoe Park, Colchester CO4 3SQ, UK

1 Introduction

During the last decade there have been significant technological advances in the development of procedures for functional expression of membrane proteins in mammalian cells. Membrane proteins are considerably more difficult to produce and purify than soluble proteins and this is reflected by the limited number of membrane-protein structures that have been solved. Obtaining large quantities of higher eukaryotic membrane proteins in functional form is further complicated due to our limited knowledge on their folding requirements. Eukaryotic membrane proteins often exhibit elaborate co/post-translational modifications including N-acetylation, N- and/or O-linked glycosylation, disulphide bond formation and fatty acylation. Mammalian cells are capable of carrying out these modifications and are also likely to have the appropriate cellular chaperones to assist in membrane-protein folding. Furthermore, the membrane phospholipid composition of mammalian cells is most likely to provide a suitable environment for promoting correct membrane-protein insertion and ensuring membrane-protein stability.

For preparation of eukaryotic membrane proteins it would seem reasonable therefore to first attempt production in a higher eukaryotic host. However, the perceived technical difficulties and high costs associated with mammalian cell culture deter many researchers and instead conventional expression systems, such as the "user-friendly" prokaryote *Escherichia coli*, or lower eukaryotes such as yeast are often tried first. *E. coli* is considered a workhorse for production of diverse proteins and benefits from decades of investment in development of bacterial genetic tools and fermentation technology. Although examples are limited, *E. coli*, has been used successfully for preparation of some eukaryotic membrane proteins as exemplified by recent work.[1,2] However, most complex membrane proteins such as G-protein-coupled receptors (GPCRs) often fail to fold correctly in these simple expression systems and, as a result, the major protein fraction is present in a misfolded non-functional form. This represents a serious limitation and requires development of refolding regimes for recovery of the target protein; a formidable task for most membrane proteins.

The merits and drawbacks of various expression systems for membrane proteins were discussed in detail elsewhere in this book. In this chapter some recent developments in mammalian cell-based expression systems will be described with emphasis on the author's research that utilizes stable mammalian cell lines for efficient production of milligram quantities of bovine rhodopsin, a GPCR.

2   Mammalian Cell Hosts and Gene Expression Vectors

Many immortalized cell lines from a variety of tissues can be obtained from repositories such as the American Type Culture Collection (ATCC) and the European distributors such as LGC Promochem. Mammalian cell lines used routinely for recombinant gene expression include COS-1/COS-7 (derived from African green monkey kidney), HEK293 (derived from human embryonic kidney) and Chinese hamster ovary (CHO) cells. The eventual choice of cell line chosen will depend on many factors including cell growth characteristics, growth medium requirements and adaptability for suspension growth, enabling scale up. It might be beneficial to perform preliminary gene expression experiments using several cell lines in order to identify the most suitable host. Gene expression optimization studies can be performed relatively rapidly by transient transfection experiments (described below). Expression levels of membrane proteins can be tested by functional assay, e.g. ligand binding, enzyme activity, signal transduction or by semi-quantitative detection using an antibody that is immunoreactive towards the target protein or towards an engineered affinity tag.

Gene expression plasmids for use in mammalian cells are geared towards the maximum production of target mRNA in stable form. In order to achieve this, some typical features of mammalian expression plasmids include: (1) a strong (usually viral origin) promoter such as the cytomegalovirus (CMV) immediate early promoter for high-level expression of the target gene, (2) a polyadenylation sequence for stabilization of the mRNA transcript, (3) a multiple cloning site (MCS) positioned between (1) and (2) to allow insertion of the target gene and (4) DNA sequences that permit replication and antibiotic selection in the cloning host *E. coli*. Some expression vectors also contain introns, short DNA sequences that are thought to increase the efficiency of mRNA transport from the nucleus to the cytosol.

3   Delivery and Maintenance of Expression Vectors in Mammalian Cells

For functional expression in mammalian cells the target gene must be delivered efficiently, and without degradation, to the nucleus of the host cell. The delivery of naked DNA is defined as transfection. For transfection to occur, host cells are treated with plasmid vector DNA in association with transfection agents (e.g. DEAE dextran, calcium phosphate or cationic lipids such as lipofectin). Expression vectors can also be delivered to host cells by other procedures, such as electroporation.

3.1   Transient Transfection

Transient transfection is a versatile procedure that permits rapid examination of a recombinant protein. Cells are transfected with the target gene followed by incubation for a limited time to allow gene expression. Accumulation of the target protein usually peaks 2–3 days post transfection and at this point the cells are collected in order to assay or purify the target protein. If the expression plasmid contains an origin of replication (such as SV40 ori) it will replicate in hosts that produce the necessary accessory factors (for SV40 ori the large SV40 T antigen). This effectively boosts gene expression by increasing gene copy number. Host cells expressing large T antigen include COS-1, COS-7 and HEK293T cells (HEK293 cells containing a chromosomal copy of the gene producing large T antigen). Transient transfection is used routinely because it is simple, fast and permits convenient side-by-side expression of many gene constructs. A good example is high-level functional expression of a synthetic rhodopsin gene[3] in COS-1 cells using DEAE-dextran as transfection agent.[3] This transient transfection procedure continues to be used for preparation and analysis of rhodopsin mutants and for expression of many other GPCRs for biochemical examination of receptor function. The major limitation of transient transfection is the difficulty in scale up for large-scale (milligram) protein production. COS-1 cells grow only as adherent monolayers and for large-scale protein production purposes, many dishes of cells are required for transfection. Handling large numbers of culture dishes is time-consuming, expensive and cumbersome. However, a recent development describes transient transfection of suspension-grown HEK293 cells.[4] This development should permit facile transient transfection for large-scale protein production.

3.2   Stable Transfection

The aim of stable transfection is permanent integration of a target gene into the chromosome of the host cell. Mammalian gene expression vectors (e.g. the pCDNA series by Invitrogen) often contain selectable antibiotic resistance markers and thus can be used for both transient transfection and for stable cell line construction. Integration of the expression plasmid occurs at random positions in the chromosome. Expression levels of the target gene will be influenced by the chromosomal site of integration (positional effects) and the number of copies of target gene stably integrated. Stable expression vectors have been designed that select for integration events into chromosomal sites that favour high-level transcription and one of them will be described here. More recent advances (Flp-In™ system, Invitrogen) now permit targeted and reproducible integration of the target gene to specific sites in the host genome mediated by DNA recombinase.[5] Such methods should enable reproducible construction of highly productive cell lines and eliminate the time-consuming screening process.

For stable cell line construction, plasmid DNA encoding the target protein is delivered to the host cell by transfection. Non-homologous recombination between the expression plasmid and the host chromosome will sometimes lead to permanent incorporation of the plasmid into the genome. The small fraction of transfected cells that undergo random plasmid integration are selected by resistance to an antibiotic marker usually present on the same plasmid. The majority of cells will be killed, whereas antibiotic resistant cells will form clonal colonies on the surface of the cell culture dish. Stable cell lines can also be constructed by co-transfection using two separate plasmids; one containing the target gene, the other containing an antibiotic resistance gene. Once established, cell lines can be stored in a frozen state or grown to provide biomass for protein purification.

The methods outlined in this chapter focus on construction of cell lines using HEK293S cells. This cell line adheres relatively loosely to the surface of cell culture dishes and adapts easily to growth in suspension and to growth in serum-free media formulations. Stable cell lines can also be constructed using other mammalian cells, with CHO cells being most widely used. CHO cells are better characterized genetically than HEK293S cells and stable cell line derivatives have been used for many years for the production of numerous commercially important recombinant proteins.[6] Techniques are available for amplification of target gene copy number in CHO cell lines and there exist many CHO cell mutants with properties designed for certain applications such as restricted N-glycosylation. Large-scale growth of CHO cells in bioreactors has also been well-established.

3.2.1   A Procedure for Stable Transfection of HEK293S Cells

The following method was used for construction of stable cell lines expressing bovine opsin and uses HEK293S cells,[7] a suspension-growth-adapted variant of HEK293 cells.[8] HEK293S cells were grown in complete Dulbecco's Modified Eagle's Medium/Ham's F-12 50/50 Mix (DMEM/F12) (DMEM/F12 base medium supplemented with heat-treated foetal bovine serum (FBS) (10% v/v), Penicillin G (100 units mL^{-1}), Streptomycin (100 µg mL^{-1}) and Glutamine (292 µg mL^{-1})). Cells were maintained at 37 °C in a 5% CO_2 humidified atmosphere and either fed or split every three days. The day prior to transfection, HEK293S cells in exponential growth phase (90% confluence) were trypsinized and plated (usually 1:10 dilution) at a density of about 1×10^6 cells/10 cm dish in 10 mL complete DMEM medium (supplemented the same as for DMEM/F12). DMEM/F12 is not used for this stage of the procedure due to potential interference by HEPES (1-piperazineethane sulfonic acid, 4-(2-hydroxyethyl)-monosodium salt) buffer in the transfection process. On the day of transfection (Day 1), highly purified plasmid DNA (30 µg) in 500 µL of CaCl$_2$ (0.25 M) was combined with 500 µL of transfection buffer (50 mM N,N-Bis [2-hydroxyethyl]-2-aminoethanesulphonic acid/250 mM NaCl/1.5 mM Na$_2$HPO$_4$, pH 7.02). This mixture was kept for 1 min at 22 °C and then added dropwise to the cell monolayer. The cells were then incubated in a humidified incubator at 35 °C in a 2% CO_2 atmosphere. After about 1 h a fine precipitate was visible upon high-power microscopic examination. The following day (Day 2), the cells were washed gently with non-supplemented DMEM/F12 medium in order to remove some of the precipitation. The cells were then trypsinized, diluted (1:10–1:100) and incubated for a further 24 h in complete DMEM/F12 medium. The following day (Day 3), the growth medium was replaced with the same medium containing the antibiotic Geneticin disulfate (G418) (1–3 mg mL^{-1}). The medium was changed every 2–3 days until G418-resistant colonies (2–3 mm diameter) appeared (2–3 weeks). Well-isolated colonies were identified by visual examination under the light microscope. These colonies were isolated by using cloning rings, trypsinized and transferred to wells in 24-well plates. These cell lines were expanded, under G418 selection, to larger cell populations. The recombinant opsin production level in each cell line was determined by signal strength using immunoblotting. The entire process of stable cell line construction takes about 4–6 weeks.

3.3   Stable Episomal Replication

Stable episomal vectors, such as those based on the Epstein-Barr virus (EBV) or bovine papilloma virus (BPV-1) replication elements, combine the features of transient and stable transfection.[9] Upon delivery to the host cell nucleus by transfection, cell populations harbouring episomal plasmids are identified by growth in the presence of an antibiotic to which the episomal plasmid confers resistance. Stable episomal plasmids segregate, along with the host chromosome, during cell division. Pools of "pseudo" stable cell lines thus generated do not require clonal selection and target gene expression is not influenced by the position of chromosomal integration, as is the case for conventional stable cell lines. Episomal cell lines thus have potential for very high-level expression because the target gene is multicopy and not influenced by chromatin context. This system has been particularly successful for expression of a serotonin receptor in milligram amounts using HEK293-EBNA cells.[9]

Attempts at establishment of HEK293S cell lines carrying an EBV-based episomal plasmid (pCEP4, Invitrogen) containing the bovine opsin was surprisingly unsuccessful (Reeves, unpublished observation). A plausible explanation for this might be that extraordinarily high-level expression of opsin-destabilized replication of the episomal vector. Another possibility was that high levels of opsin apoprotein produced were cytotoxic. Attempts to circumvent these problems by construction of a derivative of pCEP4 containing regulatory elements from the bacterial transposon Tn*10* tetracycline operon (described later) were also unsuccessful.

3.4   Viral Infection

Infection of mammalian cells using viral vectors such as Semiliki Forest virus (SFV) is an efficient method for high-level expression of eukaryotic membrane proteins such as GPCRs.[10] The SFV system can be considered analogous to the baculovirus infection-based expression system that is used for recombinant gene expression in insect cell-derived cell lines. The major advantage of the SFV system over the baculovirus system comes from the benefits afforded by mammalian cell hosts as described previously. The disadvantages of viral-based systems include the requirement for production of high-titre viral stocks and variability in infection efficiency. There is also the possibility that viral encoded proteins might dampen expression of cellular genes involved in post-translational modification and folding of the target protein. Another potential drawback of SFV expression vectors relates to health and safety risks associate with using a potential human virus pathogen.

4   HEK293S Stable Cell Lines for High-Level Expression of Eukaryotic Membrane Proteins

The following section will describe the use of HEK293S stable cell lines for production of the bovine opsin, the apoprotein of rhodopsin. HEK293S cells grow preferentially as a surface attached monolayer but will grow readily in suspension culture provided that the correct conditions are used. The following two examples describe construction of stable cell lines expressing to high levels a bovine opsin gene by using either constitutively active or regulatable CMV promoters.

4.1   Constitutive Expression

The first report of a bovine opsin gene expressed in HEK293S stable cell lines was by stable transfection of HEK293S with expression vector pRIC6.[8] The authors of this work obtained a production level of 100–200 µg rhodopsin per litre of suspension culture. This encouraging publication provided motivation for our further examination of HEK293S cells. Several different plasmid vectors were examined for stable expression of an opsin gene. The best expression levels were obtained by using pACHrhoC,[7] an expression vector derived from pACHEnc.[11] The pACHEnc vector was designed for large-scale production of secreted acetylcholinesterase from HEK293 stable cell lines. It is noteworthy that this expression vector does not contain an intron sequence in the target gene mRNA transcript. In pACHrhoC, the opsin gene is positioned downstream of the CMV promoter and the opsin mRNA transcript contains a 3′ SV40 polyadenylation (polyA) sequence. This expression vector also contains a selectable marker *neo* (neomycin resistance) that confers G418 (Geneticin) resistance to stably transfected eukaryotic cells. The neo gene is under control of the H_2L^d promoter that has relatively weak activity in HEK293S cells. HEK293S cells were transfected with pACHrhoC by calcium phosphate precipitation and stable cell lines were selected by using growth medium containing very high concentrations of G418 (1–3 mg mL^{-1}). Under these conditions cell survival is presumably dependant upon integration of the poorly expressed neo gene into highly transcriptionally active (positive positional effect) regions of the chromosome or by multiple insertions of neo gene. Proximal co-integration of the opsin gene should afford maximum potential for high-level opsin expression. This strategy led to construction of cell lines that accumulated 50 µg of opsin from cells grown to confluence in a 15 cm culture dish (about 3×10^7 cells). However, it was still necessary to screen about 10–20 clonal cell lines in order to identify stable cell lines expressing opsin at this very high level.

Stable cell lines were grown in suspension culture by inoculation of trypsinized confluent monolayers of cells directly into 2 L spinner flasks containing 500 mL of DMEM (10% FBS). The spinner flasks were stirred at very slow speed in order to minimize physical shearing forces. However, under these conditions HEK293S cells grew as large cell aggregates. Several different growth media were examined in order to optimize suspension culture growth and the best results were obtained initially using a medium designed for hybridoma growth, HB-Gro™ (Irvine Scientific) supplemented with 10% FBS. HEK293S stable cell lines grown to saturation under these conditions resulted in opsin production in the range of 2–3 mg L^{-1}. This result made it possible to purify milligram quantities of wild-type rhodopsin and similar amounts of several opsin mutants. Rhodopsin thus obtained was correctly folded and fully functional as determined by biophysical and biochemical characterization.[7]

Rhodopsin produced by HEK293S stable cell lines has been used for various applications. Milligram amounts of cytoplasmic loop rhodopsin mutants were required for examination of interactions of light-activated rhodopsin with rhodopsin kinase.[12] The constitutive expression system was next used for preparation of 6 mg of purified rhodopsin labelled with ^{15}N-lysine.[13] For isotope-labelling of rhodopsin, a calcium-free DMEM medium was formulated deficient for the amino-acids that would be introduced in stable isotope-labelled form. ^{15}N-lysine labelled rhodopsin was examined by solid-state magic angle spinning NMR spectroscopy for measurement of the distance between the Schiff base Lysine 296 retinal attachment site and the Glutamate 113 counter ion in dark-state rhodopsin. Stable HEK293S cell lines were also used for preparation of rhodopsin mutants containing single, or pairs of, reactive cysteine residues in the cytoplasmic loops.[14] Purified rhodopsin thus prepared was labelled at these positions with [13]F probes and samples were examined by solution NMR both in the dark and after light activation.

4.2 Tetracycline-Regulated Gene Expression

Construction of stable cell HEK293S lines expressing constitutively the opsin gene and mutant derivatives proved to be particularly useful for production of large quantities of rhodopsin mutants required for biophysical studies such as NMR. However, certain opsin mutants were sometimes expressed at lower levels than WT opsin and for one rhodopsin mutant of particular interest, stable cell lines could not be established. It was also evident that cell lines expressing opsin and certain opsin mutants produced less recombinant protein with increasing passage number. The instability of the wild-type opsin cell line was particularly noticeable during attempts to adapt the cells to suspension-growth in serum-free media. Indeed, the HEK293S cell line that eventually adapted to serum-free suspension growth no longer expressed opsin presumably due to outgrowth by a subpopulation of cells losing, or no longer expressing, the opsin. The most likely explanation for these unpublished observations was that high-level constitutive expression of opsin and mutant variants resulted in cytotoxicity. In order to circumvent problems associated with strong constitutive expression, tetracycline-regulated gene expression was explored.

Bacterial inducible gene expression systems have been used extensively for the production of diverse proteins such as bacteriorhodopsin and various mutants.[14] Inducible expression is a relatively recent development for gene expression in mammalian cells but has great potential for expression of potentially cytotoxic membrane proteins in stable cell lines. The most widely used inducible expression systems for mammalian cells are based upon control elements derived from the *E. coli* transposon Tn*10* tetracycline operon. The first tetracycline-inducible gene expression was described by Gossen and Bujard[15] and comprised a minimal CMV sequence containing a tetO promoter and a hybrid transactivator comprising a tetR-VP-16 fusion. More recently, Yao *et al.*[16] developed a tetracycline-inducible system that uses a full-length CMV promoter interrupted by a tetO operator sequence. The CMV-tetO promoter is repressed in cells that produce TetR protein. Full strength CMV-tetO promoter activity is obtained by addition of tetracycline to the growth medium (described later). Another example of an inducible gene expression system for mammalian cells is based on the insect moulting hormone ecdysone system.[17]

4.3   A Plasmid for Tetracycline-Regulated Expression of Opsin

The detailed methods for establishment of HEK293S stable cell lines exhibiting tetracycline-regulated gene expression have been described.[18] An outline of the procedure with supplementary information for construction and growth of inducible cell lines is described below. The decision to use the operon-based tetracycline gene regulation system, rather than the established hybrid repressor-transactivator-based system that employs a minimal CMV promoter, was made on the basis of our previous success using the full-length CMV promoter in HEK293S cells. Interruption of the CMV promoter by the short tetO operator sequence was reported not to diminish the CMV promoter strength under inducing conditions.[16] We constructed an expression vector containing vector CMV-tetO promoter and the neo gene controlled by the weak H_2L^d promoter.[18] The map of the vector (pACMV-tetO -Rho) showing these features is shown in Figure 1.

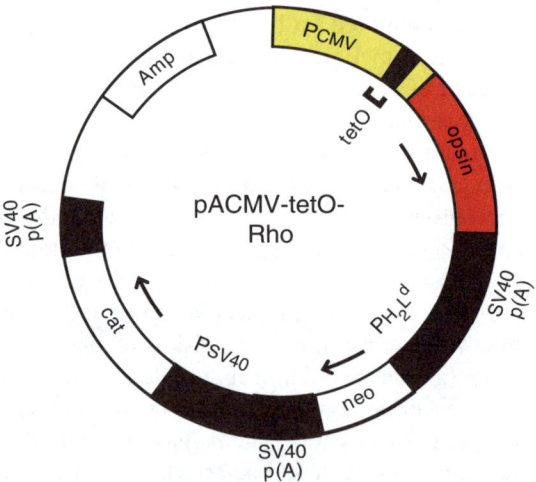

Figure 1 *Plasmid pACMV-tetO-Rho was constructed for tetracycline-inducible expression of a bovine opsin gene. The bovine opsin gene is under control of a CMV promoter containing a tetO operator sequence. The CMV-tetO promoter is inactive in HEK293S-tetR cells due to the presence of TetR repressor protein that binds to tetO. In the presence of tetracycline, binding of TetR to tetO operator sequences is prevented. This results in activation of the strong CMV-tetO promoter and expression of the opsin gene. The neo gene confers resistance to Geneticin and is controlled by the weak H_2L^d promoter. Stable cell lines are selected by growth in medium containing high concentrations of Geneticin in order to increase the likelihood of plasmid integrations into regions of the chromosome that are highly transcriptionally active*

4.3.1 Construction and Characterization of HEK293S Stable Cell Lines for Inducible Gene Expression

The use of pACMV-tetO-Rho expression vector first required construction of HEK293S cells that produced the TetR repressor protein. This HEK293S-tetR cell line was made by stable transfection of HEK293S cells with pCDNA6-tetR (Invitrogen) followed by selection with blasticidin (5 μg mL⁻¹). A population of cells resistant to blasticidin was collected and used for subsequent transfection with plasmid pACMV-tetO-Rho. Stable cell lines were now selected by using G418 (1–2 mg mL⁻¹) whereas blasticidin was discontinued. Ten isolated G418-resistant cell lines were then examined for inducible opsin expression after growth for 2-days using medium with or without inducers of the CMV-tetO promoter. A cell line that exhibited both tetracycline-regulatable and high-level expression of the opsin gene was identified by immunoblot using an anti-rhodopsin antibody. This cell line was examined in greater detail as described in Figure 2. The maximum expression level achieved was about four times higher than the maximum level ever observed using HEK293S stable cell lines constructed using pACHrhoC.[19] Nevertheless, it was still necessary to screen

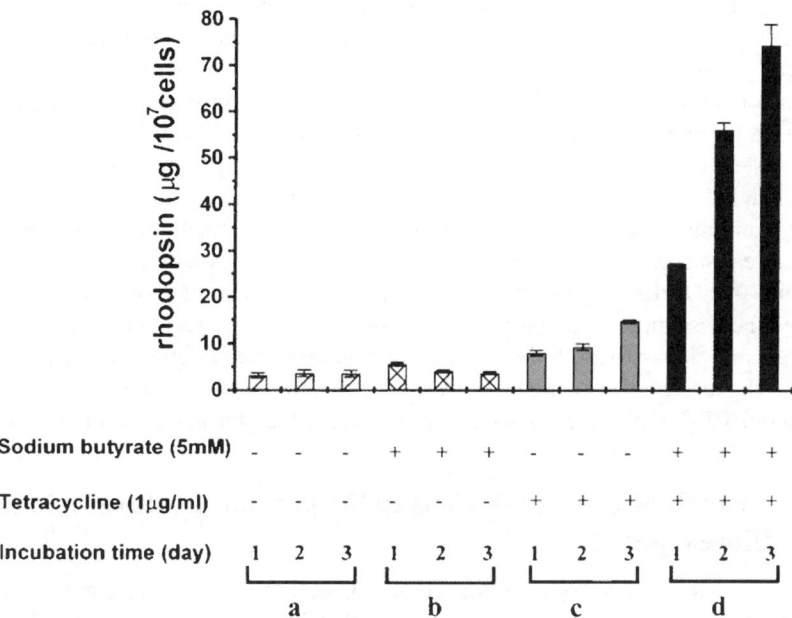

Figure 2 *Examination of conditions for optimization of inducible expression of opsin. A stable HEK293S cell line exhibiting tightly regulated expression of opsin was grown to near confluence in multiple dishes. Cells were then fed with new medium containing tetracycline and/or sodium butyrate for various times as indicated. Cells were then collected and treated with 11-cis retinal to convert opsin to rhodopsin. The total amount of rhodopsin produced under each condition was determined by a UV-visible spectroscopic assay. Maximum expression was achieved by incubation of cells in the presence of tetracycline and sodium butyrate*

several (about 10) G418-resistant colonies in order to identify a cell line expressing the opsin gene inducibly at this high level.[18]

5 Scale Up of Culture Growth and Rhodopsin Purification

Optimization of gene expression and functional protein production is the first requirement for preparation of large amounts of membrane protein using stable cell lines. However, of equal importance is the development of procedures for obtaining sufficient quantities of recombinant cells (biomass) required for subsequent protein purification. The HEK293S cell line is particularly useful because it has capacity for suspension growth. In theory, biomass yields can be improved by simply increasing culture volumes. In practice, it was necessary to optimize growth medium formulations to encourage single cell growth, implement feeding strategies and test several bioreactor formats.

Initial attempts at growing HEK293S cell lines in 500 mL suspension culture in 2 L spinner flasks while successful were often irreproducible.

Growth of larger culture volumes using 5 L spinner flasks was also unsuccessful, probably due to limitations in gaseous exchange required to maintain sufficient dissolved oxygen tension and for pH regulation. Various bioreactors formats were examined for growth of HEK293S cells. The Celligen Plus™ (New Brunswick Scientific) bioreactor equipped with a pitch blade impeller gave the most promising results and was used for optimization of suspension growth of HEK293S cell lines.[18] This bioreactor system provides very tight regulation of temperature, pH and dissolved oxygen. The pitch blade impeller provides effective mixing at low agitation speeds, thus producing minimum cell damage by shear forces. Growth medium formulations for serum-free suspension growth were also under development in the laboratory. Calcium-free DMEM supplemented with FBS (10% v/v) and Pluronic acid (0.1% v/v) formed the base medium and for further reduction of cell aggregation, dextran sulfate (300 µg mL^{-1}) was found to be a more efficient and cost-effective than heparin sulphate. The basic growth medium was enriched further by supplementation with Primatone RL-UF (0.3% w/v) that acts as a source of amino acids and small peptides.

5.1 Growth of HEK293S Cells in Suspension Culture Using a Bioreactor

Prior to growth in suspension culture, HEK293S cell monolayers were grown to confluence in 15 cm dishes containing complete DMEM supplemented with G418. For inoculation of 1.1 L of culture medium (2.2 L vessel) it was necessary to collect cells from 6 confluent 15 cm dishes. For larger volumes (3–10 L using the 14 L vessel) the number of plates was scaled up in order to provide the same inoculum size. The day before inoculation of the bioreactor the cells, at about 90–95% confluence, were fed with complete DMEM medium. On the day of bioreactor inoculation the growth medium pH and dissolved oxygen tension were brought to 7.0 and 50% respectively by sparging with a 4-gas mixture. HEK293S cells were trypsinized and collected in a separate vessel and used to inoculate the bioreactor once the pre-set conditions for temperature, pH and dO2 were reached. Growth of HEK293S cells was monitored by haemoctyometer using samples removed daily. The viable cell count at the time of inoculation (day 1) was typically about $3–5×10^5$ cells mL^{-1}. At early stages of culture growth cells were present as small chains or small clumps of 2–8 cells. At later stages of growth, typically day 5 onwards after inoculation, the cells formed larger aggregates and trypsinization was required for dispersal, enabling accurate counting. On day 5 the culture was fed by addition of glucose (10 mL of 20% w/v) and Primatone RL-UF (30 mL of 10% w/v). Induction of expression by addition of tetracycline and sodium butyrate was on day 6 by which time the cell culture reached about $5×10^6$ cells mL^{-1}. Cells were harvested by centrifugation on day 8, two days after induction of expression. The amount of rhodopsin produced by this procedure was typically 10 mg L^{-1}.[18] This level of production was also obtained for several mutants and in one case an expression level of over 20 mg L^{-1} was observed (unpublished).

5.2 Immunoaffinity Purification of Rhodopsin

Rhodopsin was purified efficiently by immunoaffinity purification after detergent solubilization of whole cells. This procedure, first described by Oprian and Khorana,[3] utilizes Sepharose-linked rho-1D4 monoclonal antibody that binds the extreme C-terminus of rhodopsin (amino acids 340–348). Elution from the immunoaffinity matrix is by competition using the corresponding synthetic nonamer peptide. A description of methods for GPCR purification is discussed in a chapter by Warne and Schertler.

6 Preparation of Eukaryotic Membrane Proteins Containing Simple N-Glycans

Recombinant membrane glycoproteins produced by HEK293S cells contain complex N-glycan, as is evident upon SDS-PAGE examination of purified opsin that migrates as a heterogeneous smear with an apparent molecular mass of 40–60 kDa.[7] The presence of complex N-glycan does not affect the biochemical or biophysical properties of rhodopsin and presumably does not affect ligand binding and signal transduction properties of other GPCRs expressed in this cell line. However, N-glycan heterogeneity was considered more likely to be problematic for growth of high quality protein crystals required for X-ray diffraction. Complex N-glycosylation can be prevented by mutagenesis of consensus glycosylation sites or by addition of inhibitors of glycosylation, such as tunicamycin, to the growth medium of cells expressing the target gene.[18,20] These strategies carry risks because complete removal of glycosylation might affect the folding or signal transduction properties of the protein as was observed for rhodopsin.[20] Complex N-glycans can also be removed by treatment with N-glycanases after purification. However, such enzymatic treatment is often inefficient for fully folded proteins and the glycanases are expensive. A more convenient solution would be biosynthesis of glycoproteins with defined N-glycans. Mutant CHO cells are available that are defective in various stages of complex N-glycan synthesis.[21] Some of these, the so-called "Lec" mutant CHO cell lines, are often used for the production of glycoproteins containing restricted N-glycan. However, no such mutant HEK293S cell lines were available.

An HEK293S mutant defective for complex N-glycan processing was made by treating HEK293S cells with a chemical mutagen and incubating surviving mutagenized cells in growth medium containing ricin.[22] Ricin enters eukaryotic cells via galactose and galactosamine moieties of complex N-glycans of cell surface glycoproteins. Cells resistant to ricin can arise by mutation of components of the complex N-glycan biosynthesis machinery. Several HEK293S cell lines were isolated that were capable of growth in medium containing ricin. These ricin-resistant cell lines were analysed by transient transfection experiments using a plasmid containing the opsin gene. After transfection and expression, opsin purified from individual ricin resistant HEK293S cell lines was examined by SDS-PAGE and visualized by silver staining. Recombinant opsin produced by one of the cell lines showed a clear absence of a trailing smear that was present in opsin produced by the parent HEK293S cell line. Detailed examination of the N-glycan present revealed that only $GlcNAc_2Man_5$ was present, indicative of complete absence of the GlcNAc transferase I (GnTI) activity. The tetracycline-inducible gene expression system was reassembled in the HEK293S (GnTI-) cell line. For unknown reasons the HEK293S (GnTI-) cell line displayed higher levels of transfection that the parent HEK293S cells. As a consequence cells had to be diluted highly after transfection in order to obtain single colonies. The HEK293S (GnTI-) cell line was also more sensitive to G418 and resistant cell lines could not be obtained using concentrations of G418 in excess of 250 μg mL^{-1}. HEK293S (GnTI-) cell lines supported tetracycline-inducible expression of opsin and several opsin mutants in amounts similar to the parent HEK293S strain. Importantly, the HEK293S (GnTI-) mutant still grew well in suspension in bioreactor culture volumes up to 10 L. Growth of the cell line and inducible expression of WT opsin in a bioreactor is shown in Figure 3. Details of this experiment are described in the Figure legend.

7 Outlook for the Use of HEK293S Tetracycline-Inducible Cell Lines for Large-Scale Preparation of Other Eukaryotic Membrane Proteins

The HEK293S expression system is now an established system for production of recombinant rhodopsin in fully functional form and in many milligram amounts. Importantly, there are now several reports that describe the use of these purified samples for analysis of rhodopsin by NMR.[12,13,23] More recently it has also been possible to grow needle crystals of recombinant rhodopsin expressed inducibly in HEK293S (GnTI-) cells (Reeves and Khorana, unpublished). This opens up the possibility of performing extensive biophysical examination of rhodopsin mutants in order to gain an understanding of receptor activation.

The HEK293S expression systems described herein should prove useful for large-scale preparation of other GPCRs and other classes of eukaryotic membrane proteins. Such investigations are underway and while there are some encouraging

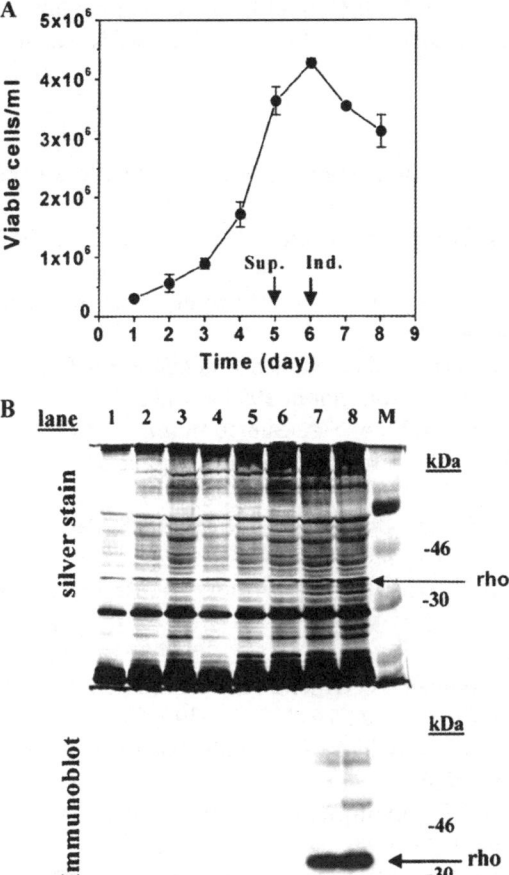

Figure 3 *Growth of an HEK293S (GnTI-) cell line expressing inducibly the bovine opsin gene. The cell line was grown in a bioreactor as described in the text. Viable cell counts (A) were recorded every day and samples were examined by SDS-PAGE (B) followed by detection using silver stain (total protein) or immunoblot (opsin specific). Cells in the bioreactor were fed on day 5 (Sup.) and opsin gene expression was induced on day 6 (Ind.) by addition of tetracycline and sodium butyrate. The appearance of opsin on day 7 and 8 only is indicated by an arrow (rho). Cells were harvested on day 8 and the total yield of rhodopsin from 1.1 L was about 6 mg*

results, it is becoming increasingly clear that bovine opsin is particularly amenable to high-level expression in various recombinant systems. Rhodopsin has evolved along with the visual system to provide exceptional sensitivity to light with low background noise. In order to achieve this it is produced naturally at high levels in the specialized rod photoreceptor cell. Furthermore, rhodopsin is extremely stable even upon solubilization in many detergents, and this helps greatly in its purification. Rhodopsin has set the standard for eukaryotic membrane protein expression and demonstrates clearly the potential that mammalian cells have for this purpose.

The author would like to thank Gobind Khorana and his lab members for all their support and encouragement during this work. This work was funded by NIH.

References

1. J.L. Baneres, A. Martin, P. Hullot, J.P. Girard, J.C. Rossi and J. Parello, *J. Mol. Biol.*, 2003, **329**, 801.
2. J.F. White, L.B. Trinh, J. Shiloach and R. Grisshammer, *FEBS Lett.*, 2004, **564**, 289.
3. D.D. Oprian, R.S. Molday, R.J. Kaufman and H.G. Khorana, *Proc. Natl. Acad. Sci. USA*, 1987, **84**, 8874.
4. Y. Durocher, S. Perret and A. Kamen, *Nucleic Acids Res.*, 2002, **30**, E9.
5. B. Sauer, *Curr. Opin. Biotechnol.*, 1994, **5**, 521.
6. R. Kunaparaju and M. Liao, N.A. Sunstrom, *Biotechnol. Bioeng.*, 2005, **91**, 670.
7. P.J. Reeves, R.L. Thurmond and H.G. Khorana, *Proc. Natl. Acad. Sci. USA*, 1996, **93**, 11487.
8. J. Nathans, C.J. Weitz, N. Agarwal, I. Nir and D.S. Papermaster, *Vis. Res.*, 1989, **29**, 907.
9. H.D. Blasey, R. Hovius, H. Vogel and A.R. Bernard, *Biochem. Soc. Trans.*, 1999, **27**, 956.
10. K. Lundstrom, *Biochim. Biophys. Acta*, 2003, **1610**, 90.
11. B. Velan, C. Kronman, A. Ordentlich, Y. Flashner, M. Leitner, S. Cohen and A. Shafferman, *Biochem. J.*, 1993, **296**, 649.
12. M. Eilers, P.J. Reeves, W. Ying, H.G. Khorana and S.O. Smith, *Proc. Natl. Acad. Sci. USA*, 1999, **96**, 487.
13. M.C. Loewen, J. Klein-Seetharaman, E.V. Getmanova, P.J. Reeves, H. Schwalbe and H.G. Khorana, *Proc. Natl. Acad. Sci. USA*, 2001, **98**, 4888.
14. S.S. Karnik, M. Nassal, T. Doi, E. Jay, V. Sagaramella and H.G. Khorana, *J. Biol. Chem.*, **262**, 9255.
15. M. Gossen and H. Bujard, *Proc. Natl. Acad. Sci. USA*, 1992, **89**, 5547.
16. F. Yao, T. Svensjo, T. Winkler, M. Lu, C. Eriksson and E. Eriksson, *Hum. Gene Ther.*, 1998, **9**, 1939.
17. D. No, T.P. Yao and R.M. Evans, *Proc. Natl. Acad. Sci. USA*, 1996, **93**, 3346.
18. P.J. Reeves, J.M. Kim and H.G. Khorana, *Proc. Natl. Acad. Sci. USA*, 2002, **99**, 13413.
19. R.L. Thurmond, C. Creuzenet, P.J. Reeves and H.G. Khorana, *Proc. Natl. Acad. Sci. USA*, 1997, **94**, 1715.
20. S. Kaushal, K.D. Ridge and H.G. Khorana, *Proc. Natl. Acad. Sci. USA*, 1994, **91**, 4024.
21. P. Stanley, S. Sallustio, S.S. Krag and B. Dunn, *Somat. Cell Mol. Genet.*, 1990, **16**, 211.
22. P.J. Reeves, N. Callewaert, R. Contreras and H.G. Khorana, *Proc. Natl. Acad. Sci. USA*, 2002, **99**, 13419.
23. A.B. Patel, E. Crocker, P.J. Reeves, E.V. Getmanova, M. Eilers, H.G. Khorana and S.O. Smith, *J. Mol. Biol.*, 2005, **347**, 803.

CHAPTER 3

Expression of Recombinant G-Protein Coupled Receptors for Structural Biology

FILIPPO MANCIA AND WAYNE A. HENDRICKSON

Howard Hughes Medical Institute and Department of Biochemistry and Molecular Biophysics, Columbia University, New York, NY 10032, USA

1 Introduction

1.1 Signal Transduction through G-protein Coupled Receptors

A cell perceives its environment through the receptor molecules embedded in the plasma membrane and endowed with a selective sensitivity toward various stimuli. Conformational changes or associations that occur when a receptor interacts with an external stimulus are transmitted to the cell interior where responses are induced, often elicited through a cascade of signal transduction events. The mechanism of signal transduction depends on molecular characteristics of the receptor, and there are several classes of receptors. Besides those linked to the downstream elements by heterotrimeric G-proteins, our subject here, there are many receptors linked to protein tyrosine kinases, ones which are linked to ion channels, and diverse receptors coupled in other ways as in the TGFβ/Smad, Notch, and Wnt systems. The stimulus detected by a receptor may be physical, *e.g.* light or an electrostatic potential, but in most cases the stimulus is a chemical ligand. Some ligands are macromolecules, others are small compounds; some are diffusible, others are associated with another cell or the extracellular matrix.

The most salient molecular characteristic of G-protein coupled receptors (GPCRs) is a pattern of seven hydrophobic segments that correspond to transmembrane α-helices; thereby GPCRs are also known as seven transmembrane (7TM) receptors. This 7TM pattern was first seen in sequences of rhodopsins[1–3] and a little later in the sequence of the hamster β2-adrenergic receptor.[4] The involvement of heterotrimeric G-proteins in signaling through 7TM receptors was first worked out for the β2-adrenergic receptor,[5] where binding of the hormone epinephrine activates Gαs which in turn stimulates adenylyl cyclase to produce the second messenger cyclic AMP. The parallel role of the G protein transducin in visual signaling, where photoactivation of rhodopsin stimulates cyclic GMP phosphodiesterase and sodium-channel closure, was discovered a little later.[6] Taken in this context, the evident homology between these two biologically disparate 7TM receptors prompted the realization that they were the founding members of the GPCR family. A flood of GPCR clonings ensued, including the first for serotonin receptors[7,8] and the discovery of the huge subfamily of odorant receptors.[9] A total of 948 GPCR genes were identified in a recent analysis of the human genome sequence,[10] up somewhat from the 616 found initially.[11] These receptors include sensors of exogenous stimuli including light and odors and others that respond to endogenous ligands ranging from cationic amines such as serotonin to peptides such as angiotensin and to proteins such as chemokines and glycoprotein hormones.

Ligand binding (or photoisomerization of retinal in the case of opsins) activates a GPCR to serve as a nucleotide exchange factor for the cognate heterotrimeric G protein. Each heterotrimer is a labile association of the GTPase component, Gα, with a Gβ:Gγ heterodimer. Gα(GDP):Gβ:Gγ dissociates to Gα(GTP) and Gβ:Gγ when stimulated by an activated GPCR, and the trimer reassociates after GTP hydrolysis. Both components are tethered to the membrane, by N-terminal myristoylation of Gα and C-terminal prenylation of Gγ, and after activation they can diffuse away from the receptor to effector sites on their membrane-associated targets. There are at least 15 different Gα proteins, 5 Gβs, and 5 Gγs[12] and different combinations are selective for specific GPCRs and for target effector molecules.[13] In particular, Gαs(GTP) stimulates adenylyl cyclase whereas Gαi(GTP) inhibits it, and Gαq(GTP) stimulates phospholipase C-β. Crystal structures have been determined for Gα proteins in various states, including a complex between Gαs(GTP) and the catalytic portion of adenylyl cyclase, of Gβ:Gγ and of heterotrimers.[14–19] GPCRs are thought to exist in equilibrium between inactive and active states, which naïvely correspond to empty and ligand-occupied receptors. The ligand-binding site is known, from studies on rhodopsin[20] and β2-adrenergic receptor,[21] to be located between helices near the center of the membrane. How activation occurs for GPCRs that bind peptide ligands, such as substance P,[22] or intact proteins, such as follicle stimulating hormone,[23] is not entirely clear; portions of such ligands might insert between transmembrane helices or else bind to interhelical loops to effect conformational changes (for a review, see Ref. 24). Changes in 7TM conformation that accompany ligand binding or activation are linked to the receptor binding of the G protein; G protein association with a receptor increases its ligand affinity. A model of G-protein coupling in GPCR activation to effector targets is given in Figure 1.

1.2   Structural and Functional Characteristics of GPCRs

The characteristic 7TM pattern of hydrophobic segments in GPCRs provides powerful constraints on the possibilities for a 3D structure. Moreover, this pattern when seen in rhodopsin was reminiscent of that in bacteriorhodopsin where the structure

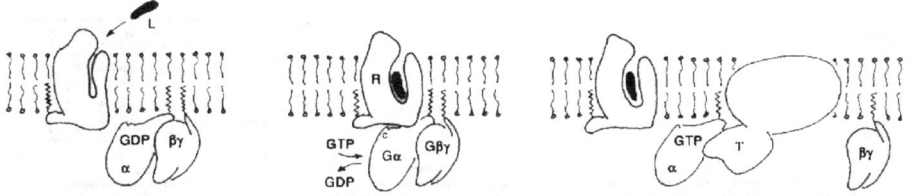

Figure 1 *Coupling of ligand (L) binding to a GPCR (R) through catalysis of GTP for GDP exchange in Gα and dissociation of free Gα(GTP) to interact with an effector target (T). Components here are approximately at scale, based on known structures of inactive rhodopsin, Gα:Gβ:Gγ, and Gα (GTP):adenylyl cyclase*

from purple membranes had shown the disposition of helices.[25] Although the sequences showed no detectable homology and these two photoreceptors have very different biochemical actions they do both use a Schiff-base linked retinal to detect light. Later, the topological connections in bacteriorhodopsin were found at a high resolution,[26] and ultimately electron crystallography showed that the helices in rhodopsin have the same topology as those in bacteriorhodopsin, although their orientation is somewhat different. Meanwhile, as sequences accumulated, conserved features were realized and comprehensive alignments were made.[27] The vast majority of GPCR sequences are homologous with rhodopsin, but there also are groups of the superfamily that are atypical. These include the secretin and metabotropic glutamate receptors.

GPCR sequence alignments reveal many features besides the 7TM pattern, and some of these are evident in the subset shown in Figure 2. Most strikingly, there is substantial conservation in the transmembrane segments. There is, however, a great variation in size as well as sequence for the N- and C-terminal segments and also for most loops. Some N-termini, *e.g.* those of glycoprotein hormone receptors, include very large domains (not shown in Figure 2). The cytoplasmic 5–6 loop is extremely variable in size. By contrast, the extracellular 2–3 and cytoplasmic 3–4 loops are relatively constant in size. The positions of several functional sites are also typically in common. These include a disulfide bridge between the N-terminus of TM3 and the 4–5 extracellular loop, N-linked glycosylation at 10–20 residues before TM1, a palmitoylation site at the end of H8, and phophorylation sites in loop 5–6 and the C-terminal segment.[28]

Structural models have also been predicted from GPCR sequences.[29,30] The best constrained model came from combining the structure of frog rhodopsin, determined at 9 Å resolution by electron microscopy of 2D crystals,[31] with an analysis of the sequences of some 500 rhodopsin-family members.[32] The result was an alpha-carbon template for the transmembrane helices for the rhodopsin family of GPCRs. Subsequently, in a landmark study, the structure of bovine rhodopsin was reported at 2.8 Å resolution,[33] and later refined to 2.6 Å resolution[34] (Figure 3). A second structure has also been reported.[35] These structures confirm predictions based on the alpha-carbon template and add rich detail on rhodopsin in the inactive, 11-*cis* retinal state.

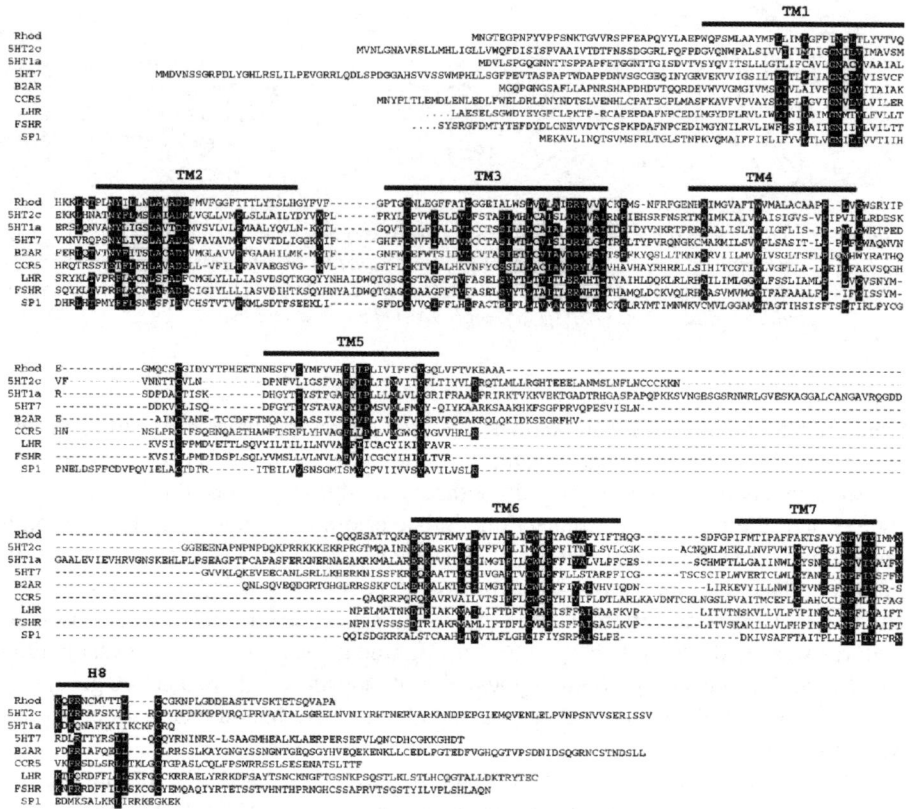

Figure 2 *Alignment based on the structure of rhodopsin of GPCR sequences. Sequences are from the following sources: Rhod, bovine rhodopsin; rat 5HT2c; human 5HT1a; mouse 5HT7; B2AR, human β2-adrenergic receptor; human CCR5; LHR, human leuteinizing hormone; FSHR, human follicle stimulating hormone; and SP1 mouse mOR28 olfactory receptor. Bars over the sequences represent transmembrane helices TM1–TM7 and C-terminal helix H8, respectively. Highlighted residues designate identities or certain close similarities*

1.3 Structure Determination of GPCRs

Integral membrane proteins present formidable, albeit not insurmountable, problems for structural analysis. There have been striking successes starting with the first result in 3Ds, by electron crystallography at 7 Å resolution, on bacteriorhodopsin[25] and the first atomic-level structure, at 3 Å resolution by X-ray crystallography, on a photosynthetic reaction center.[36] Membrane-protein structures have been determined at an accelerated pace in recent years, and many of these new structures have had a dramatic impact as in the cases of cytochrome c oxidases[37,38] and of potassium[39,40] and water channels.[41,42] Recently, the scope of successes has broadened to include several transporters,[43–47] pumps,[48,49] and other channels.[50–54] Nevertheless, the structural output on membrane proteins is a very small fraction of that for soluble macromolecules.

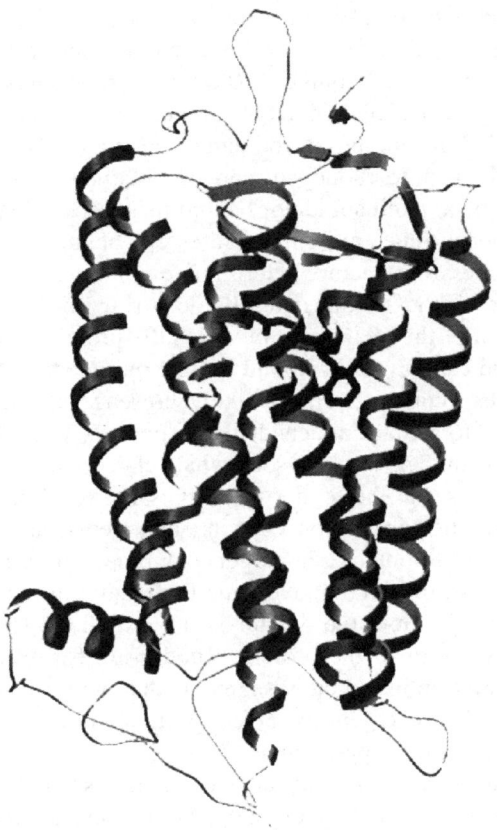

Figure 3 *Ribbon diagram of bovine rhodopsin, drawn with the extracellular side facing up, and the intracellular side facing down. The bound retinal is visible in the plane of the membrane. The figure was drawn from coordinates deposited in the Protein Data Bank (PDB) with accession number 1L9H[34]*

While membrane proteins comprise 20–30% of all proteins in both prokaryotic and eukaryotic organisms[55] they are but a fraction of a percent of those with a known structure. It is, of course, the natural association of membrane proteins with lipid bilayers that complicates their structural analysis. Once membrane-protein crystals are obtained, for example, the diffraction analysis is as straightforward as it is for aqueous soluble macromolecules. Problems that arise in the recombinant expression of membrane proteins are even more limiting than difficulties in purification and crystallization. There have been recent successes in producing recombinant bacterial proteins for analysis, but eukaryotic membrane proteins have been strikingly recalcitrant in expression at the scale needed for a structural analysis. Although there are structures of important eukaryotic membrane proteins, these have all come from natural sources except for the peripheral, single-leaflet associated cyclooxygenases[56,57] and for the very recent rat voltage-gated potassium channel.[53]

GPCRs are found only in eukaryotes, and the majority of them are present only in mammals. Essentially all GPCRs are scarce and cannot be prepared from membranes derived from natural sources. Rhodopsin represents the most notable exception as it can be isolated in large amounts and at close to 100% purity directly from purified membranes of rod outer segments.[58] The abundance and relative ease with which pure rhodopsin can be obtained from natural sources is widely believed to be a dominant factor behind its successful structure determination. High natural occurrence not only ensures an abundant supply of material, a prerequisite for successful structure determination, but it also bears two additional advantages. The first is the "ideal," naturally occurring, match between the protein and the composition of the membrane in which the protein resides. The composition of a membrane can have a profound impact on the stability of a protein that resides within it, especially, given the likely presence of specific, high-affinity lipids that are able to remain attached to the protein, once the lipid bilayer is destroyed by detergent. The second, somewhat related, advantage of using abundant natural sources is that the higher the expression level of a given membrane protein the fewer are the purification steps in a detergent-containing aqueous solution typically required to obtain a homogeneous preparation. Detergents tend not to be ideal substitutes for a lipid bilayer, and the greater the extent that one has to use these during purification, the greater is the likelihood for removal of lipids important for the stability of the protein. Rhodopsin, for example, is present at such a high concentration in rod outer segments that, with a careful preparation of these membranes, the detergent-extracted protein could be used directly for (successful) crystallization experiments.[58,59]

Unfortunately, rhodopsin represents an exception. Essentially every other GPCR is sufficiently inabundant as to defy a rhodopsin-like approach for structural studies. Moreover, GPCR systems do not exist in prokaryotes and thus bacterial homologs, exploited so effectively in structural studies of potassium channels[39] and transporters,[44,47] are not an option in this case. Appropriate recombinant expression systems are therefore needed to support these structural studies.

2 Expression of Recombinant GPCRs

2.1 Overview

GPCRs have been successfully expressed in a multitude of hosts, ranging from bacteria to mammalian cells. The expression systems currently available for the production of GPCRs have been discussed extensively in several excellent reviews.[60–63] In particular, Sarramegna and colleagues[62] provide a detailed comparison of all the expression systems that have been used for the production of GPCRs. Moreover, these authors compile a truly informative and up-to-date chart that summarizes the published results (expression levels, activity, *etc.*) of essentially every GPCR that has ever been heterologously expressed. An analysis of these data suggests that there is substantial variability (up to 100-fold) in the levels of different GPCRs produced in the same expression system. Similarly, expression levels for the same GPCR expressed in different systems can vary dramatically. In summary, given the so far scarce success of GPCR structure determination, a preferred expression system for these proteins has yet to emerge. Regardless of the problematic nature of GPCRs as targets for structural studies, an abundant supply of functional material is a prerequisite for a successful structure determination.

GPCRs, like other membrane proteins, undergo a complex and poorly understood folding and membrane-insertion mechanism. Unique properties of a cell environment that may facilitate homologous (as opposed to heterologous) expression of a GPCR include the specific lipid composition of the various membrane compartments, cell-type specific chaperones, and unique post-translational modifications including defined glycosylation, phosphorylation, sulfonation, and other covalent modifications. Following this line of thoughts, one should attempt to express the GPCR of choice in a system that matches as closely as possible its native environment. Not surprisingly, the highest expression level reported[62] for a GPCR is that of the Ste2 receptor from *Saccharomyces cerevisiae*, homologously overexpressed in *S. cerevisiae*.[64] The majority of GPCRs are from mammalian origin. Therefore, mammalian cell-based expression systems represent a likely good choice for these proteins. Although it is now routine to use mammalian protein expression systems in functional studies, based on transient-transfection experiments, their application to large-scale protein production has often been deterred by difficulties in obtaining large amounts of material, rapidly, and at a reasonable cost. A path toward the solution of these deterrents lies in the generation of stable, as opposed to transient, cell lines.

An alternative, in a way opposite, approach is to use a robust, simple, rapid, and cheap albeit "primitive" system and to optimize it to produce sufficient amounts of functional proteins. Bacterial expression systems, primarily based on the Gram-negative bacterium *Escherichia coli*, have been by far the most successful for the production of recombinant proteins for structural studies. *E. coli*-based expression of a number of GPCRs has indeed worked remarkably well, producing milligram amounts of functional receptors (for example, see Ref. 65).

These two "opposing" strategies for the expression of GPCRs, those of *E. coli*-based and stable mammalian cell-based expression systems, will be discussed in detail, analyzing their advantages and disadvantages. Other expression systems that are either an alternative to the generation of stable cell lines, such as viral infection of mammalian cells, or that fall conceptually in between these two extremes, such as those based on yeast hosts and on viral infection of insect cells will also be discussed briefly in separate paragraphs.

All of the above-mentioned expression systems can be expected to yield a functional protein. An alternative approach is to generate abundant quantities of non-functional proteins and to regain functionality by subsequent refolding, either in lipid bilayers or in detergent micelles. Expression systems that follow this approach and that have been used successfully for the production of GPCRs will also be mentioned at the end of this chapter.

2.2 Bacteria as Hosts for the Production of Functional GPCRs

High-resolution structural studies of proteins generally require large amounts of pure, properly folded material. Indeed, the advent of gene manipulation techniques for producing recombinant protein in heterologous systems is arguably the most important breakthrough of the past 30 years for structural biology, exceeding even the wondrous developments in synchrotron crystallography[66] and NMR spectroscopy.[67] Bacterial expression systems, primarily based on the Gram-negative bacterium *E. coli*, have been by far the most successful for the production of recombinant proteins for structural studies. As of mid-2005, 20,504 of the 32,643 protein structures deposited in the Protein Data Bank (PDB) had "Expression_System" records (www.rcsb.org). Of these, over 90% were produced using *E. coli*; ~3.5% were produced in yeast; ~2.5% with insect cells, and ~1.5% using mammalian cells. The remaining ~3% of these structures were determined using proteins expressed in other systems, including bacterial hosts such as *Bacillus subtilis* and cell-free expression systems. The success of bacteria-based expression systems arises from several factors including the ease with which such organisms can be genetically manipulated; the thorough understanding of their transcription and translation machinery, which has led to the ability to achieve high levels of protein expression; the rapidity of their growth; and the relatively low cost of their use.

Bacteria are unable to perform the post-translational modifications that are typical of eukaryotic membrane proteins. GPCRs undergo various modifications after protein synthesis, with glycosylation, phosphorylation, and palmitoylation being the most common. The essentiality of these modifications for function of a given receptor cannot be predicted *a priori*. For the most studied receptors, these data are typically known. Glycosylation at Asn15 of bovine rhodospin, for example, is critical for the stability of the photoactivated metarhodopsin II state, and hence for coupling to transducin.[68] This information makes rhodopsin an unsuitable candidate for bacterial expression. Many GPCRs, however, are able to function without post-translational modifications. If the protein can withstand it, the absence of these modifications eliminates a source of heterogeneity. This represents an important advantage of bacterial expression over other, eukaryotic-cell based systems.

GPCRs can be inserted in the inner bacterial membrane. The lipid composition of the bacterial inner membrane is rather different from that of eukaryotic cells.[69] Most notable is the absence of cholesterol (or any sterols for that matter) in bacteria, a major component of membranes of higher organisms, but there are other less notable but potentially as important differences. Phosphatidylserine (PS), for example, is an abundant phospholipid in mammalian cells. In *E. coli*, all the PS is converted by the enzyme phosphatidylserine decarboxylase to phosphatidylethanolamine (PE).[70] PE accumulates to become the predominant component of bacterial membranes.[69,71] In their recent review, Opekarová and Tanner[69] present conclusive examples showing that membrane proteins depend on phospholipids and sterols for their integrity and activity. Among GPCRs, the oxytocin receptor,[72] the human μ-opioid receptor,[73] and the dopamine D-1 receptor[74] have been shown to require specific lipid components for optimal function.

G-proteins do not occur naturally in bacteria. This has two effects on the expression of GPCRs in bacterial hosts. The first is that, regardless of the fact that the receptor is properly inserted in the membrane, most GPCRs will bind agonists only in a low affinity state, unless exogenous G-proteins are supplied. The second is that the system is not optimal for functional studies of the signal transduction mechanism of receptor *in vivo*. High affinity binding sites can be successfully restored, however, either by exogenous addition of purified G-proteins,[75] by addition of membranes from cells expressing G-proteins,[76] or by constructing and expressing a GPCR G-protein fusion.[77,78] In contrast to the concerns about bacteria lacking authenticity as hosts for GPCRs, the observation that bacterially expressed GPCRs are able to shift to a high-affinity state for agonists upon successful coupling to G-protein heterotrimers provides compelling support for the utility of this expression system.

Functional expression of a GPCR in *E. coli* was first shown in the late 1980s: Marullo and colleagues[79] fused the coding sequence of the human β2-adrenergic receptor to an N-terminal fragment of the β-galactosidase gene, and tested cells expressing the fusion protein with β2-adrenergic receptor specific ligands. Production of β2-adrenergic receptors was shown by the presence, on intact bacteria, of binding sites for catecholamine agonists and antagonists possessing a typical β2-adrenergic pharmacological profile. Binding and photoaffinity labeling studies performed on intact *E. coli* cells and on membrane fractions showed that these binding sites are located in the inner membrane of the bacteria. Expression levels reported in this study were too low (0.4 pmol mg^{-1} of membrane protein) to enable the use of this system for milligram-scale protein production. However, this study showed that the use of a bacterial host for the expression of GPCRs was a possibility. Protein levels improved considerably with the expression of serotonin (5-HT) receptor subtype 1a (5HT1a) as a fusion to the C-terminus of maltose-binding protein (MBP).[75] MBP, coded by the malE gene is a protein that is targeted to the bacterial periplasm.

Soon after the work on 5HT1a, Grisshammer and colleagues[80] showed that neurotensin-binding could be detected in *E. coli* membranes isolated from cells expressing the rat neurotensin receptor (NTR). The authors compared expression levels of wild-type NTR, NTR N-terminally fused to a bacterial signal peptide sequence (of enterotoxin B), and NTR N-terminally fused to the C-terminus of MBP. Expression levels, measured by radioligand-binding assay, showed a 40-fold increase with the MBP–NTR fusion (to approximately 450 binding sites per cell, equivalent to 15 pmol mg^{-1} of membrane protein), unequivocally indicating that the fusion with the malE gene was the preferred way to proceed. One of the greatest benefits of working with a bacterial system is the ease and rapidity with which an expression construct can be engineered, modified, and evaluated. Grisshammer and colleagues took advantage of this in the most thorough of ways. They systematically varied the expression construct for NTR, each time evaluating protein production levels, purification yield, and efficiency.[65,81] Following this approach, expression levels increased to a maximum of approximately 1000 copies/cell. 3 mg of functional NTR can currently be purified to homogeneity from a 20 L (100 g of cells) bacterial culture.[82] A similar approach has been applied to the expression, for example, of the rat neurokinin A receptor,[83] the human adenosine A2a receptor,[84] and the M1[85] and M2[86] muscarinic acetylcholine receptors. The strategy followed for the optimization of expression constructs for the above-mentioned examples is similar. However, the results appear to vary between different receptors, even within the same family (for example, see Ref. 87). This implies that optimization will be required for every receptor expressed in *E. coli*, and that being able to increase expression to acceptable levels cannot be guaranteed *a priori*. Analysis of the amino acid sequence of a given receptor can only provide an (not too reliable) indication of whether it will or will not express.[88]

Are there some aspects that can be generalized, and that can serve as starting guidelines for someone trying to express a GPCR in *E. coli*? Although GPCRs have been expressed without N-terminal fusion, the presence of a fusion partner seems to be advantageous, and in some cases even essential. For example, the M2 muscarinic receptor could only be expressed in an active form when fused to MBP.[86] Given its success, MBP would be the preferred choice. Other molecules used successfully include the outer membrane protein LamB,[87,89] the periplasmic protein alkaline phosphatase[90] and, surprisingly, the cytoplasmic protein β-galactosidase.[91] The role of the fusion partner is unclear. In the case of MBP, its translocation to the periplasm could help drive the topologically correct insertion of the fused GPCR into the inner membrane (Figure 4).

MBP can be removed easily from the GPCR by genetically engineering a protease recognition sequence in the linker region between the two proteins. Tobacco etch virus (TEV) protease efficiently cleaves its hepta-residue recognition sequence, engineered between the two proteins[82] (Figure 4). A stable mutant of TEV protease with enhanced catalytic activity can be produced in abundance from *E. coli*.[92] Fusion of a small and stable bacterial protein such as a truncated form of thioredoxin A (TRX) to the C-terminus of NTR leads to a triple-protein fusion construct (MBP–NTR–TRX) that appears to further stabilize the receptor and improve expression and purification yields.[65,82] A similar stabilizing effect of TRX has been observed in the bacterial expression of the rat serotonin receptor subtype 2c (5HT2c)

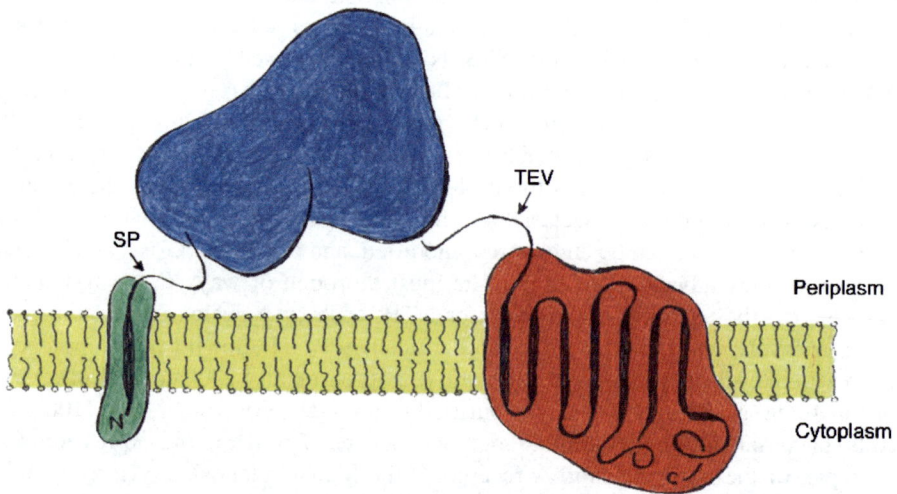

Figure 4 *A schematic drawing of MBP–GPCR fusions. MBP is shown in blue, with its signal peptide in green. The GPCR is in red. The bacterial inner membrane is drawn in yellow. SP is where the natural signal peptidase cleaves to produce the mature fusion protein. TEV is the TEV protease cleavage site utilized to separate the two proteins after expression or purification*

(Mancia and Hendrickson, unpublished results). The reasons for the beneficial effects of TRX on bacterially expressed GPCRs are unknown.

The slower the cell growth and the more attenuated the induction, the better. Therefore, growing the bacteria at low temperatures (18–25 °C) after having induced expression and using an expression plasmid bearing a weak promoter element (such as the lac wild-type promoter), are typically a good starting point and choice, respectively, for an initial experiment on a previously untested target. Finally, the need to optimize codon usage for *E. coli* expression is questionable, especially given the amount of time and effort required. Regardless, there appears to be no disadvantage in using *E. coli* strains that have been transformed to supplement the bacterium with tRNAs for rare codons. Different strains might need to be tested, as there is evidence for considerable interstrain variability in the expression levels, with DH5α and BL21 (and strains derived from this, such as Rosetta) performing the best (Mancia and Hendrickson, unpublished results).

2.3 Production of GPCRs in Stably Transfected Mammalian Cells

Despite the advantages discussed in the preceding section, bacterial systems often fail in their application to the expression of eukaryotic proteins.[93–95] Failure to achieve acceptable expression often arises from toxicity of the foreign protein or its inability to fold or be targeted properly in the bacterial cell. Such problems inevitably result in low levels of expression or protein misfolding.[93–95] Thus, despite drawbacks in efficiency, alternative expression systems based on eukaryotic hosts, have been developed for large-scale protein production. These include expression in yeast, insect cells, and mammalian cells, all of which have been used successfully in producing proteins for structure determination. In the particular case of mammalian proteins, although heterologous eukaryotic cell systems or bacteria can be effective, optimal expression of some such proteins may require mammalian host cells.

Mammalian proteins evolved in a mammalian cellular milieu, and it is understandable that both proper folding and stability may depend on this environment. Unique properties of the mammalian cell environment that may facilitate homologous expression include specific lipid compositions of the various membrane compartments, cell-type specific chaperones, and unique post-translational modifications including defined glycosylation, phosphorylation, palmitoylation, and sulfonation. Production of recombinant protein in mammalian cells can be accomplished either through transient transfection, viral infection, or through integration of expression constructs into the host genome. Each of these methods has advantages and disadvantages.

A very large number of GPCRs has been expressed in mammalian cells, predominantly for functional studies, following transient transfection protocols. Mammalian cells offer the ideal host for functional studies, as these typically require an active protein but not in large amounts. Expression levels are highly variable from one receptor to another (for a list of some examples, see Ref. 62), but functional analyses are effective nevertheless. Rapid and efficient techniques for transient transfection and site-directed mutagenesis facilitate the experiment, and the presence of a complete signaling machinery in mammalian cells provides an authentic functional readout. Thereby, numerous successes are reported in the tests of GPCR function in ligand binding, receptor activation, oligomerization, desensitization, internalization, and interaction with G-proteins. Moreover, given the presence of a complete and functional receptor–effector signaling pathway, mammalian cell-based expression of GPCRs is used routinely in assays for drug discovery directed against these important pharmaceutical targets. Although it is now routine to use mammalian protein expression systems in functional studies, their application to large-scale protein production has often been deterred by difficulty in obtaining large amounts of material, rapidly, and at a reasonable cost. When this is achieved, however, advantages ensue both for the structural work and also for functional tests of structure-inspired hypotheses. Not surprising, given the above, is the scarcity of reports in which a GPCR has been ectopically expressed in a mammalian host and then purified for structural studies. Bovine rhodopsin represents one of the very few exceptions, having been expressed to high levels (up to 10 mg L^{-1}) in human embryonic kidney 293 (HEK-293) cells, and purified to homogeneity in a single step using an immunoaffinity column.[96]

Stably transfected cells are desirable for their ability to provide a constant source of recombinant protein, but the production of stable transfectants is time consuming. To generate a cell line, the coding sequence for the gene of interest is placed under the control of a strong constitutive promoter such as the promoter element derived from cytomegalovirus (CMV).[97] The promoter can be inducible, if protein-induced toxicity to the cell should become an issue.[98] The plasmid is transfected into the host cells using standard techniques. Generation of stably producing cell lines requires integration of the expression construct into the genome of the host cell. The selection of stable integrants is typically accomplished with the use of antibiotic markers. The antibiotic marker can be present in the same or in a separate plasmid as that carrying the gene of interest. Integration is a rare event, and antibiotic-resistant cells represent a minority of the total number of cells. Expression levels will vary dramatically between these resistant cells. Heterogeneity in expression levels arises from differences in the number of integrants, and their sites of integration. Thus, a key step in harnessing such cells for protein production is the selection of those single cells that achieve the highest expression levels. Colonies grow out of single antibiotic-resistant cells. Traditionally, to screen for highly expressing cells, individual colonies are hand-picked and assayed for their levels of protein production, usually involving immunological detection. This method is time consuming and labor intensive.

A quicker and more efficient method is based on detection of the fluorescence intensity of a co-expressed marker, such as the green fluorescent protein (GFP).[99,100] Downstream of the promoter and of the coding sequence for the gene of interest, an internal ribosome entry site (IRES)[101] is followed by the coding sequence for GFP (Figure 5A). A single bicistronic messenger RNA encoding both genes is produced. The two separate proteins are then translated from the same message, and their expression levels are thereby coupled. Efficient selection of highly expressing cells can be easily achieved by monitoring fluorescence intensity of cells expressing variable amounts of GFP. The process can be accelerated further by using fluorescence-activated-cell-sorting (FACS, see Refs. 102 and 103) technology for the rapid selection of either clonal or non-clonal populations of high expressors (Figures 5B and 5C).

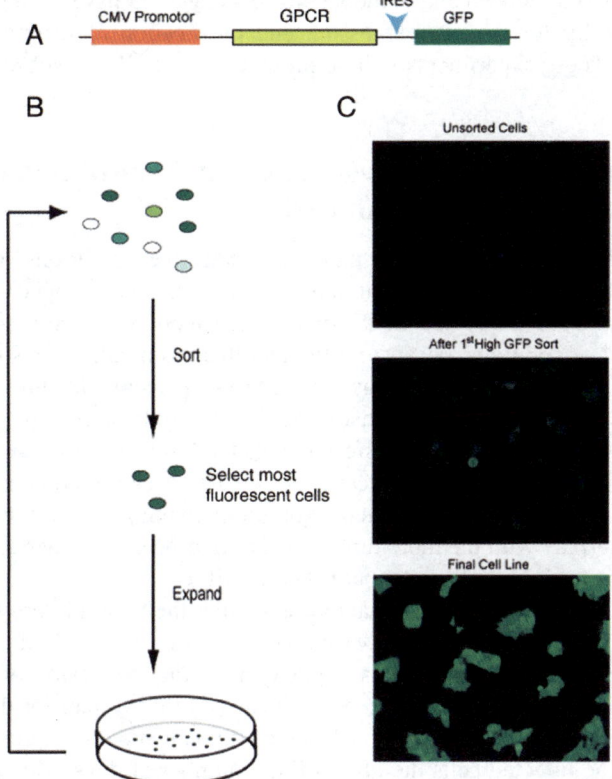

Figure 5 *Schematic representation of the GFP-selection method. (A) Scheme of the expression vector. (B) Diagram of the enrichment procedure, through successive rounds of cell sorting, and expansion of the most GFP-fluorescent cells. (C) Progression of the enrichment procedure monitored by fluorescent microscopy. The cells depicted here are HEK 293 cells expressing 5HT2c. Three stages are shown: after antibiotic selection, after the first round of sorting, and after the final round (Adapted from Ref. 106 with permission).*

The use of IRES–GFP as a binary marker for successfully transfected cells is well established. However, its use as a monitor for target protein expression levels has seen far fewer reports (for example, see Refs. 104 and 105). Stable HEK 293 cell lines expressing a considerable amount (140–160 pmol mg^{-1} of membrane protein) of the rat 5HT2c receptor have been generated using this technique.[106] The authors show clear correlation between expression levels of the GPCR and of GFP.

Many different mammalian cell lines have been used successfully for the expression of GPCRs (for a list, see Ref. 62). Baby hamster kidney (BHK) cells and African monkey (COS) cells appear to be the best suited for viral infection. HEK-293 cells and Chinese hamster ovary (CHO) cells have yielded the highest levels of GPCRs from stable lines.

GPCRs expressed in cell lines often show a highly heterogeneous pattern of glycosylation. This can constitute a severe problem for structural studies. Mutated cell lines that exhibit a reduced and homogenous glycosylation have been generated and used successfully for the overexpression and purification of recombinant bovine rhodopsin.[107] These experiments will be presented in detail in a subsequent chapter of this book.

2.4   Production of GPCRs *via* Transient Transfection or Viral Infection of Mammalian Cells

Transient expression is performed under non-selective conditions. In transiently transfected mammalian cells, protein expression levels peak around 48–72 h after transfection, and inevitably decline thereafter. Transfection on a large scale is not the most practical of solutions, because of the fact that mammalian cells tend to trans-fect more efficiently in monolayers as opposed to suspension, making scale up ardu-ous, and because of reagent expenses. As a commonly accepted general trend, expression levels of transiently transfected cells tend to be higher than that of stable cell lines. This seems not to be the case for GPCRs.[106] However, constructs can be screened easily and rapidly by transient expression, making this an excellent screen-ing tool to interplay with the more time-consuming process of generating stable cell lines (Mancia and Hendrickson, unpublished results).

Infection of cells with a recombinant virus such as the Semliki Forest virus can be extremely efficient, leading to high expression levels in the 48–72 h after infection (for a review of this system and its application to the expression of GPCRs, see Ref. 108). Excellent results have been obtained with this system for the expression of GPCRs. The negative aspects are that recombinant protein expression is again transient, as the infected cells die after a finite number of days. Moreover, the fact that viral infection typically requires the propagation of recombinant virus in a "packaging" cell line, isolation of the virus, and determination of viral titer prior to infection of the host cells, makes this approach rather time consuming, probably in the same time frame as generating a stable cell line.

2.5   Production of GPCRs in Yeast

The structure of a rat voltage-gated potassium channel has recently been solved.[53] The mammalian protein was expressed in the yeast *Pichia pastoris*.[109] Thus, a yeast-based expression system has paved the way for what is, to the best of our knowledge, the first crystal structure of a mammalian membrane protein from a recombinant source. This success should not be underestimated when choosing a system for the expression of a GPCR.

Yeasts are, in principle, excellent organisms for the expression of GPCRs. Yeasts are easy, quick, and inexpensive to grow, and they can be cultured to high density for efficient scaling up. Expression techniques are well established and offer numerous different possibilities. Moreover, these unicellular eukaryotes can potentially per-form all the post-translational modifications of mammalian cells. G-proteins are also expressed in yeast.

Glycosylation is substantially different in yeast and mammalian cells. There are differences in both the amount and the type of glycans. This can constitute a problem for those GPCRs that are sensitive to glycosylation for proper function. Moreover, there are examples of GPCRs that are not glycosylated when expressed in yeast.[110] The lipid composition of the membranes is also different in yeast and mammalian cells. Most notably, there is a reduced level of sterols in yeast. These differences can lead to the expression of GPCRs with altered ligand-binding properties.[111] Proteolysis can also be an issue that should be taken into account, although protease-deficient yeast strains can be used for expression.[110]

GPCRs have been expressed in a functional form in *S. cerevisiae*, *P. pastoris*, and *Schizosaccharomyces pombe*. Inducible as well as constitutive promoters have been used in plasmids that are either maintained in an episomal form or integrated into the host's genome. Use of a yeast leader sequence appears to be important for proper targeting to the plasma membrane. Similarly to other systems, expression levels of GPCRs in yeast are extremely variable (for a list of examples, see Refs. 62 and 112). Not surprisingly, yeast GPCRs express the best in yeast, as is the case for the Ste2 receptor in *S. cerevisiae*.[64] For a more in depth understanding of this expression system, the reader should consult some of the excellent reviews that have been written on the expression of GPCRs in yeast, and on the use of this system for functional and structural studies on these proteins.[62,112–116]

2.6   Production of GPCRs in Insect Cells

Following *E. coli* and yeast, baculovirus-based expression systems have had the greatest success in the number of structures successfully determined. In particular, expression of eukaryotic genes through the use of baculovirus-infected insect cells[117] has a long history of success (for a review of the technique, see Ref. 118). Briefly, the gene of interest is placed under the control of a strong constitutive promoter and the plasmid is inserted into the viral genome by homologous recombination. The commonly used virus is the *Autographa californica* multiply embedded nuclear polyhedrosis virus (AcMNPV). Insect cells are infected with the recombinant virus, and the expression of heterologous proteins typically peaks 48–72 h post-infection, with cell death occurring 4–5 days post-infection. Insect cells can be easily adapted to grow in suspension. Insect cells are able to perform the same post-translational modifications as those of mammalian cells, although the type and amount of glycosylation is substantially different.[119] The lipid environments are quite different in insect and mammalian cells (cholesterol levels, for example, are low in insect cells).

Insect cells have been widely used for the expression of GPCRs, yielding, once again, mixed results in terms of expression levels. There are reports in which levels are among the highest ever achieved. For an exhaustive review on the use of this system for the expression of GPCRs, the reader should consult a recent review written by Massotte.[120]

2.7   *"In Vivo"* Expression in the Eye

The amount of rhodopsin that accumulates in rod cells is extremely large. Researchers have thus begun to explore the idea of expressing other GPCRs in these compartments through the generation of appropriate transgenic animals. Eroglu and colleagues[121] were successful in expressing a *Drosophila melanogaster* metabotropic glutamate receptor in the *Drosophila* photoreceptor cells at levels (170 µg g^{-1} of fly heads, equivalent to approximately 3000 flies [Luisa Vasconcelos, personal communication]) at least 3-fold higher than those achieved with conventional baculovirus systems. However, baculovirus-infected insect cells are undoubtedly easier to scale up. Kodama and co-workers[122] have instead generated transgenic mice expressing the human endothelin receptor subtype B (hET$_B$R), fused with the C-terminal 10 residues of rhodopsin, in rod cells. Somewhat disappointingly, hET$_B$R protein levels were in the order of a 1000-fold less than rhodopsin in heterozygous animals. This approach is likely to require a better understanding of how rhodopsin is efficiently translated, folded, and transported before it can be used successfully. Moreover, more examples are necessary before the potential of this approach can be adequately assessed.

2.8   Extra-Membranous Expression Systems

All of the above-mentioned expression systems yield functional proteins. A completely different approach is to generate abundant quantities of protein outside of a membrane environment, and then to use biochemical refolding manipulations to reconstitute a properly folded and functional protein from non-functional, aggregated material. Extra-membranous expression may be possible in the bacterial cytoplasm or in cell-free systems;[123] reconstitution may be achieved either in lipid bilayers or in detergent micelles. A human leukotriene receptor has been expressed in the form of inclusion bodies in *E. coli* and successfully refolded in detergent.[124] The authors[124] showed that the refolded protein was able to bind ligand as well as interact with G-proteins.[125] An olfactory receptor fused to GST could also be accumulated in large amounts in the form of inclusion bodies when expressed in *E. coli* and refolded in detergent.[126] Recently, Ishihara and colleagues[127] were able to express several GPCRs in a cell-free expression system as fusions to thioredoxin (TRX). The insoluble proteins could be solubilized by detergents and ligand binding restored by incorporation into liposomes.[127] These approaches are currently at an early stage of development, but could potentially turn out to be extremely powerful for the field of structural biology of GPCRs.

3   Conclusions

Several viable options exist for the recombinant expression of GPCRs in sufficient abundance for structural studies. We have emphasized two alternative expression systems; one in bacteria where multiple constructs can be tested quickly and economically, and another in stable mammalian cell lines where natural modifications and functional tests can happen. What we have not stressed, but do find advantageous, is the synergy to be found in pursuing the two simultaneously. A productive interplay is then possible whereby feasibility tests are done at relatively high throughput in bacteria and functional evaluation of prime candidates can be made in a relevant mammalian setting. We find that adequate yields are possible both from MBP fusions in *E. coli* and from GFP-selected strains of HEK-293 cells. Alternative pairs of systems might also provide the respective advantages of these expression systems.

Although GPCRs have been expressed successfully in various systems and by several investigators, none of these studies has yet produced structure-quality crystals. Often, this is despite the compelling evidence of natural ligand-binding affinities in membranes or even in detergent micelles. It may be that further structural stabilization is required, which might come through the retention or reintroduction of certain lipids, through provision of physiological interactions as in dimers or in complexes with G proteins, or through complexation with engineered binding partners such as antibodies. Methods for implementing such approaches are beyond the scope of this review, but the expression methods described here surely also have relevance in these elaborated contexts.

Acknowledgments

We thank Richard Axel, Paul J. Lee, Yonghua Sun, and Risa Siegel for their precious contributions to this work and for helpful discussions. This work has been supported in part by the NIH (GM68671).

References

1. Y.A. Ovchinnikov, N.G. Abdulaev, A.E. Dergachev, A.L. Drachev, L.A. Drachev, A.D. Kaulen, L.V. Khitrina, Z.P. Lazarova and V.P. Skulachev, *Eur. J. Biochem.*, 1982, **127**, 325.
2. P.A. Hargrave, J.H. McDowell, D.R. Curtis, J.K. Wang, E. Juszczak, S.L. Fong, J.K. Rao and P. Argos, *Biophys. Struct. Mech.*, 1983, **9**, 235.
3. J. Nathans and D.S. Hogness, *Cell*, 1983, **34**, 807.
4. R.E. Diehl, R.A. Mumford, E.E. Slater, I.S. Sigal, M.G. Caron, R.J. Lefkowitz and C.D. Strader, *Nature*, 1986, **321**, 75.
5. E.M. Ross and A.G. Gilman, *J. Biol. Chem.*, 1977, **252**, 6966.
6. L. Stryer, *Annu. Rev. Neurosci.*, 1986, **9**, 87.
7. D. Julius, A.B. MacDermott, R. Axel and T.M. Jessell, *Science*, 1988, **241**, 558.
8. B.K. Kobilka, T. Frielle, S. Collins, T. Yang-Feng, T.S. Kobilka, U. Francke, R.J. Lefkowitz and M.G. Caron, *Nature*, 1987, **329**, 75.
9. L. Buck and R. Axel, *Cell*, 1991, **65**, 175.
10. S. Takeda, S. Kadowaki, T. Haga, H. Takaesu and S. Mitaku, *FEBS Lett.*, 2002, **520**, 97.
11. J.C. Venter, M.D. Adams, E.W. Myers, P.W. Li, R.J. Mural, G.G. Sutton, H.O. Smith, M. Yandell, C.A. Evans, R.A. Holt, J.D. Gocayne, P. Amanatides, R.M. Ballew, D.H. Huson, J.R. Wortman, Q. Zhang, C.D. Kodira, X.H. Zheng, L. Chen, M. Skupski, G. Subramanian, P.D. Thomas, J. Zhang, G.L. Gabor Miklos, C. Nelson, S. Broder, A.G. Clark, J. Nadeau, V.A. McKusick, N. Zinder, A.J. Levine, R.J. Roberts, M. Simon, C. Slayman, M. Hunkapiller, R. Bolanos, A. Delcher, I. Dew, D. Fasulo, M. Flanigan, L. Florea, A. Halpern,

S. Hannenhalli, S. Kravitz, S. Levy, C. Mobarry, K. Reinert, K. Remington, J. Abu-Threideh, E. Beasley, K. Biddick, V. Bonazzi, R. Brandon, M. Cargill, I. Chandramouliswaran, R. Charlab, K. Chaturvedi, Z. Deng, V. Di Francesco, P. Dunn, K. Eilbeck, C. Evangelista, A.E. Gabrielian, W. Gan, W. Ge, F. Gong, Z. Gu, P. Guan, T.J. Heiman, M.E. Higgins, R.R. Ji, Z. Ke, K.A. Ketchum, Z. Lai, Y. Lei, Z. Li, J. Li, Y. Liang, X. Lin, F. Lu, G.V. Merkulov, N. Milshina, H.M. Moore, A.K. Naik, V.A. Narayan, B. Neelam, D. Nusskern, D.B. Rusch, S. Salzberg, W. Shao, B. Shue, J. Sun, Z. Wang, A. Wang, X. Wang, J. Wang, M. Wei, R. Wides, C. Xiao, C. Yan, A. Yao, J. Ye, M. Zhan, W. Zhang, H. Zhang, Q. Zhao, L. Zheng, F. Zhong, W. Zhong, S. Zhu, S. Zhao, D. Gilbert, S. Baumhueter, G. Spier, C. Carter, A. Cravchik, T. Woodage, F. Ali, H. An, A. Awe, D. Baldwin, H. Baden, M. Barnstead, I. Barrow, K. Beeson, D. Busam, A. Carver, A. Center, M.L. Cheng, L. Curry, S. Danaher, L. Davenport, R. Desilets, S. Dietz, K. Dodson, L. Doup, S. Ferriera, N. Garg, A. Gluecksmann, B. Hart, J. Haynes, C. Haynes, C. Heiner, S. Hladun, D. Hostin, J. Houck, T. Howland, C. Ibegwam, J. Johnson, F. Kalush, L. Kline, S. Koduru, A. Love, F. Mann, D. May, S. McCawley, T. McIntosh, I. McMullen, M. Moy, L. Moy, B. Murphy, K. Nelson, C. Pfannkoch, E. Pratts, V. Puri, H. Qureshi, M. Reardon, R. Rodriguez, Y.H. Rogers, D. Romblad, B. Ruhfel, R. Scott, C. Sitter, M. Smallwood, E. Stewart, R. Strong, E. Suh, R. Thomas, N.N. Tint, S. Tse, C. Vech, G. Wang, J. Wetter, S. Williams, M. Williams, S. Windsor, E. Winn-Deen, K. Wolfe, J. Zaveri, K. Zaveri, J.F. Abril, R. Guigo, M.J. Campbell, K.V. Sjolander, B. Karlak, A. Kejariwal, H. Mi, B. Lazareva, T. Hatton, A. Narechania, K. Diemer, A. Muruganujan, N. Guo, S. Sato, V. Bafna, S. Istrail, R. Lippert, R. Schwartz, B. Walenz, S. Yooseph, D. Allen, A. Basu, J. Baxendale, L. Blick, M. Caminha, J. Carnes-Stine, P. Caulk, Y.H. Chiang, M. Coyne, C. Dahlke, A. Mays, M. Dombroski, M. Donnelly, D. Ely, S. Esparham, C. Fosler, H. Gire, S. Glanowski, K. Glasser, A. Glodek, M. Gorokhov, K. Graham, B. Gropman, M. Harris, J. Heil, S. Henderson, J. Hoover, D. Jennings, C. Jordan, J. Jordan, J. Kasha, L. Kagan, C. Kraft, A. Levitsky, M. Lewis, X. Liu, J. Lopez, D. Ma, W. Majoros, J. McDaniel, S. Murphy, M. Newman, T. Nguyen, N. Nguyen, M. Nodell, S. Pan, J. Peck, M. Peterson, W. Rowe, R. Sanders, J. Scott, M. Simpson, T. Smith, A. Sprague, T. Stockwell, R. Turner, E. Venter, M. Wang, M. Wen, D. Wu, M. Wu, A. Xia, A. Zandieh and X. Zhu, *Science*, 2001, **291**, 1304.

12. B.R. Conklin and H.R. Bourne, *Cell*, 1993, **73**, 631.
13. A.G. Gilman, *Annu. Rev. Biochem.*, 1987, **56**, 615.
14. J.P. Noel, H.E. Hamm and P.B. Sigler, *Nature*, 1993, **366**, 654.
15. D.E. Coleman, A.M. Berghuis, E. Lee, M.E. Linder, A.G. Gilman and S.R. Sprang, *Science*, 1994, **265**, 1405.
16. M.A. Wall, D.E. Coleman, E. Lee, J.A. Iniguez-Lluhi, B.A. Posner, A.G. Gilman and S.R. Sprang, *Cell*, 1995, **83**, 1047.
17. D.G. Lambright, J. Sondek, A. Bohm, N.P. Skiba, H.E. Hamm and P.B. Sigler, *Nature*, 1996, **379**, 311.
18. J. Sondek, A. Bohm, D.G. Lambright, H.E. Hamm and P.B. Sigler, *Nature*, 1996, **379**, 369.

19. J.J. Tesmer, R.K. Sunahara, A.G. Gilman and S.R. Sprang, *Science*, 1997, **278**, 1907.

20. D.D. Thomas and L. Stryer, *J. Mol. Biol.*, 1982, **154**, 145.

21. C.D. Strader, T.M. Fong, M.R. Tota, D. Underwood and R.A. Dixon, *Annu. Rev. Biochem.*, 1994, **63**, 101.

22. Y. Yokota, Y. Sasai, K. Tanaka, T. Fujiwara, K. Tsuchida, R. Shigemoto, A. Kakizuka, H. Ohkubo and S. Nakanishi, *J. Biol. Chem.*, 1989, **264**, 17649.

23. Q.R. Fan and W.A. Hendrickson, *Nature*, 2005, **433**, 269.

24. M. Berthold and T. Bartfai, *Neurochem. Res.*, 1997, **22**, 1023.

25. R. Henderson and P.N. Unwin, *Nature*, 1975, **257**, 28.

26. R. Henderson, J.M. Baldwin, T.A. Ceska, F. Zemlin, E. Beckmann and K.H. Downing, *J. Mol. Biol.*, 1990, **213**, 899.

27. W.C. Probst, L.A. Snyder, D.I. Schuster, J. Brosius and S.C. Sealfon, *DNA Cell Biol.*, 1992, **11**, 1.

28. R.J. Lefkowitz, *Nat. Cell Biol.*, 2000, **2**, E133.

29. D. Zhang and H. Weinstein, *J. Med. Chem.*, 1993, **36**, 934.

30. S. Shacham, M. Topf, N. Avisar, F. Glaser, Y. Marantz, S. Bar-Haim, S. Noiman, Z. Naor and O.M. Becker, *Med. Res. Rev.*, 2001, **21**, 472.

31. V.M. Unger, P.A. Hargrave, J.M. Baldwin and G.F. Schertler, *Nature*, 1997, **389**, 203.

32. J.M. Baldwin, G.F. Schertler and V.M. Unger, *J. Mol. Biol.*, 1997, **272**, 144.

33. K. Palczewski, T. Kumasaka, T. Hori, C.A. Behnke, H. Motoshima, B.A. Fox, I. Le Trong, D.C. Teller, T. Okada, R.E. Stenkamp, M. Yamamoto and M. Miyano, *Science*, 2000, **289**, 739.

34. T. Okada, Y. Fujiyoshi, M. Silow, J. Navarro, E.M. Landau and Y. Shichida, *Proc. Natl. Acad. Sci. USA*, 2002, **99**, 5982.

35. J. Li, P.C. Edwards, M. Burghammer, C. Villa and G.F. Schertler, *J. Mol. Biol.*, 2004, **343**, 1409.

36. J. Deisenhofer, O. Epp, K. Miki, R. Huber and H. Michel, *Nature*, 1985, **318**, 618.

37. S. Iwata, C. Ostermeier, B. Ludwig and H. Michel, *Nature*, 1995, **376**, 660.

38. T. Tsukihara, H. Aoyama, E. Yamashita, T. Tomizaki, H. Yamaguchi, K. Shinzawa-Itoh, R. Nakashima, R. Yaono and S. Yoshikawa, *Science*, 1996, **272**, 1136.

39. D.A. Doyle, C.J. Morais, R.A. Pfuetzner, A. Kuo, J.M. Gulbis, S.L. Cohen, B.T. Chait and R. MacKinnon, *Science*, 1998, **280**, 69.

40. Y. Zhou, J.H. Morais-Cabral, A. Kaufman and R. MacKinnon, *Nature*, 2001, **414**, 43.

41. D. Fu, A. Libson, L.J. Miercke, C. Weitzman, P. Nollert, J. Krucinski and R.M. Stroud, *Science*, 2000, **290**, 481.

42. T. Walz, T. Hirai, K. Murata, J.B. Heymann, K. Mitsuoka, Y. Fujiyoshi, B.L. Smith, P. Agre and A. Engel, *Nature*, 1997, **387**, 624.

43. Y. Huang, M.J. Lemieux, J. Song, M. Auer and D.N. Wang, *Science*, 2003, **301**, 616.

44. D. Yernool, O. Boudker, Y. Jin and E. Gouaux, *Nature*, 2004, **431**, 811.

45. C. Hunte, E. Screpanti, M. Venturi, A. Rimon, E. Padan and H. Michel, *Nature*, 2005, **435**, 1197.

46. J. Abramson, I. Smirnova, V. Kasho, G. Verner, H. R. Kaback and S. Iwata, *Science*, 2003, **301**, 610.

47. A. Yamashita, S. K. Singh, T. Kawate, Y. Jin and E. Gouaux, *Nature*, 2005, **437**, 215.

48. C. Toyoshima, M. Nakasako, H. Nomura and H. Ogawa, *Nature*, 2000, **405**, 647.

49. T. L. Sorensen, J. V. Moller and P. Nissen, *Science*, 2004, **304**, 1672.

50. R. Dutzler, E. B. Campbell, M. Cadene, B. T. Chait and R. MacKinnon, *Nature*, 2002, **415**, 287.

51. Y. Jiang, A. Lee, J. Chen, M. Cadene, B. T. Chait and R. MacKinnon, *Nature*, 2002, **417**, 515.

52. Y. Jiang, A. Lee, J. Chen, V. Ruta, M. Cadene, B.T. Chait and R. MacKinnon, *Nature*, 2003, **423**, 33.

53. S.B. Long, E.B. Campbell and R. Mackinnon, *Science*, 2005, **309**, 867.

54. S. Khademi, J. O'Connell III, J. Remis, Y. Robles-Colmenares, L.J. Miercke and R.M. Stroud, *Science*, 2004, **305**, 1587.

55. E. Wallin and G. von Heijne, *Protein Sci.*, 1998, **7**, 1029.

56. D. Picot, P.J. Loll and R.M. Garavito, *Nature*, 1994, **367**, 243.

57. R.G. Kurumbail, A.M. Stevens, J.K. Gierse, J.J. McDonald, R.A. Stegeman, J.Y. Pak, D. Gildehaus, J.M. Miyashiro, T.D. Penning, K. Seibert, P.C. Isakson and W. C. Stallings, *Nature*, 1996, **384**, 644.

58. T. Okada, I. Le Trong, B.A. Fox, C.A. Behnke, R.E. Stenkamp and K. Palczewski, *J. Struct. Biol.*, 2000, **130**, 73.

59. T. Okada, K. Takeda and T. Kouyama, *Photochem. Photobiol.*, 1998, **67**, 495.

60. R. Grisshammer and C.G. Tate, *Q. Rev. Biophys.*, 1995, **28**, 315.

61. C.G. Tate and R. Grisshammer, *Trends Biotechnol.*, 1996, **14**, 426.

62. V. Sarramegna, F. Talmont, P. Demange and A. Milon, *Cell. Mol. Life Sci.*, 2003, **60**, 1529.

63. K. Lundstrom, *Trends Biotechnol*, 2005, **23**, 103.

64. N.E. David, M. Gee, B. Andersen, F. Naider, J. Thorner and R.C. Stevens, *J. Biol. Chem.*, 1997, **272**, 15553.

65. R. Grisshammer and J. Tucker, *Protein Expres. Purif.*, 1997, **11**, 53.

66. W.A. Hendrickson, *Science*, 1991, **254**, 51.

67. K. Wuthrich, *Angew. Chem. Int. Ed. Engl.*, 2003, **42**, 3340.

68. S. Kaushal, K.D. Ridge and H.G. Khorana, *Proc. Natl. Acad. Sci. USA*, 1994, **91**, 4024.

69. M. Opekarova and W. Tanner, *Biochim. Biophys. Acta*, 2003, **1610**, 11.

70. D.R. Voelker, *Biochim. Biophys. Acta*, 1997, **1348**, 236.

71. W. Dowhan, *Annu. Rev. Biochem.*, 1997, **66**, 199.

72. G. Gimpl and F. Fahrenholz, *Biochim. Biophys. Acta*, 2002, **1564**, 384.

73. J. Hasegawa, H.H. Loh and N.M. Lee, *J. Neurochem.*, 1987, **49**, 1007.

74. P. Balen, K. Kimura and A. Sidhu, *Biochemistry*, 1994, **33**, 1539.

75. B. Bertin, M. Freissmuth, R.M. Breyer, W. Schutz, A.D. Strosberg and S. Marullo, *J. Biol. Chem.*, 1992, **267**, 8200.

76. B.J. Francken, J.F. Vanhauwe, K. Josson, M. Jurzak, W.H. Luyten and J.E. Leysen, *Receptors Channels*, 2001, **7**, 303.

77. R. Grisshammer and E. Hermans, *FEBS Lett.*, 2001, **493**, 101.

78. L. Stanasila, W.K. Lim, R.R. Neubig and F. Pattus, *J. Neurochem.*, 2000, **75**, 1190.
79. S. Marullo, C. Delavier-Klutchko, Y. Eshdat, A.D. Strosberg and L. Emorine, *Proc Natl. Acad. Sci. USA*, 1988, **85**, 7551.
80. R. Grisshammer, R. Duckworth and R. Henderson, *Biochem. J.*, 1993, **295**, 571.
81. J. Tucker and R. Grisshammer, *Biochem. J.*, 1996, **317**, 891.
82. J.F. White, L.B. Trinh, J. Shiloach and R. Grisshammer, *FEBS Lett.*, 2004, **564**, 289.
83. R. Grisshammer, J. Little and D. Aharony, *Receptors Channels*, 1994, **2**, 295.
84. H.M. Weiss and R. Grisshammer, *Eur. J. Biochem.*, 2002, **269**, 82.
85. E.C. Hulme and C.A. Curtis, *Biochem. Soc. Trans.*, 1998, **26**, S361.
86. H. Furukawa and T. Haga, *J. Biochem. (Tokyo)*, 2000, **127**, 151.
87. S. Marullo, C. Delavier-Klutchko, J.-G.Guillet, A. Charbit, A.D. Strosberg and L.J. Emorine, *Bio/Technology*, 1989, **7**, 923.
88. H. Kiefer, R. Vogel and K. Maier, *Receptors Channels*, 2000, **7**, 109.
89. R.A. Hill and M.N. Sillence, *Protein Expr. Purif.*, 1997, **10**, 162.
90. R.M. Lacatena, A. Cellini, F. Scavizzi and G.P. Tocchini-Valentini, *Proc. Natl. Acad. Sci. USA*, 1994, **91**, 10521.
91. S. Marullo, C. Delavier-Klutchko, Y. Eshdat, A.D. Strosberg and L. Emorine, *Proc. Natl. Acad. Sci. USA*, 1988, **85**, 7551.
92. R.B. Kapust, J. Tozser, J.D. Fox, D.E. Anderson, S. Cherry, T.D. Copeland and D.S. Waugh, *Protein Eng.*, 2001, **14**, 993.
93. F. Baneyx, *Curr. Opin. Biotechnol.*, 1999, **10**, 411.
94. S. Geisse, H. Gram, B. Kleuser and H.P. Kocher, *Protein Expr. Purif.*, 1996, **8**, 271.
95. S.C. Makrides, *Microbiol. Rev.*, 1996, **60**, 512.
96. P.J. Reeves, J. Klein-Seetharaman, E.V. Getmanova, M. Eilers, M.C. Loewen, S.O. Smith and H.G. Khorana, *Biochem. Soc. Trans.*, 1999, **27**, 950.
97. D.R. Thomsen, R.M. Stenberg, W.F. Goins and M.F. Stinski, *Proc. Natl. Acad. Sci. USA*, 1984, **81**, 659.
98. P.J. Reeves, J.M. Kim and H.G. Khorana, *Proc. Natl. Acad. Sci. USA*, 2002, **99**, 13413.
99. M. Chalfie, *Photochem. Photobiol.*, 1995, **62**, 651.
100. R.Y. Tsien, *Annu. Rev. Biochem.*, 1998, **67**, 509.
101. S. Vagner, B. Galy and S. Pyronnet, *EMBO Rep.*, 2001, **2**, 893.
102. D.W. Galbraith, M.T. Anderson and L.A. Herzenberg, *Methods Cell Biol.*, 1999, **58**, 315.
103. S.F. Ibrahim and G. van den Engh, *Curr. Opin. Biotechnol.*, 2003, **14**, 5.
104. X. Liu, S.N. Constantinescu, Y. Sun, J.S. Bogan, D. Hirsch, R A. Weinberg and H.F. Lodish, *Anal. Biochem.*, 2000, **280**, 20.
105. Y.G. Meng, J. Liang, W.L. Wong and V. Chisholm, *Gene*, 2000, **242**, 201.
106. F. Mancia, S.D. Patel, M.W. Rajala, P.E. Scherer, A. Nemes, I. Schieren, W.A. Hendrickson and L. Shapiro, *Structure (Camb)*, 2004, **12**, 1355.
107. P.J. Reeves, N. Callewaert, R. Contreras and H.G. Khorana, *Proc. Natl. Acad. Sci. USA*, 2002, **99**, 13419.

108. K. Lundstrom, *Biochim. Biophys. Acta*, 2003, **1610**, 90.

109. D.N. Parcej and L. Eckhardt-Strelau, *J. Mol. Biol.*, 2003, **333**, 103.

110. P. Sander, S. Grunewald, H. Reilander and H. Michel, *FEBS Lett.*, 1994, **344**, 41.

111. B. Lagane, G. Gaibelet, E. Meilhoc, J.M. Masson, L. Cezanne and A. Lopez, *J. Biol. Chem.*, 2000, **275**, 33197.

112. H. Reilander and H.M. Weiss, *Curr. Opin. Biotechnol.*, 1998, **9**, 510.

113. A. Celic, S.M. Connelly, N.P. Martin and M.E. Dumont, *Methods Mol. Biol.*, 2004, **237**, 105.

114. S.J. Dowell and A.J. Brown, *Receptors Channels*, 2002, **8**, 343.

115. S. Macauley-Patrick, M.L. Fazenda, B. McNeil and L.M. Harvey, *Yeast*, 2005, **22**, 249.

116. J. Minic, M. Sautel, R. Salesse and E. Pajot-Augy, *Curr. Med. Chem.*, 2005, **12**, 961.

117. G.E. Smith, M.D. Summers and M.J. Fraser, *Mol. Cell Biol.*, 1983, **3**, 2156.

118. M.J. Fraser, *Curr. Top Microbiol. Immunol.*, 1992, **158**, 131.

119. D.L. Jarvis and E.E. Finn, *Virology*, 1995, **212**, 500.

120. D. Massotte, *Biochim. Biophys. Acta*, 2003, **1610**, 77.

121. C. Eroglu, P. Cronet, V. Panneels, P. Beaufils and I. Sinning, *EMBO Rep.*, 2002, **3**, 491.

122. T. Kodama, H. Imai, T. Doi, O. Chisaka, Y. Shichida and Y. Fujiyoshi, *Exp. Eye Res.*, 2005, **80**, 859.

123. C. Klammt, F. Lohr, B. Schafer, W. Haase, V. Dotsch, H. Ruterjans, C. Glaubitz and F. Bernhard, *Eur. J. Biochem.*, 2004, **271**, 568.

124. J.L. Baneres, A. Martin, P. Hullot, J.P. Girard, J.C. Rossi and J. Parello, *J. Mol. Biol.*, 2003, **329**, 801.

125. J.L. Baneres and J. Parello, *J. Mol. Biol.*, 2003, **329**, 815.

126. H. Kiefer, J. Krieger, J.D. Olszewski, G. Von Heijne, G.D. Prestwich and H. Breer, *Biochemistry*, 1996, **35**, 16077.

127. G. Ishihara, M. Goto, M. Saeki, K. Ito, T. Hori, T. Kigawa, M. Shirouzu and S. Yokoyama, *Protein Expr. Purif.*, 2005, **41**, 27.

CHAPTER 4

The Purification of G-Protein Coupled Receptors for Crystallization

TONY WARNE AND GEBHARD F.X. SCHERTLER

MRC Laboratory of Molecular Biology, Hills Road, Cambridge CB2 2QH, UK

1 Introduction

1.1 Structural Studies of G-Protein Coupled Receptors

Structural studies of overexpressed eukaryotic membrane proteins remain a major challenge, because such work requires the preparation of milligram quantities of highly purified, fully functional, monodisperse membrane protein. Successful structural work on eukaryotic polytopic membrane proteins has been almost entirely restricted to those which are naturally abundant and can be purified from native sources. Unfortunately, the membrane proteins that are medically relevant drug targets, for example the G protein-coupled receptors (GPCRs), are of low abundance and require heterologous overexpression before structural studies can be initiated. There are already a large number of eukaryotic membrane proteins, including at least 20 GPCRs,[1] for which expression levels have been achieved that are sufficient for the preparation of milligram quantities of pure protein. The quantities of purified protein required for large-scale 3D crystallization experiments have been reduced due to the widespread implementation of nanolitre scale crystallization robots. Nevertheless, bovine rhodopsin remains the only GPCR for which there is a high-resolution structure, probably because it is naturally abundant, requires only minimal purification, and is stable in many detergents.

In this chapter we will discuss the purification of heterologously expressed GPCRs with particular emphasis on the turkey beta-adrenergic receptor (βAR). The turkey βAR has been purified in milligram quantities but has so far not produced crystals, which diffract to high-resolution. This may be attributable to some particular problems often encountered with eukaryotic membrane proteins; these include instability in detergent, delipidation through purification, post-translational heterogeneity, and inherent flexibility. In describing how these problems have been investigated and addressed, and with reference to other work, we will illustrate the crucial importance of a thorough characterization of the purified material, and the improvements to the final product, which can be gained.

1.2 The Turkey Erythrocyte Beta-Adrenergic Receptor

The turkey erythrocyte βAR, like all GPCRs, comprises seven transmembrane helices with an extracellular N-terminus and an intracellular C-terminus.[2] Based on the structure of rhodopsin, the C-terminus is now thought to contain an eighth amphipathic helix which associates with the intracellular surface of the plasma membrane and is located immediately after transmembrane helix 7.[3,4] Helix 8 is terminated by a single palmitoylated cysteine. The functionality of GPCRs can be conveniently monitored by assessing ligand binding. In the case of the turkey βAR, a radioligand binding assay utilising the βAR antagonist [³H](-) dihydroalprenolol is used to quantify expression, yields of purification steps, purity (ratio of ligand binding sites/total protein) and to monitor receptor stability by following the decline in ligand binding over time. We have overexpressed the turkey βAR in insect cells using the baculovirus system. The βAR construct, which has been the major focus of our work is truncated at the N- and C-termini and encodes amino acids 34–424 (the first transmembrane helix is predicted to start at residue 38). The construct includes a point mutation (C116L) and has a C-terminal (His)$_6$ tag. We have been able to produce functional receptor at a level of 350 pmol receptor/mg membrane protein, corresponding to about 1.5% of the total membrane protein. The receptor has been purified to apparent homogeneity in dodecylmaltoside (DDM) by a combination of metal affinity chromatography, antagonist ligand affinity chromatography, and size-exclusion chromatography. This procedure has been described previously.[5] Here, we will discuss the rationale behind the βAR construct used and the particular detergents and purification methods employed.

2 Heterogeneity of Overexpressed Receptors

2.1 Heterogeneity of GPCRs due to Post-Translational Modifications

The production of a homogenous, pure GPCR for crystallization can be complicated by the heterogeneity of a variety of post-translational modifications, which may be introduced by the host cell during overexpression. Methods employed to limit the extent and heterogeneity of modifications include the choice of expression systems, mutation or truncation of the GPCR cDNA expressed, or by biochemical modification of the purified GPCR.

2.1.1 N-Glycosylation

The N-terminus of most GPCRs contains one or more N-linked glycosylation consensus sequences (Asn-x-Ser/Thr). N-glycosylation of GPCRs is not generally thought to be essential for protein folding or function, but is usually required for trafficking of the receptor to the cell surface. During protein purification, the retention of an N-glycosylated and full-length N-terminus may help stabilize a detergent-solubilized and purified receptor. In the case of bovine rhodopsin, the native N-glycosylation did not interfere with 3D crystallization,[3,4] but glycosylation of proteins overexpressed in insect cells will be variable in extent and probably different to that found in the native receptor.[6] One method of preventing hetero-geneity of N-glycosylation in the purified protein is to avoid it altogether by over-expressing the GPCR in bacteria, such as *Escherichia coli* (*E. coli*), that are not capable of this post-translational modification. Using bacteria requires the expression of a fusion protein containing an N-terminal soluble protein for maximal expression of either correctly folded, membrane bound receptor or of misfolded, inactive receptor in inclusion bodies, which will require refolding after purification. For material expressed in eukaryotic cells, deleting or mutating the consensus sequence will prevent N-glycosylation. Deletion by N-terminal truncation is the ap-proach that we have taken with the turkey βAR, and this has actually improved expression of functional receptor. For other eukaryotic membrane proteins, N-glycosylation in extracellular loops may be essential for the folding or assembly of functional protein. In such cases, in order to maximize the expression of assem-bled, functional protein, it is likely that expression will be best in cells which are similar to those from which the native protein is obtained.[7] An interesting approach to limit the complexity of the N-glycan on a protein expressed in mammalian cells, is to utilize cells in which a glycosyl transferase has been deleted. This allows core glycosylation for correct folding and assembly but prevents complex N-glycan formation in the Golgi apparatus and should therefore eliminate heterogeneity.[8] N-Glycans can also be chemically cleaved from purified proteins. The anion exchanger AE1, purified from human erythrocytes, could only be crystallized after deglycosylation with PNGase F,[9] and the quality of crystals of bovine aquaporin-1 (AQP1) was improved by deglycosylation.[10] However, the glucose transporter Glut1 from human erythrocytes was resistant to deglycosylation with the same enzyme unless unfolded with SDS, indicating that the N-glycans were not accessi-ble to the deglycosylase.[11]

2.1.2 Palmitoylation

One or more cysteine residues located immediately after the amphipathic helix 8 may be palmitoylated when GPCRs are expressed in eukaryotic cells. Palmitoylation and subsequent depalmitoylation is a dynamic process, which is connected with receptor regulation, depalmitoylation being promoted by GPCR agonist activation.[12] Accurate quantification of the extent of this modification in receptors expressed in eukaryotic cells is difficult, and these receptors may not be 100% palmitoylated. This possible source of heterogeneity may be avoided by expression of receptors in bacteria, or by mutation of these specific cysteine residues. However, the loss of the membrane-anchoring function of the palmitate could result in some flexibility in the region of helix 8.

2.1.3   Phosphorylation

Potential phosphorylation sites are often present at the receptor C-terminus, and sometimes, on the third intracellular loop of many GPCRs. Phosphorylation is part of the normal cycle of receptor agonist-modulated activation and subsequent down-regulation. Isolation of phosphorylated GPCR species will inevitably be required for structural studies of activated receptors and receptor signalling complexes, but to obtain an initial structure, phosphorylation is probably an unnecessary and possibly harmful source of heterogeneity. Phosphorylation of GPCRs expressed in eukaryotic cells has been observed and can be assessed either by mass spectrometry[13,14] or by monitoring changes in electrophoretic mobility after incubation with a protein phosphatase, or Western blotting with anti-phosphoprotein antisera.[5,15] Our turkey βAR construct has a longer C-terminus than other GPCRs and it contains 14 serine and threonine residues, which are potential phosphorylation sites. When the turkey βAR was expressed in insect cells and purified, the band on SDS polyacrylamide gel electrophoresis was diffuse on the trailing edge, suggesting that the βAR was partially phosphorylated. This heterogeneity could not be suppressed by expression in the presence of receptor antagonist ligand, but it was reduced by including a protein phosphatase treatment in the purification. Another possible approach is the removal of minor phosphorylated species with a phosphoprotein antibody affinity column. For maximum efficiency of both dephosphorylation and affinity purification, these steps should be performed on at least partially purified receptor. Phosphorylation of overexpressed GPCRs can also be avoided by utilising bacterial expression systems or by deleting the relevant section of the C-terminus.

2.2   Other Sources of Heterogeneity

Heterogeneity of the polypeptide chain can also arise from proteolysis during either expression or purification. Partial proteolysis of the N-terminus of the turkey βAR during expression was obviated by deleting the affected section. Proteolysis during preparation of the membrane fraction and subsequent purification is a problem which can generally be avoided by working quickly, maintaining a low temperature (4 °C), and including protease inhibitors.

Another form of heterogeneity arises from the inherent structural flexibility of GPCRs that is required for their function as signalling molecules. This flexibility will inevitably give rise to conformational heterogeneity. Flexibility is not restricted to the GPCRs, as membrane transporters also show conformational changes in both helices and loop regions on substrate or inhibitor binding. Ligand, substrate or inhibitor binding may reduce flexibility and it is therefore usual to include a tightly bound ligand in the later stages of GPCR purification.

Flexibility can sometimes be eliminated by the identification of the functional core of the protein in question and it is helpful in this process to consider the structural question which is being addressed in order to trim unnecessary parts from the protein. In some cases soluble domains, such as the extracellular ligand binding domains of some GPCRs and channels, may be individually more amenable to expression and crystallization.[16–18] Proteolysis of the target protein that occurs during expression or purification may be helpful in identifying areas of flexibility, otherwise proteases can be added and products identified by mass spectrometry to identify the rigid functional core of a protein. The first potassium channel (KcsA) crystals were obtained after targeted proteolysis of 35 residues from the C-terminus,[19] and the quality of the glycerol-3-phosphate transporter (GlpT) crystals was improved by identifying a protease resistant core and expressing a new construct with a truncated C-terminus.[20] Limited proteolysis may also be useful in the identification of a flexible loop region, but it may not identify its boundaries.

Antigen-binding fragments (Fab fragments) of monoclonal antibodies have been used to stabilize flexible loops[21] and both Fab and Fv fragments have been used in the crystallization of membrane proteins to enhance the probability of crystal contacts by increasing the hydrophilic area.[22,23]

A possible explanation for the difficulty in obtaining crystals of GPCRs that diffract to high-resolution is the flexibility of some regions of the protein. The N-terminus, third intracellular loop and the C-terminus of GPCRs have all been referred to as 'intrinsically unfolded'[24] and, in most cases, these are rather extended when compared to bovine rhodopsin. The chances of successful crystallization could be increased by reducing these areas. Sequence alignment of the target receptor with bovine rhodopsin and reference to its 3D crystal structure could be used to identify the minimum receptor required to competently and stably bind its ligand, which could then allow the deletion of any sections unnecessary for functionality. In the case of the turkey βAR, the point of truncation of the N-terminus has already been indicated by proteolysis during expression. Additional deletions could include part of the third intracellular loop as well as a large section of the C-terminus. Clearly such manipulations should not affect ligand binding or agonist-induced G-protein activation if any resulting structure is to be meaningful in a biological context. Such deletions would also reduce the hydrophilic portions of the receptor (hydrophilic areas are generally thought to contribute to crystal contacts in type-II crystals), but the hydrophobic packing arrangement of bovine rhodopsin in a trigonal crystal form indicates that it is not always necessary to conserve or even increase hydrophilicity for successful 3D crystallization.[4]

3   Membrane Fractionation, Solubilization, and Detergent Selection

The first step in the purification of a membrane protein is normally the isolation of the membrane fraction. After breaking the cells, the membrane fraction is pelleted by ultracentrifugation, which separates it from the soluble proteins. This procedure results in an initial enrichment, and the resulting membrane fraction may be characterized and then stored frozen for subsequent purifications. There are numerous methods available for cell breakage, the choice depending on the host cell used for expression.[25] It is important that the effects of the chosen cell breakage method, centrifugation, methods used for resuspension, and the effects of freezing and thawing of membranes on the recovery of receptor binding sites are monitored. For example, we found that when preparing insect cell membranes containing overexpressed turkey βAR, the DIAX (Heidolph) homogenizer could be used only briefly and at low speed for the resuspension of membranes, otherwise receptor binding was greatly reduced. It has also been observed that receptor binding sites may be lost when a French pressure cell is used for cell breakage, unless the operating pressure is kept low.[26] With sensitive proteins such as GPCRs, it may be found that the best method to maximize recovery is to avoid cell breakage and membrane fractionation altogether, and directly solubilize from whole cells.[27]

The essential difference in the purification of integral membrane proteins compared to soluble proteins is the requirement for detergent to release them from the membrane and maintain them in solution. There are currently a large number of detergents available and the choice of detergent for purification is absolutely critical for success, because many detergents can denature membrane proteins resulting in inactive preparations. For relatively small integral membrane proteins like GPCRs, ideal detergents for crystallization experiments would have relatively small micelles and high critical micelle concentrations (CMCs) *e.g.* octylglucoside (OG) or nonylglucoside (NG). Unfortunately, small head groups and short aliphatic chains generally make these detergents more denaturing and their use, either as sole detergent or as an additive, is best avoided until the final stages of purification. After purification, the effect of the detergent on protein stability can be better quantified and factors that improve stability can be determined, *e.g.* addition of lipids, a receptor ligand or small amphiphiles like glycerol. This section will describe factors, which should be taken into account in deciding which detergent to use at different stages in the purification and how to optimize membrane protein stability.

3.1   Detergents for Solubilization

Large quantities of micellar detergent are required for initial solubilization to fully disrupt the membrane and form discrete micelles containing a single functional unit of a membrane protein. A mild (large, non-polar head group, C_{12} aliphatic tail), low CMC detergent can be used for solubilization, and this will be economical to use in initial purification steps due to the low working concentration required in running buffers. These detergents are also likely to maintain the receptor in a native conformation. The detergent used for solubilization is selected for its ability to effectively release receptor-binding activity from membranes and to stably maintain this activity in the solubilized state. In this respect, DDM (CMC of 0.17 mM) is one of the most popular detergents for the initial solubilization of eukaryotic membrane proteins, but polyoxyethylenes, such as $C_{12}E_8$ (CMC of 0.11 mM) and $C_{12}E_9$ should also be considered, as well as dodecanoylsucrose (CMC of 0.30 mM). These detergents all form large micelles (~80 kDa), which far exceed the size of a simple GPCR, and therefore the amount of detergent specifically bound to the receptor may be expected to exceed the mass of the protein. In the case of the purified turkey βAR, ~2.25 g of DDM are bound per gram of receptor, which corresponds to ~175 molecules of detergent bound to one βAR molecule.[5] Unfortunately, it is often not possible to use alkyl detergents that form smaller micelles, *e.g.* OG and LDAO (n-dodecyl-N, N-dimethylamine-N-oxide) because these detergents tend to be harsher than DDM. Among the GPCRs only bovine rhodopsin, stable in a range of detergents,[28] and the refolded human leukotriene (BLT1) B$_4$ receptor (stable in LDAO),[29] have so far demonstrated stability in detergents commonly thought useful for 3D crystallization. Other detergents which have been extensively used in GPCR research include digitonin and the bile-salt based detergents such as cholate and CHAPS (3-[(3-cholamidopropyl) dimethylammonio]-1-propanesulphate), often in combination.[30,31] Digitonin may not be suitable for the preparation of highly purified receptors because of its heterogeneity,[31] and it also forms unusually rigid micelles.[32] Detergents vary in their ability to solubilize lipids, and digitonin is unable to solubilize cholesterol,[33] which is essential for the activity of many receptors.[34] The bile-salt-based detergents are planar, with a hydrophobic and a hydrophilic side, and should form a less-bulky micelle than the alkyl detergents, which represents a possible advantage over alkyl detergents.

A variety of detergents may be tested for their ability to solubilize and maintain the function of the target protein. For GPCRs, solubilization of functional receptor can be determined by ligand binding, for other proteins, the efficiency of solubilization can be determined by SDS gel electrophoresis and Western blotting. Normally 2–3x more detergent (w/w) compared to the total weight of membrane protein is required for complete solubilization; this quantity can also be optimized as solubilization may be the most expensive step in a purification. Subsequent purification steps and detergent exchanges require a lower working concentration of detergent, normally 2–3x its CMC. The pH is often critical to maintain the receptor in a functional conformation, but it is also important that both pH and salt concentrations are compatible with subsequent purification protocols. The addition of salt, usually NaCl (0.1–0.4M) may be required to maximize receptor solubilization, and in rare cases there may be a requirement for a high salt concentration throughout the purification to maintain solubility. Despite the use of even the mildest, non-ionic detergents for solubilization, and the fact that delipidation of the receptor is not thought to be extensive on solubilization, stabilizing factors may still have to be included at the solubilization stage. Stabilizing factors include the addition of glycerol, cholesteryl hemisuccinate (CHS) or other lipids as well as performing all procedures at 4 °C. Some initial solubilization conditions for GPCRs are shown in Table 1.

3.2 Detergents for Final Purification Steps and Crystallization

A wide variety of detergents have been successfully used in the 3D crystallization of membrane proteins, and the relative success rates for detergents, and detergent mixtures can be obtained from lists of membrane proteins of known structure.[36] Detergents such as DDM, which form large micelles, have proved successful for the crystallization of larger membrane proteins with molecular weights equal to, or greater than, one detergent micelle. With smaller membrane proteins, specifically bound DDM might exceed the mass of the protein, and the current view is that it would be advantageous to use a detergent, which forms a smaller micelle. This may not always be possible because such detergents are generally destabilizing. One possibility is to reduce the micelle size, thus enhancing the likelihood of hydrophilic crystal contacts, by the addition of small hydrophobic organic compounds. This approach may not be so effective with mild detergents like DDM,[37] but the addition

Table 1 *Reported solubilization conditions for GPCRs*

Receptor	Host cells	Detergent % and ratio (w/w, det/protein)	Additives (%)	NaCl (M)	Notes	Ref.
βAR (turkey)	Insect (Tni)	DDM (2%, 2.0)	none	0.35	1	5
Neurotensin (rat)	E. coli	DDM (1%) and CHAPS (0.6%)	glycerol (30) CHS (0.12)	0.2	2	27
Adenosine A$_{2a}$ (human)	E. coli	DDM, UDM or DM (1%, 1.6)	glycerol (30) CHS (0.2)	0.2	3	35
Opsin (bovine)	Mammalian (HEK 293S)	NG (2%)	none	0.14	4	8

Notes:

1. The efficiency of solubilization of the turkey βAR with DDM was over 90% in the presence of >0.2 M NaCl, but at low ionic strength solubilization efficiency was only 35%. Half life of receptor ligand binding in the solubilized membranes was several weeks.

2. To avoid membrane preparations by cell breakage and ultracentrifugation, which caused a loss of receptor binding sites, intact cells were solubilized; therefore no value for the detergent: protein ratio is given. This approach results in a viscous solubilizate as total cellular protein is released and this can only be applied at a low flow rate to the first column or applied in batch. A detergent mixture and the use of two additives were necessary for efficient solubilization and subsequent receptor stability.

3. Receptor stability in the solubilisate, as indicated by half-life of the receptor's ligand binding, was 40–130 days (DDM), 26 days (UDM), and 7 days (DM). Half-lives in UDM and DM, but not in DDM were improved by addition of CHS. The addition of CHS was essential to maintain receptor stability during subsequent purification.

4. Solubilization from whole cells after addition of the retinal ligand, which extends the detergent stability of the unliganded receptor to allow the use of NG.

Abbreviations used in Table 1: DDM, dodecylmaltoside; UDM, undecylmaltoside; DM, decylmaltoside; CHS, cholesteryl hemisuccinate; CHAPS, 3-[(3-cholamidopropyl) dimethylammonio]-1-propanesulphate; NG, nonylglucoside.

of a second detergent with a shorter alkyl chain is more likely to reduce micellar size. The problem can also be circumvented by extending the hydrophilic area of the membrane protein by creating a membrane protein-soluble protein fusion or by binding an antibody fragment, raised against hydrophilic loops of the membrane protein.

The choice of detergent is also crucial in maintaining the appropriate proportion of specifically bound lipid. It has been demonstrated, by phospholipid determination, that detergents with shorter aliphatic tails are more effective in delipidating membrane proteins than milder detergents with longer aliphatic tails.[9] Excessive delipidation of membrane proteins may result in an increased tendency to form aggregates and loss of functional activity.

Bile-salt based detergents have not been used extensively in large-scale purification and there has not so far been any successful crystallization of any membrane protein solubilized exclusively in such a detergent, although they have been used as additives in crystallization.[38–40] We have found the turkey βAR to be completely inactivated by CHAPS, but other GPCRs may be more stable in this detergent,[41] and there are also many new synthetic bile-salt detergents available which should all form relatively small micelles. These could also be tried, either alone or in combination with alkyl detergents.

There are a number of newly developed detergents, which may be considered for GPCR purification. The Cymal® range (maltoside detergents with a cyclohexyl aliphatic tail) may be sufficiently mild for detergent-sensitive proteins and should have a reduced micellar size compared to the maltosides with linear aliphatic tails. Fos-choline® detergents have lipidic head groups but simple aliphatic tails. Although this concept sounds attractive, we have found that these detergents, even with C_{12} and C_{14} aliphatic tails, destabilize turkey βAR antagonist binding compared to DDM. Derivatives of Fos-choline® with cyclohexyl aliphatic chains are also available.

In order to accomplish a complete detergent exchange, the turkey βAR is bound to a ligand affinity column, where it is washed with several column volumes of buffer containing the replacement detergent. We have used this method for exchanges into UDM, DM, and decylthiomaltoside for crystallization experiments, after having determined the suitability of these detergents by dilution. The variety of detergents now available is so large, that rather than carrying out a full detergent exchange by binding a protein to a column, it is perhaps easier to initially determine suitability of a detergent by dilution of the purified protein into the new detergent. This technique is rapid and can be performed on a small scale, but it will only give an indication whether a complete exchange will be tolerated. For example, the turkey βAR tolerates exposure to NG when diluted into it, as ligand binding is maintained, indicating its suitability as a secondary detergent in combination with DDM, but a complete detergent exchange into NG resulted in the precipitation of the receptor. It is therefore important, in order to perform such dilution experiments, to use purified protein, which is sufficiently concentrated to allow an appropriate dilution into the new detergent. It is also important that protein concentration has been achieved without concentration of non-protein containing detergent micelles by using appropriate molecular weight cut-off concentrators,[42] otherwise an excess of the initial detergent may protect the protein from contact with the secondary detergent. For the turkey βAR purified in DDM, we have found it essential to use sample concentrators with a molecular weight cut-off of 100 kDa.

4 Purification

The techniques required for purifying a membrane protein are usually dictated by the level of overexpression. Generally, the expression of GPCRs is about 0.1–1mg per litre of culture, although βAR, rhodopsin, and a few others are expressed at higher levels. The turkey βAR overexpressed using the baculovirus expression system represents 1.5% of total membrane protein. In contrast the neurotensin receptor, expressed in *E. coli*, represents 0.15% of total protein, but this has proved sufficient for large-scale purification[27] and structural studies by solid state NMR.[43] Low expression levels necessitate high purification factors and the implementation of purification protocols with maximal yields. An additional problem is that some expression systems may generate a large proportion of non-functional protein, which can be detergent-solubilized and could be co-purified with functional receptor. In the case of the turkey βAR overexpressed in insect cells the proportion of non-functional receptor may be ~80%. This can be removed by using a ligand affinity column in the purification procedure.

4.1 Use of Purification Tags and Fusions

The method of choice, as with recombinant soluble proteins, is to express the membrane protein as a fusion with one or more affinity tags to assist purification, as only the use of highly specific affinity steps will provide the high purification factors required. An affinity tag purification method for membrane proteins should have the following properties: (i) the method should not be affected by the presence of detergent micelles, preferably even at the 1–2% detergent concentrations employed during initial membrane protein solubilization; (ii) the tag or elution conditions should not adversely affect the protein; (iii) it should allow quantitative and selective adsorption from extremely dilute samples; (iv) the tag should not adversely affect expression; (v) the system will ideally be applicable to a variety of proteins. Many affinity tag methods are available, and the principles underlying their use have been reviewed elsewhere.[44] The large-scale purification of overexpressed membrane proteins is a developing area, and details of yields obtained with affinity tags other than polyhistidine are still sparse, although the application of a variety of tags in the expression and purification of the rat neurotensin receptor has been described in detail.[45] We have summarized the properties of some of the fusion tags, which have been used for membrane protein purification in Table 2.

The addition of small affinity tags does not normally have any effect on receptor ligand binding properties.[52,53] However, the addition of tags will often affect expression levels.[54] For example, increasing the length of polyhistidine tags from $(His)_6$ to $(His)_{10}$ may in some cases reduce expression[55] and gains in recovery and purity that accompany increased tag length must therefore be offset against any reduction in expression. It is possible that the addition of other tags at either terminus will reduce receptor expression, and this effect will vary with tag, position, and expression

Table 2 *Affinity tags used in GPCR research*

Tag	Matrix	Ref.	Notes
Polyhistidine	Ni^{2+}-NTA, Talon (Co^{2+}), Ni^{2+}-Sepharose	46	Recovery 40–60% with turkey βAR, depending on matrix. Purification moderate, requires a second affinity step. Elution conditions harsh, high imidazole concentration should be quickly reduced. Histidine can be used for milder elution at lower concentration.
Strep II	Strep-tactin	47	Affinity 1 μM, therefore poor binding and recovery unless target protein concentration is high, otherwise excellent purification. An initial metal affinity step can be used to achieve a more concentrated sample suitable for further purification via the Strep II tag with high recovery.
FLAG	Anti-FLAG M1 resin	48–50	Excellent purification, mild binding, and elution conditions. Tag encodes an enterokinase cleavage site.
Bio-tag	Avidin	45	Tag 13 aa, therefore cleavage is preferable, biotinylation by expression host required.
Maltose-binding	Amylose	26	Excellent purification, may be sensitive to high detergent concentration and may therefore work protein best as a second purification step. Attachment of a large tag may introduce flexibility or destabilize the receptor. It has been recommended that maltoside detergents should be avoided as they may interfere with binding.[26]
1D4	1D4 sepharose	51	Excellent purification, very mild peptide elution, yield can be variable and binding conditions require optimization.

system. All of these factors need to be optimized to minimize any deleterious effect on expression levels.

The yield of membrane proteins in affinity tag purification steps may be low, possibly due to the partial burial of the tag in the detergent micelle. Improvements in recovery have been observed when histidine tags are lengthened,[46] and the insertion of a flexible linker and protease cleavage site might also improve accessibility. For the purification of histidine-tagged proteins by metal affinity chromatography a variety of matrices is available, which we tested, finding an inverse relationship between purity and recovery. As a second affinity step is available for the turkey βAR, we compromise purity for increased recovery. Recovery can also be improved by loading in batch or at low flow-rate. For efficient use of laboratory time, column loading steps, elution, and peak collection can all be automated and run overnight.[5,27] High salt concentrations, which often help initial solubilization, are beneficial with metal affinity chromatography as they reduce non-specific protein binding, making IMAC an ideal first purification step.

Large affinity tags such as maltose-binding protein (MBP) and glutathione-S-transferase (GST) have been used as an aid to expression of GPCRs in *E. coli*, but are not generally used for purification. It has been reported that it is not possible to purify MBP-receptor fusion proteins, solubilized in DDM, using amylose affinity resins. Dodecanoylsucrose, which due to the ester linkage of the aliphatic tail is a heterogenous mixture of three isomers, has been used instead.[26] We were able to purify a MBP-turkey βAR fusion protein from solubilized membranes containing 2% DDM, by amylose affinity. It may be that micellar maltoside does not interfere with binding of the MBP to amylose and that the presence of 0.17 mM monomeric maltoside (*i.e.* at the CMC of DDM) is not a problem, as long as the fusion protein concentration is relatively high (Warne *et al.*, unpublished results).

The MBP fusion approach could also be used as a means of extending the hydrophilicity of GPCRs to make them more amenable to 3D crystallization.[26] The use of MBP fusions has assisted in the crystallization of some soluble proteins,[56] and in all of these successful cases the linker region between the MBP and the target protein has been altered with the aim of reducing flexibility in this region. However, especially after the attachment of a large fusion partner, or indeed any protein modification, it is essential to determine whether protein stability or function has in any way been affected before proceeding further.

The use of the Bio-tag, Strep II tag, FLAG tag, and 1D4 epitope tag are all highly recommended for their high specificity of binding and the mild conditions for binding and elution required. The Bio-tag is one of the lesser-used tags; a 13 amino acid biotinylation target is fused to the protein of interest, and the protein, which is biotinylated in vivo can then be purified using a monomeric avidin column.[57] To ensure good recovery of the neurotensin receptor using the Bio-tag system, it was necessary to co-express a biotin ligase with the neurotensin receptor in *E. coli* to ensure maximal biotinylation of the receptor; subsequently a 60-fold purification was achieved using a monoavidin affinity column.[45] We have found the Strep II tag inefficient (10% yield) when used as a first purification step for the turkey βAR because of the low receptor concentration in the solubilizate, and we would recommend that it be used after a first capture step, *e.g.* metal affinity chromatography, which will also concentrate the receptor. The use of antibody affinity columns, such as anti-FLAG M1 and 1D4 can effect remarkable, one-step purifications. The 1D4 antibody recognizes the last eight C-terminal residues of bovine rhodopsin, and overexpressed rhodopsin bound to a 1D4 antibody resin can be selectively eluted with a rhodopsin peptide, with excellent recovery and purification.[51] As far as being useful in the purification of recombinant rhodopsins, the 1D4 epitope has been added to the C-terminus of a number of other overexpressed visual pigments and GPCRs. However, yields have proved to be variable, with poor recovery of the β_2AR despite its longer C-terminus compared to rhodopsin.[58]

4.2 Removal of Tags and Fusions

With soluble proteins and some of the more highly expressed prokaryotic membrane proteins, a protease recognition sequence to allow removal of the tag is often included, as the removal of a tag may prove useful in order to improve crystal quality. This may remove a stretch of unstructured residues, e.g. a $(His)_6$ or $(His)_{10}$ tag. However, it should be noted that histidine tags are often placed at the receptor C-terminus in order to maintain expression[54] and that protease recognition sites will generally leave between four and six additional residues when placed at the C-terminus. In cases where it is necessary to add two tags in order to effect sufficient purification, it is clear that removal would be preferable. With proteins, which are not so well-expressed, there is perhaps a reluctance to perform a cleavage step. This is because the inclusion of such a step will lengthen and further complicate the purification procedure, and if the cleavage step is not efficient, the already low yield might be further reduced. However, it has been shown that TEV protease cleavage can proceed efficiently at 4 °C, resulting in the quantitative overnight cleavage of the N-terminal MBP and C-terminal thioredoxin-$(His)_{10}$ fusions from the neurotensin receptor.[27]

4.3 Ligand Affinity Chromatography

The heterologous overexpression of receptors may result in the generation of a proportion of non-functional, misfolded receptor, which must be separated from functional receptor. It has proved possible, by careful adjustment of 1D4 immunoaffinity chromatography conditions, to separate misfolded rhodopsin mutants, which were unable to bind the retinal chromophore, from functional receptor.[59] However, ligand affinity chromatography is the method of choice for this essential step. Unfortunately, ligands with appropriate binding affinities (<0.5 μM) may not be readily available for less well-characterized GPCRs or cannot easily be linked to a support matrix. With the turkey βAR, affinity chromatography has been carried out by binding, functional receptor to the antagonist ligand alprenolol linked to a Sepharose matrix. It has been found that competitive elution of receptor from the matrix with free ligand (100 μm) can be performed at 4 °C and the receptor is eluted as a sharp peak provided 0.35 M NaCl is present in the elution buffer.[5] Others utilising the same affinity resin to purify the human β_2AR eluted bound receptor with free antagonist by warming the column to 37 °C for 1 h;[49] this is a risky procedure with a receptor which is best maintained at low temperature for maximum stability. With the human adenosine A2a receptor, a temperature of 23 °C was used to effect elution from an agonist affinity column and, despite high ionic strength, elution of receptor was very slow.[35] The difficulties sometimes experienced in eluting GPCRs from ligand affinity columns mean that ligand affinity is probably best utilized as a second purification step. Column size can then be reduced, and a difficult and lengthy elution step may be minimized, and excessive dilution of the receptor can be avoided.

While ligand affinity chromatography will remove non-functional turkey βAR and is essential for further purification, only 50% of the functional receptor can be recovered in the eluate and the final purified turkey βAR exhibits only 50% of the theoretical maximum ligand binding activity, despite having bound to the ligand affinity column.[5] The reasons for this are not clear, although it seems to be an observation, which is peculiar to the βARs. In contrast, both the neurotensin receptor and the adenosine A_{2a} receptor, when eluted from their ligand affinity columns, yield purified receptors that bind ligand at close to 100% of the theoretical maximum.[27,35]

4.4 Final Purification Steps before Crystallization and Assembly of Complexes

The final purification step of turkey βAR was size exclusion chromatography, which was used to remove high molecular weight aggregates. Purification of membrane proteins by size exclusion is generally not very effective because the molecular weight of each membrane protein is increased by the weight of the detergent bound. For example, the βAR has a calculated molecular weight of 46 kDa but migrates on size exclusion chromatography at 120 kDa due to bound DDM. Baseline separation of soluble (globular) protein peaks is possible where there is a factor of 2 difference in molecular weight, but the differences in apparent molecular weight for membrane proteins are reduced because of the bound detergent. However, size exclusion is still useful for the production of a monodisperse preparation. The elution of a sharp, symmetric peak should be observed for a pure protein, even in a detergent micelle, and this is an excellent diagnostic of a high-quality purification procedure.

While separation of different membrane proteins by size exclusion may prove difficult, the binding of detergent will enhance separations of membrane proteins from soluble proteins. The hydrophilic area of membrane proteins can be extended by the binding of antibody fragments raised against hydrophilic loops of the membrane protein, and the preparation and isolation of such a complex may be performed immediately prior to crystallization. We have used size exclusion to separate the turkey βAR with a bound Fab fragment from excess Fab (50 kDa). A clear isolation of the βAR-Fab complex was easily achieved, as the weight of this complex is increased from the predicted value of 98 kDa to 185 kDa, due to the detergent bound to the receptor component of the complex (Figure 1).

Size exclusion chromatography will generally result in a five-fold dilution of the applied protein sample, and this will necessitate a further concentration step to achieve a concentration sufficient for crystallization (5–20 mg/mL). Concentration is usually achieved with centrifugal concentration devices and it is important to determine whether these generate further high molecular weight aggregates. If this is the case, ultracentrifugation of the sample prior to crystallization may be sufficient

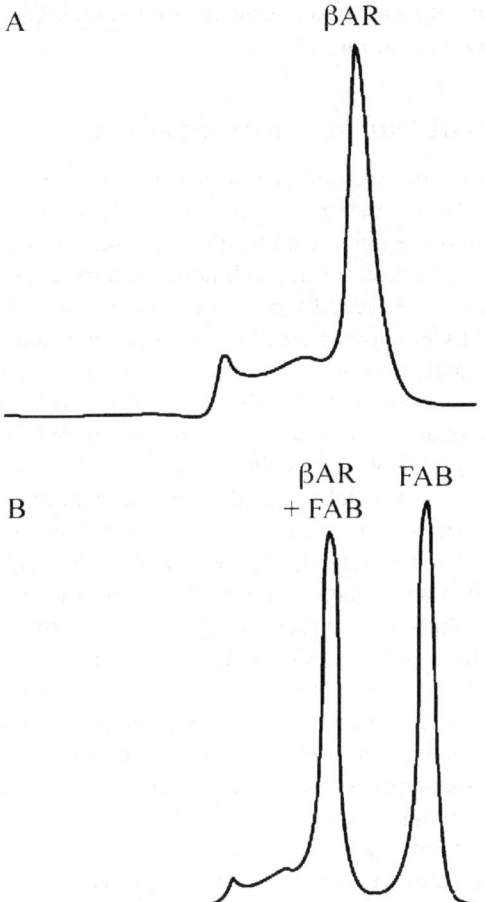

Figure 1 *Preparation of the turkey βAR-Fab complex by size exclusion chromatography on a Superdex HR10/30 column (elution volumes in brackets). A: βAR alone (13 mL); B: βAR-Fab (11.9 mL), excess Fab (15.4 mL). The Fab fragment was prepared by proteolysis of the monoclonal antibody TBA2–354–74 using papain*

to remove aggregates, or a further chromatography step (*e.g.* affinity resin or ion exchange) could be implemented to concentrate the protein. This should be performed with the minimum amount of column matrix, eluting with a steep gradient and dialysing the sample to reduce salt if necessary. The binding of membrane proteins without large hydrophilic domains to an ion exchange matrix may be weak, although binding in the early stages of purification can be stronger as it can be mediated by receptor-associated charged phospholipids, which may be lost with further purification. A Mono Q anion-exchange chromatography step has been used for the final purification of bovine rhodopsin prior to crystallization, and the receptor was eluted with only 50 mM NaCl, demonstrating weak binding. This step was essential for delipidation, detergent exchange, separation of aggregated species, and also indicated the potential of using anion exchange for the separation of phosphorylated species, which bound more strongly.[60]

4.5   Lipid Content During Purification

One of the major problems encountered during purification is the reduction in the amount of lipid bound to the membrane protein, which occurs by the partitioning of protein-bound lipid into free detergent micelles. This may have a destabilising effect on the GPCR, resulting in reduced ligand binding of a receptor or aggregation. The stripping of lipid may be reduced by reducing washing steps on columns, keeping protein concentrated by employing steep elution gradients and keeping the column size to a minimum, and of course, using an appropriate detergent. The working concentration of detergent in column buffers can also be reduced to just above the CMC, which will maintain the protein in its micelle, but reduce lipid partition into non-protein-containing detergent micelles. Washing of columns with the target protein bound is essential for the removal of contaminating polypeptides, but the quest for absolute polypeptide purity may have to be compromised in order to maintain optimal bound lipid. Alternatively, lipid may be added back to the GPCR during or after purification, after prior determination of the specific lipids, which enhance stability of ligand binding or help prevent aggregation, or by reference to the lipid composition of the native bilayer. An example of the effect of delipidation through over-purification is the aggregation of the *E. coli* glycerol-3-phosphate transporter (GlpT) when a third chromatography step was introduced to remove two co-purifying polypeptides.[20] The aggregation problem was circumvented by the optimization of expression conditions to suppress the expression of the contaminants, thus avoiding the need for a third chromatography step.[61]

The quantity of co-purifying bound lipid can also be of crucial importance to the success of 2D crystallization experiments. Purified bovine rhodopsin is crystallized in two dimensions by removing the detergent by dialysis and relying solely on endogenous, co-purifying lipid to form the lipid bilayer.[62] During purification, the rhodopsin is bound to a concanavalin A column, and if this is not washed extensively, too much lipid is present and the reconstituted rhodopsin is disordered in large vesicles. With optimal washing using a concanavalin A column of the appropriate size for the given amount of protein, 2D crystals are formed. Further purification by anion exchange results in rhodopsin that is too delipidated and detergent removal causes aggregation as opposed to 2D crystallization. However, the delipidated rhodopsin prepared with anion exchange was used to produce 3D rhodopsin crystals that diffracted to 2.65 Å resolution.[60]

5   Final Quality Control, Monitoring Protein Stability, Aggregational State, Lipid, and Bound Detergent

Crystallization experiments may last for several days, if not weeks, and are unlikely to succeed if the protein is destabilized by inappropriate choice of detergent, poor control of detergent concentration, pH, salt concentration, or temperature. It is essential to determine both the limitations of each parameter and also to look for positive influences, such as the addition of glycerol or lipids. Detergent-solubilized GPCRs can be assayed directly for ligand-binding activity using radiolabelled ligands and the stability can be monitored over a period of weeks under different conditions. The determination of specifically bound detergent and an estimation of the molecular weight of the protein in its micelle, either by gel filtration or analytical ultracentrifugation, is useful in order to determine its aggregational state. The ideal sample for crystallization will be a concentrated, completely monodisperse receptor that maintains both its function and monodispersity. Important parameters affecting functional stability and monodispersity are exposure to different detergents and detergent mixtures, the detergent concentration, extent of delipidation, temperature, pH, ionic strength, and additives. Monodispersity can be ascertained by performing analytical gel filtration on the purified protein incubated in different detergents and, in order to accelerate this analysis, incubation may be carried out at room temperature.[61] Reduction in peak size of the eluted protein compared to the control indicates aggregation, with aggregates having been removed by prior centrifugation. The results of a comprehensive analysis of the effect of detergents on the transporter GlpT are illustrated in Figure 2. A protein assay on the supernatant after centrifugation or SDS gel electrophoresis could also demonstrate the same effect. Another method that has been used to monitor protein monodispersity/aggregation is

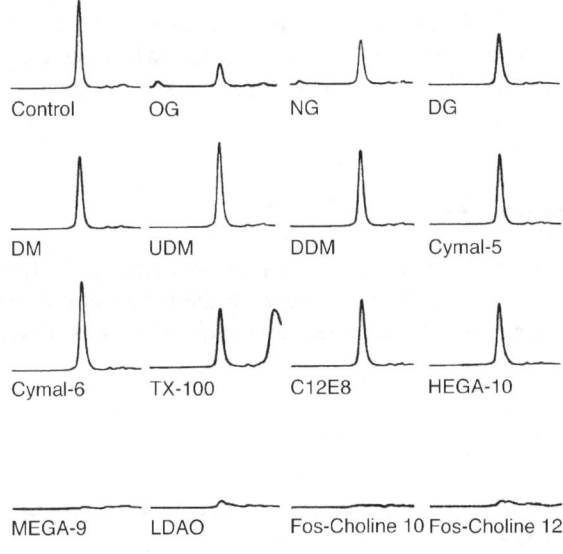

Figure 2 *Oligomeric state, stability and monodispersity of the glycerol-3-phosphate transporter (GlpT) in different detergents. Size exclusion chromatography was performed after 2 h incubation at room temperature. MEGA-9, LDAO and Fos-choline detergents caused aggregation. Incubation with glucoside detergents resulted in variable aggregation which was dependent on alkyl chain length. DG; decylglucoside. Reprinted with permission from Ref. 61. Copyright (2001) American Chemical Society.*

dynamic light scattering although this is difficult because there is already a mixed population of sizes with detergent-only and protein-detergent micelles. Electron microscopy of samples for crystallization has also proved effective in sample quality control and has also indicated the importance of a final ultracentrifugation step in improving crystal quality.[63]

The analysis of lipid specifically bound to the purified protein by thin layer chromatography (TLC),[64] in conjunction with a phosphorus assay,[65] may determine where delipidation occurs and can be helpful in the standardization of protein preparations. More important, however, is to recognize the effects of delipidation and to take remedial action. If purification is so demanding that minimization of washing steps and detergent exposure are not effective in conserving lipid, or if the expression host does not supply the correct lipid, then addition of lipid during the purification is necessary. If the protein is sensitive in this respect, it follows that crystallization methods which are mediated by potentially stabilizing lipid additions, such as 2D crystallization and the newer bicelle[66] or lipidic cubic phase[67] crystallization methods should be considered in addition to standard vapour diffusion methods, which might also be enhanced by the addition of lipid.[68]

6   Conclusions

The purification of mammalian integral membrane proteins is extremely demanding due to the multitude of factors that can adversely affect the stability of the protein. An appreciation of solubilized membrane proteins as a complex and variable assemblage of protein, detergent, and lipid is essential when devising a purification strategy to produce fully functional, monodisperse, stable membrane proteins ideal for crystallization.

Acknowledgments

We would like to thank Dr. C.G. Tate for his critical reading of the manuscript and Professor Paul Hargrave and Dr. D. Dugger for the production of monoclonal antibodies. We thank Dr. Elliot Ross for helping us in the initiation of our studies of the turkey βAR.

References

1. V. Sarramegna, F. Talmont, P. Demange and A. Milon, *Cell Mol. Life Sci.*, 2003, **60**(8), 1529–1546.
2. Y. Yarden, H. Rodriguez, S.K. Wong, D.R. Brandt, D.C. May, J. Burnier, R.N. Harkins, E.Y. Chen, J. Ramachandran, A. Ullrich and E.M. Ross, *Proc. Natl. Acad. Sci. USA*, 1986, **83**(18), 6795–6799.
3. K. Palczewski, T.Kumasaka, T. Hori, C.A. Behnke, H. Motoshima, B.A. Fox, I. Le Trong, D.C. Teller, T. Okada, R.E. Stenkamp, M. Yamamoto and M. Miyano, *Science*, 2000, **289**, 739–745.

4. J. Li, P.C. Edwards, M. Burghammer, C. Villa and G.F. Schertler, *J. Mol. Biol.*, 2004, **343**, 1409–1438.
5. T. Warne, J. Chirnside and G.F. Schertler, *Biochim. Biophys. Acta*. 2003, **1610**(1), 133–140.
6. C.G. Tate, J. Haase, C. Baker, M. Boorsma, F. Magnani, Y. Vallis and D.C. Williams, *Biochim. Biophys. Acta*. 2003, **1610**(1), 141–153.
7. C.G. Tate, *FEBS Lett.*, 2001, **504**(3), 94–98.
8. P.J. Reeves, N. Callewaert, R. Contreras and H.G. Khorana, *Proc. Natl. Acad. Sci. USA*, 2002, **99**(21), 13419–13424.
9. M. Lemieux, R.A. Reithmeier and D.N. Wang, *J. Struct. Biol.*, 2002, **137**(3), 322–332.
10. H. Sui, P.J. Walian, G. Tang, A. Oh and B.K. Jap, *Acta. Crystallogr. D. Biol. Crystallogr.*, 2000, **56**(9), 1198–1200.
11. J.M. Boulter and D.N. Wang, *Protein. Expr. Purif.*, 2001, **22**(2), 337–348.
12. B. Mouillac, M. Caron, H. Bonin, M. Dennis and M. Bouvier, *J. Biol. Chem.*, 1992, **267**(30), 21733–21737.
13. M. Trester-Zedlitz, A. Burlingame, B. Kobilka and M. von Zastrow, *Biochemistry*, 2005, **44**(16), 6133–6143.
14. J. Godovac-Zimmermann, V. Soskic, S. Poznanovic and F. Brianza, *Elecrophoresis*, 1999, **20**(4–5), 952–961.
15. K. Bjorklof, K. Lundstrom, L. Abuin, P.J. Greasley and S. Cotecchia, *Biochemistry*, 2002, **41**(13), 4281–4291.
16. W.N. Zagotta, N.B. Olivier, K.D. Black, E.C. Young, R. Olson and E. Gouaux, *Nature*, 2003, **425**(6954), 200–205.
17. A.P. West Jr., L.L. Llamas, P.M. Snow, S. Benzer and P.J. Bjorkman, *Proc. Natl. Acad. Sci. USA*, 2001, **98**(7), 3744–3749.
18. Q.R. Fan and W.A. Hendrickson, *Nature*, 2005, **433**(7023), 269–277.
19. D.A. Doyle, J.M. Cabral, R.A. Pfuetzner, A. Kuo, J.M. Gulbis, S.L. Cohen, B.T. Chait and R. MacKinnon, *Science*, 1998, **280**(5360), 69–77
20. M.J. Lemieux, J. Song, M.J. Kim, Y. Huang, A. Villa, M. Auer, X.D. Li and D.N. Wang, *Protein Sci.*, 2003, **12**(12), 2748–2756.
21. P.D. Kwong, R. Wyatt, J. Robinson, R.W. Sweet, J. Sodroski and W.A. Hendrickson, *Nature*, 1998, **393**(6686), 648–659.
22. R. Dutzler, E.B. Campbell and R. MacKinnon, *Science*, 2003, **300**(5616), 108–112.
23. C. Ostermeier, S. Iwata, B. Ludwig and H. Michel, *Nat. Struct. Biol.*, 1995, **2**(10), 842–846.
24. V.P. Jaakola, J. Prilusky, J.L. Sussman and A. Goldman, *Protein Eng. Des. Sel.*, 2005, **18**(2), 103–110.
25. G. von Jagow, T.A. Link, H. Schägger in *Membrane Protein Purification and Crystallization, a Practical Guide*, C. Hunte, G. von Jagow and H. Schägger (eds), 2nd edn, Academic Press, London, 2003.
26. W. Hampe, R.H. Voss, W. Haase, F. Boege, H. Michel and H. Reiländer, *J. Biotechnol.*, 2000, **77**(2–3), 219–234.
27. J.F White, L.B. Trinh, J. Shiloach and R. Grisshammer, *FEBS Lett.*, 2004, **564**(3), 289–293.

28. W.J. DeGrip, *Methods Enzymol.*, 1982, **81**, 256–265.
29. J.L. Baneres, A. Martin, P. Hullot, J.P. Girard, J.C. Rossi and J. Parello, *J. Mol. Biol.*, 2003, **329**(4), 801–814.
30. P. Park, C.S. Sum, D.R. Hampson, H.H. Van Tol and J.W. Wells, *Eur. J. Pharmacol.*, 2001, **421**(1), 11–22.
31. A. Parini and R.M. Graham, *Anal. Biochem.*, 1989, **176**(2), 375–381.
32. S. Kopanchuck and A. Rinken, *Proc. Estonian. Acad. Sci. Chem.*, 2001, **50**(4), 229–240.
33. P. Banerjee, J.B. Joo, J.T. Buse and G. Dawson, *Chem. Phys. Lipids.*, 1995, **77**(1), 65–78.
34. M. Opekarova and W. Tanner, *Biochim. Biophys. Acta.* 2003, **1610**(1), 11–22.
35. H.M. Weiss and R. Grisshammer, *Eur. J. Biochem.*, 2002, **269**(1), 82–92.
36. http://www.mpibp-frankfurt.mpg.de/michel/public/memprotstruct.html.
37. C. Hunte and H. Michel in *Membrane Protein Purification and Crystallization, a Practical Guide*, C. Hunte, G. von Jagow and H. Schägger (eds), 2nd edn, Academic Press, London, 2003.
38. J. Abramson, I. Smirnova, V. Kasho, G. Verner, H.R. Kaback and S. Iwata, *Science.*, 2003, **301**(5633), 610–615.
39. Z. Liu, H. Yan, K. Wang, T. Kuang, J. Zhang, L. Gui, X. An and W. Chang, *Nature.*, 2004, **428**(6980), 287–292.
40. A.W. Roszak, T.D. Howard, J. Southall, A.T. Gardiner, C.J. Law, N.W. Isaacs and R.J. Cogdell, *Science.*, 2003, **302**(5652), 1969–1972.
41. V. Harvey, J. Jones, A. Misra, A.R. Knight and K. Quirk, *Eur. J. Pharmacol.*, 2001, **431**(2), 171–177.
42. A. Urbani and T. Warne, *Anal. Biochem.*, 2005, **336**(1), 117–124.
43. S. Luca, J.F. White, A.K. Sohal, D.V. Filippov, J.H. van Boom, R. Grisshammer and M. Baldus, *Proc. Natl. Acad. Sci. USA*, 2003, **100**(19), 10706–10711.
44. K. Terpe, *Appl. Microbiol. Biotechnol.*, (2003), **60**(5), 523–533.
45. J. Tucker and R. Grisshammer, *Biochem. J.*, 1996, **317**(3), 891–899.
46. R. Grisshammer and J. Tucker, *Protein Expr. Purif.*, 1997, **11**(1), 53–60.
47. S. Voss and A. Skerra, *Protein Eng.*, 1997, **10**(8), 975–982.
48. B.K. Kobilka, *Anal. Biochem.*, 1995, **231**(1), 269–271.
49. C. Reinhart, H.M. Weiss and H. Reiländer, in *Membrane Protein Purification and Crystallization, a Practical Guide*, C. Hunte, G. von Jagow and H. Schägger (eds), 2nd edn, Academic Press, London, 2003.
50. Z.Gao, A.S. Robeva and J. Linden, *Biochem. J.*, 1999, **338**(3), 729–736.
51. D.D. Oprian, R.S. Molday, R.J. Kaufman and H.G. Khorana, *Proc. Natl. Acad. Sci. USA.*, 1987, **84**(24), 8874–8878.
52. D. Massotte, L. Baroche, F. Simonin, L. Yu, B. Kieffer and F. Pattus, *J. Biol. Chem.*, 1997, **272**(32), 19987–19992.
53. D. Massotte, C.A. Pereira, Y. Pouliquen and F. Pattus, *J. Biotechnol.*, 1999, **69**(1), 39–45.
54. D. Massotte, *Biochim. Biophys. Acta.* 2003, **1610**(1), 77–89.
55. A.K. Mohanty and M.C. Wiener, *Protein Expr. Purif.,* 2004, **33**(2), 311–325.
56. D.R. Smyth, M.K. Mrozkiewicz, W.J. McGrath, P. Listwan and B. Kobe, *Protein Sci.*, 2003, **12**(7), 1313–1322.

57. K.L. Tsao, B. DeBarbieri, H. Michel and D.S. Waugh, *Gene.*, 1996, **169**(1), 59–64.
58. D.D. Oprian, A.B. Asenjo, N. Lee and S.L. Pelletier, *Biochemistry.* 1991, **30**(48), 11367–11372.
59. K.D. Ridge, Z. Lu, X. Liu and H.G. Khorana, *Biochemistry.* 1995, **34**(10), 3261–3267.
60. P.C. Edwards, J. Li, M. Burghammer, J.H. McDowell, C. Villa, P.A. Hargrave and G.F. Schertler, *J. Mol. Biol.*, 2004, **343**(5), 1439–1450.
61. M. Auer, M.J. Kim, M.J. Lemieux, A. Villa, J. Song, X.D. Li and D.N. Wang *Biochemistry.* 2001, **40**(22), 6628–6635.
62. A. Krebs, C. Villa, P.C. Edwards and G.F. Schertler, *J. Mol. Biol.*, 1998, **282**(5), 991–1003.
63. R. Horsefield, V. Yankovskaya, S. Tornroth, C. Luna-Chavez, E. Stambouli, J. Barber, B. Byrne, G. Cecchini and S. Iwata, *Acta. Crystallogr. D. Biol. Crystallogr.*, 2003, **59**(3), 600–602.
64. D.L. Vaden, V.M. Gohil, Z. Gu and M.L. Greenberg, *Anal. Biochem.*, 2005, **338**(1), 162–164.
65. P.S. Chen, T.Y. Toribara and H. Warner, *Anal. Chem.*, 1956, **28**, 756–1758.
66. S. Faham and J.U. Bowie, *J. Mol. Biol.*, 2002, **316**(1), 1–6.
67. G. Rummel, A. Hardmeyer, C. Widmer, M.L. Chiu, P. Nollert, K.P. Locher, I.I. Pedruzzi, E.M. Landau and J.P. Rosenbusch, *J. Struct. Biol.*, 1998, **121**(2), 82–91.
68. H. Zhang, G. Kurisu, J.L. Smith and W.A. Cramer, *Proc. Natl. Acad. Sci. USA*, 2003, **100**(9), 5160–5163.

CHAPTER 5

An Introduction to Detergents and Their Use in Membrane Protein Studies

FABIEN WALAS, HIROYOSHI MATSUMURA AND BEN LUISI

Department of Biochemistry, University of Cambridge, Cambridge CB2 1GA, UK

1 Introduction

The cells from all organisms are enveloped by lipidic membranes, which act as protective barriers with restricted permeability. Similar membranes also partition the interiors of eukaryotic cells into specialized compartments. Although they vary greatly in their detailed chemical composition, cellular membranes are all composed of amphipathic lipids that can spontaneously self-assemble into sheets that are precisely two molecules in width. The lipids usually associate within this molecular bilayer by positioning their hydrophobic hydrocarbon moiety into the interior, and exposing their hydrophilic groups to the solvent [Figure 1(a)]. The membrane bilayers are somewhat like liquid crystals, and the degree of fluidity of the interior and exposed surface are governed by the torsional freedom of the hydrocarbon constituent and, in the case of phospholipids, electrostatic repulsion of the clustered charged groups.

Embedded in the lipid bilayers are "membrane" proteins whose diverse biological roles include enzymatic processing, signal transduction and transport of molecules across the bilayer. Membrane proteins contain two distinct domains: a hydrophilic surface exposed to the aqueous environment and a hydrophobic region buried within the hydrocarbon core of the lipid membrane bilayer [Figure 1(b)]. To study their structure and function, membrane proteins are often isolated from their natural lipidic environment and placed into an analogous milieu that mimics simultaneously both hydrophilic and hydrophobic aspects of the membrane. To date, the greatest success in isolating membrane proteins has made use of surfactants known as detergents, which can disrupt cell membranes to free and solubilize membrane-bound proteins.[1–4]

(a)   (b)

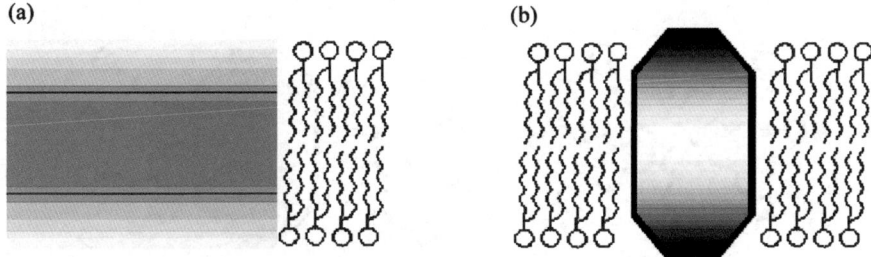

Figure 1 *(a) A schematic representation of a cross-section through a lipid bilayer (courtesy of Prof. Stephen White). The membrane bilayer is highly dynamic. The grey area represents the mean hydrocarbon density, while the red lines indicate a rough boundary of this hydrophobic core (left-hand side). The circles represent the negatively charged phosphate "head" groups. The zigzagging lines represent the two hydrophobic hydrocarbon chain "tails" (right-hand side).The phospholipid tails create a hydrophobic environment within the membrane, and the charged phosphate groups face the aqueous environment. The tails are generally acyl chains, but can include occasional double bonded carbons. Naturally occurring membranes are chemically heterogeneous, and the different types of lipids probably cluster into the so-called "rafts".[80] (b) A membrane protein is shown embedded in a lipid bilayer*

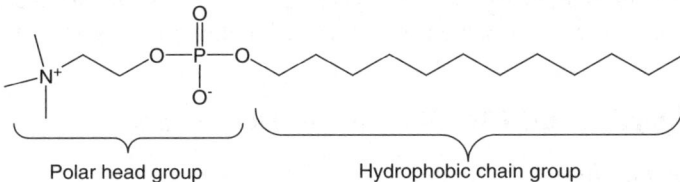

Polar head group   Hydrophobic chain group

Figure 2 *Schematic structure of the DPC detergent. DPC is a synthetic detergent often used in the studies of proteins by nuclear magnetic resonance (NMR). The phospho group has di-ester linkages to choline, $(CH_3)_3N(CH_2)_2OH$, and to dodecyl (or lauryl) alcohol, $CH_3(CH_2)_{11}OH$*

2 Physical Properties of Detergents Used in Membrane Protein Studies

Detergent molecules possess typically two distinct parts: a polar hydrophilic group and a hydrophobic carbon chain group (Figure 2), referred respectively as the "head" and "tail" in accordance with their orientations to the solvent. Due to their amphipathic nature, detergents placed in aqueous media can spontaneously form spherical assemblies, known as "micelles" [Figure 3(a)] in which their polar groups are solvent exposed, while their hydrocarbon tails pack to form a hydrophobic core.

The main purpose of using detergent molecules is to provide a favourable environment that allows membrane protein extraction from the lipid bilayer and solubilization. The detergent imitates the lipid membrane by masking the hydrophobic surface of the protein, and generates a soluble protein–detergent complex (PDC) that can then be manipulated for structural and functional characterization.[5–7]

(a)   (b)

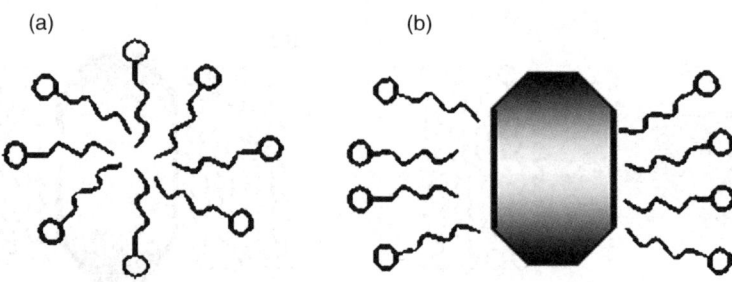

Figure 3 *(a) Idealized structure of a detergent micelle and (b) a schematic representation of a protein-detergent complex (PDC)*

Not all detergents are suitable for use with membrane proteins. Certain detergents cause denaturing of the native protein conformation or even complete unfolding, which results in the loss of biological function. The so-called mild detergents are non-denaturing and bind to the membrane protein, allowing the solubilization in aqueous buffers [Figure 3(b)].[8-10] However, a very small number of mild detergents can nonetheless inactivate a whole variety of eukaryotic integral membrane proteins. Thus, to investigate a protein of interest, a number of different detergents might be explored with different properties. Detergents are commonly organized according to their physical properties, which will be described in the following section.

2.1 Properties and Classification of Detergents

2.1.1 Properties of Detergents

An important detergent characteristic is its ability to form micelles. This property is quantified by the concentration at which the detergent forms micelles, which is known as the critical micelle concentration (CMC).[5] All detergent molecules added to a solution already at its CMC will be incorporated into detergent micelles. Above the CMC, the detergent monomer concentration is independent of the total detergent concentration. In general, CMC values are a guide to detergent hydrophobic binding strengths. The CMC of a detergent can be measured by surface tension, light scattering, iso-thermal calorimetric or dye-binding experiments.

Another characteristic of detergents is the aggregation number, which designates the average number of monomers in a micelle. The aggregation number may be determined experimentally from the micellar molecular weight. Classical light scattering, fluorescence quenching, sedimentation equilibrium, dynamic light scattering and small-angle X-ray scattering experiments can establish a micellar molecular weight. The CMC and aggregation number are both highly dependent on physical factors such as temperature, pH, ionic strength, and detergent homogeneity and purity.

At low temperatures, detergents may separate into two phases and form a cloudy suspension. As the temperature increases, the suspension clears to form monomers if the concentration is below the CMC, or micelles if the concentration is above the CMC. At this temperature point, the detergent exists predominantly in the micellar form. The temperature at which the monomer reaches the CMC is called the critical micellar temperature (CMT). The temperature at which all the three phases (crystalline, micellar and monomeric) exist is called as the Krafft point. At this temperature, the detergent solution turns clear and the concentration of the detergent reaches its CMC. For most detergents, the Krafft point is identical to the CMT. At a particular temperature above the CMT, non-ionic detergents (see Section 2.1.3) become cloudy and undergo phase separation to yield a detergent-rich layer (Figure 4). The cloud point corresponds to the temperature above which this phase-separation occurs. The cloud point phenomenon interferes with applications that require optical clarity, but can be used as an advantage in removing a detergent from aqueous solution.

Another useful measure of detergent property is the hydrophile–lipophile balance (HLB), which can predict surfactant activity based on the semi-empirical analysis of chemical structure. HLB is calculated by summing positive values assigned for individual hydrophilic groups and negative values assigned for hydrophobic constituents. The surfactant properties have been scaled and range between 1 and 40 according to the HLB system, introduced by Griffin.[11] The HLB of a detergent appears to correlate with its ability to solubilize membrane proteins. Detergents with HLB in the lower end of the range (12–15) are best for extracting and solubilizing integral membrane proteins.[12–15]

Detergents used for biological applications may be conveniently sorted according to the nature of their hydrophilic groups, which can be ionic, non-ionic or zwitterionic. Examples of each are described below.

2.1.2 Ionic Detergents

Ionic detergents possess a head group with a net charge, which can be either negative or positive. The most well-known and commonly used detergent in biochemical

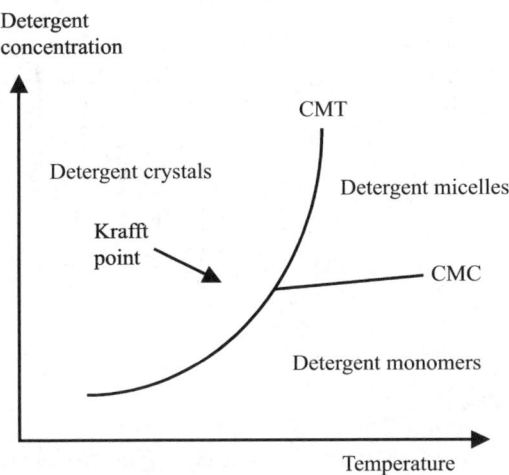

Figure 4 *Schematic representation of the temperature-composition phase diagram for detergent solutions*

experiments is the anionic detergent sodium dodecyl sulfate (SDS), which contains a 12-carbon acyl chain and a sulfate head group [Figure 5(a)]. This compound is regarded as a "strong" and "denaturing" detergent because it can disrupt hydrophobic interactions in the core of proteins, leading to their unfolding.

Examples of anionic detergents that are sometimes used for membrane protein studies are bile acid salts, such as cholate and deoxycholate, which function as lipid emulsifiers [Figure 5(b)]. These compounds contain a rigid hydrophobic steroidal group carrying a carboxyl group at the end of the short alkyl chain. Cetyl trimethyl-ammonium bromide (CTAB) is one example of a cationic detergent carrying a positively charged trimethyl-ammonium group [Figure 5(c)].

2.1.3   Non-Ionic Detergents

Non-ionic detergents have been widely used for functional and structural studies on membrane proteins. They are considered to be milder detergents compared to ionic detergents, which as a benefit do not denature the membrane protein upon extraction and solubilization. Non-ionic detergents contain uncharged hydrophilic head groups and chemically distinct tail groups. Unlike ionic detergents, salts have a minimal effect on their micellar sizes. Triton X-100 exemplifies the non-ionic class of detergents [Figure 6(a)]: this compound has a neutral head group containing polyoxyethylene and conjugated aromatic rings forming its hydrophobic portion. Conjugated aromatic rings are often a disadvantage as they absorb light in the ultra-violet region, thus interfering with spectrophotometric monitoring of a protein at 280 nm.

Other popular choices of non-ionic detergents used in membrane protein studies are the alkyl–sugar detergents, such as *n*-dodecyl-β-D-maltoside (DDM) and *n*-octyl-β-D-glucopyranoside (OG) [Figures 6(b) and (c)]. Alkyl detergents have been

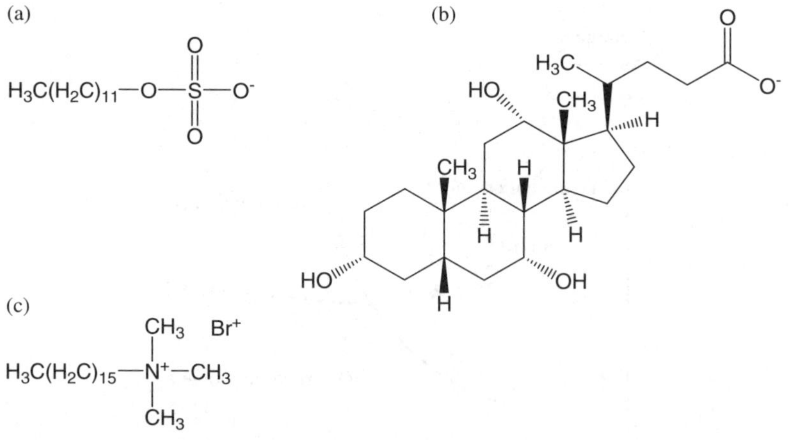

Figure 5 *Examples of ionic detergents: (a) dodecyl sulfate, (b) cholate and (c) CTAB. In micelles, the ionic detergents experience repulsion between the charged polar groups, and their aggregation is therefore strongly affected by ionic strength*

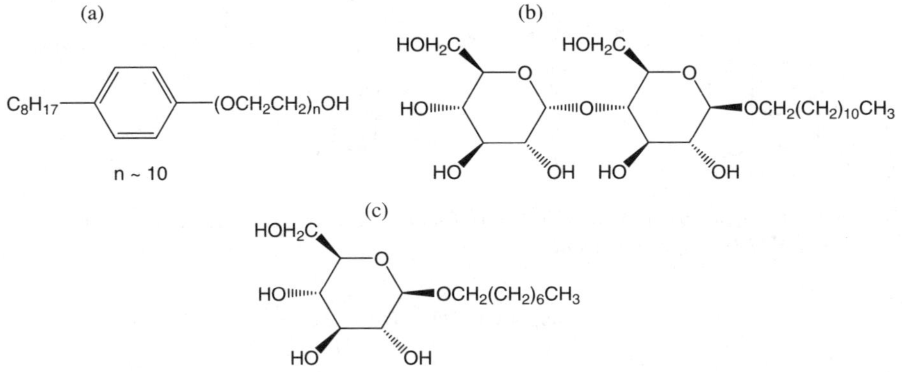

(a)

C_8H_{17}—⬡—$(OCH_2CH_2)_nOH$

n ~ 10

(b)

$OCH_2(CH_2)_{10}CH_3$

(c)

$OCH_2(CH_2)_6CH_3$

Figure 6 *Examples of non-ionic detergents: (a) chemical structure of Triton X-100, (b) DDM and (c) OG*

successfully employed for the isolation and the crystallization of membrane proteins. In comparison with other detergents, they have good spectral properties, which make them convenient for assessing the protein stability and concentration during the purification process. There are several variations of alkyl head groups available (*e.g.* glucoside or maltoside), each of them containing different combinations of carbohydrate chains. Their physicochemical properties such as CMC differ. These differences often prove vital in establishing a particular purification protocol for a specific membrane protein. The optimal choice of detergent for solubilizing a particular membrane protein is found by trial-and-error experiments.

2.1.4 Zwitterionic Detergents

Zwitterionic agents have the combined properties of ionic and non-ionic detergents. These detergents are of interest in the biological study of membrane proteins due to their non-denaturing characteristics, and they are usually used for membrane protein solubilization and purification. However, the overall charge of zwitterionic detergents depends on the pH of the buffer solution, which may make them less appropriate for purification procedures based on ion-exchange chromatography methods. The most widely used zwitterionic detergents are CHAPS (3-[(3-Cholamidopropyl) dimethylammonio]-1-propanesulfonate) and CHAPSO (3-[(3-Cholamidopropyl) dimethylammonio]-2-hydroxy-1-propanesulfonate) (Figure 7).

2.1.5 Amphipols

Amphipols are a new type of surfactants that have been created to allow the manipulation of membrane proteins in aqueous solution.[16–19] Their amphiphilic character arises from the bristle-like arrangement of hydrophilic and hydrophobic groups on an acyl backbone (Figure 8). They bind avidly to the transmembrane surface of the protein. Membrane proteins in complex with amphipols remain in their stable native state and are water soluble without requirement of supplementary detergent.

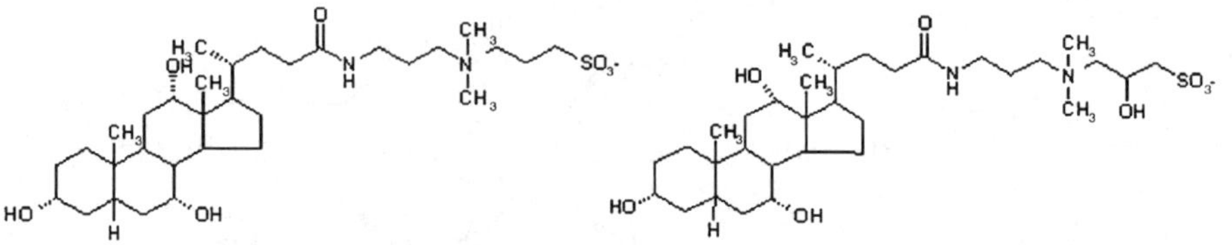

Figure 7 *Chemical structure of CHAPS (left-hand side) and its hydroxylated derivative, CHAPSO (right-hand side)*

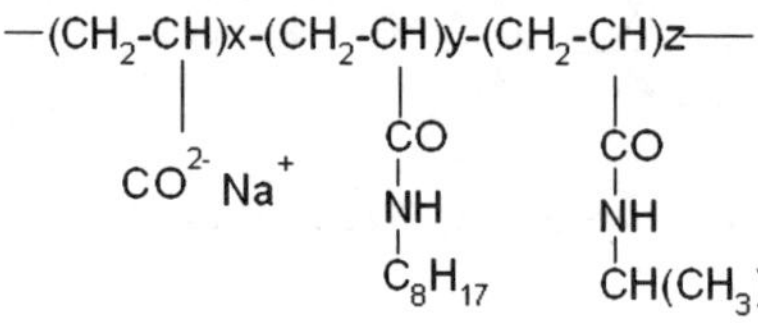

$$-(CH_2-CH)x-(CH_2-CH)y-(CH_2-CH)z-$$

with substituents:
$CO^{2-} Na^+$ on the x unit;
CO, NH, C_8H_{17} on the y unit;
CO, NH, $CH(CH_3)_2$ on the z unit

Figure 8 *Chemical structure of the amphipol A8-35. x, y and z correspond to 35%, 25% and 40%, respectively*

2.2 Lipopeptide Detergents

Lipopeptide detergents (LPDs) are amphiphile agents that have been designed to mimic the lipid membrane geometry.[20,21] LPDs consist of a peptide scaffold that supports two alkyl chains, one anchored to each end of an α-helix (Figure 9).

Like other detergents, LPDs have the ability to self-assemble into small micelles in aqueous solution. These assemblies are similar in size to OG or dodecylphosphocholine (DPC) micelles.[22] Another interesting attribute of LPDs is that they can dissolve lipids into small mixed-compound micelles. However, the most important application of these lipopeptides is their effectiveness for solubilizing and stabilizing membrane proteins.[21] Moreover, they have proved to be non-denaturing detergents that preserve the membrane protein structure in solution for extended periods of time. The LPD structure is expected to maintain the acyl chain orientation by forming a belt around the protein of interest. Therefore, their intrinsic properties make them very attractive for membrane protein structural studies.

2.3 Supplements and Additives for Detergents

A number of small chemical compounds have been found to be useful as a supplement to detergents for the manipulation and crystallization of membrane proteins. They might be envisaged as filling in the space in the imperfect fit of the detergent around the hydrophobic portion of the membrane protein. For example, compounds such as benzamidine and 1,2,3-heptanetriol have proven helpful for crystallization of membrane proteins.[23–28] Other examples of useful additives are the so-called non-detergent sulfobetaines (NDSBs). NDSBs have the advantage of being a non-denaturing substitute for harsh solubilizing agents and can be helpful to enhance the extraction yields of proteins. However, NDSBs cannot replace detergents during extraction. Nevertheless, these have proven to be very effective when used in conjunction with detergents for solubilization,

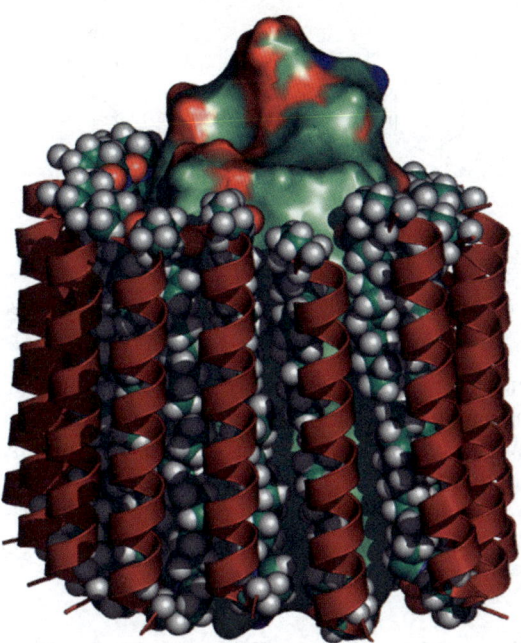

Figure 9 *Schematic representation of a LPD-12/membrane protein complex courtesy of Dr. Gil Privé). LPD-12 formula: Ac-AO(C$_{12}$)AEAAEKAAKYAAEAAEKAAKAO(C$_{12}$)A-NH$_2$*

purification, *in vitro* refolding, manipulation and crystallization of membrane proteins.[29] NDSBs are zwitterionic molecules, soluble in water, and do not significantly alter the pH or viscosity of biological buffers. As an added advantage, they can easily be removed by dialysis since they do not form micelles. As an example, NDSB201 (3-(1-pyridino)-1-propane sulfonate) [Figure 10(a)] has been reported to prevent protein aggregation and facilitate the refolding of chemically and thermally denatured proteins.[30,31] NDSB211 (dimethyl-(2-hydroxyethyl)-(3-sulfopropyl)-ammonium) is useful for protein solubilization, including halophilic proteins [Figure 10(b)].

3 Extraction and Purification Procedure Using Common Detergents

3.1 Choice of Detergent

The first point to be taken into consideration when using a detergent is the ionic form of its hydrophilic group. Anionic and cationic detergents typically affect protein structure to a greater extent than the neutral and zwitterionic detergents. The degree of perturbation varies with the individual protein and the particular detergent. Ionic detergents are also more sensitive to pH, ionic strength, chemical properties of the counter-ion, and can interfere with ion-exchange chromatography and other charge-based analytical methods. Alternatively, most non-ionic detergents are non-denaturing, but are less effective at disrupting protein aggregation. Zwitterionic detergents

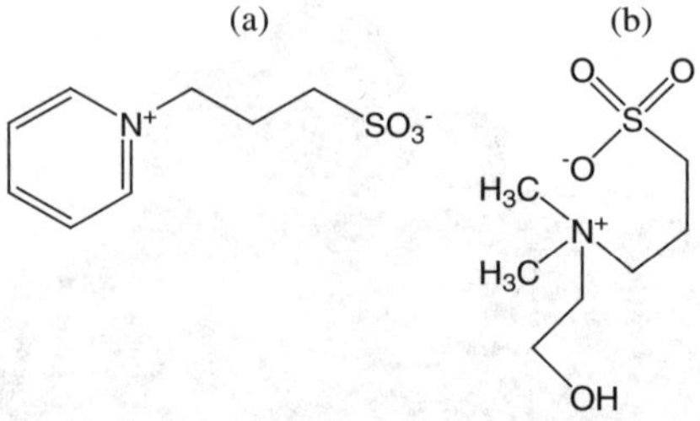

Figure 10 *Chemical structure of two representative NDSBs: (a) NDSB201 and (b) NDSB211*

uniquely offer some intermediate class properties that are superior to the other detergent types for some applications.

Non-ionic or zwitterionic detergents are the most commonly used for protein extraction from the lipid bilayer.[32–35] The most successfully used detergents for crystallization are non-ionic alkyl-sugars with a C_8–C_{12} alkyl chains such as OG, DDM and octa-(ethyleneglycol)-monododecylether ($C_{12}E_8$). The longer chain detergents usually form larger micelles around the protein.

3.2 Purification of Membrane Proteins in the Presence of Detergents

3.2.1 Strategy and Method

The protein purification method appears to be the most critical factor for determining the success of membrane protein crystallization. The procedure requires finding an adequate way to disrupt the membrane and solubilize the membrane protein of interest in its active form and native conformation.[36–38] Parameters such as temperature, pH, ionic strength, along with the determination of the appropriate detergent or mixture of detergents need to be defined and should be assessed by biophysical characterization[39,40] or biological assays.

The stability of a membrane protein will dictate the choice of detergent for solubilization, purification and crystallization. A detergent that is effective for solubilization may not be the optimum choice for studying the purified membrane protein.[41] The nature of the detergent needs to be taken into consideration because it may also reduce or abolish the biological activity of the native protein.[28,36,37,42] Overpurification may have the effect of damaging the protein sample by dissociating loosely bound but structurally important subunits and lipid molecules. Thus, some purification procedures may involve the addition of natural lipids during the isolation of membrane proteins.[43]

The selected detergent should possess a CMC value that is in the range of 0.5–50 mM. If the CMC value is less than 0.5 mM, it may be difficult to exchange the detergent with another more suitable detergent and if it is greater that 50 mM, it will probably require very high concentrations in the extraction and the purification process.[42] Usually, several different detergents are tested during an experimental survey.[44] Excess detergent is normally employed in solubilization of membrane proteins to ensure complete dissolution of the membrane and to provide a large number of micelles such that only one protein molecule is present per micelle.

Purification techniques for membrane proteins are similar to those for soluble proteins and commonly involve chromatographic techniques. However, membrane proteins tend to have strong interactions with chromatography media due to their hydrophobic nature, thus resulting in lower column efficiency. Affinity chromatography is the easiest choice for purifying recombinant membrane proteins, which have been fused with a specific affinity tag.[45] Other methods such as iso-electric focusing or size exclusion chromatography are alternatively used.[46]

3.2.2 Detergent Exchange or Removal

Excess detergent may need to be removed for further membrane protein biochemical characterization, or in order to minimize the contribution of numerous detergent micelles, which may bias the crystallization process or lead to the formation of detergent crystals. Several methods are effective in the removal of excess detergent, including dialysis, hydrophobic adsorption, ion-exchange[47] or size exclusion chromatography.

When detergent solutions are diluted below the CMC, the micelles are dispersed into monomers. The size of the monomers is usually an order of magnitude smaller than that of the micelles, and can thus be easily removed by dialysis. For detergents with high CMC values, dialysis is usually the preferred choice because the detergent is readily removed from detergent–protein complexes, whereas a detergent with low CMC dialyses away very slowly.

The hydrophobic adsorption method takes advantage of the detergent's ability to bind to hydrophobic resins. This technique is effective for most detergents. Several commercial kits are available for this purpose (*e.g.* BioBeads (BioRad)). If the adsorption of the protein to the resin is of concern, the resin can be included in a dialysis buffer and the protein dialysed. However, this usually requires an extended period of dialysis.

The ion-exchange method is frequently used as a chromatography technique for the purification of proteins. This method is based on the interaction between charged solute molecules and oppositely charged moieties covalently linked to the adsorption matrix. For example, when non-ionic detergents are used, conditions can be set up so that the PDCs are adsorbed on the ion-exchange resin and the protein-free detergent micelles pass through. Adsorbed protein can be washed and eluted by changing either the ionic strength, the pH, or by the combination of both. Alternatively, the protein can be extensively washed and eluted with a detergent belonging to a different class, thus replacing the detergent used during extraction and solubilization.

The gel filtration method is routinely used for water-soluble proteins. However, the behaviour of membrane proteins solubilized by detergent on gel filtration columns is different owing to the effect of detergent binding to the proteins. The PDC will have a bigger size than the protein alone, thus migrating differently. Size exclusion chromatography takes advantage of the differences in size between PDCs, detergent–lipid and homogenous detergent micelles.

4 Use of Detergents in Membrane Protein Crystallization

4.1 Introduction

Crystallization of membrane proteins remains a challenge in the field of structural biology due to the difficulty in obtaining sufficient material and moreover in finding the right detergent for isolating, purifying and crystallizing the protein of interest. These different features make it often difficult to grow well-ordered 3D crystals that are required for X-ray crystallography. As an alternative, the preparation of 2D crystals of membrane proteins for electron microscopy studies can provide important structural information and help to understand protein–detergent interactions.[48–52] Other biophysical techniques such as small-angle X-ray scattering, Fourier-transform infrared spectroscopy and solution or solid-state NMR have also yielded structural information on membrane proteins.

Membrane protein crystals are usually obtained with standard crystallization techniques and conditions that are routinely used for soluble proteins.[42] Other techniques were alternatively developed where added lipids are used during the crystallization process such as in the so-called lipid cubic phase method.[53–58]

The solubilization of integral membrane proteins with detergents produces PDCs, where in some cases, lipid molecules remain tightly bound to the membrane protein.[54] The most commonly used precipitants to achieve super-saturation of the membrane protein solution and thereby nucleation for the crystal growth are polyethylene glycols (PEG), polyethylene glycol mono-methylether (PEG-MME), ammonium sulfate and phosphate salts. The effects of the precipitating agent PEG upon these protein–detergent interactions have been examined, focusing on the detergent system used to crystallize the bacterial outer membrane protein OmpF porin.[59,60] The results show that the precipitating agent may have important effects on the detergent phase diagram (Figure 11) and in the nucleation of membrane protein crystals. Small amphiphilic molecules (such as those described in Section 2.3) are often employed as additives during the crystallization trial.[23,61]

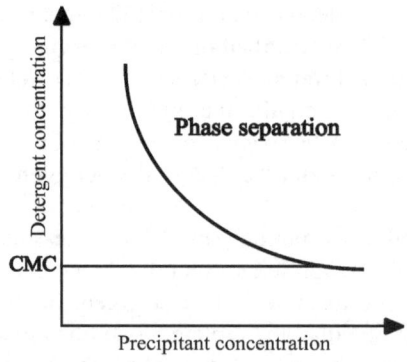

Figure 11 *Detergent phase diagram showing the solubility of a protein in solution as a function of the detergent concentration*

The vapour diffusion method with sitting or hanging drops is most frequently applied in crystallization surveys. Data collection and structure determination are in general the same as with soluble proteins. Flash cooling is often used to enhance crystal stability in the X-ray beam or for trapping reaction intermediates.[62,63] In some cases establishing cryo conditions for membrane proteins may require screening a variety of conditions. This might be due to the presence of detergents micelles in the sample and high solvent content in the crystals.

4.2 Membrane Protein Crystallization in Lipid Cubic Phase

The lipid cubic phase method uses a complex mixture of solvent, lipid, protein and other amphiphile components as the hosting medium for membrane protein crystallization.[64–67] The cubic phase provides a lipid bilayer into which the protein presumably reconstitutes and from which protein crystals nucleate and grow (see central portion of Figure 12). The solutions used to spontaneously form the protein-enriched cubic phase often contain significant amounts of detergents that were employed initially to purify and to solubilize the membrane protein. By virtue of their surface activity, detergents have the potential to impact on the phase properties of the *in meso* system and, by extension, the outcome of the crystallization process. Detergents are used to solubilize membrane proteins and are likely to be integrated into the cubic medium with the target protein. Depending on the identity and concentration of the detergent, the cubic phase, which is membranous, may be solubilized or destabilized in such a way as to render it unsuitable as a crystal growing system.

4.3 Crystal Lattice Organization

Membrane protein crystals may be classified according to the organization of the crystal lattice.[68–70] Figure 12 shows a schematic representation of the type I and type II lattice packing often observed in membrane proteins crystals.

Type I crystals are composed of stacks of 2D membrane protein layers, where there are interactions between the hydrophobic domains of each protein membrane layer. Due to the difficulties in creating both hydrophobic and polar interactions during the crystallization process, it is often complex to obtain large type I crystals. However, a method using a lipidic cubic matrix (see Section 4.2) was recently reported to form type I crystals.[71] This technique seems useful for hydrophobic membrane proteins, since it allows direct contact between molecules owing to the hydrophobic surface. Another advantage of this method is that the crystals obtained have a low solvent content and usually diffract well. So far, this technique has been used to crystallize the membrane proteins of bacteriorhodopsin,[55] halorhodopsin[72] and sensory rhodopsin II.[62,73]

The most common type of membrane protein crystal obtained is the type II crystal. Type II crystals are grown using the same techniques and precipitants (*e.g.* PEG or ammonium sulfate) that are applied to soluble proteins. However, the presence of detergents has two impeding influences on the formation of the crystals. The first problem is that protein–protein contacts, essential for crystal lattice formation, can be mediated through the hydrophilic surfaces of the membrane protein. The second is that the detergent molecules covering the hydrophobic region prevent specific protein–protein

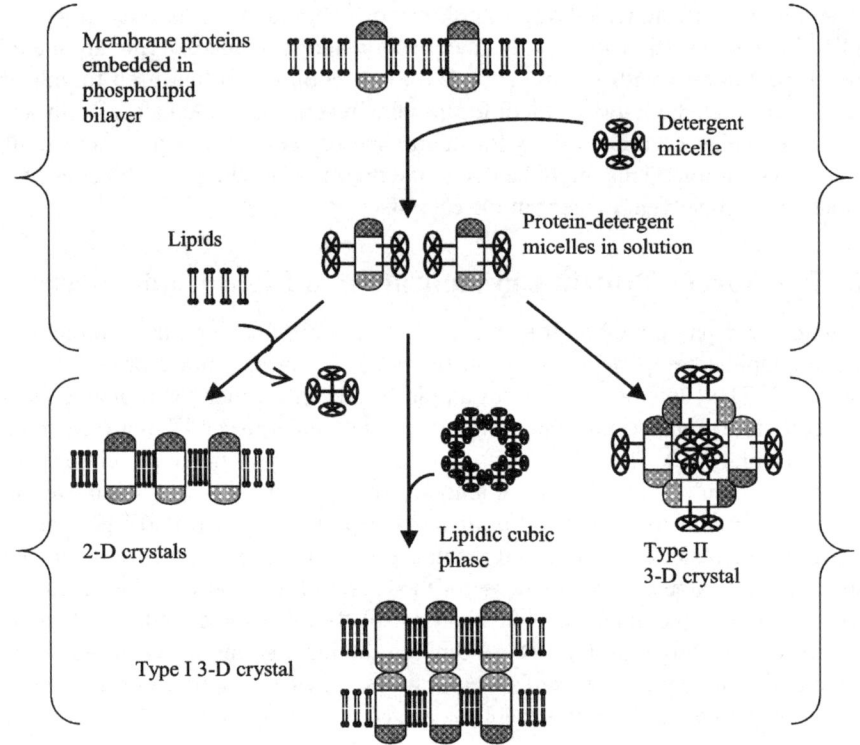

Figure 12 *Schematic representation of crystal lattice packing observed for membrane proteins*

contacts. Due to these characteristics, type II crystals often have a very high solvent content (65–80%) and very few protein–protein contacts in the crystals, resulting in poor crystal quality.

The choice of detergent is therefore clearly essential to obtain the best possible quality crystals. In summary, hydrophobic and detergent-mediated contacts are important in the type I crystals, where molecules form 2D crystal layers (hydrophobic, detergent and lipid interactions between the molecules) which are then stacked by polar interactions. High-quality type II crystal lattices are formed predominantly by polar interactions. The hydrophobic surfaces do not contribute significantly to the lattice formation. This type of interaction is observed in those structures where molecules possess relatively large protein domains outside the membrane. Regular molecule shapes make it easier to form a regular arrangement of molecules in the crystal lattice. Membrane proteins that form multimers whose shape is regular and symmetric have a good chance to crystallize.

4.4 Example of Detergent Interactions with β-Sheet Membrane Proteins

In the following sections, we present some representative examples of protein–detergent and protein–lipid interactions as observed from X-ray and neutron diffraction.

The examples are grouped into cases of β-barrel membrane proteins, found exclusively in the outer membrane of Gram-negative bacteria and some organelles, and α-helical membrane proteins. Despite the difference in secondary structure exposed, these different classes of membrane proteins make analogous interactions with the lipids or detergents.

4.4.1 Crystal Structure of VceC, an Outer Membrane Protein from Vibrio Cholerae

VceC is a channel-like outer membrane protein and a component of the energy-driven drug efflux machinery in the pathogen *Vibrio cholerae*. The crystal structure of VceC has been solved at 1.8 Å resolution.[74] Interestingly, VceC trimers pack in laminar arrays in the crystal [Figure 13(a)]. This packing brings the exposed hydrophobic surfaces of the β-barrel together in the same plane and mimics the organization of a planar membrane ("type I crystal packing" as described above). Within this membrane-like layer, OG molecules are found to nestle between two adjacent β-barrel domains. Two such detergent molecules line up end-to-end and span the height of the porin-like domain as though they were lipids in the two leaflets of a bilayer membrane [Figure 13(b)]. The detergent molecules make also both hydrophobic interactions with the protein through their aliphatic chains and a number of hydrogen bond interactions via their sugar moieties [Figure 13(c)].

4.4.2 Detergent Organization in Crystals of Monomeric Outer Membrane Phospholipase A

The organization of the detergent OG in the outer membrane phospholipase A crystals (OMPLA) has been determined by neutron diffraction.[75] Large crystals were soaked in stabilizing solutions, each containing a different water/deuterium (H_2O/D_2O) contrast ratio. From the neutron diffraction at five contrast points, the detergent micelle structure around the protein was resolved at 12 Å resolution. Rings of detergent could be observed covering the hydrophobic β-barrel surfaces of the protein molecules (Figure 14). These detergent belts are fused to neighbouring detergent rings forming a continuous 3D network throughout the crystal. The thickness of the detergent layer around the protein varies from 7 to 20 Å. These results indicate a role for detergent coalescence in lattice organization.

4.5 Example of Detergent and α-Helical Type Membrane Protein Contact

4.5.1 Crystal Structure of Rotor Rings

The Na-ATPase rotor ring from the vacuolar-type (V-type) ATPase of *Enterococcus hirae*, which is homologous to other V-ATPases and F-ATPases, was recently solved at 2.1 Å.[76] The protein crystallizes in a type II crystal-packing manner [Figure 15(a)]. The protein consists of ten NtpK subunits, composed of four transmembrane

Figure 13 *(a) Packing of VceC in the crystal lattice. The unit cell is represented by a red square. (b) Laminar packing of VceC in sheets that mimic the lipid bilayer. Two octyl-glucoside molecules resemble a bilayer and are related by crystallographic symmetry. (c) The octyl-β-glucoside molecule at the interface between symmetry-related protomers (inside a 2 F_o–F_c electron density map, contoured at 1.5σ) (PDB code: 1YC9)*

helices each, in which a Na$^+$ ion is bound between helices 2 and 4. Electron density lying close to the inside surface has been interpreted as phosphatidylglycerol (PG) head groups, which is the most abundant (57.1%) phospholipid in *E. hirae*. Because the most common fatty acyl chain in *E. hirae* phospholipids is C_{16}, 1,2-dipalmitoyl-phosphatidylglycerol (DPPG) and 1,2-dipalmitoyl-glycerol (DPG) moieties were modelled into the upper and lower features, respectively [Figure 15(b)]. DDM detergent molecules have also been allocated in the model, and shown to be interacting closely with the protein.

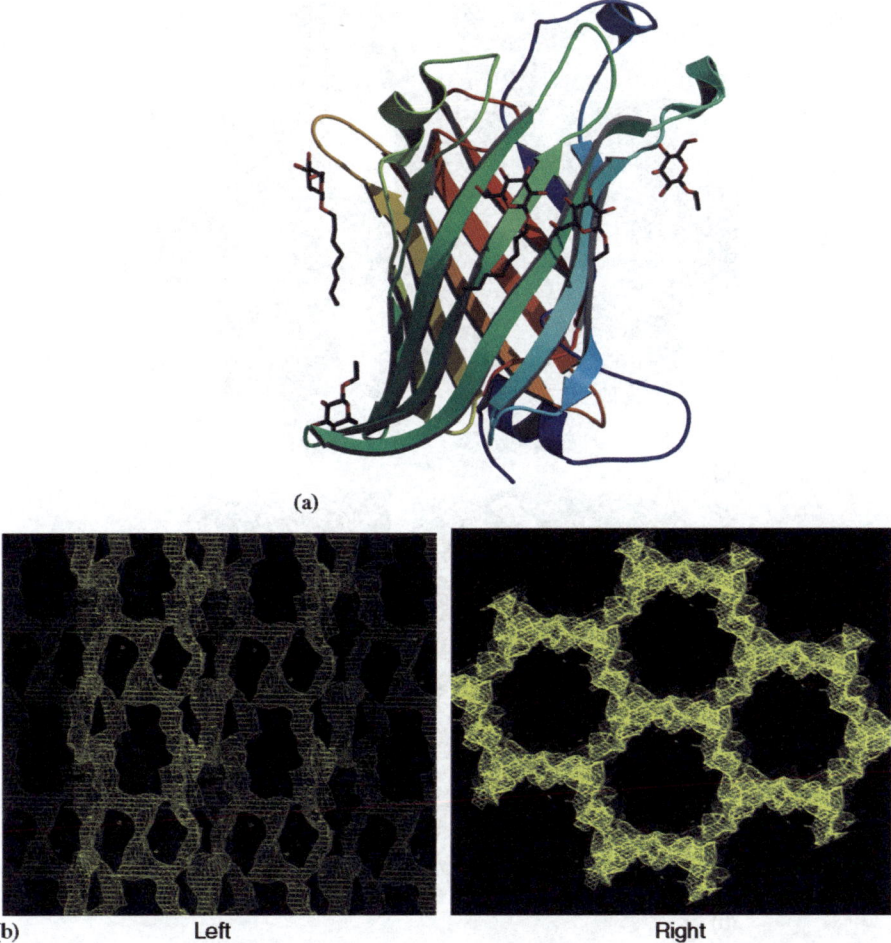

Figure 14 *(a) Crystal structure of OMPLA (PDB code: 1QD5) and (b): OG detergent ring around OMPLA protein, side view and top view, respectively (courtesy of Dr. Arjan Snijder)*

4.5.2 Structure of Bovine Rhodopsin in a Trigonal Crystal Form

The structure of bovine rhodopsin was determined at 2.65 Å resolution.[77–79] The structure reveals the presence of lipid and detergent molecules (Figure 16). An ordered detergent molecule is seen wrapped around the kink in transmembrane helix H6, stabilizing the structure around the potential hinge in H6.

4.6 A Synopsis of Detergent–Protein Interactions in Crystals

From the examples described above and other structures available, a few points can be made about the distribution of detergents around membrane proteins. Detergents interact closely and directly with membrane proteins, forming a belt- or ring-like environment

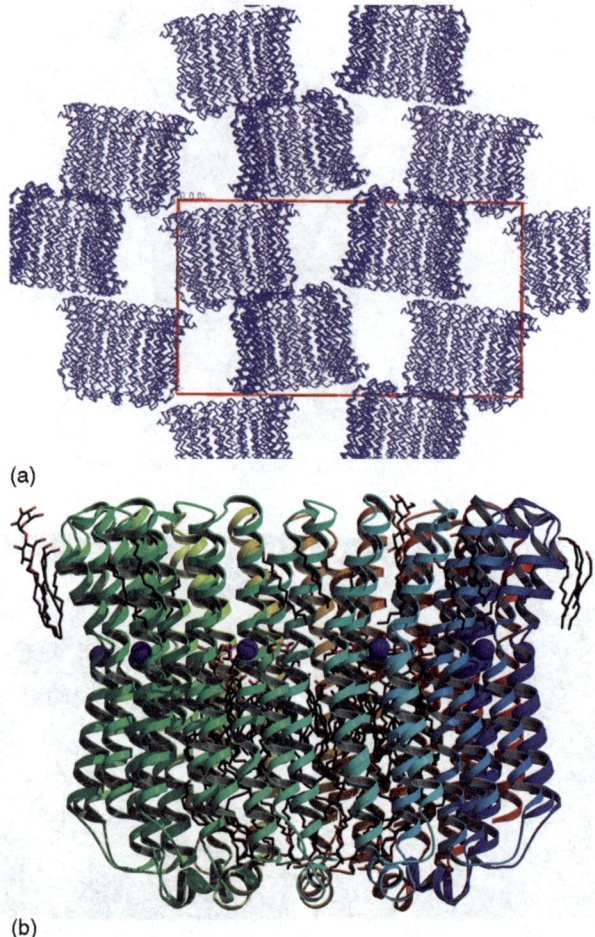

(a)

(b)

Figure 15 *(a) Crystal packing and (b) localization of lipid and detergent molecules (in black (carbon atoms) and red (oxygen atoms)) in the V-type Na-ATPase rotor ring (PDB code: 2BL2) (top view and side view, respectively). The unit cell is represented by a red square in (a)*

that surrounds the protein and more or less mimics the constitutive lipid bilayer. Our analysis separated the helical proteins and the β-sheet proteins, but in fact these two groups appear to interact with the detergent acyl groups in a similar fashion despite presenting surfaces with different organization of protruding side chains. Most of the contacts between detergents and proteins are of hydrophobic moieties of the detergent fitting into small hydrophobic grooves on the protein surface made by the protruding amino acid side chains. These contacts make favourable van der Waals interactions.

The nature of the detergent does not seem to affect nor correlate with the crystal-packing type. Amphiphilic compounds and other small additives used in conjunction with one or a mixture of different classes of detergents (long plus small aliphatic chain for example) could make all the difference in the ability to successfully crystallize membrane proteins.

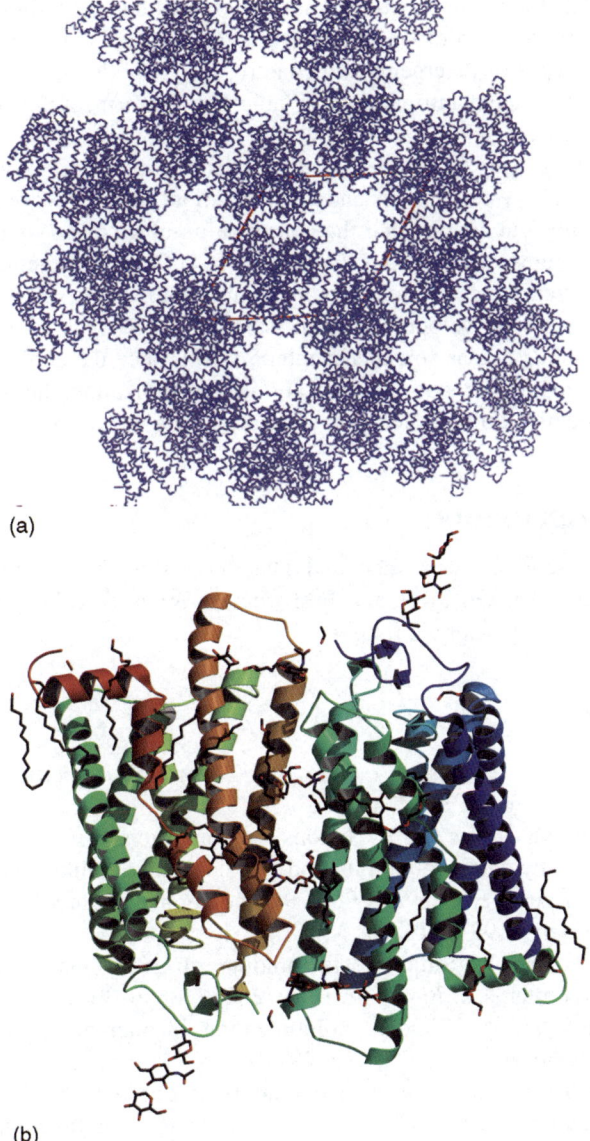

(a)

(b)

Figure 16 *(a) Crystal packing and (b) structure of the bovine rhodopsin dimer: localization of bound lipid and detergent molecules (in black and red) in the bovine rhodopsin (PDB code: 1GZM) (side view and top view, respectively). The unit cell is represented by a red square in (a)*

5   Conclusion

Detergents used for biological studies in general, and for structural characterization in particular, are essential for the isolation of membrane proteins. In the process of purification, membrane proteins have to be isolated from cell membranes and brought into stable aqueous solution. Thus, detergents have the ability to encapsulate membrane proteins in order to create protective protein–detergent complexes. Many different types of detergents and chaotropic agents that can solubilize membrane proteins in aqueous buffers are available, but fewer meet the key criterion of preserving the protein biological activity. Also, the detergent should be easily separable by chromatography techniques (if neutral detergents are used) or by dialysis in order to pursue biophysical studies and crystallization trials. However, the preservation of such biological activity and the conservation of secondary/ternary structure of membrane proteins still remain a challenging task. Often detergents will denature proteins at elevated concentrations. At the same time, a lower concentration will reduce protein yield. Development of an isolation protocol is a highly empirical process that requires extensive experimentation. But, once the choice of detergents has been optimized in protocols of purification and stabilization, there is tremendous scope for biochemical, biophysical and structural characterization.

Acknowledgements

We thank Dr. Hendrick van Veen and Dr. Chris Tate for helpful discussions, Dr. Arjan Snijder, Dr. Gil Privé and Prof. Steven White for kindly providing the images to make the appropriate figures.

References

1. A. Helenius, D.R. McCaslin, E. Fries and C. Tanford, Properties of detergents. *Meth. Enzymol.*, 1979, **56**, 734–749.
2. A. Helenius, M. Sarvas and K. Simons, Asymmetric and symmetric membrane reconstitution by detergent elimination. Studies with Semliki-Forest-virus spike glycoprotein and penicillinase from the membrane of *Bacillus licheniformis*. *Eur. J. Biochem.*, 1981, **116**, 27–35.
3. A. Helenius and K. Simons, The binding of detergents to lipophilic and hydrophilic proteins. *J. Biol. Chem.*, 1972, **247**, 3656–3661.
4. A. Helenius and K. Simons, Solubilization of membranes by detergents. *Biochim. Biophys. Acta*, 1975, **415**, 29–79.
5. J.M. Neugebauer, Detergents: an overview. *Meth. Enzymol.*, 1990, **182**, 239–253.
6. L.M. Hjelmeland and A. Chrambach, Solubilization of functional membrane proteins. *Meth. Enzymol.*, 1984, **104**, 305–318.
7. L.M. Hjelmeland, Solubilization of native membrane proteins. *Meth. Enzymol.*, 1990, **182**, 253–264.
8. L.M. Hjelmeland, The design and synthesis of detergents for membrane biochemistry. *Meth. Enzymol.*, 1986, **124**, 135–164.
9. L.M. Hjelmeland, Removal of detergents from membrane proteins. *Meth. Enzymol.*, 1990, **182**, 277–282.
10. M. Seigneuret, J.M. Neumann and J.L. Rigaud, Detergent delipidation and solubilization strategies for high-resolution NMR of the membrane protein bacteriorhodopsin. *J. Biol. Chem.*, 1991, **266**, 10066–10069.

11. W.C. Griffin, Classification of surface-active agents by HLB. *J. Soc. Cosmet. Chem.*, 1949, **1**, 311–326.

12. D.W. Nicholson and W.C. McMurray, Triton solubilization of proteins from pig liver mitochondrial membranes. *Biochim. Biophys. Acta*, 1986, **856**, 515–525.

13. E. Slinde and T. Flatmark, Effect of the hydrophile–lipophile balance of non-ionic detergents (Triton X-series) on the solubilization of biological membranes and their integral b-type cytochromes. *Biochim. Biophys. Acta*, 1976, **455**, 796–805.

14. D.R. Storm, S.O. Field and J. Ryan, The HLB dependency for detergent solubilization of hormonally sensitive adenylate cyclase. *J. Supramol. Struct.*, 1976, **4**, 221–231.

15. J.N. Umbreit and J.L. Strominger, Relation of detergent HLB number to solubilization and stabilization of D-alanine carboxypeptidase from *Bacillus subtilis* membranes. *Proc. Natl. Acad. Sci. USA*, 1973, **70**, 2997–3001.

16. J.L. Popot, E.A. Berry, D. Charvolin, C. Creuzenet, C. Ebel, D.M. Engelman, M. Flotenmeyer, F. Giusti, Y. Gohon, Q. Hong, J.H. Lakey, K. Leonard, H.A. Shuman, P. Timmins, D.E. Warschawski, F. Zito, M. Zoonens, B. Pucci and C. Tribet, Amphipols: polymeric surfactants for membrane biology research. *Cell Mol. Life Sci.*, 2003, **60**, 1559–1574.

17. C. Tribet, R. Audebert and J.L. Popot, Amphipols: polymers that keep membrane proteins soluble in aqueous solutions. *Proc. Natl. Acad. Sci. USA*, 1996, **93**, 15047–15050.

18. B.M. Gorzelle, A.K. Hoffman, M.H. Keyes, D.N. Gray, D.G. Ray and C.R. Sanders, Amphipols can support the activity of a membrane enzyme. *J. Am. Chem. Soc.*, 2002, **124**, 11594–11595.

19. Y. Gohon, G. Pavlov, P. Timmins, C. Tribet, J.-L. Popot and C. Ebel, Partial specific volume and solvent interactions of amphipol A8-35. *Anal. Biochem.*, 2004, **334**, 318–334.

20. S. Zhang, D.M. Marini, W. Hwang and S. Santoso, Design of nanostructured biological materials through self-assembly of peptides and proteins. *Curr. Opin. Chem. Biol.*, 2002, **6**, 865–871.

21. C.L. McGregor, L. Chen, N.C. Pomroy, P. Hwang, S. Go, A. Chakrabartty and G.G. Privé, Lipopeptide detergents designed for the structural study of membrane proteins. *Nature Biotechnol.*, 2003, **21**, 171–176.

22. B. Lorber, J.B. Bishop and L.J. DeLucas, Purification of octyl beta-D-glucopyranoside and re-estimation of its micellar size. *Biochim. Biophys. Acta*, 1990, **1023**, 254–265.

23. M.A. Rosenow, J.C. Williams and J.P. Allen, Amphiphiles modify the properties of detergent solutions used in crystallization of membrane proteins. *Acta Crystallogr. D Biol. Crystallogr.*, 2001, **57**, 925–927.

24. S.M. Yu, D.T. McQuade, M.A. Quinn, C.P. Hackenberger, M.P. Krebs, A.S. Polans and S.H. Gellman, An improved tripod amphiphile for membrane protein solubilization. *Protein Sci.*, **9**, 2518–2527.

25. P. Nollert, Membrane protein crystallization in amphiphile phases: practical and theoretical considerations. *Prog. Biophys. Mol. Biol.*, 2005, **88**, 339–357.

26. S.M. Prince, T.D. Howard, D.A. Myles, C. Wilkinson, M.Z. Papiz, A.A. Freer, R.J. Cogdell and N.W. Isaacs. Detergent structure in crystals of the integral

membrane light-harvesting complex LH2 from *Rhodopseudomonas acidophila* strain 10050. *J. Mol. Biol.*, 2003, **326**, 307–315.

27. R.J. Cogdell, N.W. Isaacs, A.A. Freer, T.D. Howard, A.T. Gardiner, S.M. Prince and M.Z. Papiz, The structural basis of light-harvesting in purple bacteria. *FEBS Lett.*, **555**, 35–39.

28. H. Michel, Crystallization of membrane proteins. *Trends Biochem. Sci.*, 1983, **8**, 56–59.

29. L. Vuillard, D. Madern, B. Franzetti and T. Rabilloud, Halophilic protein stabilization by the mild solubilizing agents nondetergent sulfobetaines. *Anal. Biochem.*, 1995, **230**, 290–294.

30. L. Vuillard, T. Rabilloud and M.E. Goldberg, Interactions of non-detergent sulfobetaines with early folding intermediates facilitate *in vitro* protein renaturation. *Eur. J. Biochem.*, 1998, **256**, 128–135.

31. L. Vuillard, C. Braun-Breton and T. Rabilloud, Non-detergent sulphobetaines: a new class of mild solubilization agents for protein purification. *Biochem. J.*, 1995, **305**(Pt 1), 337–343.

32. M. Bannwarth and G.E. Schulz, The expression of outer membrane proteins for crystallization. *Biochim. Biophys. Acta*, 2003, **1610**, 37–45.

33. F.J. Sharom, Characterization and functional reconstitution of the multidrug transporter. *J. Bioenerg. Biomembr.*, 1995, **27**, 15–22.

34. K. Ohlendieck, Extraction of membrane proteins. *Meth. Mol. Biol.*, 2004, **244**, 283–293.

35. I. Lehner, M. Niehof and J. Borlak, An optimized method for the isolation and identification of membrane proteins. *Electrophoresis*, 2003, **24**, 1795–1808.

36. R.M. Garavito, D. Picot and P.J. Loll, Strategies for crystallizing membrane proteins. *J. Bioenerg. Biomembr.*, 1996, **28**, 13–27.

37. C. Ostermeier and H. Michel, Crystallization of membrane proteins. *Curr. Opin. Struct. Biol.*, 1997, **7**, 697–701.

38. H. Peters, C. Schmidt-Dannert, L. Cao, U.T. Bornscheuer and R.D. Schmid, Purification and reconstitution of an integral membrane protein, the photoreaction center of *Rhodobacter sphaeroides*, using synthetic sugar esters. *Biotechniques*, 2000, **28**, 1214–1219.

39. F.M. Goni and A. Alonso, Spectroscopic techniques in the study of membrane solubilization, reconstitution and permeabilization by detergents. *Biochim. Biophys. Acta*, 2000, **1508**, 51–68.

40. S. Tornroth, V. Yankovskaya, G. Cecchini and S. Iwata, Purification, crystallisation and preliminary crystallographic studies of succinate: ubiquinone oxidoreductase from *Escherichia coli. Biochim. Biophys. Acta*, 2002, **1553**, 171–176.

41. R. Newman, Crystallization and structure analysis of membrane proteins. *Meth. Mol. Biol.*, 1996, **56**, 365–387.

42. M. Caffrey, Membrane protein crystallization. *J. Struct. Biol.*, 2003, **142**, 108–132.

43. A.M. Seddon, P. Curnow and P.J. Booth, Membrane proteins, lipids and detergents: not just a soap opera. *Biochim. Biophys. Acta*, 2004, **1666**, 105–117.

44. J. Barber, Membrane proteins. Detergent ringing true as a model for membranes. *Nature*, 1989, **340**, 601.

45. A.K. Mohanty and M.C. Wiener, Membrane protein expression and production: effects of polyhistidine tag length and position. *Protein Expr. Purif.*, 2004, **33**, 311–325.

46. M.C. Wiener, A pedestrian guide to membrane protein crystallization. *Methods*, 2004, **34**, 364–372.

47. J. Van Ede, J.R. Nijmeijer, S. Welling-Wester, C. Orvell and G.W. Welling, Comparison of non-ionic detergents for extraction and ion-exchange high-performance liquid chromatography of Sendai virus integral membrane proteins. *J. Chromatogr.*, 1989, **476**, 319–327.

48. D. Levy, G. Mosser, O. Lambert, G.S. Moeck, D. Bald, and J.L. Rigaud, Two-dimensional crystallization on lipid layer: a successful approach for membrane proteins. *J. Struct. Biol.*, **127**, 44–52.

49. L. Hasler, J.B. Heymann, A. Engel, J. Kistler and T. Walz, 2D crystallization of membrane proteins: rationales and examples. *J. Struct. Biol.*, 1998, **121**, 162–171.

50. J.P. Rosenbusch, A. Lustig, M. Grabo, M. Zulauf and M. Regenass, Approaches to determining membrane protein structures to high resolution: do selections of subpopulations occur? *Micron*, 2001, **32**, 75–90.

51. J. Rigaud, M. Chami, O. Lambert, D. Levy and J. Ranck, Use of detergents in two-dimensional crystallization of membrane proteins. *Biochim. Biophys. Acta*, 2000, **1508**, 112–128.

52. D. Levy, M. Chami and J.L. Rigaud, Two-dimensional crystallization of membrane proteins: the lipid layer strategy. *FEBS Lett.*, 2001, **504**, 187–193.

53. E. Pebay-Peyroula, R. Neutze and E.M. Landau, Lipidic cubic phase crystallization of bacteriorhodopsin and cryotrapping of intermediates: towards resolving a revolving photocycle. *Biochim. Biophys. Acta*, 2000, **1460**, 119–132.

54. E. Pebay-Peyroula and J.P. Rosenbusch, High-resolution structures and dynamics of membrane protein–lipid complexes: a critique. *Curr. Opin. Struct. Biol.*, 2001, **11**, 427–432.

55. E.M. Landau and J.P. Rosenbusch, Lipidic cubic phases: a novel concept for the crystallization of membrane proteins. *Proc. Natl. Acad. Sci. USA*, 1996, **93**, 14532–14535.

56. P. Nollert, H. Qiu, M. Caffrey, J.P. Rosenbusch and E.M. Landau, Molecular mechanism for the crystallization of bacteriorhodopsin in lipidic cubic phases. *FEBS Lett.*, 2001, **504**, 179–186.

57. P. Nollert, J. Navarro and E.M. Landau, Crystallization of membrane proteins in cubo. *Meth. Enzymol.*, 2002, **343**, 183–199.

58. V. Cherezov, J. Clogston, Y. Misquitta, W. Abdel-Gawad and M. Caffrey, Membrane protein crystallization in meso: lipid type-tailoring of the cubic phase. *Biophys. J.*, 2002, **83**, 3393–3407.

59. C. Hitscherich Jr., V. Aseyev, J. Wiencek and P.J. Loll, Effects of PEG on detergent micelles: implications for the crystallization of integral membrane proteins. *Acta Crystallogr. D Biol. Crystallogr.*, 2001, **57**, 1020–1029.

60. C. Hitscherich Jr., J. Kaplan, M. Allaman, J. Wiencek and P.J. Loll, Static light scattering studies of OmpF porin: implications for integral membrane protein crystallization. *Protein Sci.*, 2000, **9**, 1559–1566.

61. M.A. Rosenow, D. Brune and J.P. Allen, The influence of detergents and amphiphiles on the solubility of the light-harvesting I complex. *Acta Crystallogr. D Biol. Crystallogr.*, 2003, **59**, 1422–1428.

62. H. Luecke, B. Schobert, J.K. Lanyi, E.N. Spudich and J.L. Spudich, Crystal structure of sensory rhodopsin II at 2.4 angstroms: insights into color tuning and transducer interaction. *Science*, 2001, **293**, 1499–1503.

63. T. Tsukihara and S. Yoshikawa, Crystal structural studies of a membrane protein complex, cytochrome *c* oxidase from bovine heart. *Acta Crystallogr. A*, 1998, **54**, 895–904.

64. Y. Misquitta and M. Caffrey, Detergents destabilize the cubic phase of monoolein: implications for membrane protein crystallization. *Biophys. J.*, 2003, **85**, 3084–3096.

65. Y. Misquitta, V. Cherezov, F. Havas, S. Patterson, J.M. Mohan, A.J. Wells, D.J. Hart and M. Caffrey. Rational design of lipid for membrane protein crystallization. *J. Struct. Biol.*, **148**, 169–175.

66. C. Sennoga, A. Heron, J.M. Seddon, R.H. Templer and B. Hankamer, Membrane-protein crystallization in cubo: temperature-dependent phase behaviour of monoolein-detergent mixtures. *Acta Crystallogr. D Biol. Crystallogr.*, 2003, **59**, 239–246.

67. Y. Qutub, I. Reviakine, C. Maxwell, J. Navarro, E.M. Landau and P.G. Vekilov. Crystallization of transmembrane proteins in cubo: mechanisms of crystal growth and defect formation. *J. Mol. Biol.*, **343**, 1243–1254.

68. H. Stahlberg, A. Engel and A. Philippsen, Assessing the structure of membrane proteins: combining different methods gives the full picture. *Biochem. Cell Biol.*, 2002, **80**, 563–568.

69. J. Torres, T.J. Stevens and M. Samso, Membrane proteins: the 'wild west' of structural biology. *Trends Biochem. Sci.*, 2003, **28**, 137–144.

70. M.C. Wiener, A.S. Verkman, R.M. Stroud and A.N. van Hoek, Mesoscopic surfactant organization and membrane protein crystallization. *Protein Sci.*, 2000, **9**, 1407–1409.

71. G. Katona, U. Andreasson, E.M. Landau, L.E. Andreasson and R. Neutze, Lipidic cubic phase crystal structure of the photosynthetic reaction centre from Rhodobacter sphaeroides at 2.35 Å resolution. *J. Mol. Biol.*, 2003, **331**, 681–692.

72. M. Kolbe, H. Besir, L.O. Essen and D. Oesterhelt, Structure of the light-driven chloride pump halorhodopsin at 1.8 Å resolution. *Science*, 2000, **288**, 1390–1396.

73. A. Royant, P. Nollert, K. Edman, R. Neutze, E.M. Landau, E. Pebay-Peyroula and J. Navarro. X-ray structure of sensory rhodopsin II at 2.1-Å resolution. *Proc. Natl. Acad. Sci. USA*, 2001, **98**, 10131–10136.

74. L. Federici, D. Du, F. Walas, H. Matsumura, J. Fernandez-Recio, K.S. McKeegan, M.I. Borges-Walmsley, B.F. Luisi and A.R. Walmsley. The crystal structure of the outer membrane protein VceC from the bacterial pathogen *Vibrio cholerae* at 1.8 Å resolution. *J. Biol. Chem.*, 2005, **280**, 15307–15314.

75. H.J. Snijder, P.A. Timmins, K.H. Kalk and B.W. Dijkstra, Detergent organisation in crystals of monomeric outer membrane phospholipase A. *J. Struct. Biol.*, 2003, **141**, 122–131.

76. T. Murata, I. Yamato, Y. Kakinuma, A.G. Leslie, J.E. Walker, Structure of the rotor of the V-Type Na+-ATPase from *Enterococcus hirae. Science*, 2005, **308**, 654-659.
77. J. Li, P.C. Edwards, M. Burghammer, C. Villa and G.F. Schertler, Structure of bovine rhodopsin in a trigonal crystal form. *J. Mol. Biol.*, 2004, **343**, 1409–1438.
78. T. Okada and K. Palczewski, Crystal structure of rhodopsin: implications for vision and beyond. *Curr. Opin. Struct. Biol.*, 2001, **11**, 420–426.
79. K. Palczewski, T. Kumasaka, T. Hori, C.A. Behnke, H. Motoshima, B.A. Fox, I. Le Trong, D.C. Teller, T. Okada, R.E. Stenkamp, M. Yamamoto, and Miyano. Crystal structure of rhodopsin: A G protein-coupled receptor. *Science*, 2000, **289**, 739–745.
80. A.I. Magee and I. Parmryd, Detergent-resistant membranes and the protein composition of lipid rafts. *Genome Biol.*, 2003, **4**, 234.

Web-Based Resources

http://blanco.biomol.uci.edu/Membrane_Proteins_xtal.html
http://moose.bio.ucalgary.ca/Downloads/
http://www.elmhurst.edu/~chm/vchembook/558micelle.html
http://cellbio.utmb.edu/cellbio/membrane_intro.htm

Section 2

Methods for Structural Characterization of Membrane Proteins

Section 2

Methods for Structural Characterisation of Inorganic Phases

CHAPTER 6

Solution NMR Approaches to the Structure and Dynamics of Integral Membrane Proteins

JOHN H. BUSHWELLER[a,b], TOMASZ CIERPICKI[a] AND YUNPENG ZHOU[a]

[a]Department of Molecular Physiology and Biological Physics, University of Virginia, Charlottesville, VA 22908, USA
[b]Department of Chemistry, University of Virginia, Charlottesville, VA 22908, USA

1 Introduction

Although a wealth of structural information on soluble proteins has been generated in the recent past, the determination of structures of integral membrane proteins has lagged far behind. This is a reflection of the significant challenges associated with structural studies of this class of proteins. One of the most exciting recent developments in solution NMR has been the successful application of these methods to the determination of the structures of a number of integral membrane proteins in detergent micelles. Application of transverse relaxation optimized spectroscopy (TROSY)-based experiments for uniformly deuterated proteins[1] has extended the size limit for studies of membrane protein-detergent complexes beyond 100 kDa[2] and made possible structure determinations of moderate-sized integral membrane proteins. The number of successful applications of high-resolution NMR for integral membrane proteins is growing and numerous new systems are being characterized.[2–4] Current NMR methodology can now routinely define backbone structures of moderate-sized β-barrel proteins.[5–7] Proteins containing α-helical membrane spanning motifs have, thus far, proven more challenging targets for structural analysis. Until recently, only a handful of small helical polypeptides, limited to the presence of two membrane spanning fragments, had been structurally characterized by NMR.[8–10] The recent resonance assignment of the α-helical 40 kDa trimeric protein diacylglycerol kinase (DAGK),[2] preliminary NMR data for the human vasopressin V2 receptor,[4] a G-protein-coupled receptor, and the recent structure determination of the putative membrane protein Mistic,[11] all bode well for future success in this area. In addition, recent solution NMR studies of the dynamics of two membrane proteins[12,13] have clearly demonstrated the power of these methods to provide unique insights into the function of integral membrane proteins.

2 Protein Production and Optimization for NMR Studies

2.1 Protein Production

High-level production of functional membrane proteins has proven to be challenging. Because NMR studies of these proteins require the use of quite costly triple-labeled (^{2}H,^{13}C,^{15}N) protein, the first step in NMR-based structural studies of membrane proteins is the development of a cost-effective approach for such high-level protein production. Both bacterial expression and cell-free synthesis have been employed for the production of membrane proteins, whereas eukaryotic expression has not thus far due to the high cost of labeling in such systems.

Bacterial expression has proven effective for a number of systems. In particular, expression of β-barrel membrane proteins into inclusion bodies followed by refolding has provided a powerful approach to achieve expression and exchange of the amide ND moieties to NH during the refolding process, a necessary prerequisite for collection of amide NH-based NMR data. This has been used successfully for structural studies of OmpA, OmpX, and PagP[5–7] and has recently been reviewed.[14] To date, there has not been a successful demonstration of the refolding of an α-helical membrane protein with more than 2 transmembrane helices for NMR studies. However, a number of α-helical membrane proteins with more than 2 transmembrane helices have been successfully expressed and labeled in *Escherichia coli* including Mistic, DAGK, and human vasopressin V2 receptor.[2,4,11] Although it remains to be demonstrated that Mistic is definitively an integral membrane protein, the approaches used with that system for the solution structure determination are illustrative of the methods that will be required. As has been reviewed elsewhere,[15,16] membrane protein expression has proven quite sensitive to the constructs used, cells employed, and the culture conditions. In our recent work on the integral membrane protein DsbB, we have been able to achieve an 8-fold increase in the expression level of membrane-inserted protein by optimization of the cell line employed, temperature of induction, and the concentration of IPTG used for induction (Zhou and Bushweller, unpublished results). This was necessary to make labeling of the protein cost-effective. Recently, Sanders and co-workers[4] showed that by lowering the induction temperature from 37 °C to 12 °C, human vasopressin V2 receptor could be expressed in the membrane and labeled at a level of 5 mg L^{-1}. While low-temperature induction has worked for several systems, it remains to be demonstrated how general this approach will be.

Cell-free expression of proteins has the distinct advantage of avoiding the cellular toxicity that can be associated with membrane protein expression. In terms of NMR studies, it also has the distinct advantage of using very small quantities of labeled components, making it very cost-effective for labeling. In addition, the use of cell-free expression makes it possible to employ selective labeling patterns[17] to aid in the assignment process. Indeed, recent successes in the cell-free expression and labeling of SugE (a 4-transmembrane helix with small multidrug resistance transporter), TehA (a multidrug transporter with 10 putative transmembrane helices), and YfiK (a cysteine transporter with 6 putative transmembrane helices)[18] as well as of EmrE (a 4-transmembrane helix with small multidrug resistance transporter)[19] suggest that it will be a powerful approach for future studies of integral membrane proteins.

2.2 Sample Optimization

Successful structural studies using NMR spectroscopy require membrane protein samples to be stable, monodisperse, and free of chemical exchange-mediated degradation of the NMR spectra. In addition, functional assays are critical to establish that the protein is active under the conditions being employed for the structural studies. To meet all these criteria, it is necessary to optimize temperature, pH, ionic strength, and, most importantly, the identity and concentration of the detergent used. Solution NMR approaches require relatively rapid tumbling of the protein in solution to avoid unfavorable linewidth in the NMR spectra, therefore membrane proteins must be extracted from the lipid bilayer and solubilized in the detergent for these studies. Different detergents form different sizes of protein-detergent complexes. Smaller protein-detergent complexes tumble more rapidly, yielding narrower linewidths and higher quality spectra. In addition, different membrane proteins will behave differently in different detergents with regard to activity, stability, monodispersity, and quality of the NMR data. As studies carried out to date demonstrate, there is not a single magical detergent that works for all proteins. Therefore, a detergent screen is essential to identify the best detergents to optimize the sample behavior.[3] Lipopeptides and amphipols (amphipathic polymers) provide alternative ways to maintain membrane proteins stable in solution, as demonstrated with the outer membrane proteins OmpA and PagP, and DAGK.[20–22] Organic solvent systems have also been used to solubilize membrane proteins.[9,23] As the organic solvent does not form a micelle structure, the overall tumbling of the protein is more rapid, resulting in narrower linewidths and higher quality NMR data. However, these conditions are a vast departure from the native environment, resulting in concerns about the validity of structures determined under these conditions. In addition, the use of organic solvents often precludes any form of functional assay to verify activity, so it is not clear how generally applicable this approach will be.

Our recent experience with the integral membrane protein DsbB (a disulfide oxidoreductase in the *E. coli* inner membrane) is illustrative with regard to this process. Using a simple fluorescence-based assay for activity, we screened a panel of 93 detergents to identify those in which the protein retained enzymatic activity. Subsequently, detergents that retained activity were analyzed for their ability to stabilize the protein at elevated temperature for extended periods of time. Following incubation at 40 °C for 24 h, the sample was assayed for activity. Using these criteria, a small set of detergents were identified that could be employed for the NMR studies. Protein was then analyzed by size-exclusion chromatography in these detergents to assay for aggregation and by ^{15}N-^{1}H TROSY-HSQC spectra to assess the quality of the NMR data. Detergents were examined at several concentrations using these approaches to identify the best possible conditions for the NMR spectroscopy (Zhou and Bushweller, unpublished results).

Although this approach may yield high-quality spectra, there have been numerous observations of chemical exchange broadening in NMR spectra of integral membrane proteins. This may just reflect the change from the more "structured" bilayer environment to the less "structured" micelle environment. Alternatively, this may reflect a significant structural plasticity that is common to these proteins and likely plays a critical role in their function. While changes in temperature can be used to manipulate such exchange behavior, such as has been done for PagP,[13] we anticipate that this will not always be sufficient. In these cases, modifications to the protein will be necessary in order to quench the exchange behavior and make it possible to collect high-quality NMR data. In the case of DAGK, a single point mutation increased the thermal inactivation half-life from 5.7 min to beyond 40 min at 55 °C in n-octylglucoside, while retaining enzymatic activity[24] and dramatically improved the quality of the NMR data.[2] Alternatively, homologs of the target protein can be examined to identify one with more favorable NMR behavior. Those with shorter loops or with higher thermal stability could provide a decrease in the conformational exchange behavior resulting in higher quality spectra. Indeed, the NMR spectra of OmpX, which has shorter extracellular loops, are significantly better than those of the homologous OmpA, which has significantly longer loops.[6]

3 NMR Methodology for the Study of Integral Membrane Proteins

Membrane proteins represent a challenging problem for NMR spectroscopy due to their significant size. Even moderate-sized membrane proteins, *ca.* 20 kDa, form large protein-detergent micelles with molecular weights of 50–60 kDa.[5–7] Furthermore, the size of relatively small oligomeric membrane protein-detergent complexes extends beyond 100 kDa.[2] Although still challenging, successful studies of membrane proteins by NMR are feasible and critically depend on appropriate labeling strategies and carefully optimized experiments performed using state-of-the art instrumentation. The strategy for structure determination of membrane proteins is illustrated in Figure 1.

3.1 High Level Deuteration and Assignment Strategies Using TROSY-Based Experiments

Assignment of chemical shifts in $^{13}C/^{15}N$-labeled proteins can be accomplished using a large set of triple resonance experiments (Table 1).[25] However, the short transverse relaxation times of nuclei involved in transfer of magnetization in these large proteins results in very poor signal-to-noise ratios for these standard triple resonance experiments. The major source of relaxation for ^{13}C nuclei is dipolar relaxation mediated by the attached ^1H.[26] The most efficient method to minimize this effect is to replace protons with deuterons.[27] Since ^2H has an ~6.5-fold lower gyromagnetic ratio, deuteration of protein molecules significantly reduces the transverse

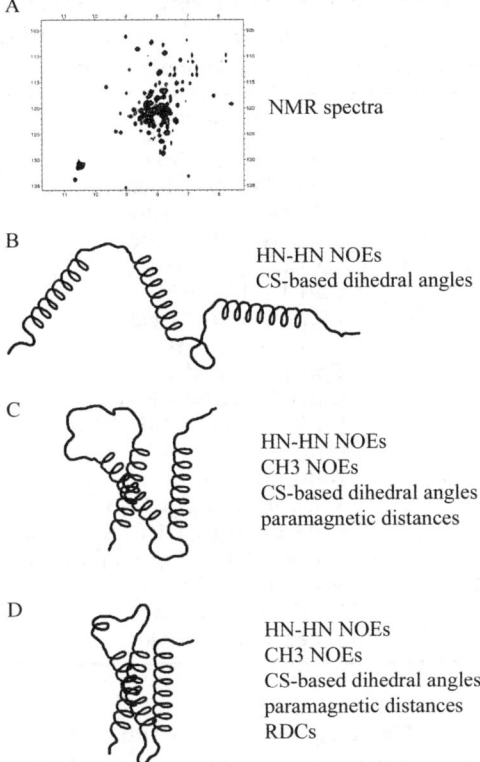

A

NMR spectra

B

HN-HN NOEs
CS-based dihedral angles

C

HN-HN NOEs
CH3 NOEs
CS-based dihedral angles
paramagnetic distances

D

HN-HN NOEs
CH3 NOEs
CS-based dihedral angles
paramagnetic distances
RDCs

Figure 1 *Steps in the structure determination of membrane proteins by NMR spectroscopy: (A) resonance assignment, (B) secondary structure determination, (C) global fold determination, and (D) refinement with dipolar couplings*

relaxation rates of ^{13}C nuclei.[28–30] Broadening of carbon resonances resulting from the ^{13}C-^{2}H couplings can be effectively removed by deuterium decoupling.[30]

Further improvement toward the structure determination of large molecules was achieved by development of TROSY.[1,31] For certain nuclei (amide ^{15}N-^{1}H and aromatic ^{13}C-^{1}H), interference of dipole–dipole and chemical shift anisotropy (CSA) relaxation mechanisms at high-magnetic fields can be used to minimize transverse relaxation. This methodology is being routinely used for studying proteins in the 25–50 kDa range. High resolution NMR spectra can be obtained for even larger proteins with molecular masses over 100 kDa with application of cross-correlated relaxation-enhanced polarization transfer (CRINEPT) and cross-correlated relaxation-induced polarization transfer (CRIPT).[32,33] A set of TROSY-based triple resonance experiments for assignment of backbone resonances in large deuterated proteins is shown in Table 1. These experiments utilize the TROSY effect for amide groups and rely on magnetization that is created and detected using HN nuclei. In this way, chemical shifts of amide protons can be correlated with Cα, Cβ, and CO of two sequential residues in a series of NMR experiments.[34–36]

Table 1 *Family of triple resonance experiments for assignment of backbone and side chain methyl groups. All NH-detected experiments are TROSY-based. CM and HM are methyl carbon and methyl protons, respectively*

Experiment name	Type of correlations	References
3D HNCO	CO(i-1), N(i), HN(i)	34
3D HN(CA)CO	CO(i), N(i), HN(i) and CO(i-1), N(i), HN(i)	35
3D HNCA	Cα(i), N(i), HN(i) and Cα(i-1), N(i), HN(i)	34
3D HN(CO)CA	Cα(i-1), N(i), HN(i)	35
3D HN(CA)CB	Cβ(i), N(i), HN(i) and Cβ(i-1), N(i), HN(i)	35
3D HN(COCA)CB	Cβ(i-1), N(i), HN(i)	35
4D HNCACO	Cα(i), CO(i), N(i), HN(i)	37
4D HNCOCA	Cα(i-1), CO(i-1), N(i), HN(i)	37
3D (H)C(CO)NH-TOCSY	CM(i), N(i), HN(i)	51
3D H(C)(CO)NH-TOCSY	HM(i), N(i), HN(i)	51
3D HMCM[CG]CBCA	Val: $C\gamma H_3$(i), Cβ(i), Cα(i) Ile: $C\delta 1H_3$(i), C1γ(i), Cβ(i), Cα(i) Leu: $C\delta H_3$(i), Cγ(i), Cβ(i), Cα(i)	52
3D HMCM(CBCA)CO	Val: $C\gamma H_3$(i), CO(i)	52
3D HMCM(CGCBCA)CO	Ile: $C\delta 1H_3$(i), CO (i) Leu: $C\delta H_3$(i), CO(i)	52

Application of TROSY-based experiments was sufficient to yield assignments for several integral membrane proteins including OmpA, OmpX, PagP, and DAGK.[2,5–7] If necessary, further improvement in resolution can be obtained using four-dimensional experiments, such as HNCACO and HNCOCA.[37] This may be indispensable for assignment of α-helical proteins with significantly higher chemical shift degeneracy.[2]

3.2 Carbon Detected Experiments: Breaking the Limit of Sensitivity

A valuable alternative for the assignment of deuterated proteins is the application of experiments involving direct detection of ^{13}C nuclei.[38] This would overcome the rapid relaxation of protons by utilizing the more slowly relaxing ^{13}C nuclei and yield high-resolution spectra for large proteins. However, such approaches are limited by the very low sensitivity of carbon-detected experiments and require substantial improvement in spectrometer hardware. The sensitivity of NMR spectrometers is primarily limited by thermal noise generated by the signal detection hardware, so a significant gain in sensitivity can be achieved by lowering the temperature of the coil and preamplifier. This led to the development of cryogenically cooled probes that increase the signal-to-noise ratio of NMR spectrometers up to 3–4 fold.[39]

Application of cryogenic probes has made it possible to directly detect ^{13}C nuclei in proteins and paved the way for development of new types of heteronuclear NMR experiments.[39] Recently, a new strategy for chemical shift assignment relying on protonless experiments was proposed.[40,41] A family of 2D and 3D heteronuclear experiments have been developed and tested for the 32 kDa protein superoxide dismutase.[41] Development of 2D ^{13}C TOCSY and its application for nearly complete assignment of side chain carbons was demonstrated for the 44 kDa trimeric chorismate mutase,[38] demonstrating the utility of this approach for side chain assignments as well.

3.3   Use of Methyl Protonation to Increase the Number of Nuclear Overhauser Effect-Derived Distance Constraints

Structure determination of proteins using NMR spectroscopy typically relies heavily on distance information obtained from measurement of the nuclear Overhauser effect (NOE). In fully protonated and perdeuterated proteins there are, on average, 15.7+/−2.0 and 2.5+/−0.4 protons located within a 5 Å radius.[42] In perdeuterated proteins, distance measurements are restricted to HN–HN interactions that are frequently not sufficient to determine the global fold of the molecule.[42] In order to increase the number of NOE-based distances, protonated methyl groups of Val, Leu, and Ile (δ1) can be selectively incorporated into the proteins.[43] The rationale for choosing methyl protonation is based on dispersion of the spectra, location of Leu, Ile, and Val side chains in the hydrophobic core of proteins, and relatively narrow signals due to fast rotation about the methyl symmetry axis.[44] It was shown that measurement of NOEs involving methyl groups has a significant impact on structure quality and makes it possible to obtain global folds for large proteins.[45–47] Because the side chains of these amino acids predominantly reside in the bilayer environment in membrane proteins, it is not clear that this approach will have the same utility for membrane proteins as it does for soluble proteins. The most common strategy to incorporate methyl groups in a highly deuterated background involves labeling of methyl groups of Val, Leu, and Ile (δ1). This is achieved by adding [3-^{2}H] ^{13}C α-ketoisovalerate and [3,3-^{2}H$_2$] ^{13}C α-ketobutyrate to the growth medium prior to induction of *E. coli* cells.[48] This method results in over 90% labeling of methyl groups without CH$_2$D and CHD$_2$ isotopomers.

Basic experiments for assignment of methyl groups are summarized in Table 1. The first class of experiments is based on transfer of magnetization from methyl groups to backbone Cα and subsequently through CO to NH for detection.[49] Application of (H)C(CO)NH-TOCSY and H(C)(CO)NH-TOCSY spectra was successfully used to assign maltose-binding protein[50] and OmpX.[51] The disadvantage of this labeling strategy is the presence of two methyl groups that can degrade magnetization transfer due to two possible pathways.[52] In order to overcome this problem a second labeling strategy was developed leading to only one ^{13}CH$_3$ group.[52] This labeling strategy can be combined with assignment based on COSY experiments, resulting in a sensitivity gain. The most recent modifications to experiments for methyl assignment involve the use of "out-and-back" magnetization transfer from methyl spins to aliphatic carbons or carbonyl and back to methyl protons for detection (Table 1).[52] This method is substantially more sensitive and can be applied to larger systems.

3.4 Application of Electron-Nuclear Relaxation for Long Range Distances

It has long been recognized that paramagnetic centers in protein molecules lead to significant broadening of adjacent resonances.[53] This effect is caused by electron-nuclear relaxation that originates from a strong interaction between free electrons and nuclei. Enhancement of relaxation can be quantitatively analyzed and provide distance information between the paramagnetic center and the affected nuclei.[54] Unlike the NOE, this approach makes it possible to obtain longer range distance restraints (up to 25 Å).[55]

Paramagnetic relaxation enhancement (PRE) can provide critical restraints to determine the global fold of large deuterated proteins.[11,56] Distance information can be obtained from experimentally measured proton relaxation times and solving the Solomon–Bloembergen equation.[55,57,58] The most sensitive probe of paramagnetic relaxation is *via* the proton transverse relaxation (T2).[55] The first, more rigorous, approach requires measurement of proton relaxation times for diamagnetic and paramagnetic samples.[57,59] The second method is based on quantification of paramagnetic broadening of HSQC spectra and estimation of PRE from the ratio of signal intensities of paramagnetic and spin-labeled samples.[55,56]

Paramagnetic centers can be introduced into proteins using various approaches: (i) utilizing natural metal binding sites in metalloproteins,[60] (ii) engineering of metal-binding sites,[61] and (iii) site-directed spin labeling.[56,62] The last method is well suited for membrane proteins and is frequently used to characterize protein structures using EPR spectroscopy.[63] It is based on modification of cysteine residues using a spin-labeled reagent, such as the nitroxide-containing MTSL.[56,62] Nitroxides are convenient as Cys residues can be introduced at many sites in the protein and thereby a large set of PRE data can be generated. Another method relies on introduction of an amino terminal Cu(II) or Ni(II) binding motif (ATCUN) by appending an NH_2-X-X-His tripeptide sequence at the N-terminus of the protein.[57]

3.5 Residual Dipolar Couplings

A powerful approach to improve the accuracy of membrane protein structures is based on the application of residual dipolar couplings (RDCs).[64,65] RDCs contain information about orientation of internuclear vectors relative to the external magnetic field[66,67] that can be readily introduced into structure calculations.[68,69] Measurement of RDCs is feasible for protein molecules that are partially oriented in solution.[66,67] Spontaneous alignment in the magnetic field is very weak.[70] Modification of a diamagnetic protein by introduction of a paramagnetic tag can be used to achieve modest alignment in the magnetic field.[61] Typically it is necessary to introduce weak alignment of protein molecules *via* a properly selected medium[71] for these measurements. The most suitable media for integral membrane proteins are polyacrylamide gels,[72] due to their inert nature with respect to the detergent present in these samples. Sample preparation is relatively straightforward and can be produced using various methods.[73–75] The most generally applicable method, to date, relies on compressed charged copolymer gels and was recently used to measure RDCs for OmpA.[74]

Measurement of RDCs is achieved *via* a comparison of measured coupling constants in the aligned as well as in the unaligned state. Numerous two and three-dimensional experiments have been designed to measure a variety of coupling constants.[76,77] This commonly includes three types of backbone couplings: $^1J_{HN}$, $^1J_{NCO}$, and $^1J_{COC\alpha}$ that can be accurately measured for large proteins using a set of TROSY-based HNCO experiments.[77] A very promising recent development is the demonstration of dipolar couplings for side chains using directly detected ^{13}C spectroscopy.[78] Dipolar couplings can be readily utilized in structure calculations using standard software for molecular dynamics with simulated annealing, such as CNS [68] and Xplor-NIH.[69] A significant improvement in structure accuracy can be achieved by simultaneous use of RDCs measured from more than one alignment.[79]

Measurements of RDC for polypeptides embedded in detergent micelles are more difficult due to the size of the protein-detergent complex. To date, structure refinement using RDCs has been demonstrated for the small α-helical peptide HIV-1 gp41,[80] the single transmembrane α-helical proteins Vpu[81] and Pf1 coat protein,[82] and recently to MerF with two transmembrane helices.[83]

4 Solution NMR Structures of Helical Integral Membrane Proteins

Helical membrane proteins constitute the majority of integral membrane proteins. This class of proteins presents significant challenges for structure determination by NMR due to the extensive chemical shift degeneracy (overlap) typical of highly helical proteins and the lack of inter-helical NOEs to define the overall fold of the protein. Due to the long distances between amide NH moieties in the helical elements in this class of proteins, few, if any, inter-helical NOEs can be obtained from perdeuterated samples. The combination of these two challenges has limited structure determinations to small and structurally simple membrane proteins, mostly containing two or less transmembrane helices. With the development of alternative techniques to obtain inter-helical constraint information, particularly the use of electron – nuclear relaxation methods, more complex systems have been tackled including the recent determination of the structure of the putative integral membrane protein Mistic.[11] The structures of several helical membrane proteins determined using NMR spectroscopy are shown in Figure 2.

4.1 F1Fo ATP Synthase Subunit c from *E. coli*

F1Fo ATP synthase subunit c was the first membrane protein with two transmembrane helices whose structure has been determined by solution NMR.[84] F1Fo ATP synthase uses the proton gradient across the membrane to drive ATP synthesis. The membrane embedded Fo portion contains a rotor assembled with 12 subunit c monomers, which couples the rotation of the ring to the movement of protons across the membrane. The structure of subunit c showed that it consists of two long transmembrane helices and a short cytoplasmic loop. There is a conserved Asp in the middle of the C-terminal helix that mediates proton transfer across the membrane. Protonation and ionization of this highly conserved Asp was shown to result in a

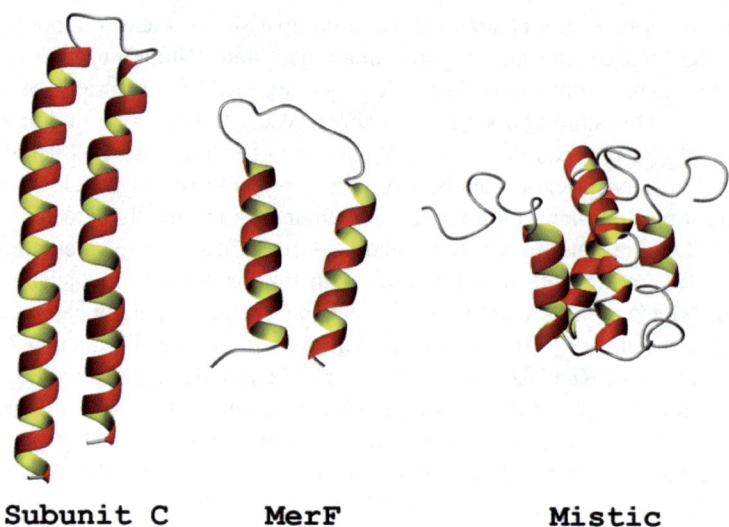

Subunit C   MerF   Mistic

Figure 2 *Ribbon representations of the structures of three α-helical integral membrane proteins solved using solution NMR methods*

rotation of the C-terminal helix, which was proposed to drive the rotation of the rotor ring.[9] The NMR structures of protonated and ionized subunit c were carried out in chloroform/methanol/water (4:4:1) solution at pH 5 and pH 8. Due to the short correlation time for subunit c in chloroform/methanol/water, it was possible to employ fully protonated samples for resonance assignment and NOE data collection. Thus, inter-helical distance information was readily obtained from NOEs to side chain protons. Indeed, more than 100 interhelical NOEs were assigned unambiguously for protonated subunit c. Standard structure calculations using NOEs, $^{3}J_{HNH\alpha}$ coupling constants, and hydrogen bond restraints resulted in high-quality structures for the pH 5 and pH 8 structures. The use of a mixed solvent system significantly reduces the overall correlation time for the protein as a result of the reduced viscosity of this solvent system and also the lack of micelle formation that occurs in detergent solutions. This provides a large improvement in relaxation behavior and makes it possible to use protonated samples and standard NMR methods for membrane protein studies. While there is no question that these improvements are dramatic, concerns remain as to whether such solvent systems can adequately mimic the bilayer environment. Indeed, the crystal structure of the rotor ring from *Ilyobacter tartaricus* shows a different conformation for the C-terminal helix of subunit c,[85] which may reflect crystal packing or oligomerization effects but could also reflect the influence of the organic solvent.

4.2 MerF

MerF is an 81 amino acid bacterial mercury transport protein, which contains two transmembrane helices connected by a short loop. The structure of MerF was solved using dipolar couplings and chemical shift-based dihedral angles in SDS micelles.[83] Because of the well-known propensity of SDS to induce helix formation, it is not entirely clear that the structure in SDS will recapitulate what is found in the bilayer. Two different constructs of MerF in SDS micelles were weakly aligned in stressed polyacrylamide gels. Backbone ^{1}H-^{15}N RDCs derived from these two different alignments were used to fit a dipolar wave,[86] which determines the boundaries and relative orientation of the two helices and generates supplemental dihedral restraints in this region. The dipolar coupling and dihedral angle data are sufficient to determine the conformation of the helices. For structure calculations, a radius of gyration term that reflects the packing tendency of secondary structure elements was used to pack these two helices together. The final MerF structure was determined to high resolution. While the utility of this approach for larger helical membrane protein remains to be demonstrated, this structure clearly illustrates the utility of dipolar couplings in the determination of the structures of helical membrane proteins.

4.3   Mistic

The structure determination of the protein Mistic[11] has provided a useful illustration of the approach necessary for NMR structure determination of multiple transmembrane helix proteins. Although it remains to be demonstrated that Mistic is definitively an integral membrane protein, it still serves as a useful example of how to apply solution NMR methods to the structure determination of α-helical membrane proteins. Indeed, the structure of Mistic displays no α-helices long enough to span the membrane and the protein has a hydrophobic interior and hydrophilic exterior, neither of which is found in typical integral membrane proteins. Mistic was overexpressed in *E. coli* membranes and the sample was prepared in LDAO micelles. Light-scattering data and backbone ^{15}N T1, T2 measurements demonstrated that it is a monomer and provided an estimate of the size of the protein-detergent complex (22 kDa). Partial deuteration with full ^{13}C/^{15}N labeling was employed to acquire full backbone sequential assignments and partial side chain assignments. Backbone ^{13}C chemical shift derived dihedral angle restraints, NOEs, and hydrogen bond restraints were used to determine the conformation for each of the four helices. Long-range distance constraints (487) derived from 5 different paramagnetic spin labels and 29 long-range NOEs from methyl or aromatic side chains to backbone amide protons were used to accurately determine the fold of the protein with a backbone rmsd of 1.0 Å. The structure determination of Mistic clearly demonstrates that the challenge of identifying long-range distance information that is critical for determination of the fold can be solved through the introduction of spin labels and the measurement of electron-nuclear relaxation effects.

5   Solution NMR Structures of β-Barrel Membrane Proteins

Small monomeric β-barrel proteins are attractive targets for NMR structure determination. Unlike α-helical proteins, the global fold is much easier to determine due to the abundance of long-range H^N-H^N NOEs that can be detected between residues in neighboring β–strands. Indeed, solution NMR has been successfully applied for the structure determination of several β-barrel membrane proteins (see Figure 3)

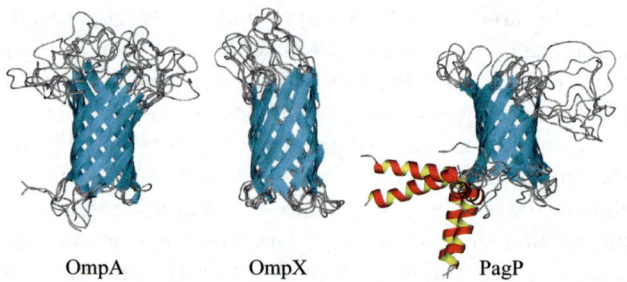

OmpA   OmpX   PagP

Figure 3 *Ribbon representations of the structures of three β-barrel integral membrane proteins solved using solution NMR methods. Overlay of 10 conformers shown for each structure to illustrate the increased mobility observed for the extracellular loops*

including OmpA, OmpX, and PagP.[5–7] All the three proteins possess common features including excellent stability in small detergents (*e.g.* dodecylphosphocholine (DPC) and β-octylglucoside (β-OG)), high solubility (up to 2 mM), and long–term stability at elevated temperatures (up to 50 °C for OmpA).

5.1 OmpA

One of the major proteins found in the *E. coli* outer membrane is the 35 kDa OmpA. OmpA has a predominantly structural function in maintaining the proper shape of Gram-negative bacteria, however it can also function as an ion channel.[87] OmpA consists of a 19 kDa transmembrane domain followed by a 16 kDa globular periplasmic domain. The transmembrane domain of OmpA can be overexpressed in *E. coli* as inclusion bodies in high amounts and efficiently refolded into various detergents. High-quality spectra obtained for deuterated OmpA in DPC[5] or dihexanoylphosphatidylcholine (DHPC)[6] micelles paved the way for structural studies using NMR. Application of TROSY-based triple resonance experiments yielded assignment of 138 out of 177 residues in OmpA.[5] Complete assignment of the protein was not possible due to complex dynamics resulting in extensive signal broadening for extracellular loops and the appearance of multiple peaks for numerous residues in OmpA.[5] The interpretation of NOESY spectra for deuterated and fractionally protonated protein samples yielded 91 distances, including 49 H^N–H^N and 42 H^N-H^α NOEs. The global fold of OmpA was obtained based on distance restraints, backbone dihedral angles based on chemical shift analysis, and 58 interstrand hydrogen bonds.[5] Despite relatively sparse data, the fold of the eight stranded OmpA β-barrel is very similar to that seen in the X-ray structure.[88]

5.2 OmpX

Another example of a β-barrel integral membrane protein whose structure has been solved by NMR is 16.5 kDa OmpX. Although the function of OmpX remains unclear, the protein is involved in cell adhesion of *E. coli* cells.[89] The NMR studies were facilitated by high-quality NMR spectra of OmpX refolded into DHPC. Application of TROSY-based experiments yielded nearly complete backbone assignments.[6] The initial structure of OmpX was determined based on 107 H^N-H^N NOEs and 140 chemical shift based dihedral angles.[6] The global fold was similar to the X-ray structure of OmpX; however the coordinate precision for β-barrel residues was quite low with a backbone rmsd of about 3.1 Å. In order to increase the accuracy and precision of the structure, additional NOEs were collected from a methyl protonated protein sample. This yielded a substantial increase in the number of NOEs from 107 to 526[90] resulting in improvement of the precision of β-barrel residues (from 3.1 to 1.42 Å rmsd for backbone atoms of all NMR conformers). An additional improvement to 1.17 Å was achieved by supplementing the data with 34 interstrand hydrogen bonds.[90]

5.3 PagP

The bacterial outer membrane enzyme PagP is a transacylase that catalyzes transfer of a palmitate chain from phospholipid to lipid A.[91] Its relatively small size and efficient refolding facilitated NMR studies and structure determination in two different detergents: n-octyl-β-*D*-glucoside (β-OG) and DPC.[7] Nearly complete assignment of backbone chemical shifts was obtained with only a small number of resonances in the loops broadened beyond detection. The structure of PagP in DPC was solved based on 147 H^N-H^N NOE based distances, 74 hydrogen bond restraints, and 234 dihedral angles.[7] The fold of PagP is very similar to OmpA and OmpX, consisting of an eight-stranded β-barrel with an additional N-terminal α-helix. The core of the barrel is very well defined in the NMR structure (backbone rmsd = 0.9 Å), while the loops are disordered. The position of the β-helix could not be obtained from NMR data, however it is most likely located on the surface of the membrane. Very similar structures of PagP have been obtained in both DPC and β-OG detergents.

Recently, PagP was reconstituted in CYFOS-7, a detergent that supports enzymatic activity.[13] It was found that under these conditions the protein exists in equilibrium between two states: relaxed and tense. Comparison of chemical shifts for the two states shows that major structural changes occur in the large extracellular loop and adjacent regions of the β-barrel. These results clearly highlight the potential effects of the detergent environment on both integral membrane protein structure and activity.

6 Solution NMR Characterization of Membrane Protein Dynamics

The functions of integral membrane proteins as ion channels, transporters of ligands (iron, vitamin B12, *etc.*) across the membrane, and transducers of information across the membrane (GPCRs) require significant structural plasticity. Namely, significant conformational changes are associated with all of these processes. The necessary conformational heterogeneity is difficult to characterize by X-ray crystallography, which typically only provides information about a single conformation that is readily crystallizable. Indeed, the use of mutations or antibodies has been essential for successful crystallization of this class of proteins, resulting in a "freezing" of the protein into a single conformation, which is favorable for crystallization. However, this also results in the loss of information on the dynamics of the protein, which is clearly critical for function. Solution NMR spectroscopy has been used very effectively to study protein dynamics.[92–95] In the context of membrane proteins, such studies will be imperative to gain a complete picture of the mechanism. Indeed, recent efforts to analyze the dynamics of the integral membrane proteins OmpA and PagP have provided tantalizing evidence of the power of NMR spectroscopy to elucidate mechanism in membrane proteins.

6.1 OmpA

OmpA is a structural protein found in high abundance in the outer membrane. OmpA has a role in bacterial conjugation and as a receptor for bacteriophages and colicins,[96–98] OmpA has also been shown to function as an ion channel.[99,100] This latter function provided an interesting opportunity to explore the role of dynamics in channel function. [15]N backbone relaxation measurements (T1, T2, [1]H-^{15}N NOE) at multiple field strengths have been employed to characterize the backbone dynamics of the protein.[12] The data showed significantly increased mobility for the extracellular loops in OmpA, similar to results for the β-barrel membrane protein PagP,[13] suggesting this may be a common property of the β-barrel class of membrane proteins. In addition, a number of residues with motion on the μs–ms timescale were identified. Interestingly, a significant number of these were located in the center of the OmpA β-barrel. This region of the structure has been shown to have a collection of charged residues inside the barrel, which participate in a significant hydrogen-bonded network. Indeed, both the crystal structure[88] and NMR solution structure of OmpA[5] show a total occlusion in this region of the barrel, which would make passage of any ion impossible. It is clear that the dynamics of the amino acids in this region must be such as to periodically permit the passage of ions. An MD simulation of OmpA[101] has also suggested a role for specific amino acids identified by the dynamics studies in the channel function of OmpA, providing independent corroboration. These results have suggested specific point mutations in the barrel to assess their role in channel function. Similar to results from NMR dynamics studies of enzyme function,[102,103] these results clearly demonstrate the power of this approach to elucidate the role of various amino acids in channel function, which will be generally applicable to all ion channels.

6.2 PagP

Kay and co-workers[7] have recently described both the structure and the dynamics[13] of the integral membrane protein PagP. PagP catalyzes the transfer of the *sn*-1 palmitate chain from phospholipids to lipopolysaccharide in Gram-negative bacteria and is a critical determinant of virulence. Structure determinations of PagP by NMR[7] and X-ray crystallography[104] employed detergents in which PagP was not enzymatically active. The identification of a bound detergent molecule in the barrel in the X-ray crystal structure suggested that the detergents used could penetrate the barrel and thereby inhibit activity. Subsequently, a more bulky detergent (CYFOS-7) was

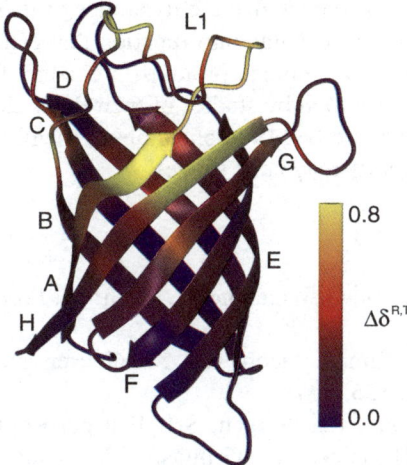

Figure 4 *Ribbon diagram of PagP color according to the difference in chemical shift between the relaxed (R) and tense (T) states (reprinted with permission from Hwang et al., PNAS, 2004, **101**, 9618–9623)*

employed and PagP was found to be active in this detergent. This provided conditions under which the role of dynamics in PagP function could be assessed. Using NMR spectroscopy, it was shown that PagP exists in solution in equilibrium between two states (see Figure 4). At modest temperature (25 °C), these two 2 states were readily distinguished whereas at the higher temperatures used for the structural studies the corresponding peaks coalesced into a single set of peaks, which were exchange-broadened. Careful measurements using ^{15}N relaxation data made it possible to characterize the rate of conversion between the two states as well as the relaxation behavior of each state individually. These measurements showed the major species, referred to as the R (relaxed) state, to be quite flexible, whereas the minor species, referred to as the T (tense) state, displayed dynamic behavior consistent with a much more rigid ordered structure. Based on this data, Kay and co-workers proposed a model for enzymatic activity in which the more flexible R state provides the necessary plasticity for substrate binding and the more rigid T state is the conformation necessary to effect catalysis. Again here, the extraordinary power of NMR to probe the dynamics of the protein is providing powerful insights into the mechanism of action of the protein.

7 Future Directions

The development of approaches to determine the folds of both β-barrel and α-helical membrane proteins now makes it possible for solution NMR to make a substantial contribution to the structural biology of membrane proteins. Indeed, recent successes for both classes of proteins provide a tantalizing view of future prospects in this area. Significant challenges remain, however, particularly in going from fold determination to high-resolution structures with well-characterized side chain structural information. The ability to characterize the dynamics of these proteins promises to provide unique and powerful insights into function that cannot be obtained by any other methods. The role of membrane proteins in channel function, transport, and signaling can only be understood by studies of their dynamics.

Indeed, the future of solution NMR applications to membrane proteins appears so bright, the NMR community may need shades!

References

1. K. Pervushin, R. Riek, G. Wider and K. Wuthrich, *Proc. Natl. Acad. Sci. USA*, 1997, **94**, 12366.
2. K. Oxenoid, H.J. Kim, J. Jacob, F.D. Sonnichsen and C.R. Sanders, *J. Am. Chem. Soc.*, 2004, **126**, 5048.
3. R.D. Krueger-Koplin, P.L. Sorgen, S.T. Krueger-Koplin, I.O. Rivera-Torres, S.M. Cahill, D.B. Hicks, L. Grinius, T.A. Krulwich and M.E. Girvin, *J. Biomol. NMR*, 2004, **28**, 43.
4. C. Tian, R.M. Breyer, H.J. Kim, M.D. Karra, D.B. Friedman, A. Karpay and C.R. Sanders, *J. Am. Chem. Soc.*, 2005, **127**, 8010.
5. A. Arora, F. Abildgaard, J.H. Bushweller and L.K. Tamm, *Nat. Struct. Biol.*, 2001, **8**, 334.
6. C. Fernandez, C. Hilty, S. Bonjour, K. Adeishvili, K. Pervushin and K. Wuthrich, *FEBS Lett.*, 2001, **504**, 173.
7. P.M. Hwang, W.Y. Choy, E.I. Lo, L. Chen, J.D. Forman-Kay, C.R. Raetz, G.G. Prive, R.E. Bishop and L.E. Kay, *Proc. Natl. Acad. Sci. USA*, 2002, **99**, 13560.
8. K.R. MacKenzie, J.H. Prestegard and D.M. Engelman, *Science*, 1997, **276**, 131.
9. V.K. Rastogi and M.E. Girvin, *Nature*, 1999, **402**, 263.
10. M.F. Mesleh, S. Lee, G. Veglia, D.S. Thiriot, F.M. Marassi and S.J. Opella, *J. Am. Chem. Soc.*, 2003, **125**, 8928.
11. T.P. Roosild, J. Greenwald, M. Vega, S. Castronovo, R. Riek and S. Choe, *Science*, 2005, **307**, 1317.
12. L.K. Tamm, F. Abildgaard, A. Arora, H. Blad and J.H. Bushweller, *FEBS Lett.*, 2003, **555**, 139.
13. P.M. Hwang, R.E. Bishop and L.E. Kay, *Proc. Natl. Acad. Sci. USA*, 2004, **101**, 9618.
14. M. Bannwarth and G.E. Schulz, *Biochim. Biophys. Acta*, 2003, **1610**, 37.
15. D.N. Wang, M. Safferling, M.J. Lemieux, H. Griffith, Y. Chen and X.D. Li, *Biochim. Biophys. Acta*, 2003, **1610**, 23.
16. S. Eshaghi, M. Hedren, M.I. Nasser, T. Hammarberg, A. Thornell and P. Nordlund, *Protein. Sci.*, 2005, **14**, 676.
17. J. Shi, J.G. Pelton, H.S. Cho and D.E. Wemmer, *J. Biomol. NMR*, 2004, **28**, 235.
18. C. Klammt, F. Lohr, B. Schafer, W. Haase, V. Dotsch, H. Ruterjans, C. Glaubitz and F. Bernhard, *Eur. J. Biochem.*, 2004, **271**, 568.
19. Y. Elbaz, S. Steiner-Mordoch, T. Danieli and S. Schuldiner, *Proc. Natl. Acad. Sci. USA*, 2004, **101**, 1519.
20. C.L. McGregor, L. Chen, N.C. Pomroy, P. Hwang, S. Go, A. Chakrabartty and G.G. Prive, *Nat. Biotechnol.*, 2003, **21**, 171.

21. B.M. Gorzelle, A.K. Hoffman, M.H. Keyes, D.N. Gray, D.G. Ray and C.R. Sanders, *J. Am. Chem. Soc.*, 2002, **124**, 11594.

22. M. Zoonens, L.J. Catoire, F. Giusti and J.L. Popot, *Proc. Natl. Acad. Sci. USA*, 2005, **102**, 8893.

23. M. Schwaiger, M. Lebendiker, H. Yerushalmi, M. Coles, A. Groger, C. Schwarz, S. Schuldiner and H. Kessler, *Eur. J. Biochem.*, 1998, **254**, 610.

24. Y. Zhou and J.U. Bowie, *J. Biol. Chem.*, 2000, **275**, 6975.

25. M. Sattler, J. Schleucher and C. Griesinger, *Prog. Nucl. Mang. Reson. Spectrosc.*, 1999, **34**, 93.

26. A. Allerhand, D. Doddrell, V. Glushko, D.W. Cochran, E. Wenkert, P.J. Lawson and F.R. Gurd, *J. Am. Chem. Soc.*, 1971, **93**, 544.

27. D.M. LeMaster and F.M. Richards, *Biochemistry*, 1988, **27**, 142.

28. D.M. Kushlan and D.M. LeMaster, *J. Biomol. NMR*, 1993, **3**, 701.

29. R.A. Venters, B.T. Farmer II, C.A. Fierke and L.D. Spicer, *J. Mol. Biol.*, 1996, **264**, 1101.

30. S. Grzesiek, J. Anglister, H. Ren and A. Bax, *J. Am. Chem. Soc.*, 1993, **115**, 4369.

31. K. Pervushin, *Q. Rev. Biophys.*, 2000, **33**, 161.

32. J. Fiaux, E.B. Bertelsen, A.L. Horwich and K. Wuthrich, *Nature*, 2002, **418**, 207.

33. R. Riek, J. Fiaux, E.B. Bertelsen, A.L. Horwich and K. Wuthrich, *J. Am. Chem. Soc.*, 2002, **124**, 12144.

34. M. Salzmann, K. Pervushin, G. Wider, H. Senn and K. Wuthrich, *Proc. Natl. Acad. Sci. USA*, 1998, **95**, 13585.

35. M. Salzmann, G. Wider, K. Pervushin, H. Senn and K. Wuthrich, *J. Am. Chem. Soc.*, 1999, **121**, 844.

36. M. Salzmann, K. Pervushin, G. Wider, H. Senn and K. Wuthrich, *J. Biomol. NMR*, 1999, **14**, 85.

37. D. Yang and L.E. Kay, *J. Am. Chem. Soc.*, 1999, **121**, 2571.

38. A. Eletsky, O. Moreira, H. Kovacs and K. Pervushin, *J. Biomol. NMR*, 2003, **26**, 167.

39. H. Kovacs, D. Moskau and M. Spraul, *Prog. Nucl. Mang. Reson. Spectrosc.*, 2005, **46**, 131.

40. I. Bertini, B. Jimenez and M. Piccioli, *J. Magn. Reson.*, 2005, **174**, 125.

41. W. Bermel, I. Bertini, L. Duma, I.C. Felli, L. Emsley, R. Pierattelli and P.R. Vasos, *Angew. Chem. Int. Ed. Engl.*, 2005, **44**, 3089.

42. K.H. Gardner and L.E. Kay, *Annu. Rev. Biophys. Biomol. Struct.*, 1998, **27**, 357.

43. K.H. Gardner and L.E. Kay, *Annu. Rev. Biophys. Biomol. Struct.*, 1998, **27**, 357.

44. L.K. Nicholson, L.E. Kay, D.M. Baldisseri, J. Arango, P.E. Young, A. Bax and D.A. Torchia, *Biochemistry*, 1992, **31**, 5253.

45. G.A. Mueller, W.Y. Choy, D. Yang, J.D. Forman-Kay, R.A. Venters and L.E. Kay, *J. Mol. Biol.*, 2000, **300**, 197.

46. C. Fernandez, C. Hilty, G. Wider, P. Guntert and K. Wuthrich, *J. Mol. Biol.*, 2004, **336**, 1211.

47. V. Tugarinov, W.Y. Choy, V.Y. Orekhov and L.E. Kay, *Proc. Natl. Acad. Sci. USA*, 2005, **102**, 622.

48. N.K. Goto, K.H. Gardner, G.A. Mueller, R.C. Willis and L.E. Kay, *J. Biomol. NMR*, 1999, **13**, 369.
49. V. Tugarinov, P.M. Hwang and L.E. Kay, *Annu. Rev. Biochem.*, 2004, **73**, 107.
50. K.H. Gardner, X. Zhang, K. Gehring and L.E. Kay, *J. Am. Chem. Soc.*, 1998, **120**, 11738.
51. C. Hilty, C. Fernandez, G. Wider and K. Wuthrich, *J. Biomol. NMR*, 2002, **23**, 289.
52. V. Tugarinov and L.E. Kay, *J. Am. Chem. Soc.*, 2003, **125**, 13868.
53. P.G. Schmidt and I.D. Kuntz, *Biochemistry*, 1984, **23**, 4261.
54. P.A. Kosen, R.M. Scheek, H. Naderi, V.J. Basus, S. Manogaran, P.G. Schmidt, N.J. Oppenheimer and I.D. Kuntz, *Biochemistry*, 1986, **25**, 2356.
55. J.R. Gillespie and D. Shortle, *J. Mol. Biol.*, 1997, **268**, 158.
56. J.L. Battiste and G. Wagner, *Biochemistry*, 2000, **39**, 5355.
57. L.W. Donaldson, N.R. Skrynnikov, W.Y. Choy, D.R. Muhandiram, B. Sarkar, J.D. Forman-Kay and L.E. Kay, *J. Am. Chem. Soc.*, 2001, **123**, 9843.
58. I. Solomon and N. Bloembergen, *J. Chem. Phys.*, 1956, **25**, 261.
59. J. Iwahara, D.E. Anderson, E.C. Murphy and G.M. Clore, *J. Am. Chem. Soc.*, 2003, **125**, 6634.
60. D. Bentrop, I. Bertini, M.A. Cremonini, S. Forsen, C. Luchinat and A. Malmendal, *Biochemistry*, 1997, **36**, 11605.
61. C. Ma and S.J. Opella, *J. Magn. Reson.*, 2000, **146**, 381.
62. V. Gaponenko, J.W. Howarth, L. Columbus, G. Gasmi-Seabrook, J. Yuan, W.L. Hubbell and P.R. Rosevear, *Protein Sci.*, 2000, **9**, 302.
63. E. Perozo, L.G. Cuello, D.M. Cortes, Y.S. Liu and P. Sompornpisut, *Novartis Found. Symp.*, 2002, **245**, 146.
64. J.H. Prestegard, H.M. Al-Hashimi and J.R. Tolman, *Q. Rev. Biophys.*, 2000, **33**, 371.
65. A. Bax, *Protein Sci.*, 2003, **12**, 1.
66. J.H. Prestegard, J.H. Al-Hashimi and J.R. Tolman, *Q. Rev. Biophys.*, 2000, **33**, 371.
67. A. Bax, G. Kontaxis and N. Tjandra, *Methods Enzymol.*, 2001, **339**, 127.
68. A.T. Brunger, P.D. Adams, G.M. Clore, W.L. DeLano, P. Gros, R.W. Grosse-Kunstleve, J.S. Jiang, J. Kuszewski, M. Nilges, N.S. Pannu, R.J. Read, L.M. Rice, T. Simonson and G.L. Warren, *Acta Crystallogr. D Biol. Crystallogr.*, 1998, **54**, 905.
69. C.D. Schwieters, J.J. Kuszewski, N. Tjandra and G. Marius Clore, *J. Magn. Reson.*, 2003, **160**, 65.
70. N. Tjandra, J.G. Omichinski, A.M. Gronenborn, G.M. Clore and A. Bax, *Nat. Struct. Biol.*, 1997, **4**, 732.
71. A. Bax and N. Tjandra, *J. Biomol. NMR*, 1997, **10**, 289.
72. H.J. Sass, G. Musco, S.J. Stahl, P.T. Wingfield and S. Grzesiek, *J. Biomol. NMR*, 2000, **18**, 303.
73. J.J. Chou, S. Gaemers, B. Howder, J.M. Louis and A. Bax, *J. Biomol. NMR*, 2001, **21**, 377.
74. T. Cierpicki and J.H. Bushweller, *J. Am. Chem. Soc.*, 2004, **126**, 16259.
75. D.H. Jones and S.J. Opella, *J. Magn. Reson.*, 2004, **171**, 258.

76. M. Ottiger, F. Delaglio and A. Bax, *J. Magn. Reson.*, 1998, **131**, 373.
77. D. Yang, R.A. Venters, W.Y. Choy and L.E. Kay, *J. Biomol. NMR*, 1999, **14**, 333.
78. B. Vogeli, H. Kovacs and K. Pervushin, *J. Am. Chem. Soc.*, 2004, **126**, 2414.
79. G.M. Clore, M.R. Starich, C.A. Bewlwy, M. Cai and J. Kuszewski, *J. Am. Chem. Soc.*, 1999, **121**, 6513.
80. J.J. Chou, J.D. Kaufman, S.J. Stahl, P.T. Wingfield and A. Bax, *J. Am. Chem. Soc.*, 2002, **124**, 2450.
81. S.H. Park, A.A. Mrse, A.A. Nevzorov, M.F. Mesleh, M. Oblatt-Montal, M. Montal and S.J. Opella, *J. Mol. Biol.*, 2003, **333**, 409.
82. S. Lee, M.F. Mesleh and S.J. Opella, *J. Biomol. NMR*, 2003, **26**, 327.
83. S.C. Howell, M.F. Mesleh and S.J. Opella, *Biochemistry*, 2005, **44**, 5196.
84. M.E. Girvin, V.K. Rastogi, F. Abildgaard, J.L. Markley and R.H. Fillingame, *Biochemistry*, 1998, **37**, 8817.
85. T. Meier, P. Polzer, K. Diederichs, W. Welte and P. Dimroth, *Science*, 2005, **308**, 659.
86. M.F. Mesleh and S.J. Opella, *J. Magn. Reson.*, 2003, **163**, 288.
87. A. Arora, D. Rinehart, G. Szabo and L.K. Tamm, *J. Biol. Chem.*, 2000, **275**, 1594.
88. A. Pautsch and G.E. Schulz, *J. Mol. Biol.*, 2000, **298**, 273.
89. K. Otto, J. Norbeck, T. Larsson, K.A. Karlsson and M. Hermansson, *J. Bacteriol.*, 2001, **183**, 2445.
90. C. Fernandez, C. Hilty, G. Wider, P. Guntert and K. Wuthrich, *J. Mol. Biol.*, 2004, **336**, 1211.
91. L. Guo, K.B. Lim, C.M. Poduje, M. Daniel, J.S. Gunn, M. Hackett and S.I. Miller, *Cellular*, 1998, **95**, 189.
92. L.E. Kay, *J. Magn. Reson.*, 2005, **173**, 193.
93. J.G. Kempf and J.P. Loria, *Cell. Biochem. Biophys.*, 2003, **37**, 187.
94. R. Ishima and D.A. Torchia, *Nat. Struct. Biol.*, 2000, **7**, 740.
95. A.G. Palmer III, *Chem. Rev.*, 2004, **104**, 3623.
96. G. Ried and U. Henning, *FEBS Lett.*, 1987, **223**, 387.
97. L. Van Alphen, L. Havekes and B. Lugtenberg, *FEBS Lett.*, 1977, **75**, 285.
98. R. Morona, C. Kramer and U. Henning, *J. Bacteriol.*, 1985, **164**, 539.
99. A. Arora, D. Rinehart, G. Szabo and L.K. Tamm, *J. Biol. Chem.*, 2000, **275**, 1594.
100. N. Saint, C. El Hamel, E. De and G. Molle, *FEMS Microbiol. Lett.*, 2000, **190**, 261.
101. P.J. Bond, J.D. Faraldo-Gomez and M.S. Sansom, *Biophys. J.*, 2002, **83**, 763.
102. D. Kern, E.Z. Eisenmesser and M. Wolf-Watz, *Methods Enzymol.*, 2005, **394**, 507.
103. D. Kern and E.R. Zuiderweg, *Curr. Opin. Struct. Biol.*, 2003, **13**, 748.
104. V.E. Ahn, E.I. Lo, C.K. Engel, L. Chen, P.M. Hwang, L.E. Kay, R.E. Bishop and G.G. Prive, *Embo. J.*, 2004, **23**, 2931.

CHAPTER 7

Membrane Proteins Studied by Solid-State NMR

ADAM LANGE AND MARC BALDUS

Solid-state NMR, Department for NMR-based Structural Biology, Max-Planck-Institute for Biophysical Chemistry, Am Fassberg 11, 37077 Göttingen, Germany

1   Introduction

Nuclear magnetic resonance (NMR) probes the interaction of microscopic, nuclear magnetic moments (such as ^{1}H, ^{13}C, or ^{15}N spins) with a static magnetic field (B_0) and an external radio frequency (rf) field.[1] Solid-state NMR refers to applications where molecular size or chemical environment prohibit fast molecular tumbling in solution. As a result, the size and the orientation dependence of the nuclear spin interactions require the use of specialized instrumentation and methodology to study molecular structure and dynamics at atomic resolution. While such experiments are not influenced by protein size or the degree of structural order, applications in structural biology were, for a long time, hampered by spectral resolution and sensitivity. During the last decade, access to high-field NMR instruments, advancements in solid-state NMR methodology and novel sample preparation and isotope labelling techniques have significantly expanded the utility of solid-state NMR spectroscopy for biomolecular applications.

In the following chapter, we will discuss principles and recent applications of solid-state NMR spectroscopy in the context of membrane proteins. We introduce the two principal techniques to achieve high-resolution solid-state NMR conditions and summarize recent studies aiming at the determination of membrane protein structure and dynamics and their interactions with ligands. For more detailed information, the interested reader is referred to a series of recent reviews.[2–4]

2   Sample Preparation and Methodology

2.1   Isotope Labelling and Solid-State NMR Sample Preparation

Unless molecular size or mobility allow for the direct use of ^{1}H evolution and detection periods, isotope labelling (including ^{13}C, ^{15}N or ^{2}H) is mandatory if molecular structure is to be investigated by solid-state NMR techniques. Depending on the application of interest, chemical synthesis or a variety of expression systems are employed. In the case of protein expression in cell cultures, uniform isotope labelling is easily achieved using uniformly labelled starting media such as uniformly ^{13}C labelled glucose and $^{15}NH_4Cl$ (see, for example, Ref. 5,6) that can be obtained from commercial sources. Expression and purification efficiency are strong determinants on whether isotope-labelling is a viable option for the system of interest. More advanced labelling patterns can be generated if specifically labelled precursors, amino acids or growth media are supplied.[6] In the case of membrane proteins, the purified, uniformly labelled protein of interest is usually reconstituted into model membranes which can be studied as liposomes, possibly macroscopically oriented on solid glass or polymer supports.[7,8]

In general, spectral resolution is critical for solid-state NMR applications. As a result, two complementary experimental strategies have evolved for structural investigations of membrane proteins (Figure 1). The first, and perhaps most generally applicable method, relates to Magic Angle Spinning (MAS[9]) where the sample of interest is put in a sample container of typically 12–80 µL volume and rapidly spun (*i.e.*, 1–32 kHz) around the so-called magic angle in the static magnetic field. Under these conditions, the spectral broadening due to the anisotropic nature of the sample is minimized and high-resolution spectroscopy is possible. If membrane proteins are

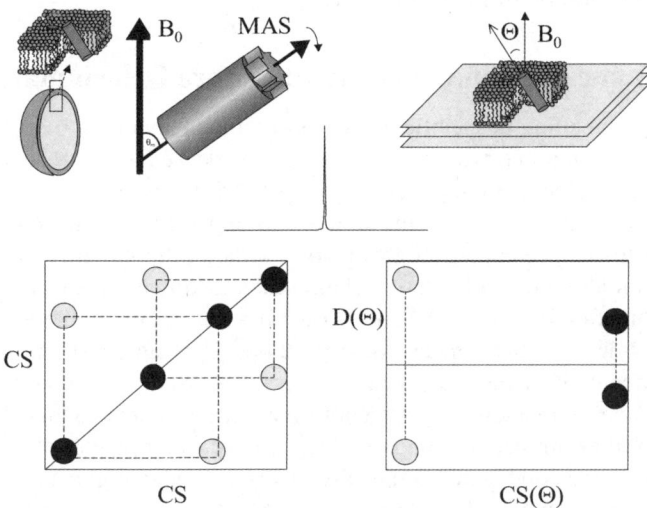

Figure 1 *The two principal experimental techniques to achieve high-resolution solid-state NMR conditions: Magic-Angle-Spinning (MAS) and macroscopic sample orientation. For MAS, the sample is spun rapidly around a rotation axis that is inclined by $\theta_m = 54.7°$ with respect to the static magnetic field. Proteoliposomes may correspond to small unilamellar vesicles in the case of MAS (left) or oriented lipid bilayers put on solid glass or polymer supports (right). Two-dimensional correlation experiments exist to study molecular structure using chemical shifts (CS) as a spectroscopic marker. In oriented systems, the measured values of CS and dipolar coupling (D) not only encode local structural topology but also the overall molecular arrangement (related to the tilt angle Θ) in the membrane*

under investigation, experiments can be conducted on frozen, detergent-solubilized protein, proteoliposomes and two-or three-dimensional nanocrystals.[10] In all cases, the sample of interest is transferred into the NMR rotor and experiments are conducted at variable temperature and MAS rate. As with other biophysical techniques, the choice of lipid and the molar ratio of protein to lipid are factors that, within the limits of spectroscopic sensitivity, are usually determined by biological activity.

Alternatively, membrane peptides or proteins can be macroscopically oriented in the magnetic field (Figure 1). Again, the anisotropic nature of a solid-state NMR spectrum is minimized and high-resolution spectroscopy becomes possible. There are two methods for mechanically aligning lipid bilayers[11]: deposition from organic solvents followed by evaporation and lipid hydration, and fusion of unilamellar reconstituted lipid vesicles with the glass or polymer surface. Maintaining a constant hydration level of the sample is critical. For this reason, stacked glass plates are in general placed in thin polymer films that achieve heat-sealing and hence stable sample hydration. In addition, bicelles have been used to study membrane proteins by NMR.[12] They represent molecular aggregates composed of long-chain phospholipids (such as 1,2-dimyristoyl-sn-3-glycerophosphocholine (DMPC)) and either short-chain lipids or surfactants. The long-chain lipids are organized into planar bilayers with the short-chain lipids arranged in a rim surrounding the bilayer edges. Bicellar solutions are lyotropic liquid crystalline solutions and can form a nematic phase that aligns in the magnetic field. The orientation of these bicelles has been shown to be affected by the addition of lanthanide ions.

2.2   Resonance Assignments and Structure Determination

Unless isotope-labelling is specific and/or assumptions about the overall secondary structure of the membrane protein of interest are made, sequential resonance assignments must be obtained. For applications using MAS, the NMR chemical shifts (CS, Figure 1) can be readily related to the amino acid sequence if multi-dimensional correlation experiments are conducted that invoke scalar or dipolar interactions to transfer polarization along the polypeptide chain. Such information is usually generated using a combination of $(^{13}C,^{15}N)^{13}$ and $(^{13}C,^{13}C)^{14}$ correlation methods. Since backbone CS provide a sensitive means of local dihedral angle constraints, these resonance assignments provide an easy and powerful instrument to assess polypeptide secondary structure. In addition to the conformation-dependent chemical shift,[15] distances (encoded as dipolar couplings D) can be used to refine the local polypeptide conformation in the solid state. In the case of macroscopic alignment, the chemical shift is a strong function of both local protein topology and the protein orientation (Θ) in the membrane (Figure 1). To detect the latter information in a reliable manner, tailored (often referred to PISEMA[16]-type) correlation methods have been developed. These two- and three-dimensional techniques are usually applied to a series of selectively ^{15}N-labelled polypeptides.[4] Of course, block-, pattern- or amino-acid-selective labelling (see, for example, Ref. 17) can also be applied under MAS conditions to reduce spectral overlap.

The most direct instrument to establish orientational constraints in solid-state NMR is the measurement of through-space $(^{13}C,^{13}C)$, $(^{13}C,^{15}N)$ or $(^{1}H,^{1}H)$ distances.[3,18]

Additional structural information becomes available if paramagnetic interactions[19] and hydrogen exchange is probed or if macroscopically oriented membrane proteins are studied under MAS conditions.[7,8,20] Such NMR parameters could furthermore be supplemented by an analysis of the hydrophobic mismatch,[21] which may strongly correlate with membrane protein topology.[22]

Once a sufficient number of local and overall orientational constraints have been obtained, a conventional structure calculation routine can be used to construct a local molecular structure or 3D membrane protein structure from solid-state NMR data. The number and precision of the experimentally derived solid-state NMR parameters determine the accuracy of the resulting 3D structure.

3 Applications

3.1 Membrane Protein Structure

MAS has long been employed to obtain site-specific structural information in membrane proteins such as bacteriorhodopsin or rhodopsin (for recent reviews, see for example Ref. 23, 24). In addition, MAS-based solid-state NMR studies were also conducted using selectively labelled membrane peptides[25] or peptides reconstituted into deuterated model lipids.[26,27] Such experimental conditions can also easily be modified for the case of macroscopically oriented systems.[20,28] To date, most complete 3D molecular arrangements of membrane embedded peptides have been obtained using macroscopic alignment methods. The first structure reported using solid-state NMR methods relates to Gramicidin A (PDB entry: 1MAG,[29]). Additional studies of membrane-embedded peptides relate to the M2 channel-lining segment from the nicotinic acetylcholine receptor (1CEK,[30]), the Fd bacteriophage coat protein in lipid bilayer membranes (1MZT,[31]), the closed state structure of M2 protein H+ channel (1NYJ,[32]), a peptide segment of the 6th *trans*-membrane domain of the *Saccharomyces cerevisiae* α-factor receptor (1PJD,[33]) and the structure of the channel-forming *trans*-membrane domain of virus protein "U" (Vpu) from HIV-1 (1PJE,[34]). These structures are summarized in Figure 2. Uniformly [^{13}C,^{15}N] labelled peptides studied recently by high-resolution MAS-based correlation techniques include neurotensin,[35] mastoparan-X[36] and a variety of fusion peptides.[37] Irrespective of whether measured in the gel or liquid–crystalline phase, secondary CS can report on local backbone conformation.[8,37,38]

To study membrane protein conformation or conformational changes, selective labelling by amino acid types has been used for a variety of membrane proteins. For example, Thompson *et al.* conducted solid-state NMR studies of bacterial chemoreceptors to measure helix–helix distances as a probe of local tertiary structure and structural changes that may mediate *trans*-membrane signalling.[39] Smith and coworkers used a combination of receptor and ligand binding to examine the coupling of retinal isomerization to the activation of rhodopsin.[40]

3.2 Ligand Binding to Membrane Proteins

G-protein-coupled receptors (GPCRs)[41] and ion channels[42] represent important membrane protein classes that are involved in cellular activity. They are known to

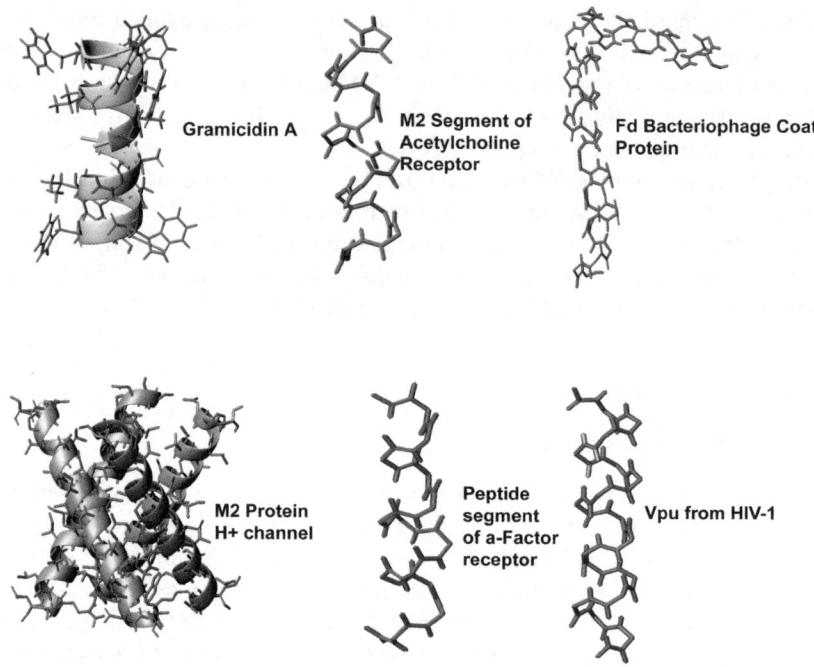

Gramicidin A

M2 Segment of
Acetylcholine
Receptor

Fd Bacteriophage Coat
Protein

M2 Protein
H+ channel

Peptide
segment
of a-Factor
receptor

Vpu from HIV-1

Figure 2 *Structures determined by solid-state NMR in oriented bilayers. See text or Ref. 4 for further details*

interact with a variety of physical stimuli or chemical reagents such as ions, ATP, or polypeptides. Because spectroscopic investigations of the ligand conformation do not require isotope-labelling of the membrane protein, solid-state NMR techniques are often ideally suited to investigate the ligand structure in the receptor-bound form in an efficient manner. For more than a decade, MAS solid-state NMR experiments have been conducted on selectively ^{13}C or ^{2}H labelled ligands to study ligand structure and dynamics (see, for example, Ref. 43).

Advancements in instrumentation and methodology now permit us to extend such investigations to uniformly labelled ligands. Using a uniformly ^{13}C-labelled retinal chromophore of rhodopsin,[44] changes in ^{1}H and ^{13}C CS in MAS solid-state NMR spectra were, for example, recently interpreted in terms of conformational rearrangements due to retinal-receptor interactions. In the case of polypeptides, CSs can be readily used to infer the backbone conformation of the bound ligand. Such a strategy was employed to determine the backbone conformation of neurotensin bound to its G-protein-coupled receptors NTS-1[45] and to monitor the folding pathway of neurotensin starting from an aqueous solution to a receptor-bound form.[35] Earlier, an NMR study in the presence of detergent had reported NT(8–13) CS changes upon receptor interaction.[46]

Moreover, solid-state NMR studies of neurotoxin II bound to nicotinic acetylcholine receptor have been reported.[47] Our group has used solid-state NMR to elucidate the structural details of K+ ion channel inhibition by Kaliotoxin (KTX), which has been shown to block voltage-dependent eukaryotic Kv (*e.g.*, Shaker and Kv1.3) channels in high affinity and with a 1:1 stoichiometry.[48] Using a KcsA–Kv1.3 chimeric channel that binds KTX with high affinity, we have found that residues shown to be important for high-affinity toxin binding exhibit significant solid-state NMR CS changes compared to free, solid phase KTX.[49] An analysis of the secondary and tertiary structure of bound KTX in complex with KcsA–Kv1.3 suggests that the backbone structure of KTX is preserved upon channel binding and that structural differences predominantly relate to toxin side chains.[50]

Furthermore, ligand/membrane protein interactions can be monitored, not only by using a labelled ligand in combination with an unlabelled membrane protein, but also by direct isotope labelling of the membrane protein (when isotope labelling is possible). Figure 3 shows a 2D ^{13}C spin diffusion spectrum of uniformly [^{13}C, ^{15}N] labelled KcsA–Kv1.3 potassium channel in complex with non-labelled KTX. The functional KcsA–Kv1.3 potassium channel is a homotetramer. Most of the 173

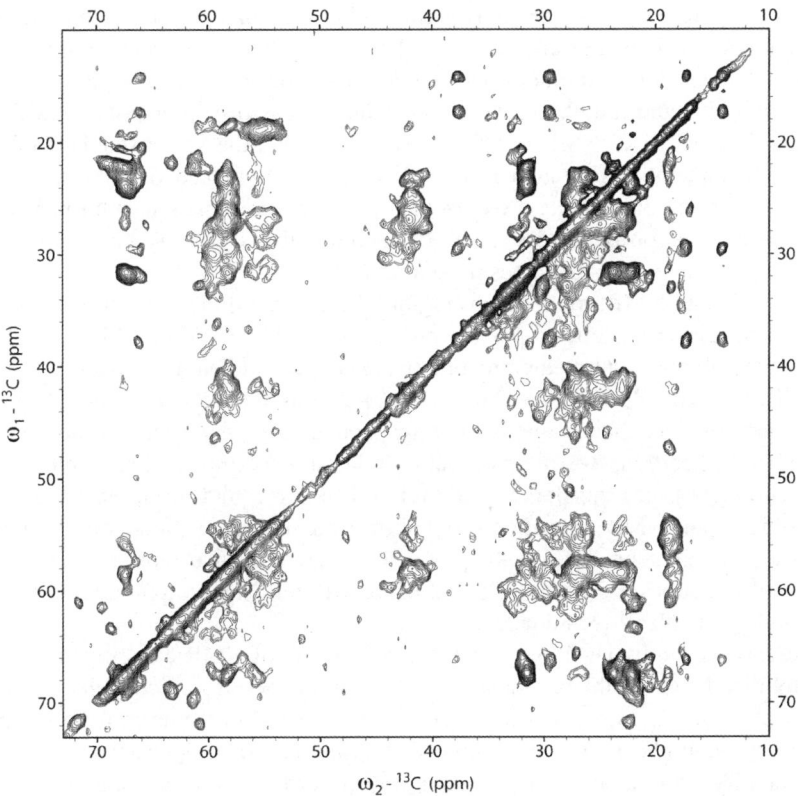

Figure 3 *2D ^{13}C spin diffusion spectrum of uniformly [^{13}C, ^{15}N] labelled KcsA–Kv1.3 potassium channel in complex with non-labelled Kaliotoxin. The experiment was conducted on a 600 MHz spectrometer using 9.375 kHz MAS and a longitudinal mixing time of 80 ms. The probe temperature was set to –25 °C (total acquisition time: 23 h). The spectrum reveals line widths well below 1 ppm and shows that the channel is in a well-defined mostly α-helical conformation*

residues in each subunit are in an α-helical conformation. Thus, the corresponding solid-state NMR correlations cluster in the spin diffusion spectrum. On the other hand, some 22 residues in the pore- and selectivity-filter region are not in an α-helical conformation and show well-separated signals. Their resonances exhibit widths well below 1 ppm, which indicates that the ion channel is in a well-defined conformation. Pore and selectivity region are known to constitute the binding site for KTX. Because the corresponding resonances are largely resolved, a variety of techniques could be applied for the structural characterization of the binding pocket. By comparing solid-state NMR results for KcsA–Kv1.3 with and without bound KTX, residues that are key players in the interaction could be identified.[50]

3.3   Membrane Protein Dynamics

For a long time, NMR has provided information on structural propensities and dynamics of partially disordered proteins in solution.[51,52] The occurrence of such unstructured regions is surprisingly common in functional proteins[53] and relevant in crucial areas such as transcriptional regulation, translation and cellular signal transduction.[53,54] As for membranes and membrane-associated proteins, the increase in molecular size and tumbling rate can prohibit the use of solution-state NMR techniques. Instead, MAS NMR has been shown to extend the use of standard solution-state NMR methods to characterize molecular structure and dynamics.[24,27,55] For larger membrane-interacting proteins, the molecular tumbling rate is further reduced and the application of solid-state NMR methods that explicitly take into account anisotropic interactions becomes mandatory.[4,23,24,56]

In solid-state NMR, motional averaging of the anisotropic interactions and modulation of relaxation times report on protein dynamics.[51,57] Under MAS, the topology of membranes and membrane proteins has been studied using one-dimensional ^{2}H, ^{13}C, ^{15}N and ^{31}P NMR experiments (see, for example, Refs. 4, 23, 24, 56, 58, 59), paramagnetic quenchers,[60] and by performing deuterium–hydrogen exchange experiments.[61] In addition, two-dimensional correlation experiments that employ dipolar (*i.e.*, through-space) transfers[62,63] and rely on the profound influence of molecular dynamics upon NMR relaxation[58,63,64] have been suggested. Since dipolar interactions are largely suppressed in the presence of fast molecular motion, protein structure and dynamics are often analyzed using different sample preparation methods and solid-state NMR techniques.

We have recently developed a complementary set of NMR experiments to detect immobilized and flexible segments of a membrane protein studied under MAS conditions.[65] Our experiments were demonstrated on the monomeric form of the 52 residue protein phospholamban (AFA–PLN) that plays an important role in cardiac contractility[66] and modulates the active transport of calcium into the lumen of the sarcoplasmic reticulum (SR) by the Ca–ATPase (SERCA). As indicated in Figure 4, tailored (^{1}H,^{13}C) and (^{13}C,^{13}C) correlation experiments suggest that a significant population of monomeric phospholamban exists in liposomes that is characterized by an α-helical *trans*-membrane segment and a cytoplasmic domain that exhibits a high degree of structural disorder. While these findings are in agreement with other biophysical[67] and biochemical[68] results, the existence of a further population of

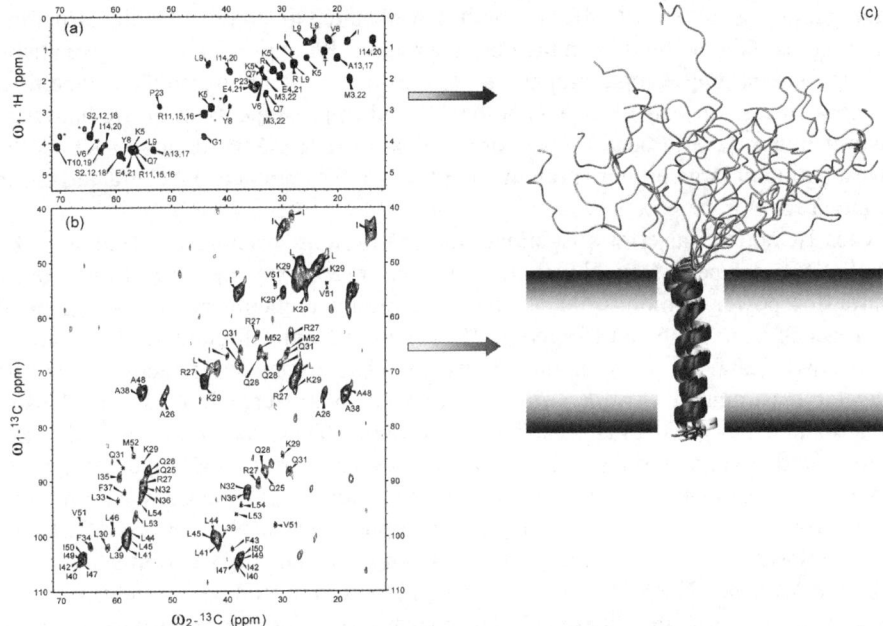

Figure 4 *Results of an HC–INEPT experiment (a) and a double quantum CC experiment (b) on a uniformly labelled sample of AFA–phospholamban (lipid/peptide = 20:1). Experiments were conducted at 600 MHz using MAS rates of 9 kHz and 7.5 kHz for Figure 4(a) and (b), respectively. Contributions from natural abundance lipid background are marked '*' in (a). (c) Structural model of AFA-phospholamban in lipid bilayers as seen by 2D MAS NMR. While the trans-membrane α-helix is buried in the membrane, the cytoplasmic N terminus exhibits a high degree of molecular disorder and is in close contact to an aqueous environment. An ensemble of 15 structures was selected to represent the molecular conformation of the PLN monomer. Figure produced using MOLMOL[73]*

membrane-associated phospholamban that contains an α-helical cytoplasmic domain as suggested by solution-state NMR[69] cannot be ruled out at present. Additional MAS-based correlation experiments that are sensitive to different motional regime are planned to elucidate these aspects in further detail.

Monitoring through-space and through-bond polarization transfer in two complementary sets of NMR experiments as shown in Figure 4, not only provides access to the study of structure and dynamics under physiological conditions but also simplifies the spectroscopic analysis. Ongoing studies in our laboratory show that the concept discussed not only can be applied to proteoliposomes containing the 52-residue polypeptide phospholamban but can also be extended to study protein dynamics in larger membrane proteins or protein fibrils.

4 Conclusions

Solid-state NMR has recently made considerable progress in the structural study of polypeptides and proteins. In addition to recent reports in the field of protein aggregation,[70] solid-state NMR is a method well suited to study molecular structure, dynamics and ligand binding in membrane proteins. Examples have been given here. Advances regarding sample preparation (for example, including modular labelling, *in vitro* expression and intein technology[71]) and improvements in NMR hardware instrumentation could open up new areas of solid-state NMR research such as the investigation of large protein–protein complexes or the complete 3D characterization of larger membrane proteins.

Clearly, additional efforts to streamline 3D protein structure determination by MAS NMR are necessary. Novel concepts may make extensive use of the rapidly increasing power of bioinformatics and computational chemistry. Already, general approaches have appeared to predict 3D structure from a limited set of solution-state NMR data. In parallel, the ability to relate molecular structure to NMR detectable parameters by *ab initio* quantum chemistry calculations and Density Functional Theory (DFT) approaches is improving. These techniques already have empowered novel material science applications of solid-state NMR and will greatly expand the use of MAS solid-state NMR for the study of polypeptides and proteins. With these advancements at hand, further applications in the context of *in vivo* structural studies may become possible. Complementary to solution-state methods,[72] MAS-based NMR may, for example, allow for the study of ligand-binding to membrane receptors in a dense cellular environment. Complementary to crystallographic methods and solution-state NMR techniques, MAS solid-state NMR methods hence provide a powerful tool to study protein structure, folding and function under biologically relevant conditions.

Acknowledgements

We thank our group members and collaborators for their contributions to the research described here. Financial support by the DFG and Volkswagen foundation is gratefully acknowledged. AL thanks the Fonds der Chemischen Industrie for a Ph.D. fellowship.

References

1. R.R. Ernst, G. Bodenhausen and A. Wokaun, *Principles of Nuclear Magnetic Resonance in One and Two dimensions*, Clarendon Press, Oxford, 1987.
2. L.K. Thompson, *Curr. Opin. Struct. Biol.*, 2002, **12**, 661–669.
3. S. Luca, H. Heise and M. Baldus, *Acc. Chem. Res.*, 2003, **36**, 858–865.
4. S.J. Opella and F.M. Marassi, *Chem. Rev.*, 2004, **104**, 3587–3606.
5. M. Kainosho and T. Tsuji, *Biochemistry*, 1982, **21**, 6273–6279; D.C. Muchmore, L.P. McIntosh, C.B. Russell, D.E. Anderson and F.W. Dahlquist, *Method Enzymol.*, 1989, **177**, 44–73.
6. N.K. Goto and L.E. Kay, *Curr. Opin. Struct. Biol.*, 2000, **10**, 585–592; L.Y. Lian and D.A. Middleton, *Prog. Nucl. Magn. Reson. Spectrosc.*, 2001, **39**, 171–190.
7. C. Glaubitz and A. Watts, *J. Magn. Reson.*, 1998, **130**, 305–316.

8. O.C. Andronesi, J.R. Pfeifer, L. Al-Momani, S. Özdirekcan, D.T.S. Rijkers, B. Angerstein, S. Luca, U. Koert, J.A. Killian and M. Baldus, *J. Biomol. NMR*, 2004, **30**, 253–265.
9. E.R. Andrew, A. Bradbury and R.G. Eades, *Nature*, 1958, **182**, 1659.
10. R.W.Martin and K.W. Zilm, *J. Magn. Reson.*, 2003, **165**, 162–174.
11. A.A. de Angelis, D.H. Jones, C.V. Grant, S.H. Park, M.F. Mesleh and S.J. Opella, *Method. Enzymol.*, T.L. James (ed), 2005, **394**, 350–382.
12. C.R. Sanders, B.J. Hare, K.P. Howard and J.H. Prestegard, *Prog. Nucl. Magn. Reson. Spectrosc.*, 1994, **26**, 421–444.
13. M. Baldus, *Prog. Nucl. Magn. Reson. Spectrosc.*, 2002, **41**, 1–47.
14. K. Seidel, A. Lange, S. Becker, C.E. Hughes, H. Heise and M. Baldus, *Phys. Chem. Chem. Phys.*, 2004, **6**, 5090–5093.
15. D.S. Wishart and B.D. Sykes, *Methods Enzymol.*, 1994, **239**, 363–392.
16. R. Jelinek, A. Ramamoorthy and S.J. Opella, *J. Am. Chem. Soc.*, 1995, **117**, 12348–12349.
17. C.E. Hughes and M. Baldus, *Annu. Rep. NMR spectrosc.*, 2005, **55**, 121–158.
18. R.G. Griffin, *Nat. Struct. Biol.*, 1998, **5**, 508–512; N.C. Nielsen, A. Malmendal and T. Vosegaard, *Mol. Membr. Biol.*, 2004, **21**, 129–141.
19. G. Grobner, C. Glaubitz and A. Watts, *J. Magn. Reson.*, 1999, **141**, 335–339.
20. C. Sizun and B. Bechinger, *J. Am. Chem. Soc.*, 2002, **124**, 1146–1147.
21. S.H. White and W.C. Wimley, *Annu. Rev. Biophys. Biomolec. Struct.*, 1999, **28**, 319–365.
22. S.H. Park and S.J. Opella, *J. Mol. Biol.*, 2005, **350**, 310–318.
23. S.O. Smith, K. Aschheim and M. Groesbeek, *Q. Rev. Biophys.*, 1996, **29**, 395–449.
24. J.H. Davis and M. Auger, *Prog. Nucl. Magn. Reson. Spectrosc.*, 1999, **35**, 1–84.
25. O.B. Peersen, S. Yoshimura, H. Hojo, S. Aimoto and S.O. Smith, *J. Am. Chem. Soc.*, 1992, **114**, 4332–4335; D.J. Hirsh, J. Hammer, W.L. Maloy, J. Blazyk and J. Schaefer, *Biochemistry*, 1996, **35**, 12733–12741.
26. J.H. Davis, M. Auger and R.S. Hodges, *Biophys. J.*, 1995, **69**, 1917–1932.
27. M. Bouchard, J.H. Davis and M. Auger, *Biophys. J.*, 1995, **69**, 1933–1938; W.Y. Zhang, E. Crocker, S. McLaughlin and S.O. Smith, *J. Biol. Chem.*, 2003, **278**, 21459–21466.
28. C. Glaubitz, *Concepts Magn. Resonance*, 2000, **12**, 137–151.
29. R.R. Ketchem, W. Hu and T.A. Cross, *Science*, 1993, **261**, 1457–1460.
30. S.J. Opella, F.M. Marassi, J.J. Gesell, A.P. Valente, Y. Kim, M. Oblatt-Montal and M. Montal, *Nat. Struct. Biol.*, 1999, **6**, 374–379.
31. F.M. Marassi and S.J. Opella, *Protein Sci.*, 2003, **12**, 403–411.
32. J.F. Wang, S. Kim, F. Kovacs and T.A. Cross, *Protein Sci.*, 2001, **10**, 2241–2250.
33. K.G. Valentine, S.F. Liu, F.M. Marassi, G. Veglia, S.J. Opella, F.X. Ding, S.H. Wang, B. Arshava, J.M. Becker and F. Naider, *Biopolymers*, 2001, **59**, 243–256.
34. S.H. Park, A.A. Mrse, A.A. Nevzorov, M.F. Mesleh, M. Oblatt-Montal, M. Montal and S.J. Opella, *J. Mol. Biol.*, 2003, **333**, 409–424.
35. H. Heise, S. Luca, B.L. de Groot, H. Grubmuller and M. Baldus, *Biophys. J.*, 2005, **89**, 2113–2120.

36. T. Fujiwara, Y. Todokoro, H. Yanagishita, M. Tawarayama, T. Kohno, K. Wakamatsu and H. Akutsu, *J. Biomol. NMR*, 2004, **28**, 311–325.
37. M.L. Bodner, C.M. Gabrys, P.D. Parkanzky, J. Yang, C.A. Duskin and D.P. Weliky, *Magn. Reson. Chem.*, 2004, **42**, 187–194.
38. P. Barre, O. Zschornig, K. Arnold and D. Huster, *Biochemistry*, 2003, **42**, 8377–8386.
39. O.J. Murphy, F.A. Kovacs, E.L. Sicard and L.K. Thompson, *Biochemistry*, 2001, **40**, 1358–1366.
40. A.B. Patel, E. Crocker, M. Eilers, A. Hirshfeld, M. Sheves and S.O. Smith, *Proc. Natl. Acad. Sci. USA*, 2004, **101**, 10048–10053.
41. U. Gether and B.K. Kobilka, *J. Biol. Chem.*, 1998, **273**, 17979–17982; T.H. Ji, M. Grossmann and I.H. Ji, *J. Biol. Chem.*, 1998, **273**, 17299–17302.
42. B. Hille, *Ionic Channels of Excitable Membranes,* 3 edn, Sinauer Associates Inc., Sunderland, MA, 2001; O. Pongs, *FEBS Lett.*, 1999, **452**, 31–35; K. George Chandy, H. Wulff, C. Beeton, M. Pennington, G.A. Gutman and M.D. Cahalan, *Trends Pharmacol. Sci.*, 2004, **25**, 280–289.
43. V. Copie, A.E. McDermott, K. Beshah, J.C. Williams, M. Spijkerassink, R. Gebhard, J. Lugtenburg, J. Herzfeld and R.G. Griffin, *Biochemistry*, 1994, **33**, 3280–3286; A. Watts, *Curr. Opin. Biotechnol.*, 1999, **10**, 48–53; J. Herzfeld and J.C. Lansing, *Annu. Rev. Biophys. Biomolec. Struct.*, 2002, **31**, 73–95; M. Eilers, W.W. Ying, P.J. Reeves, H.G. Khorana and S.O. Smith, *Method Enzymol.*, 2002, **343**, 212–222.
44. A.F.L. Creemers, S. Kiihne, P.H.M. Bovee-Geurts, W.J. DeGrip, J. Lugtenburg and H.J.M. de Groot, *Proc. Natl. Acad. Sci. USA*, 2002, **99**, 9101–9106.
45. S. Luca, J.F. White, A.K. Sohal, D.V. Filippov, J. H. van Boom, R. Grisshammer and M. Baldus, *Proc. Natl. Acad. Sci. USA*, 2003, **100**, 10706–10711.
46. P.T.F. Williamson, S. Bains, C. Chung, R. Cooke and A. Watts, *FEBS Lett.*, 2002, **518**, 111–115.
47. L. Krabben, B.J. van Rossum, F. Castellani, E. Bocharov, A.A. Schulga, A.S. Arseniev, C. Weise, F. Hucho and H. Oschkinata, *FEBS Lett.*, 2004, **564**, 319–324.
48. S. Grissmer, A.N. Nguyen, J. Aiyar, D.C. Hanson, R.J. Mather, G.A. Gutman, M.J. Karmilowicz, D.D. Auperin and K.G. Chandy, *Mol. Pharmacol.*, 1994, **45**, 1227–1234; J. Aiyar, J.M. Withka, J.P. Rizzi, D.H. Singleton, G.C. Andrews, W. Lin, J. Boyd, D.C. Hanson, M. Simon, B. Dethlefs, C.L. Lee, J.E. Hall, G.A. Gutman and K.G. Chandy, *Neuron*, 1995, **15**, 1169–1181.
49. A. Lange, S. Becker, K. Seidel, K. Giller, O. Pongs and M. Baldus, *Angew. Chem.-Int. Edit.*, 2005, **44**, 2089–2092.
50. A. Lange, K. Giller, S. Hornig, M.F. Martin-Eauclaire, O. Pongs, S. Becker and M. Baldus, 2006, submitted.
51. A.G. Palmer, J. Williams and A. McDermott, *J. Phys. Chem.*, 1996, **100**, 13293–13310.
52. H.J. Dyson and P.E. Wright, *Chem. Rev.*, 2004, **104**, 3607–3622.
53. A.K. Dunker, C.J. Brown, J.D. Lawson, L.M. Iakoucheva and Z. Obradovic, *Biochemistry*, 2002, **41**, 6573–6582.
54. H.J. Dyson and P.E. Wright, *Nat. Rev. Mol. Cell Bio.,* 2005, **6**, 197–208.

55. E. Oldfield, J.L. Bowers and J. Forbes, *Biochemistry*, 1987, **26**, 6919–6923; D. Huster, K. Kuhn, D. Kadereit, H. Waldmann and K. Arnold, *Angew. Chem.-Int. Edit.*, 2001, **40**, 1056–1057; O. Soubias, V. Reat, O. Saurel and A. Milon, *J. Magn. Reson.*, 2002, **158**, 143–148.

56. T.A. Cross and S. J. Opella, *Curr. Opin. Struct. Biol.*, 1994, **4**, 574–581; J.M. Griffiths and R.G. Griffin, *Anal. Chim. Acta*, 1993, **283**, 1081–1101.

57. J.W. Peng and G. Wagner, *Method. Enzymol.*, T.L. James and N.J. Oppenheimer (eds), 1994, **239**, 563–596.

58. M.A. Keniry, H.S. Gutowsky and E. Oldfield, *Nature*, 1984, **307**, 383–386.

59. B.A. Lewis, G.S. Harbison, J. Herzfeld and R.G. Griffin, *Biochemistry*, 1985, **24**, 4671–4679; J.L. Bowers and E. Oldfield, *Biochemistry*, 1988, **27**, 5156–5161; S. Tuzi, A. Naito and H. Saito, *Biochemistry*, 1994, **33**, 15046–15052; M. Engelhard, S. Finkler, G. Metz and F. Siebert, *Eur. J. Biochem.*, 1996, **235**, 526–533; F. Creuzet, A. McDermott, R. Gebhard, K. Vanderhoef, M.B. Spijkerassink, J. Herzfeld, J. Lugtenburg, M.H. Levitt and R.G. Griffin, *Science*, 1991, **251**, 783–786; A.J. Mason, G.J. Turner and C. Glaubitz, *FEBS J.*, 2005, **272**, 2152–2164.

60. L.R. Brown, W. Braun, A. Kumar and K. Wuthrich, *Biophys. J.*, 1982, **37**, 319–328; J. Villalain, *Eur. J. Biochem.*, 1996, **241**, 586–593; S. Tuzi, J. Hasegawa, R. Kawaminami, A. Naito and H. Saito, *Biophys. J.*, 2001, **81**, 425–434.

61. C.L. Tian, P.F. Gao, L.H. Pinto, R.A. Lamb and T.A. Cross, *Protein Sci.*, 2003, **12**, 2597–2605.

62. J.D. Gross, D.E. Warschawski and R.G. Griffin, *J. Am. Chem. Soc.*, 1997, **119**, 796–802; D. Huster, X.L. Yao and M. Hong, *J. Am. Chem. Soc.*, 2002, **124**, 874–883.

63. K.K. Kumashiro, K. Schmidt-Rohr, O.J. Murphy, K.L. Ouellette, W.A. Cramer and L.K. Thompson, *J. Am. Chem. Soc.*, 1998, **120**, 5043–5051.

64. D.A. Torchia, *Ann. Rev. Biophys. Bioeng.*, 1984, **13**, 125–144.

65. O.C. Andronesi, S. Becker, K. Seidel, H. Heise, H.S. Young and M. Baldus, *J. Am. Chem. Soc.*, 2005, **127**,12965–12974.

66. D.H. MacLennan and E.G. Kranias, *Nat. Rev. Mol. Cell Bio.*, 2003, **4**, 566–577.

67. C.B. Karim, T.L. Kirby, Z.W. Zhang, Y. Nesmelov and D.D. Thomas, *Proc. Natl. Acad. Sci. U.S.A.*, 2004, **101**, 14437–14442; J.H. Li, Y.J. Xiong, D.J. Bigelow and T.C. Squier, *Biochemistry*, 2004, **43**, 455–463.

68. L.G. Reddy, L.R. Jones, S.E. Cala, J.J. Obrian, S.A. Tatulian and D.L. Stokes, *J. Biol. Chem.*, 1995, **270**, 9390–9397; Y. Kimura, K. Kurzydlowski, M. Tada and D.H. MacLennan, *J. Biol. Chem.*, 1996, **271**, 21726–21731; C.B. Karim, C.G. Marquardt, J.D. Stamm, G. Barany and D.D. Thomas, *Biochemistry*, 2000, **39**, 10892–10897; Y. Kimura, M. Asahi, K. Kurzydlowski, M. Tada and D.H. MacLennan, *FEBS Lett.*, 1998, **425**, 509–512; N.A. Lockwood, R.S. Tu, Z.W. Zhang, M.V. Tirrell, D.D. Thomas and C.B. Karim, *Biopolymers*, 2003, **69**, 283–292; T. Toyofuku, K. Kurzydlowski, M. Tada and D.H. Maclennan, *J. Biol. Chem.*, 1994, **269**, 3088–3094; D.H. MacLennan, M. Abu-Abed and C. Kang, *J. Mol. Cell. Cardio.*, 2002, **34**, 897–918.

69. E.E. Metcalfe, J. Zamoon, D.D. Thomas and G. Veglia, *Biophys. J.*, 2004, **87**, 1205–1214.

70. A.T. Petkova, R.D. Leapman, Z. Guo, W.-M. Yau, M.P. Mattson and R. Tycko, *Science*, 2005, **307**, 262–265; C. Ritter, M.-L. Maddelein, A.B. Siemer, T. Luhrs, M. Ernst, B.H. Meier, S.J. Saupe and R. Riek, *Nature*, 2005, **435**, 844–848; H. Heise, W. Hoyer, S. Becker, O.C. Andronesi, D. Riedel and M. Baldus, *Proc. Natl. Acad. Sci. USA*, 2005, **102**, 15871–15876.

71. D. Staunton, J. Owen and I.D. Campbell, *Acc. Chem. Res.*, 2003, **36**, 207–214.

72. Z. Serber, L. Corsini, F. Durst and V. Dotsch, *Method. Enzymol.*, T.L. James (ed), 2005, **394**, 17–41.

73. R. Koradi, M. Billeter and K. Wüthrich, *J. Mol. Graph.*, 1996, **14**, 51–60.

CHAPTER 8

Assessing Structure and Dynamics of Native Membrane Proteins

W. KUKULSKI, T. KAUFMANN, T. BRAUN, H. RÉMIGY,
D. FOTIADIS AND A. ENGEL

M.E. Müller Institute for Structural Biology, Biozentrum, University of Basel,
Klingelbergstrasse 70, 4056 Basel, Switzerland

Abstract

Membrane proteins are studied in their native state by reconstituting two-dimensional crystals in the presence of lipids and assessing their atomic structure by electron crystallography and the surface dynamics by atomic force microscopy (AFM).

1 Introduction

More than 30% of eukaryotic proteomes represent membrane proteins.[1] They fulfill key functions such as energy conversion, solute transport, secretion, and signal transduction, and exhibit characteristic dimensions of 5–10 nm. A wide range of membrane protein-related diseases and their central role in cell–cell communication might explain the fact that 30% of all drug targets are G-protein-coupled receptors (GPCRs).[2] While the structures of more than 10,000 soluble proteins are solved, the number of membrane protein structures is smaller than 100. The lack of structural information is related to the instability of membrane proteins in a detergent-solubilized state. Two-dimensional (2D) crystals of membrane proteins reconstituted in the presence of lipids provide an environment that allows the function and native structure of membrane proteins to be assessed. Electron crystallography is used to establish the 3D structure. A modern electron microscope transfers information at atomic resolution and allows the amplitude and phase of diffracted electrons to be determined. 2D crystallization methods are steadily improving and can now yield highly ordered crystals of several microns in size, which allow structure determination at 3 Å or better.[3,4] Atomic force microscopy (AFM) produces images with an outstanding signal-to-noise ratio and addresses single molecules under native conditions, keeping the sample in buffer solution. Progress in sample preparation and instrumentation has led to topographs that reveal sub-nanometer details and surface dynamics of membrane proteins.[5] In addition, AFM is the ideal method to directly image the topography of native membranes.[6,7] This chapter presents methods to grow 2D crystals, to record them by cryo-electron microscopy and to extract the structural information by digital image processing. AFM techniques are discussed and relevant results presented. The chapter is intended to provide a compact description of the current possibilities offered by these methods and their power is demonstrated by the presented examples.

2   Assembly of 2D Crystals

2D crystals consisting of membrane proteins and lipids are produced by different methods.[8] The first method induces a regular packing of a highly abundant protein in its native membrane by eliminating interspersed lipids using lipases,[9] or specific detergents.[10] This is the most gentle 2D crystallization (2DX) method, because it does not require solubilization of the membrane protein, but it is unfortunately not generally applicable.

The second method reconstitutes the purified membrane protein into a lipid bilayer at high protein density.[11] Detergent-solubilized proteins are mixed with solubilized lipids and the detergent is removed. This results in the formation of protein aggregates in the worst case, and in the progressive formation of proteoliposomes with large 2D crystalline regions in the best case. The respective affinities between the components of the ternary mixture dictate the progress of the reconstitution process. Ideally, a starting condition should be established where mixed detergent-protein and mixed detergent-lipid micelles have exchanged their constituents to the extent that the mixture consists mainly of ternary detergent-protein-lipid micelles.

The third method concerns the reconstitution of the membrane proteins at the air–water interface by attaching the solubilized membrane protein to an active lipid monolayer prior to detergent removal.[12] In this process, membrane proteins are concentrated at the monolayer, brought into a planar configuration and finally squeezed together during detergent removal. This approach is useful for membrane proteins that are present in small amounts and are stably solubilized only in low CMC detergents.[13]

All methods (Figure 1) have in common that the detergent is brought below its CMC to foster assembly of a bilayer, into which the membrane protein should integrate. Generally used methods to bring the detergent concentration below the CMC include dialysis,[14] adsorption of the detergent to Bio-Beads,[15] and dilution of the ternary mixture.[16] In all methods the amount of interspersed lipid must be minimized to ensure regular interactions between the membrane proteins. The pertinent interactions depend on the shape and surface charges of the components. For a given protein, the lipid-detergent mixture, pH, counter ions, and temperature must all be optimized. In addition, the concentration, the ratio of the respective components, and the detergent removal rate are critical. This gives a multidimensional parameter space that needs to be experimentally sampled, a similar task to that carried out in 3D crystallization screens. The difficulty of such experiments is the management of the screens and the

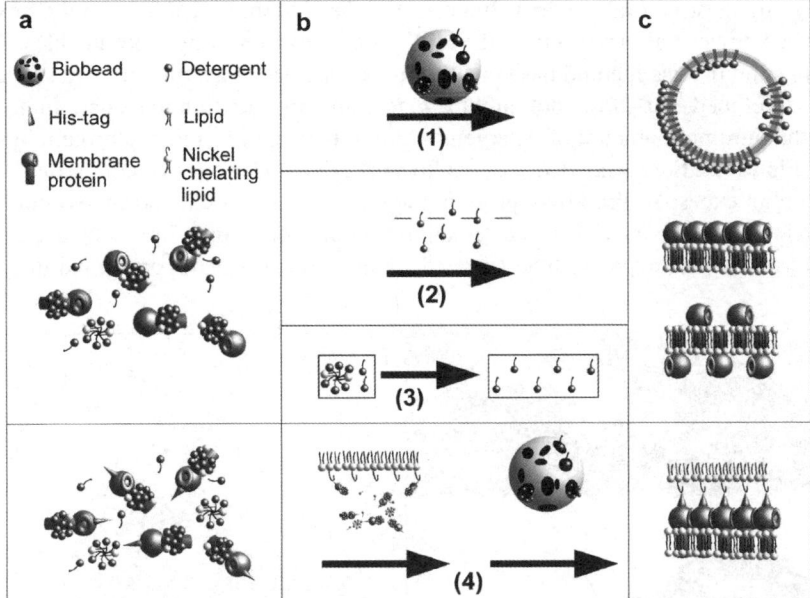

Figure 1 *2DX methods. (a) Starting conditions: detergent solubilized proteins are mixed with lipids/detergent micelles and equilibrated to achieve partitioning of lipids to the protein moiety. (b) Detergent removal using (1) Biobeads,[59] (2) dialysis,[19] (3) dilution,[16] and (4) Biobeads combined with a monolayer at the air–water interface, which binds His-tagged proteins.[12] (c) The resulting 2D crystals can exist as a vesicular or a sheet-like assembly, or they are trapped at the air–water interface*

assessment of results. With 2DX the latter is particularly cumbersome because 2D crystals cannot be detected by light microscopy and screening by electron microscopy is time consuming. Nevertheless, 2DX is well worth the effort, since 2DX methods are likely to foster perfect reconstitution and possibly highly ordered 2D crystals of a functional membrane protein, provided that the protein remains in its native, properly folded state during the solubilization and isolation steps.

Critical in the 2D crystallization process is the time point when the lipid molecules replace the detergent molecules to stabilize the membrane protein.[17] The adsorption–desorption kinetics of the detergent molecule are not necessarily the same for the mixed protein-detergent or the lipid-detergent micelles, respectively. As indicated above, the chance for the protein to interact with the appropriate lipids early enough increases when the crystallization cocktail can be incubated before detergent removal. If the interactions between the detergent-protein and detergent-lipid complexes are not properly balanced during the reconstitution, nonspecific aggregation occurs and the protein is lost.

In general, membrane proteins are destabilized upon solubilization,[13] especially when short-chain, high CMC detergents are used. Smaller but harsher detergents are an advantage for dialysis-driven 2DX, while low CMC and thus mild detergents are not easily removed. Thus, the choice of detergent is critical: there is a fine balance between disruption of the membrane to solubilize a membrane protein and preserving its structural integrity. Importantly, the amount of detergent in the mixture must be known at all times during the 2DX process for achieving reproducible results. A rapid and precise method based on the shape analysis of a sitting drop has recently been developed.[18] It allows not only the detergent concentration to be determined but also the minimum amount of detergent required to keep a membrane protein in solution (Figure 2). Such knowledge is key to start 2DX under optimal conditions.

With an excess of lipid over protein, the protein is mainly incorporated into lipid bilayers, similar to its native state. In an excess of protein over lipid, however, some of the protein aggregates, likely in a denatured form. An important parameter is

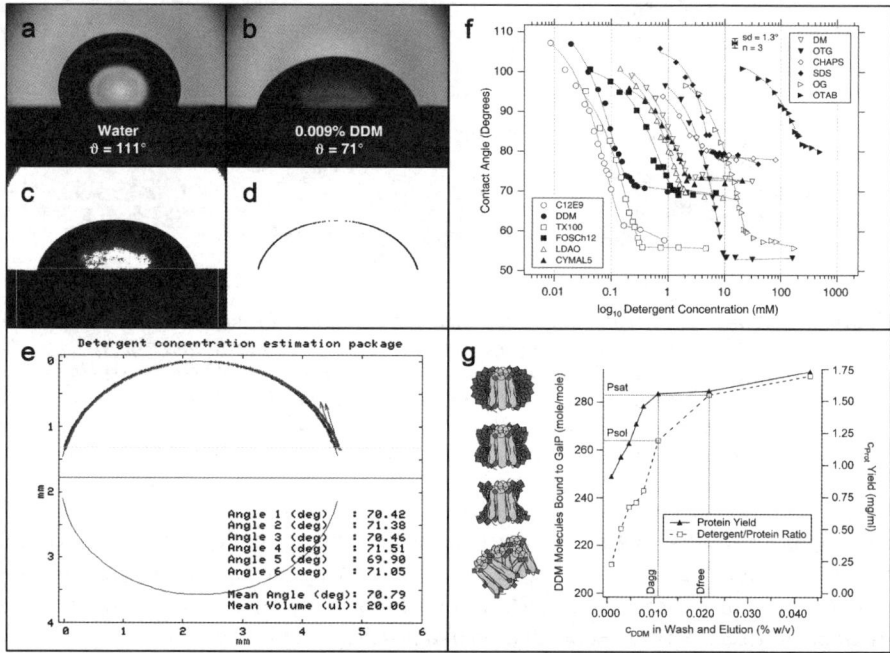

Figure 2 *Detergent concentration determination. Sideviews of 20 μl droplets deposited on Parafilm M containing no detergent or 0.009% (w/v) dodecyl-β,D-maltoside (DDM) are displayed in (a) and (b) respectively. Image processing steps are further outlined for image (b). (c) Image with applied threshold and baseline. (d) Extracted droplet contour. (e) Output file from GNUPLOT displaying contact angles and mean volume. (f) Semi-logarithmic plot of the detergent concentration vs. experimental contact angles for all calibrated detergents. (g) Ni-NTA affinity chromatography with galactose/proton symporter of E. coli (GalP) using washes of different DDM concentrations. The amount of DDM bound to each GalP monomer is monitored together with the protein yield from the column. Based on the assumption that the protein is saturated with detergent after solubilization (P_{sat}), one would expect that when lowering the detergent concentration in the wash, the detergent-to-protein ratio (DPR) would decrease before the protein elution yield decreases. This is the range where excess detergent molecules are drawn off the protein without affecting its solubility (between D_{free} and D_{agg}). When a critical DPR is reached (P_{sol}) the protein yield starts to decrease too, indicating that part of the protein aggregates and that there is not enough detergent to keep all the protein soluble*

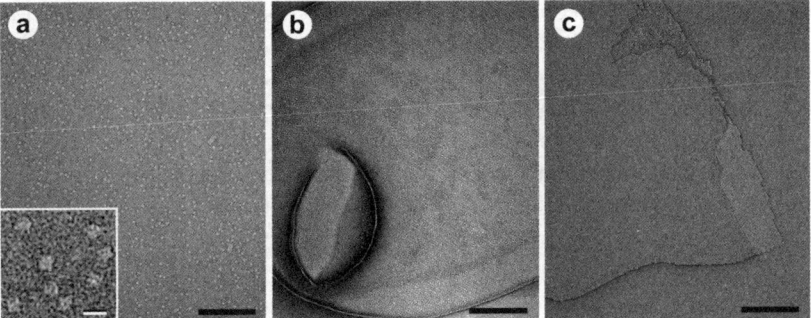

Figure 3 *EM analysis of protein sample quality and 2DX results by negative staining of the plant aquaporin SoPIP2;1. (a) Detergent-solubilized SoPIP2;1 particles. In the inset at higher magnification, the square shape of the particles indicates their tetrameric state. (b) Vesicles consisting of rows that contain alternating SoPIP2;1 tetramers from their extracellular and intracellular side. (c) Double-layered crystal sheets of SoPIP2;1 tetramers. The two sheets face each other with the extracellular side, which exhibits a higher contrast compared to the cytoplasmic side on the surface of the double layer, as seen at surface ruptures. Scale bars in (a), (b), and (c) represent 200 nm. Scale bar in the inset represents 10 nm*

therefore the lipid-to-protein ratio (LPR), which should be low enough to promote crystal contacts between protein molecules, but not so low that the protein is lost in aggregation.[19] In addition to an appropriate lipid and detergent combination, 2DX of many membrane proteins depends on additives such as divalent ions, among them Mg^{2+} having a special role, likely as a result of its interaction with the lipids.[11]

As an example, the 2DX of the plant aquaporin SoPIP2;1 is illustrated in Figure 3.[20] SoPIP2;1 was expressed both as *His-* and non-tagged protein in the methylotrophic yeast *Pichia pastoris*. Both clones express SoPIP2;1 at similarly high levels. When reconstituted into proteoliposomes and exposed to an osmotic gradient, recombinant SoPIP2;1 shows efficient water channel activity. Both *His-* and non-tagged SoPIP2;1 yielded homogeneous preparations of purified protein tetramers free of aggregates (Figure 3a), and both could be reconstituted into several crystal forms. Non-tagged SoPIP2;1 crystallized in two forms, tubular crystals and sheets. Tubular vesicles exhibited a specific surface texture, resulting from up- and down-oriented tetramers that are packed into alternating rows (Figure 3b), but such crystals were anisotropically ordered, thus not suitable for high-resolution structural analysis. Membrane sheets, however, that were mostly double-layered, were well ordered in all directions, exhibited p4 symmetry and a lattice constant of 65 Å (Figure 3c).

3 Electron Microscopy

3.1 Image Formation

Electrons are elastically scattered by the nuclei of the atoms, which are orders of magnitude heavier than the moving electrons. Electrons are inelastically scattered by the inner- and outer-shell electrons, to which they transmit a fraction of their kinetic energy. Whereas elastic electrons contribute to the coherent axial bright-field image that carries the high-resolution information on the 3D arrangement of the sample atoms, the inelastic electrons carry interesting chemical information. Importantly, inelastic scattering is directly related to the beam-induced specimen degradation. Here, we consider the coherent phase contrast image formation, since the elastically scattered electrons alone contribute to a high-resolution image. A thin object comprising only light elements (such as 2D membrane protein crystals) is a weak-phase object

$$t(x,y) = 1 + i\varphi(x,y), \quad \varphi(x,y) < \frac{\pi}{4} \tag{1}$$

The amplitude distribution in the image plane is the coherent superposition of the unscattered wave and the elastically scattered waves, which is conveniently described as the convolution of the function $t(x,y)$ describing the object with the point spread function $h(x,y)$ describing the electron optical system

$$a(x,y) = h(x,y)((1 + i\varphi(x,y)) \tag{2}$$

Intensities $|a(x,y)|^2$ are recorded on the film. Omitting small quadratic terms, the image is thus written as

$$|a(x,y)|^2 = 1 - 2h_i(x,y)\varphi(x,y) \tag{3}$$

The imaginary part $h_i(x,y)$ is the inverse Fourier transform of the phase contrast transfer function (CTF)

$$h_i(x,y) = FT^{-1}[A(p)\sin(\pi(C_s\lambda^3 p^{4/2} + \Delta f\lambda p^2))] \tag{4}$$

where $A(p)$ the envelope of the CTF, C_s the spherical aberration constant, Δf the defocus, λ the electron wavelength (about 0.02 Å for 300 kV electrons), and p the distance from the origin in the reciprocal space.

Figure 4 shows the CTF for weak-phase objects. The contrast is weak when the microscope is operated close to focus because the prominent low-resolution features of the specimen are transferred with small amplitude. The contrast can be enhanced by moving out of focus, because the phase difference between the scattered and unscattered electrons becomes $\pi/2$ or $\pi(n + 1/2)$. The phase shift of electrons scattered elastically by an atom is proportional to the coulomb-potential of this atom. Therefore, the ensemble of all electrons singly scattered by a specimen produces a projection of its coulomb potential, which is dominated by the atom's nuclei rather than the electron shells as in X-ray crystallography.

The power spectrum of an axial bright field image of a 2D crystal (Figure 5) reveals (i) the discrete spots representing the crystal information (see image processing), (ii) the CTF, *i.e.* bright concentric rings with gaps in between (Thon rings), and (iii) the envelope modulation decreasing the signal intensity at higher resolutions. The decrease of contrast toward high resolution (envelope function) results

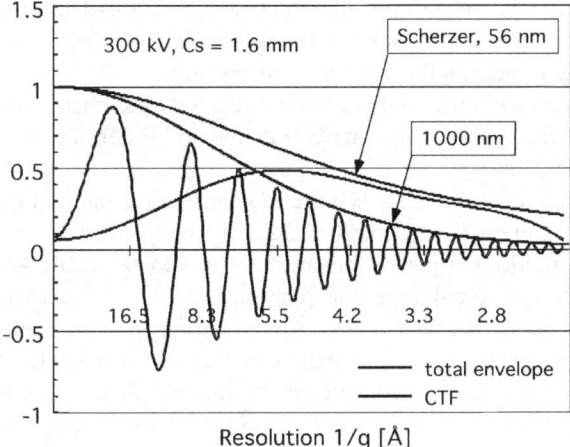

Figure 4 *CTF of a magnetic lens. The spherical aberration can be partially compensated by operating the lens at Scherzer focus. For a 300 kV instrument having a $C_s = 1.6$ mm, the Scherzer focus is at $\Delta f = 56$ nm (underfocus). Under this condition spatial frequencies are transferred with identical sign up to a resolution better than 3 Å. Contrast variations having spatial frequencies below $(15 \text{ Å})^{-1}$ are strongly attenuated, hence micrographs recorded at Scherzer focus exhibit weak contrast. A stronger contrast is obtained by further underfocusing the lens. In this case, the fine structures are transmitted with opposite signs. The envelope function resulting from the source size and stability of the electron gun limits the contrast transfer at high resolution*

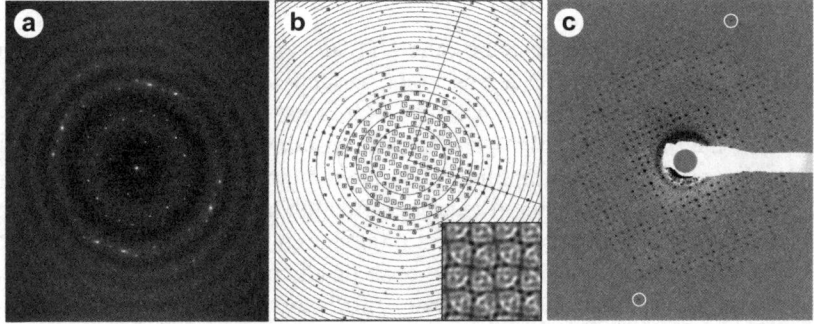

Figure 5 *Calculated diffraction of an electron micrograph and electron diffraction of 2D crystals. (a) Calculated power spectrum ($|modulus|^2$ of the Fourier transform) of a high-resolution image of a 2D crystal of SoPIP2;1. The periodic arrangement of the protein in the 2D crystal leads to a periodic signal in the recorded image. Therefore, the power spectrum shows discrete sharp spots in which the structural information of the protein is concentrated. Furthermore, noise (all information outside the spots) and the CTF (see Figure 4) are visible.(b) Interpreted calculated Fourier transform, where numbers indicate the signal strength of diffraction spots and the zero crossings of the fitted CTF are drawn. (c) Electron diffraction of a highly ordered 2D crystal, which is not affected by the CTF but does not carry phase information. Diffraction spots at the periphery of the pattern represent a resolution of $(3 \text{ Å})^{-1}$*

from the partial incoherence of the electron beam. Modern electron microscopes are equipped with field emission guns exhibiting a high degree of coherence so that the contrast decrease is acceptable even at high resolution. The Thon rings with alternating positive and negative contrast are the result of the phase shift introduced by the electron optical system. This phase shift has to be corrected for correct image interpretation (see below).

The great advantage to directly acquire the phase information out to atomic scale resolution in a modern field-emission electron microscope is diluted by several experimental difficulties. First, the instrument has to be stable and installed in a field- and vibration-free environment, prerequisites that are not routinely reached. Second, beam-induced damage is not only changing the specimen structure but it also leads to specimen charging. Since these charges act like an electrostatic lens, the focus changes during image recording. If the sample plane is perpendicular to the optical axis, the overall effect is quite small: the focus change occurring during image acquisition may have an influence only at very high resolution. However, when the sample is tilted for collecting the 3D information (see below), the electrostatic lens building up during irradiation introduces an image shift, a problem that cannot always be solved satisfactorily.[21] Third, a further optical defect is related to the changing focus for tilted specimens. Here, the point spread function is not space invariant, and the commonly used CTF correction is only an approximate measure to eliminate the phase distortions of the electron optical system. Last but not least the depth of focus must be considered and it limits the thickness of the object to be imaged at high resolution. This limitation is not of concern if thin samples such as 2D crystals are studied with a 300 kV instrument.

3.2   Electron Diffraction

Electron diffraction is not affected by the CTF, the envelope function, or specimen charging, but it requires the 2D crystals to be perfectly ordered and of sufficient size. Electron diffraction is much more effective for the collection of high-resolution information than imaging, although the phase information is not retrieved (Figure 5c). The directly measured amplitudes can be combined with the phases from the images during image processing. Electron diffraction is not an absolute requirement for determining a structure, but it allows a fast judgment of the crystal quality and helps correcting the CTF. Importantly, electron diffraction of highly ordered 2D crystals provides suitable high-resolution information for molecular replacement methods.[3]

3.3   Specimen Preparation

The major advantage of EM over X-ray diffraction is its ability to directly record an image and not just a diffraction pattern of the specimen under investigation. Two major problems have to be overcome to reach high-resolution 3D structures: first, the strong interaction of electrons with matter leads to rapid sample destruction and, second, the 2D crystals have to be prepared such that the high-resolution structure is preserved within the vacuum of the electron optical system. The same answer was found for both problems by one technique: cryo-electron microscopy.[22] With this technique, crystals are adsorbed onto a thin carbon-layer, surplus liquid is blotted away and the sample is frozen very quickly in liquid ethane. By this preparation method the crystals are embedded in a thin layer of amorphous ice or in a layer of carbohydrates substituting water, which conserves fine structures of the 2D crystal. As a result of the low temperature of the specimen (at least liquid nitrogen temperature, 90 K), the sample is less susceptible to beam-induced damage, allowing images to be recorded at electron doses of 5 $e^-/\text{Å}^2$. If the sample is cooled down to liquid helium temperature (4.2 K), electron doses of 20 $e^-/\text{Å}^2$ are possible without intolerable sample damage.[23]

To screen the results of 2DX experiments, the most efficient preparation method is to embed the sample in a heavy metal salt solution and let it dry in air. In spite of non-physiological pH and ionic strength, this approach is able to preserve the sample structure to a resolution in the nanometer (nm) range that suffices to evaluate the sample quality in a first screen. Negatively stained samples are inspected at room temperature, and the high contrast and beam resistance provided by the heavy metal salt greatly facilitate the identification of appropriate 2DX conditions (Figure 3).

3.4 Data Processing

Images recorded by cryo-electron microscopy have a very low signal-to-noise ratio. Therefore the structure of a 2D crystal can only be seen after image processing. In the power spectrum, the crystallinity of the sample is manifested in discrete spots containing the structural information of the periodically arranged protein (Figure 5a). In contrast to X-ray crystallography, not only the amplitude of the diffraction pattern can be measured but also the corresponding phase, which both are read out from the calculated Fourier transform, from which the averaged unit cell is calculated (Figure 5b, inset). Such information cannot be generated from the electron diffraction pattern obtained from identical crystals (Figure 5c).

The steps involved in extracting the full information from the projection of a 2D crystal recorded by high-resolution cryo-EM are described in Figure 6. Step 1 concerns calculating the Fourier transform of a crystal image (Figure 6a) and displaying the modulus of the transform (Figure 6b). Diffraction peaks are indexed (manually or automatically) to define the crystal lattice. In step 2, this lattice is used to generate a filter, which is zero except for small windows centered at lattice line crossings. The Fourier transform is then multiplied by the filter. This Fourier peak filtering step eliminates most of the noise (Figure 6c). Back-transformation, step 3, yields the averaged crystal image, which reveals some structural features (Figure 6d, inset). Crystal defects such as distortions, lattice defects, or mosaicity become clearly visible. To determine lattice distortions, the cross-correlation of the image (Figure 6a) with a small reference selected in step 4 (Figure 6e) is calculated and the peak-search in step 5 detects the real positions of the unit cells (Figure 6f, inset). Unit cell positions are compared to the theoretical lattice vectors in step 6, and a field of shift-vectors describing the crystal-distortions is calculated (Figure 6g). This information is used to interpolate the original image along curved lattice lines to unbend the distorted image of the 2D crystal. However, before the unbending can be applied, the CTF correction must be implemented. This involves the sign changes required to elimi-nate the phase jump introduced by the optical system. In addition, amplitude atten-uation introduced by the CTF is corrected using a Wiener filter. By the subsequent real-space unbending procedure (step 7) lattice distortions are eliminated and the dif-fraction-spots become sharper (see insets in Figure 6h). This combination of crys-tallographic methods in Fourier space and image processing methods in real-space allows the resolution to be improved by a factor of two,[24] and provides the best esti-mate of the unit cell-density distribution (Figure 6i). While the initial reference is a small area of the original image housing a few unit cells, the refinement (step 9) is executed using the first average or even the combined average from other images.

To get a 3D structure, the 2D crystal has to be tilted in the electron microscope so that 'side-views' of the protein are recorded (Figure 7a). These images are Fourier peak-filtered and unbent in the same way as the untilted ones. However, the CTF correction is complicated by the fact that the point spread function is not space invariant, requiring an additional step to take the effects of the tilt into account.[24] The information from different images is then combined in the Fourier space according to the central section theorem, shifting the phase origin of each projection to the ori-gin of the Fourier space. The amplitudes and phases of the crystals are concentrated on lattice lines in the z^*-direction (Figure 7b). Since there is no repetition in the z-direction of the crystal, the lines are sampled continuously in the Fourier space and are interpolated to obtain a regularly spaced 3D sampling. The 3D map is then cal-culated by the 3D back-transform of interpolated values (Figure 7c). Note that not all of the Fourier space can be sampled as result of the maximal tilt angle (around 60°) of the sample, thus leaving a cone of missing data. Therefore, 3D maps obtained by electron crystallography exhibit always a better resolution perpendicular to than along the optical axis.

To interpret a 3D map obtained by cryo-EM in terms of an atomic protein structure is challenging, even if data of better than 4 Å are available. In a first step the macro-molecular fold must be determined and the backbone traced through the map, which can be achieved by visual inspection. Structural clues can further be obtained from bioinformatics methods: extensive sequence alignments and the analysis of correlated mutations led to valuable insights in the case of aquaporin-1 (AQP1).[25] In the case of α-helical proteins, automated procedures have been developed to determine both the location and direction of individual helices in the map.[26] When used in conjunction with constraints from the sequence (*e.g.* the maximal length of a loop), macromolec-ular folds can be unambiguously derived. Once the fold has been determined, an ini-tial model of the backbone structure is generated. What follows are rounds of manual or semi-automated model (re-) building and refinement. The structures of bacteri-orhodopsin (bR)[4,24,27] and AQP1[28,29] have been solved by electron crystallography in this way and were subsequently confirmed by X-ray crystallography. These successes demonstrate that even at a resolution of around 3.5 Å, a cryo-EM map contains suf-ficient structural information to uniquely define the atomic structure. The practical challenge is to find this optimal structure in the high-dimensional search space. The available tools are for a substantial part the same as used in X-ray crystallography. In particular, molecular replacement has successfully been used to exploit the informa-tion from electron diffraction data to determine the structure of AQP0.[3] In contrast to

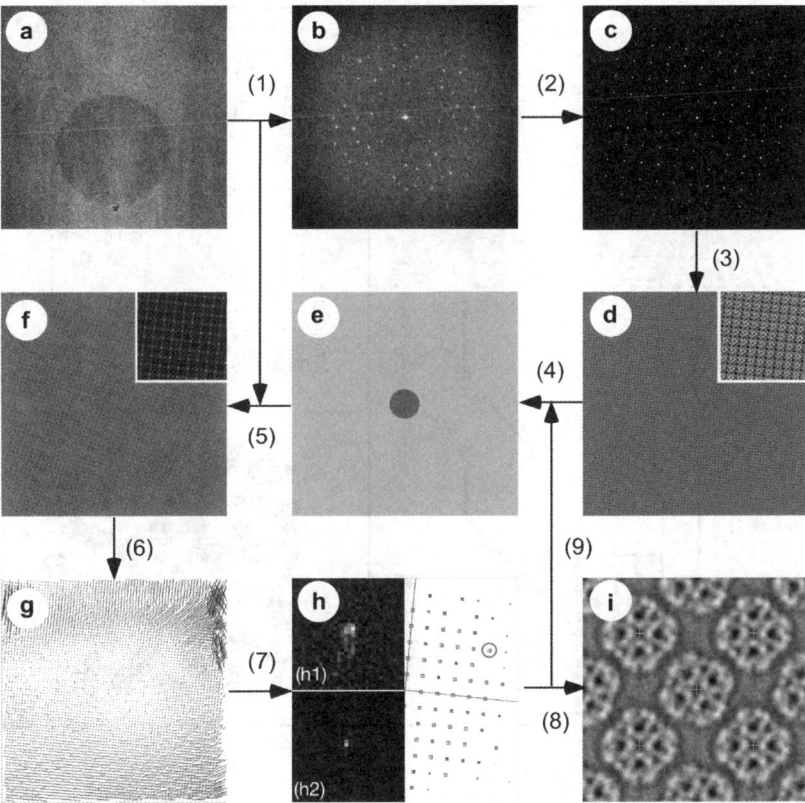

Figure 6 *Image processing of a 2D crystal (for details see text). The raw image (a) Fourier transformed and the crystal lattice of one layer is indexed in the power spectrum (b) of the raw image. For the Fourier peak filtering (c) the spots containing all the crystal information of this crystal layer are transferred to the filtered image, whereas the amplitudes outside the masked area (containing the other crystal layer and noise) are set to 0. The back-transform (d) reveals the packing of the crystal (inset). To unbend the 2D crystal, a reference (e) is selected and the cross-correlation function with the raw image is calculated. The cross-correlation (f) reveals the positions of the unit cells, from which distortion vectors are calculated (g). This information is then used to interpolate the raw image along curved lattice lines to unbend the crystal. Importantly, this step is only executed after the CTF correction. As a result, the spots of the power spectrum are better focused (h). In inset h1 peak 5,3 (indicated with a circle) is depicted before unbending and in h2 after unbending. Amplitudes and phases of the spots are combined with the data of other crystals to yield a final projection map. In this case the result is the 3.7 Å map of GlpF (i) revealing the typical tetrameric structure of an aquaglyceroporin.[60] This map can also be used to generate a synthetic reference for refining the unbending procedure*

X-ray crystallography, however, experimental phase information is available in the case of EM. This advantage is not yet fully exploited and requires the development of novel automated techniques specifically targeted at model building and refinement of EM data.

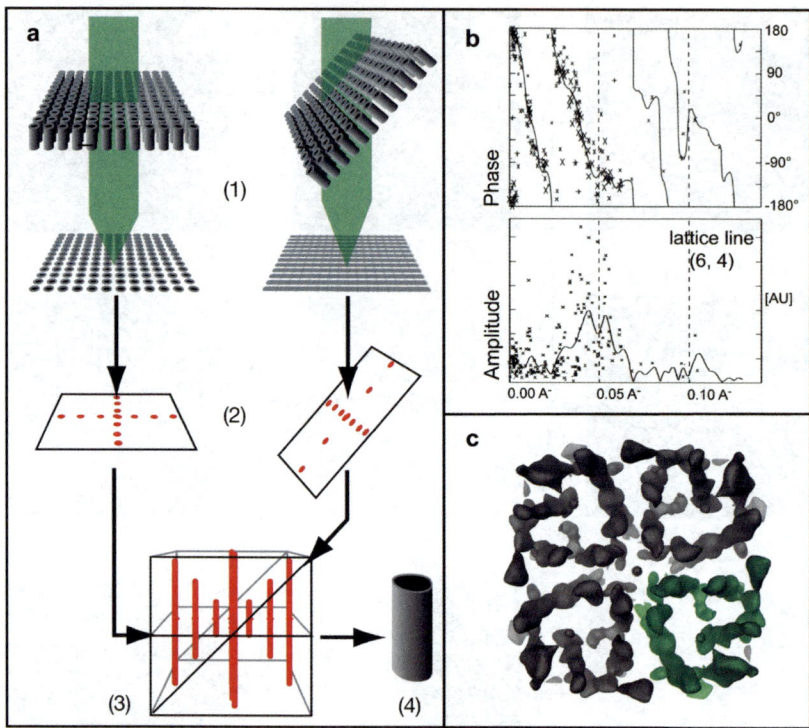

Figure 7 *3D reconstruction of membrane proteins by 2D crystallography. (a) A 3D recon-struction from 2D crystals requires projections to be recorded at different tilt angles (1). Images are Fourier filtered and processed as described in Figure 6 (2), and the Fourier transforms are combined in the 3D Fourier space according to the central section theorem (3). The discrete orders in the Fourier transform from the crystal are aligned in continuous lattice lines, since sampling is not periodic in the z* direction. Lattice lines are regularly interpolated to sample the 3D Fourier space on a cubic raster. Back-transformation of the combined data finally leads to the rep-resentation of the 3D unit cell (4). (b) Amplitude and phase values of lattice line 6,4 extend to a z* resolution of (7 Å)$^{-1}$. The plotted curve indicates the interpolation of the lattice line. (c) Cytosolic view of the 3D map of SoPIP2;1 calculated from 156 electron micrographs*

4   Atomic Force Microscopy

4.1   Image Formation

In the AFM[30] a sharp stylus at the end of a flexible cantilever is raster-scanned by a piezo device over the sample surface submerged in buffer solution. Cantilever deflections measured at a resolution of about 1 Å are used to determine the surface contour of the sample, exploiting a sensitive servo system to minimize the force applied to the stylus. Biomolecular forces lie between a few pN (1 pN = 10^{-12} N), as generated, *e.g.* by myosin, and 250 pN, as required to dissociate the streptavidin-biotin complex. It is important to ensure that imaging forces between stylus and sam-ple are of similar magnitude, because higher forces would lead to sample distortion or even disruption. Hence, while the tip travels over corrugations, the feedback of a fast servo-loop drives the piezo scanner perpendicular to the x–y raster plane to minimize the deviation of the cantilever deflection from a preset value. In this way, the cantilever tip contours the sample surface at a preset force.

4.2 Sample Preparation

Sample preparation methods for raster-scanning samples in buffer solutions all have the objective to attach the specimen firmly to a support. Immobilization is required for the contact-imaging mode, since even small lateral forces exerted by the tip tend to push away the structure to be imaged. Lateral forces are largely eliminated by the dynamic imaging mode, where the tip is oscillated vertically, lifting it away from the sample surface most of the time. Biological membranes are sufficiently large that they can simply be adsorbed to freshly cleaved mica. Adjusting the electrolyte concentration and the pH of the buffer solution facilitates this: the pH dictates the surface charge of the membrane, while the electrolyte concentration determines the thickness of the Debye layer.[31–34] Since most biological membranes have a negative surface charge at neutral pH, the repulsive electrostatic force resulting from the negative surface charge of mica prevents adsorption unless the Debye layer is thin compared to the decay length of van der Waals forces. Therefore, the critical concentration of a KCl solution for adsorbing biological membranes to mica is 50–100 mM. Once a membrane is absorbed, the gap between sample and support will decrease until van der Waals attractions and osmotic pressure, resulting from the electrolytes in the gap, are in equilibrium. Because that gap is small, solutes cannot quickly diffuse away and it is possible to exchange the adsorption buffer with a buffer optimized for imaging without desorbing the membranes.

4.3 Optimized Imaging Conditions

Electrostatic and van der Waals interactions generate the major forces interacting between AFM stylus and biological sample when imaging in buffer solution.[32] Force–distance curves, which are acquired by approaching the sample with the stylus while measuring its deflection, reveal the nature of these forces. Adjusting the electrolyte concentration and pH of the buffer can minimize these forces. In addition, AFM feedback parameters must be optimized to ensure distortion free imaging,[35] and cantilevers exhibiting a force constant of about 0.1 N/m must be used. Under such conditions, topographs of protein surfaces revealing details with a lateral resolution of 5 Å and a vertical resolution of 1 Å can be recorded routinely. Importantly, such topographs exhibit an outstanding signal-to-noise ratio, enabling recognition of single molecule structural features with unprecedented clarity.

Figure 8 shows topographs of both cytosolic and extracellular surfaces of purple membranes.[36] Purple membrane is a native 2D crystal composed of bR trimers and lipids.[37] The cytosolic (Figure 8a) and extracellular (Figure 8b) surfaces of bR in its most native state reveal the structural details of bR trimers (see broken circles in Figure 8). The calculated 2D power spectrum of the extracellular side (Figure 8b) is displayed in Figure 8c. Sharp spots extend beyond the $(1\ nm)^{-1}$ resolution limit

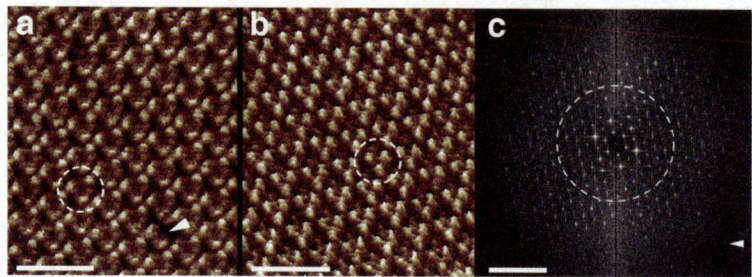

Figure 8 *High-resolution AFM topographs of the cytoplasmic (a) and the extracellular sur-*
face (b) of bR. Single bR trimers are indicated by broken circles. In (a) a missing
bR monomer in the bR trimer is indicated by an arrowhead. (c) Calculated power
spectrum from the extracellular side of bR (b). The broken circle represents the
(1 nm)⁻¹ resolution limit. The diffraction spot (6,9) corresponds to a resolution limit
of 0.41 nm (see arrowhead). Imaging buffers: 20 mM Tris-HCl (pH 7.8), 150 mM
KCl (cytoplasmic side of bR), and 20 mM Tris-HCl (pH 7.8), 150 mM KCl, 25 mM
MgCl₂ (extracellular side of bR). Scale bars represent: 15 nm (a and b) and (1 nm)⁻¹
(c). Z-scale ranges: 1.2 nm (a) and 0.8 nm (b)

(broken circle), *e.g.* spot (6,9) which corresponds to a lateral resolution of 0.41 nm.
Since many grey levels are resolved in this image of the purple membrane, whose
extracellular surface exhibits a corrugation of 0.6 nm, the vertical resolution
achieved in this case is better than 0.1 nm.

4.4 Imaging Native Membranes

The best way to study membrane proteins under native or near-native conditions is
to directly prepare and observe native membranes in buffer solution. The AFM is
currently the only instrument that allows these conditions in combination with a resolution better than 2 nm. Recently, topographs of the native bacterial photosynthetic
membrane of *Rhodopseudomonas viridis* have revealed the molecular organization
of the photosynthetic core complex[38] and its changes during light-adaptation.[7]
Moreover, AFM images of disk membranes of murine retina have elucidated the
native conformation of rhodopsin molecules, which are arranged in rows of dimers[6]
(Figure 9a), in contrast to findings from early optical studies. These AFM data were
further confirmed by blue native-polyacrylamide gel electrophoresis (Figure 9b) and
electron microscopy of detergent-solubilized rhodopsin dimers (Figure 9c).[39] The
importance of this observation is remarkable since for a long time cross-linking and
pharmacokinetic studies have suggested that GPCR are working as dimers.[40–48]
Because rhodopsin is the only GPCR with a known structure,[49] the AFM data now
provide a solid basis to model a GPCR dimer and to discuss possible interactions
with the cognate proteins, such as arrestin and the G-protein heterotrimer.[50,51]

4.5 Nanodissection

The examples shown were all recorded at minimal forces (<100 pN) to prevent
deformation of the molecules, which would cause a reduction of resolution.

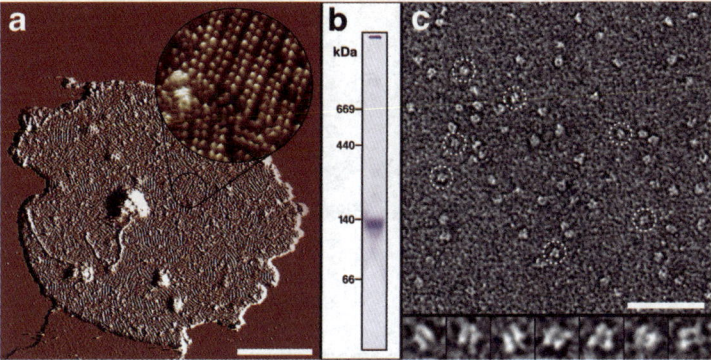

Figure 9 *(a) Atomic force microscopy of native disk membranes isolated from mice. The deflection image and the surface topography (inset) reveals rows of rhodopsin dimers densely packed in paracrystalline arrays. The diameter of the round inset is 68 nm and the z-range 1.6 nm. Imaging buffer: 20 mM Tris-HCl (pH 7.8), 150 mM KCl, and 25 mM MgCl₂. (b) Blue native-polyacrylamide gel electrophoresis (BN-PAGE) of dark-adapted disk membranes from bovine: rhodopsin migrates as dimer when solubilized in dodecylmaltoside (DDM). Prior BN-PAGE, disk membranes were solubilized in 0.3% DDM at a protein concentration of ~0.6 mg/ml. (c) Transmission electron microscopy of negatively stained DDM-solubilized disk membranes. The selected rhodopsin dimers which are marked by broken circles were magnified and are displayed in the gallery. The frame size of the magnified particles in the gallery is 10.4 nm. Scale bars represent: 250 nm (a) and 50 nm (c)*

However, at higher forces the stylus may be used as nanoscalpel to disrupt supramolecular assemblies.[52,53] In this case, the force applied to the tip is increased to 0.75–10 nN, depending on the dissection to be achieved. Quite small forces (typically 1 nN) suffice to separate stacked layers of membranes or 2D crystals.[53–55] Even smaller forces and repeated scanning at high magnification allow extrinsic proteins that are specifically complexed to an integral membrane protein to be pushed away.[36,38,56,57] Such nanodissections are of great interest to unveil otherwise hidden surfaces and to estimate interaction forces between subunits in a complex.

4.6 Image Processing

Image processing allows a variety of parameters describing the mechanical properties of a membrane protein surface to be extracted. First, the individual molecular images need to be aligned laterally and angularly. While not required for highly ordered 2D crystals, this can be accomplished to sub-nanometer accuracy with randomly oriented single particles as result of the high signal-to-noise ratio of topographs recorded by AFM. As illustrated for the bacterial aquaporin, AqpZ, average measurements and the related standard deviation of the height measurement are calculated pixel by pixel from aligned single molecular topographs (Figure 10). Calculating two independent averages from all odd and all even numbered single particles allows the resolution to be determined by comparing the Fourier coefficients at a given resolution of these two averages. This yields the phase residual (Figure 10b) and the spectral signal-to-noise

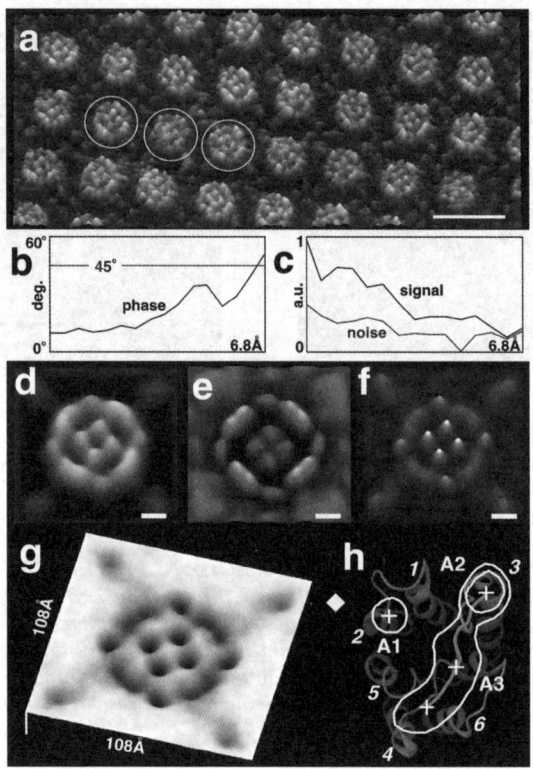

Figure 10 *Image analysis of the bacterial aquaporin (AqpZ) surface.[61] Topographs were recorded in 10 mM Tris-HCl, pH 7.8, 100 mM KCl. (a) High-resolution raw data AFM image of an AqpZ 2D-crystal. Circles indicate individual extracellular protein surfaces imaged at a loading force to the tip of ~80 pN (scale bar: 100 Å; full gray scale: 10 Å). (b) Phase residual analysis from averaging 1447 topographs of the AqpZ extracellular surface. The Nyquist frequency is 1/(6.8 Å). (c) Spectral signal-to-noise ratio analysis from averaging the 1447 AqpZ topographs. The Nyquist frequency is 1/(6.8 Å). (d) Average topograph calculated from a total of 1447 aligned tetramers (scale bar: 10 Å). (e) Standard deviation map corresponding to average (d) (scale bar: 10 Å). (f) Position probability map calculated from the peak position distribution of 1447 aligned tetramers (scale bar: 10 Å). (g) Energy landscape as function of protrusion peak position (full image size: 108 Å; ΔkT=6.6). (h) Overlay of the FWHM outline and a model of the AQP1 structure.[28] The crosses indicate the highest probability positions, the red line the top probability tracing line of loop C. Transmembrane helices are numbered in italics (helices 1 and 2 connected by loop A (FWHM: A1); helices 3 and 4 connected by loop C (FWHM: A3); helices 5 and 6 are connected by loop E which folds back into the water channel). The diamond outside the monomer indicates the four-fold axis symmetry center*

ratio (Figure 10c) as function of resolution. These analyses both indicate the lateral resolution to be 0.7 nm. The vertical scale of Figure 10e documents that vertical measurements can be better than 0.1 nm provided that the sample is sufficiently stiff, implying that larger variations reflect sample flexibility. Clearly, vertical fluctuations will emerge from lateral displacements of protruding loops that connect the helices. Therefore, it is advantageous to characterize variations in the protein surface structure by calculating the probability $p(x,y)_d$ to find a certain domain at a certain position $(x,y)_d$ by mapping the corresponding peak positions of all individual AqpZ tetramers, as illustrated by Figure 10f. This map is readily converted to a free energy landscape $G(x,y)_d$ using Boltzmann's law (Figure 10g).

$$G(x,y)_d = -k_B T \ln \{p(x,y)_d\} \tag{5}$$

where $k_B T$ is the thermal energy, which is $4.1 \ 10^{-21}$ Joule or 4.1 pN nm at room temperature. Interpretation of this map is facilitated by the 3D structure of AQP1: the long loop C spanning the entire molecule is likely to be flexible – although it is fixed to the C-terminal end of helix 3, whereas loop A is kept fixed by helices 1 and 2 (Figure 10h).

5 Conclusion and Perspectives

Electron and AFM are powerful tools to analyze structure, dynamics, and function of membrane proteins. These tools allow membrane proteins to be studied in their native environment, where their function is preserved.[58] Technical problems concerning instrumentation, data acquisition technology and image processing still remain to be solved. However, there is no fundamental limitation that would prevent achieving atomic scale insights into functional proteins embedded in a lipid bilayer.

Future developments concern crystallization robots akin to those now available for 3D crystallization. Combined with automated EM sample screening methods, this will solve the bottleneck of 2D crystallization. Commercial electron microscopes now provide the prerequisites for routine acquisition of atomic scale 3D data. Image processing has to be developed to a level akin to that currently available in X-ray crystallography. AFM instrumentation for biological applications has not been strongly developed since a decade. Small cantilevers, highly sensitive deflection measurement and high-speed scanners are only slowly emerging in commercial instruments. However, such progress will be available and routinely applied to the study of membrane proteins being only a matter of time. The real bottleneck thus is the necessity to express the membrane protein of interest in sufficient quality and quantity for structural analyses.

References

1. I. Ubarretxena-Belandia and D.M. Engelman, Helical membrane proteins: diversity of functions in the context of simple architecture. *Curr. Opin. Struct. Biol.*, 2001, **11**, 370–376.
2. A. Wise, K. Gearing and S. Rees, Target validation coupled receptors. *Drug Discov. Today*, 2002, **7**, 235–246.
3. T. Gonen, P. Sliz, J. Kistler, Y. Cheng and T. Walz, Aquaporin-0 membrane junctions reveal the structure of a closed water pore. *Nature*, 2004, **429**, 193–197.
4. K. Mitsuoka, T. Hirai, K. Murata, A. Miyazawa, A. Kidera, Y. Kimura and Y. Fujiyoshi, The structure of bacteriorhodopsin at 3.0 Å resolution based on

electron crystallography: implication of the charge distribution. *J. Mol. Biol.*, 1999, **286**, 861–882.

5. A. Engel and D.J. Müller, Observing single biomolecules at work with the atomic force microscope. *Nat. Struct. Biol.*, 2000, **7**, 715–718.

6. D. Fotiadis, Y. Liang, S. Filipek, D.A. Saperstein, A. Engel and K. Palczewski, Atomic-force microscopy: rhodopsin dimers in native disc membranes. *Nature*, 2003, **421**, 127–128.

7. S. Scheuring and J.N. Sturgis, Chromatic adaptation of photosynthetic membranes. *Science*, 2005, **309**, 484–487.

8. T. Braun and A. Engel, Two-dimensional electron crystallography. *Nature Encyclopedia of Life Sciences*, 2005. http://www.els.net

9. C.A. Mannella, Phospholipase-induced crystallization of channels in mitochondrial outer membranes. *Science*, 1984, **224**, 165–166.

10. V.M. Unger, N.M. Kumar, N.B. Gilula and M. Yeager, Expression, two-dimensional crystallization, and electron cryo-crystallography of recombinant gap junction membrane channels. *J. Struct. Biol.*, 1999, **128**, 98–105.

11. B.K. Jap, M. Zulauf, T. Scheybani, A. Hefti, W. Baumeister, U. Aebi and A. Engel, 2D crystallization: from art to science. *Ultramicroscopy*, 1992, **46**, 45–84.

12. D. Levy, M. Chami and J.L. Rigaud, Two-dimensional crystallization of membrane proteins: the lipid layer strategy. *FEBS Lett.*, 2001, **504**, 187–193.

13. R.M. Garavito and S. Ferguson-Miller, Detergents as tools in membrane biochemistry. *J. Biol. Chem.*, 2001, **276**, 32403–32406.

14. A. Engel, A. Hoenger, A. Hefti, C. Henn, R.C. Ford, J. Kistler and M. Zulauf, Assembly of 2-D membrane protein crystals – dynamics, crystal order, and fidelity of structure analysis by electron microscopy. *J. Struct. Biol.*, 1992, **109**, 219–234.

15. J.L. Rigaud, G. Mosser, J.J. Lacapere, A. Olofsson, D. Levy and J.L. Ranck, Bio-beads: an efficient strategy for two-dimensional crystallization of membrane proteins. *J. Struct. Biol.*, 1997, **118**, 226–235.

16. H.W. Remigy, D. Caujolle-Bert, K. Suda, A. Schenk, M. Chami and A. Engel, Membrane protein reconstitution and crystallization by controlled dilution. *FEBS Lett.*, 2003, **555**, 160–169.

17. M. Dolder, A. Engel and M. Zulauf, The micelle to vesicle transition of lipids and detergents in the presence of a membrane protein: towards a rationale for 2D crystallization. *FEBS Lett.*, 1996, **382**, 203–208.

18. T.C. Kaufmann, A. Engel and H.-W. Rémigy, A novel method for detergent concentration determination. *Biophys. J.*, 2006, **90**, 310–317.

19. L. Hasler, J.B. Heymann, A. Engel, J. Kistler and T. Walz, 2D crystallization of membrane proteins: rationales and examples. *J. Struct. Biol.*, 1998, **121**, 162–171.

20. W. Kukulski, A.D. Schenk, U. Johanson, T. Braun, B.L. de Groot, D. Fotiadis, P. Kjellbom and A. Engel, The 5 Å structure of heterologously expressed plant aquaporin SoPIP2;1. *J. Mol. Biol.*, 2005, **350**, 611–616.

21. N. Gyobu, K. Tani, Y. Hiroaki, A. Kamegawa, K. Mitsuoka and Y. Fujiyoshi, Improved specimen preparation for cryo-electron microscopy using a symmetric carbon sandwich technique. *J. Struct. Biol.*, 2004, **146**, 325–333.

22. J. Dubochet, M. Adrian, J.-J. Chang, J.-C. Homo, J. Lepault, A.W. McDowall and P. Schultz, Cryo-electron microscopy of vitrified specimens. *Q. Rev. Biophys.*, 1988, **21**, 129–228.
23. Y. Fujiyoshi, The structural study of membrane proteins by electron crystallography. *Adv. Biophys.*, 1998, **35**, 25–80.
24. R. Henderson, J.M. Baldwin, T.A. Ceska, F. Zemlin, E. Beckmann and K.H. Downing, Model for the structure of bacteriorhodopsin based on high-resolution electron cryo-microscopy. *J. Mol. Biol.*, 1990, **213**, 899–929.
25. J.B. Heymann and A. Engel, Structural clues in the sequences of the aquaporins. *J. Mol. Biol.*, 2000, **295**, 1039–1053.
26. B.L. de Groot, J.B. Heymann, A. Engel, K. Mitsuoka, Y. Fujiyoshi and H. Grubmüller, The fold of human aquaporin 1. *J. Mol. Biol.*, 2000, **300**, 987–994.
27. Y. Kimura, D.G. Vassylyev, A. Miyazawa, A. Kidera, M. Matsushima, K. Mitsuoka, K. Murata, T. Hirai and Y. Fujiyoshi, Surface of bacteriorhodopsin revealed by high-resolution electron crystallography. *Nature*, 1997, **389**, 206–211.
28. K. Murata, K. Mitsuoka, T. Hirai, T. Walz, P. Agre, J.B. Heymann, A. Engel and Y. Fujiyoshi, Structural determinants of water permeation through aquaporin-1. *Nature*, 2000, **407**, 599–605.
29. B.L. de Groot, A. Engel and H. Grubmüller, A refined structure of human aquaporin-1. *FEBS Lett.*, 2001, **504**, 206–211.
30. G. Binnig, C.F. Quate and C. Gerber, Atomic force microscopy. *Phys. Rev. Lett.*, 1986, **56**, 930–933.
31. D.J. Müller, M. Amrein and A. Engel, Adsorption of biological molecules to a solid support for scanning probe microscopy. *J. Struct. Biol.*, 1997, **119**, 172–188.
32. D.J. Müller and A. Engel, The height of biomolecules measured with the atomic force microscope depends on electrostatic interactions. *Biophys. J.*, 1997, **73**, 1633–1644.
33. D.J. Müller, M. Amrein and A. Engel, Adsorption of biological molecules to a solid support for scanning probe microscopy. *J. Struct. Biol.*, 1997, **119**, 172–188.
34. D.J. Müller, A. Engel and M. Amrein, Preparation techniques for the observation of native biological systems with the atomic force microscope. *Bioelectrochem. Bioenerg.*, 1997, **12**, 867–877.
35. D.J. Müller, D. Fotiadis, S. Scheuring, S.A. Müller and A. Engel, Electrostatically balanced subnanometer imaging of biological specimens by atomic force microscope. *Biophys. J.*, 1999, **76**, 1101–1111.
36. D. Fotiadis and A. Engel, High-resolution imaging of bacteriorhodopsin by atomic force microscopy. *Methods Mol. Biol.*, 2004, **242**, 291–303.
37. A.E. Blaurock and W. Stoeckenius, Structure of the purple membrane. *Nat. New Biol.*, 1971, **233**, 152–155.
38. S. Scheuring, J. Seguin, S. Marco, D. Levy, B. Robert and J.L. Rigaud, Nanodissection and high-resolution imaging of the *Rhodopseudomonas viridis* photosynthetic core complex in native membranes by AFM. *Proc. Natl. Acad. Sci. USA*, 2003, **100**, 1690–1693.
39. K. Suda, S. Filipek, K. Palczewski, A. Engel and D. Fotiadis, The supramolecular structure of the GPCR rhodopsin in solution and native disc membranes. *Mol. Membr. Biol.*, 2004, **21**, 435–446.

40. S. Angers, A. Salahpour and M. Bouvier, Dimerization: an emerging concept for G protein-coupled receptor ontogeny and function. *Annu. Rev. Pharmacol. Toxicol.*, 2002, **42**, 409–435.

41. M. Bai, Dimerization of G-protein-coupled receptors: roles in signal transduction. *Cell Signal.*, 2004, **16**, 175–186.

42. Z.J. Cheng, K.G. Harikumar, E.L. Holicky and L.J. Miller, Heterodimerization of type A and B cholecystokinin receptors enhance signaling and promote cell growth. *J. Biol. Chem.*, 2003, **278**, 52972–52979.

43. S.P. Lee, B.F. O'Dowd, R.D. Rajaram, T. Nguyen and S.R. George, D2 dopamine receptor homodimerization is mediated by multiple sites of interaction, including an intermolecular interaction involving transmembrane domain 4. *Biochemistry*, 2003, **42**, 11023–11031.

44. J.F. Mercier, A. Salahpour, S. Angers, A. Breit and M. Bouvier, Quantitative assessment of beta 1- and beta 2-adrenergic receptor homo- and heterodimerization by bioluminescence resonance energy transfer. *J. Biol. Chem.*, 2002, **277**, 44925–44931.

45. M.C. Overton, S.L. Chinault and K.J. Blumer, Oligomerization, biogenesis, and signaling is promoted by a glycophorin A-like dimerization motif in transmembrane domain 1 of a yeast G protein-coupled receptor. *J. Biol. Chem.*, 2003, **278**, 49369–49377.

46. C.D. Rios, B.A. Jordan, I. Gomes and L.A. Devi, G-protein-coupled receptor dimerization: modulation of receptor function. *Pharmacol. Ther.*, 2001, **92**, 71–87.

47. F. Trettel, S. Di Bartolomeo, C. Lauro, M. Catalano, T.M. Ciotti and C. Limatola, Ligand-independent CXCR2 dimerization. *J. Biol. Chem.*, 2003, **278**, 40980–40988.

48. O.S. Soyer, M.W. Dimmic, R.R. Neubig and R.A. Goldstein, Dimerization in aminergic G-protein-coupled receptors: application of a hidden-site class model of evolution. *Biochemistry*, 2003, **42**, 14522–14531.

49. K. Palczewski, T. Kumasaka, T. Hori, C.A. Behnke, H. Motoshima, B.A. Fox, I. Le Trong, D.C. Teller, T. Okada, R.E. Stenkamp, Masaki Yamamoto and Masashi Miyano, Crystal structure of rhodopsin: a G protein-coupled receptor. *Science*, 2000, **289**, 739–745.

50. Y. Liang, D. Fotiadis, S. Filipek, D.A. Saperstein, K. Palczewski and A. Engel, Organization of the G protein-coupled receptors rhodopsin and opsin in native membranes. *J. Biol. Chem.*, 2003, **278**, 21655–21662.

51. S. Filipek, K.A. Krzysko, D. Fotiadis, Y. Liang, D.A. Saperstein, A. Engel and K. Palczewski, A concept for G protein activation by G protein-coupled receptor dimers: the transducin/rhodopsin interface. *Photochem. Photobiol. Sci.*, 2004, **3**, 1–12.

52. D. Fotiadis, S. Scheuring, S.A. Müller, A. Engel and D.J. Müller, Imaging and manipulation of biological structures with the AFM. *Micron*, 2002, **33**, 385–397.

53. F.A. Schabert, C. Henn and A. Engel, Native *Escherichia coli* OmpF porin surfaces probed by atomic force microscopy. *Science*, 1995, **268**, 92–94.

54. D. Fotiadis, L. Hasler, D.J. Müller, H. Stahlberg, J. Kistler and A. Engel, Surface tongue-and-groove contours on lens MIP facilitate cell-to-cell adherence. *J. Mol. Biol.*, 2000, **300**, 779–789.

55. D.J. Müller, G.M. Hand, A. Engel and G.E. Sosinsky, Conformational changes in surface structures of isolated connexin 26 gap junctions. *EMBO J.*, 2002, **21**, 3598–3607.

56. D. Fotiadis, D.J. Müller, G. Tsiotis, L. Hasler, P. Tittmann, T. Mini, P. Jenö, H. Gross and A. Engel, Surface analysis of the photosystem I complex by electron and atomic force microscopy. *J. Mol. Biol.*, 1998, **283**, 83–94.

57. C.A. Siebert, P. Qian, D. Fotiadis, A. Engel, C.N. Hunter and P.A. Bullough, Molecular architecture of photosynthetic membranes in *Rhodobacter sphaeroides*: the role of PufX. *EMBO J.*, 2004, **23**, 690–700.

58. T. Walz, B.L. Smith, M.L. Zeidel, A. Engel and P. Agre, Biologically active 2-dimensional crystals of aquaporin chip. *J. Biol. Chem.*, 1994, **269**, 1583–1586.

59. J.L. Rigaud, B. Pitard and D. Levy, Reconstitution of membrane proteins into liposomes: application to energy-transducing membrane proteins. *Biochim. Biophys. Acta*, 1995, **1231**, 223–246.

60. T. Braun, A. Philippsen, S. Wirtz, M.J. Borgnia, P. Agre, W. Kühlbrandt, A. Engel and H. Stahlberg, The 3.7 Å projection map of the glycerol facilitator GlpF: a variant of the aquaporin tetramer. *EMBO reports*, 2000, **1**, 183–189.

61. S. Scheuring, D.J. Müller, H. Stahlberg, H.A. Engel and A. Engel, Sampling the conformational space of membrane protein surfaces with the AFM. *Eur. Biophys. J.*, 2002, **31**, 172–178.

CHAPTER 9

State-of-the-Art Methods in Electron Microscopy, including Single-Particle Analysis

VINZENZ M. UNGER

Department of Molecular Biophysics and Biochemistry, Yale School of Medicine, PO Box 208024, New Haven, CT 06520-8024, USA

1   Introduction

January 13, 1968: on this day an article, authored by De Rosier and Klug from the Medical Research Council Laboratory of Molecular Biology at Cambridge, appeared in the journal *Nature*,[1] describing how electron microscopic images can be used to reconstruct the three-dimensional (3D) structure of biological samples. The key discovery described in this chapter lies in the realization that electron microscopic images are true projections, representing structural information from all levels within a thin biological sample. Moreover, the diffraction patterns that can be calculated from a set of projections showing the molecule from different directions, all share a common origin in reciprocal space or in other words, the Fourier transform of each projection represents a central section through the 3D Fourier transform of the molecule. While each of these projections and their transforms are 2D, the simple analogy shown in Figure 1 makes it easy to appreciate that information about the 3D structure of the object can be obtained by combining information from images/transforms that represent different views of the object.

This discovery, which can be regarded as the "birth" of molecular microscopy, happened at a time that also saw the first X-ray crystal structures of proteins emerge.[2,3] Yet, only X-ray crystallography quickly developed into a powerful tool for structural analysis of biological samples, while another 22 years passed before the first near atomic structure was solved by electron microscopy.[4] During this period, a silent revolution took place that transformed electron microscopy from a shadow existence into a vibrant and rapidly developing method which has just begun to

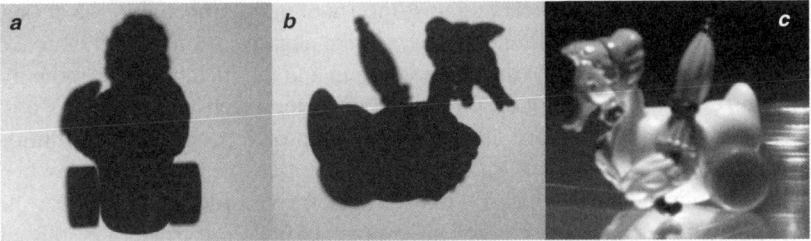

Figure 1 *3D structures and their 2D projections. Electron microscopic images are 2D projections of the imaged object, similar, but not identical, to optical projections. Nevertheless, the principle of how 3D structures can be reconstructed from 2D projections can be gleaned from the simple example given here. Panels a and b are projections of an umbrella carrying duck (panel c), seen from above (panel a) and from the side (panel b). Neither of the two projections reveals all the detail about the object. For instance, the top view does not tell that the object is a duck, nor does the apparently curved feature on the left side of the view give away that the duck carries an umbrella. Both features can clearly be recognized in the side view. However, this view does not reveal the presence of the wheels, whose projections are more revealing in the top view. From this simple example it follows that 3D structures can be obtained from projections if a sufficiently large number of different views has been obtained and if the geometric relation between the different projections are known*

assume a pivotal role in the structural analysis of soluble and membrane-embedded macromolecular complexes. This chapter touches on some of the technical hurdles that needed to be overcome to establish molecular microscopy as a useful tool; outlines the concepts of the two most commonly used electron microscopic approaches, electron crystallography and single-particle analysis; and illustrates how these approaches can be applied to structure determination of membrane proteins.

2   Sample Preparation

Although conceptually simple, structure determination by electron microscopy is complicated by the necessity to protect the sample from the vacuum of the microscope column and the damaging effects of the electron beam. In the early days, coating the sample with a film of heavy metal salts, such as uranyl-acetate or phosphotungstic acid, was the only solution to these problems. This approach, which is also known as negative staining, is very efficient and still has many uses. However, the stain cannot penetrate inside tightly packed protein molecules and, therefore, images of negatively stained samples represent a projection of the regions from which stain is excluded. In most cases, this stain-exclusion pattern coincides with the molecular envelope of the biological sample. Consequently, this approach limits the resolution that can be obtained and, therefore, this hurdle needed to be overcome before electron microscopy could begin to unfold its full potential. A surprisingly simple solution to the problem came from studies on the archebacterial membrane protein bacteriorhodopsin, which spontaneously forms well-ordered 2D arrays. In their quest to determine the structure of bacteriorhodopsin, Henderson and Unwin discovered that the sample could be protected against the vacuum by embedding the crystalline membranes in a layer of the sugar, glucose. Unlike negative stains, the chemical composition of glucose is similar to that of the protein and the membrane (carbon, hydrogen, oxygen), which makes glucose transparent to the electron beam. This allowed Henderson and Unwin to record data to a resolution of 7 Å, at which resolution the packing arrangement of bacteriorhodopsin's seven transmembrane helices became visible.[5] At about the same time, Taylor and Glaeser recorded the first near atomic electron diffraction data of a biological sample. Using thin, frozen crystals of the enzyme catalase, this groundbreaking work provided proof that electron microscopy is capable of resolving near atomic detail in biological samples.[6] Both discoveries represented large leaps forward. However, the close chemical match between small molecules such as glucose or trehalose and tannin, which were later used in other cases, acts like a camouflage, and attenuates the signal at low resolution. This effect did not interfere much with the analysis of crystalline samples, such as catalase and bacteriorhodopsin, because in these cases, the structural information is contained in the discrete and strongly amplified diffraction signal that emerges from a regular lattice containing thousands of molecules. However, if structure reconstruction relies on the combination of projections from individual molecules, commonly referred to as "particles," any reduction in contrast caused by small molecules such as glucose, detergents (in the case of membrane proteins), or even the presence of a continuous carbon support film makes detection of individual-particle projections more difficult, if not impossible. To become a generally applicable method, the field of electron microscopy therefore required yet another breakthrough. The advancement came from the work of Dubochet and co-workers who were the first to successfully use rapid freezing to turn thin layers of water into amorphous, solid films.[7,8] This method, which is also known as vitrification (Figure 2), opened the door to the structure determination of noncrystalline specimens because embedding in vitreous water not only improves contrast, but also allows single particles to be suspended in the holes of a holey carbon support film used to mount such samples on the electron microscope grid (Figure 3). Moreover, the ability to replace small embedding molecules, such as glucose, by water allows biological macromolecules to be preserved with minimal perturbation, and creates an opportunity to control the composition of the embedding medium (pH, salt, detergents, ligands, etc.). It, therefore, is not surprising that vitrification has become the most widely used approach to preserve samples for high-resolution macromolecular microscopy. An extensive discussion of the background and applications of sample vitrification can be found in an excellent review by Dubochet *et al.*[9]

3 Low-Dose Microscopy

3.1 Sample Holders

Although it is key to the success of molecular microscopy, the ability to prepare frozen-hydrated samples also created new challenges. For instance, vitreous water is metastable and, therefore, frozen-hydrated samples need to be maintained at liquid nitrogen temperatures (or colder) at all times after freezing. Consequently, special

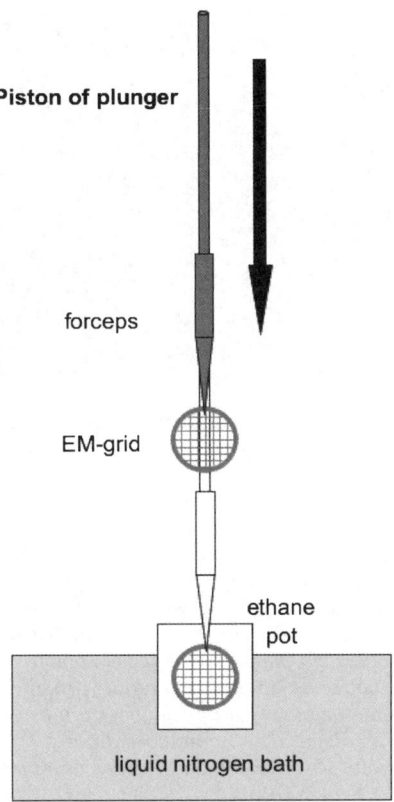

Figure 2 *Principle of sample vitrification. Rapid freezing of thin aqueous layers results in vit-rification of the sample. In this state, water exists as an amorphous solid rather than adopting a crystalline structure like in regular ice. In practice, vitrification is accomplished by first applying sample to an electron microscope grid that is mounted on a plunger by removable forceps. After removing excess sample by blot-ting with a piece of filter paper, the piston of the plunger is released which causes rapid transfer (indicated by the arrow) of the grid into a pot filled with liquid ethane. Upon entering the ethane, freezing occurs within microseconds, which is too fast for crystalline ice to form. By dislodging the forceps from the piston, the grid can then be transferred and stored in liquid nitrogen until use for the actual data collection session*

transfer stations and sample holders needed to be invented to allow samples to stay cold during transfer into the microscope and data collection. The whole procedure of observing unstained samples in a frozen state has since become known as cryo-electron microscopy (cryoEM). The most straightforward solution to make cryoEM a reality was to integrate a dewar into the part of the sample holder that remains out-side the microscope column. This design, which is still widely used, is simple and efficient (Figure 4). However, the presence of the additional dewar makes the holder very sensitive to acoustic interference, and even minute thermal imbalances between the holder and the mounted sample, or the holder and the microscope column, cause

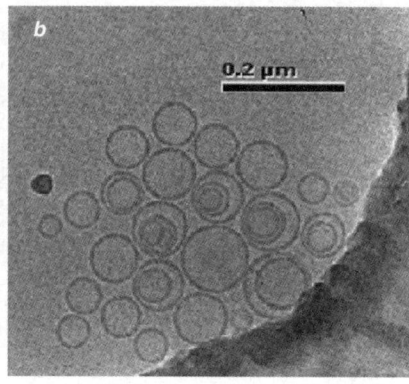

Figure 3 *Comparison of negative stain and frozen-hydrated samples. In this example, lipo-somes were chosen as a test object. Panel a shows a field from a negatively stained sample deposited on a continuous carbon support film. The individual liposomes appear as round or polygonal features, separated by dark lines where different lipo-somes touch each other. Panel b shows a sample of the same liposome preparation. However, in this case the liposomes were vitrified over a holey carbon film. The image shows a group of liposomes that were trapped in a hole but close to the edge of the carbon film (lower right corner). Notably, all liposomes appear perfectly round in the vitrified sample. This is because, unlike negative staining, no osmotic stress and mechanic deformation are induced upon freezing. Moreover, the vitrified sample reveals that some liposomes contained smaller liposomes trapped inside. This feature cannot be observed in negative stain because the stain envelope only accounts for the volume that is excluded from the liposomes. The comparison clearly demonstrates why vitrification is necessary if internal detail of a biological macromolecule is to be resolved*

sample movements that preclude imaging of near atomic resolution detail. To put it into perspective: if high-resolution detail is to be preserved in images, sample move-ment must be limited to less than 1 Å ($= 1 \times 10^{-10}$ m) during an exposure that typ-ically takes about 1 s. The requirement for such extreme stability, which can be upset by as little as a whisper or a $1°$ change in room temperature, has been the reason for continued efforts to develop more stable cryostages.

A very different and extremely efficient technical solution was developed by Fujiyoshi and co-workers who build the first liquid helium cooled "top-entry stage."[10] In their design, the electron microscope grid is mounted on a separate cartridge, which through a mechanical mechanism is introduced and released into the micro-scope column. This stage design provides extreme sample stability because the car-tridge and grid become parts of the column and no longer interact directly with the environment outside the column. Yet, this design has the disadvantage that the orien-tation of the cartridge with respect to the electron beam cannot be changed. This inability to continuously tilt the cartridge inside the column is a limiting factor for applications such as electron crystallography, where the crystal lattice constrains the orientation of the molecules. Because of this, a third type of cryostage has been developed over recent years that solves the tilting problem by combining elements of both top- and side-entry stages. Just as top-entry stages, these hybrid stages eliminate

Traditional Side Entry Cold Stage

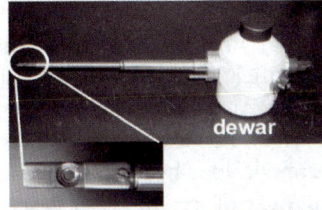

Specimen Cartridges for
Use in Top Entry Stage

Figure 4 *Comparison of cryostage components. An example for a traditional side-entry cold stage is shown (top). In this design, the tip holding the electron microscope grid (see enlarged area) is connected to a dewar at the far end of the holder. A matching transfer station (not shown) allows the grid to be mounted under liquid nitrogen. By filling the dewar with nitrogen, the sample tip is held at a temperature of about −180 °C. In contrast, top-entry stages rely on the use of specimen cartridges as those shown in the lower part of the figure. The grids are mounted onto these cartridges under nitrogen and the assembly is then transferred into the microscope column where the cartridge is positioned within a cooling stage. In contrast to side-entry stages, which are mounted in goniometers that provide continuous tilt capability, top-entry cartridges are machined to hold the grid at a fixed angle (0° and 45° in this example). This is a disadvantage for some applications and has prompted the development of stages whose design uses components of both side- and top-entry stages*

sample movements, commonly referred to as "drift," and thereby overcome one of the most significant factors limiting resolution in electron microscopic images.

3.2 Radiation Damage and Low-Dose Imaging

Although of tremendous importance, stage stability is only one of the factors limiting the use of electron microscopy for structure determination of biological samples. An even more fundamental hurdle originates from a very simple fact: organic matter is rapidly destroyed by its interaction with an electron beam. Consequently, the advent of unstained samples that no longer were embedded in a coat of negative stain required the development of new imaging protocols known as "low-dose microscopy." As implied by its name, the key purpose of low-dose imaging is to minimize radiation-induced damage to the sample by minimizing the electron dose that is delivered prior to and during recording of an image. Just like taking a picture with an optical camera, taking low-dose images of biological samples involves three separate steps: (1) searching for a region of interest, (2) focusing, and (3) exposure.

To put the three operations of low-dose microscopy into perspective: finding suitable targets for data collection is a formidable challenge because if the molecules/crystals were the size of a person, then one would need to find these targets in an area roughly the size of California. In practical terms, this is accomplished by setting the microscope to a very low magnification. Under these conditions, the outlines of large objects, such as crystals, can be detected within the embedding layer of ice. However, individual molecules/complexes are often too small to be visible at low magnifications. In these cases, imaging is done blindfolded by simply choosing regions that at low magnification show a clean and thin layer of ice. In either case, pooling electrons from a very large area allows searching to be carried out at minute electron doses that cause almost no damage to the sample. Unfortunately, this is not true for the second step when the objective lens needs to be focused. This step is performed at high magnifications (typically >200,000-fold), which delivers a large electron dose to the illuminated area. Fortunately, damage to the sample can be avoided because the electron optics allow the focusing step to be performed on an area that is adjacent to the region that will be recorded in the image. Once focus has been found, the lens is then defocused before the picture is taken. At first, this last step is counterintuitive. Why would one throw the lens out of focus again before taking a picture? The roots of the answer lie in the fact that just as light microscopy, electron microscopy is a phase contrast microscopy. In contrast to light microscopy however, no generally useful electron phase plates have yet been developed that generate contrast by uniformly maximizing interference between the diffracted and undiffracted wavefronts. Without such a phase plate, the only way to generate contrast is by defocusing the beam with respect to the sample. This generates contrast by virtue of the interference that is caused by pathlength differences between the undiffracted and diffracted wavefronts. Unfortunately, relying on pathlength differences makes transfer of information by the objective lens a complex function of (1) the electron wavelength, (2) diffraction angle, (3) spherical aberration of the lens, and (4) the actual defocus chosen for the exposure. This notorious function, which is also known as contrast transfer function (CTF), distorts the image and needs to be corrected for (see Section 4.2.3 and Figure 12). In practice, images are always recorded at underfocus because the impact of the CTF on the data is easier to account for at underfocus than at overfocus. For this very reason, focus needs to be found first because only then can the lens be adjusted to record the image at a defined amount of underfocus. This brings us to the last step: recording of the actual picture. When working at liquid nitrogen temperatures, an electron dose of ~ 10 electrons per Å^2 per s offers the best compromise between accumulating useful signal over noise.[11] At this dose, the electron flux is so small that the electrons pass through the sample one at a time. Yet even at this dose, the high-resolution information in the sample is destroyed during the time it takes to record the image. For exactly this reason, only a single useful exposure of any one area can be taken. This is quite different from negatively stained samples where the shell formed by the negative stain withstands radiation quite easily, allowing higher doses to be used. As a consequence, images of negatively stained and unstained samples show large differences in contrast. This is illustrated in Figure 5, which emphasizes that structure determination by electron microscopic approaches relies on computational methods to average signal from hundreds of thousands of molecules. This aspect will be covered in the next two sections, which will deal with the concepts of and some of the differences between the two most widely used image processing approaches: single-particle reconstruction and electron crystallography.

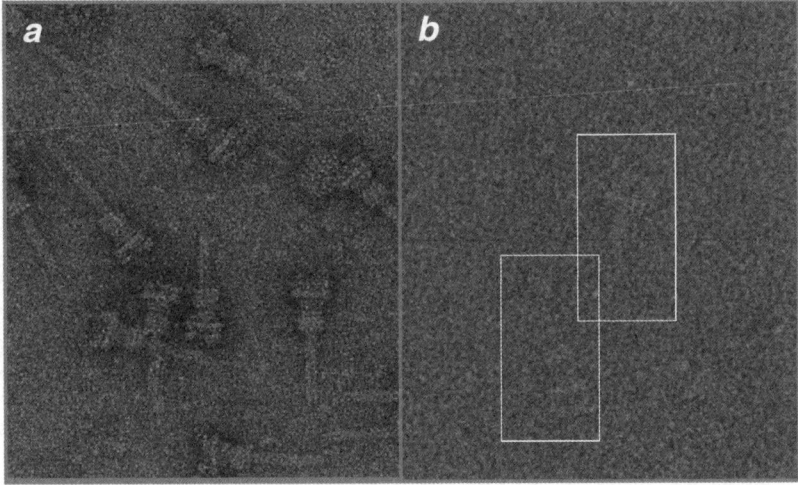

Figure 5 *Low-dose imaging of vitrified samples. Low-dose imaging of unstained samples results in images with very low contrast. This is illustrated by a direct comparison between a negatively stained sample (panel a) and an unstained sample (panel b) of a type III secretion needle complex.[34] In this particular case, the presence of detergent and a continuous carbon support film also contributes to the low contrast by matching the density of protein complexes. The projections of two particles have been boxed in the image of the unstained sample. Yet, even with this guide, the particle images can barely be detected above the background. Notably, these complexes are >4 MDa in size, emphasizing that detection of particles of small membrane proteins can easily become the limiting factor*

4   Applications of CryoEM

4.1   Single-Particle Approaches

The ability to determine structures of biological macromolecules directly from images of individual molecules is a distinct advantage of cryoEM over other structural approaches for mostly three reasons: (1) samples need not be crystalline, (2) only small amounts of material are required to prepare each grid (a few microliters at protein concentrations between 0.1 and 1 mg mL^{-1}), and (3) impurities as well as structural heterogeneity are acceptable within limits. In the case of membrane proteins however, these advantages are muted by just as many specific difficulties: first, most membrane proteins are small. This makes detection in low-dose images difficult, if not impossible, if the molecular weight of the protein/complex is lower than ~250 kDa. Second, membrane proteins require detergents to stay in solution. Because of their chemical composition, detergent micelles match the contrast of the protein, and the detergent-induced reduction in surface tension makes sample vitrification over holey carbon films difficult. Both these factors add to the challenge of identifying individual-particle projections in the low-dose images of unstained preparations (see Figure 5 for instance). It, therefore, is not surprising that single-particle reconstructions of membrane proteins rarely advance past analysis of negatively stained samples.[12–15] However, there are few exceptions and, therefore, the basic concepts of single-particle approaches will briefly be outlined here (for detailed treatments see Refs. 16 and 17).

As shown in Figure 1, the underlying principle of single-particle reconstruction exploits the fact that the microscope generates projections of the molecules. If the sample adopts random orientations on the support film or within holes of a holey carbon film, these projections evenly sample views from all directions, which allows the 3D structure to be determined. Toward this goal, each single-particle project goes through four stages: (1) a data set, consisting of thousands of individual-particle projections, must be compiled; (2) particle projections are then sorted and averaged to improve the signal-to-noise ratio for each distinct view; (3) the orientations of averaged views with respect to each other need to be determined; and (4) the structure needs to be calculated. Steps (2) through (4) are then iterated several times until the parameters determined for each projection no longer change. The following is a brief description of the steps involved.

4.1.1   Generating a Data Set

Regardless of the image processing approach taken, the first step is to extract single-particle projections from images. If the images were recorded on film, then the micrographs need to be digitized first. However, recent improvements in size and quality of digital cameras make digital data collection an increasingly popular alternative to taking images on film. In either case, starting from the images, suitable particle projections (i.e., those showing an intact particle and no ice contamination or other artifacts) are identified and boxed off into separate, smaller images that constitute the raw data (Figure 6). Over recent years, a number of algorithms have been developed to assist with particle picking to various degrees (for recent review see Ref. 18). However, particles of membrane proteins are often only barely visible above the background which causes (semi)automatic picking algorithms to fail or to erroneously pick large numbers of false positives. Therefore, it often is preferable to manually select particles, despite a certain danger to bias particle selection, which can turn out to be detrimental at later stages.

4.1.2   Particle Classification

Individual-particle projections are too noisy to allow a meaningful structure to be determined (see Figure 5). This limitation can be overcome by averaging. Of course, averaging makes sense only if all the component projections represent the same view, as otherwise detail will not be reinforced but will add to a random array of gray levels. Thus, once a set of randomly oriented particle projections has been extracted from the images, the task becomes to sort the particle images into groups that represent identical views of the molecule/complex. This step, which is also known as classification, can be accomplished in different ways. The most intuitive approach toward classification is "projection matching." In this method, a model is generated and projections, representing different views of the model, are calculated. Each particle image in the data set is then compared to all the reference projections and

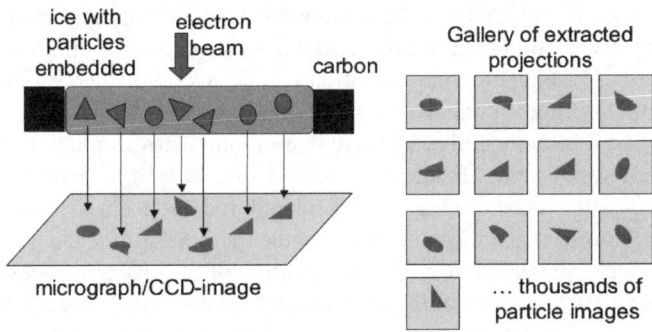

Figure 6 *Principle of particle picking. This schematic illustrates how a raw data set is generated in the case of a single-particle project. In the first step, low-dose images of randomly oriented particles, suspended in holes or over continuous carbon, are recorded on either film or with a digitial camera. If recorded on film, the micrographs need to be digitized for further processing, a step that is not necessary if digital images are recorded. In either case, images usually contain several particle projections. Suitable projections (undamaged particles, no contamination) are then selected and written to smaller image files, each containing the data for only a single particle. After collecting several thousands of particle images, the projections are then subjected to classification and calculation of class-averages*

finally gets assigned to the class corresponding to the reference projection that provides the closest match. This approach is quite efficient for sorting particle projections into groups, and often works even if nothing or only little is known about the actual structure of the molecule/complex. How is this possible? The answer to this question lies in the fact that initial particle alignments are dominated by the overall shape of the molecules/complexes rather than by fine structural detail. Consequently, the starting model can be very rough. In some cases, a simple geometric shape like a cylinder of roughly the right dimensions or a combination of shapes such as spheres, cylinders, and toroids may suffice to get the process started. The structure resulting from such an initial sorting process will leave much room for improvements. However, it will contain some features that are characteristic for the actual molecule that were not present in the original starting model. If that is the case, then the structure can be vastly improved by iterative refinement in which the actual, and steadily improving, structure replaces the synthetic starting model for calculation of the reference projections. An inherent danger of this approach is that it can result in model bias, especially if the particles lack symmetry.[19] For this reason, a different approach, known as multivariate statistical analysis (MSA), has become increasingly popular to generate the initial alignment parameter for the images in the data set. MSA treats images as vectors, where each of the image's pixels represents one dimension. For instance, if the particle projection were contained in a box of 64×64 pixels, then MSA would treat this image as a 4096-dimensional vector. The key idea underlying this approach is that the gray level value of each pixel within the box is a composite of information stemming from the molecule/complex and noise accumulated during the low-dose exposure. Although difficult to picture, images representing the same view of the molecule/complex will have similar gray levels for each "dimension" (pixel) and, therefore, the endpoints of their corresponding vectors will lie closely together within the multidimensional space defined by the size of the individual images. Consequently, classification can be accomplished by defining clusters within this *n*-dimensional space. A major advantage of this approach is that it does not rely on a reference, that is, there is no input as to what one is looking for.

Regardless of how classification is carried out, once the particles have been grouped a so-called class-average is calculated for each class. In the averaging process only features that are shared between the members of the class are enhanced, leading to a much clearer, less noisy representation of the respective view of the molecule/complex than was present in the individual projections. An example, showing a field of particles and three class-averages for the "base" of a type III secretion complex, is shown in Figure 7. For simplicity, the example is shown for a negatively stained specimen, but the improvement in the signal-to-noise ratio, brought about by averaging, can nevertheless be appreciated.

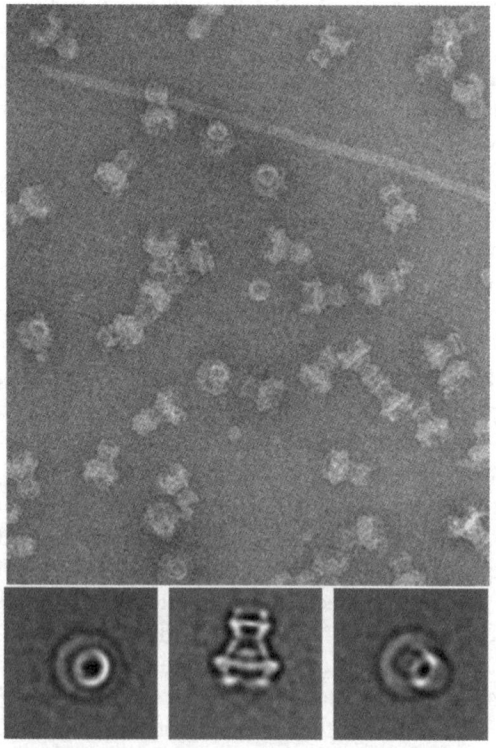

Figure 7 *Class-averages. The noisy nature of the individual-particle projections requires extensive averaging to retrieve the full information for each particle view. This figure shows a field of negatively stained particles representing the base of the type III secretion needle complex shown in Figure 5, and three selected class-averages that were obtained by MSA classification of the data set. Even in this case of a negatively stained sample, averaging many particle images for each view clearly improves the signal-to-noise ratio*

4.1.3 Euler Angle Determination

The next step toward structure calculation consists in determining how the different projections are related to each other. Specifically, the task consists in figuring out the three angles, also known as Euler angles, that describe how the particles are oriented with respect to a common *xyz*-coordinate system. Determination of Euler angles is the most critical step during single-particle reconstruction because the assignments determine how the class-averages will be combined into a 3D structure. In other words, inclusion of "bad" particle projections in any one class-average will only blur the average. However, errors in the assignment of the Euler angles will cause hundreds, if not thousands, of particles to be added to the final structure in an incorrect way, which will result in a significant distortion of the reconstructed volume. As for classification, several approaches have been developed to determine particle orientations. The majority of these approaches exploit the fact that each pair of projections (or their calculated Fourier transforms) shares a line along which the data are identical. This property is known as "common lines" and identifies determination of Euler angles as a combinatorial problem, the most difficult part of which is to find sets of angles for three different projections, such that each of the projections shares a common line with each of the remaining two projections. Once this initial set has been found, the frame becomes fixed and a solution can be determined for all remaining projections as long as they originate from the same molecule/complex. The latter remark is meant to emphasize that it is at the stage of Euler angle determination where structural and conformational heterogeneity comes into play. While heterogeneous populations can always be classified, especially when using reference-free methods such as MSA, determination of Euler angles may fail because no solution exists to the common lines problem. Thus, failure to obtain clear Euler angle assignments at any stage of structure refinement invariably spells trouble. However, such failure is unlikely, especially during early rounds of refinement where the resolution of the structure/model is very low. Because of this, modest heterogeneity may go undetected, particularly if the molecule/complex has an asymmetric shape, leading to an incorrect structure. Unfortunately, methods and criteria for structure/data validation are still subjects of debate and consequently, single-particle reconstructions should be treated with a healthy amount of skepticism, particularly if the particles lack symmetry and if the claimed resolution is high.

4.1.4 Structure Calculation

Once the Euler angles are known, the structure can be calculated by combining the information from the different views (Figure 8). This is conveniently done in reciprocal space, where the Fourier Transforms of the class-averages can be combined to recreate the 3D transform of the molecule/complex. The real-space volume is then obtained by calculating the inverse transform and by choosing a suitable threshold for surface rendering. With regard to the latter, it should be emphasized that detergent bound to the protein cannot be distinguished from the protein itself at low resolutions. Consequently, at low resolutions (worse than ~15 Å) the contribution of the

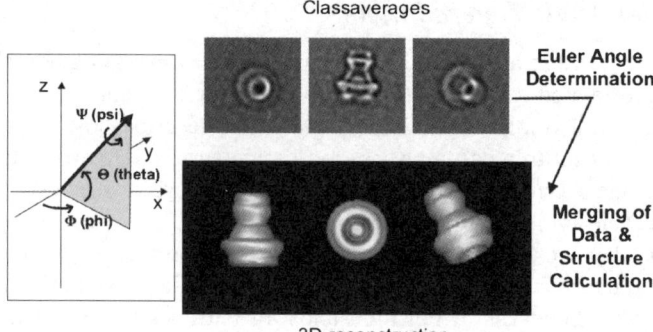

Figure 8 *Euler angles and structure calculation. Once class-averages have been calculated, the last step consists in the assignments of the three real-space angles that describe the orientation of the particle on the electron microscope grid when the image was recorded. Knowledge of these angles defines the geometric relation between the different class-averages and hence allows their information to be combined in a meaningful way. A structure can be calculated if sufficiently different views have been obtained. The number of views required depends on the size and symmetry of the complex*

detergent to the overall molecular weight needs to be considered when determining the contouring threshold. This detail is often overlooked and can give rise to artifacts in the appearance of the final volume.

After a first data-based model has been obtained, the structure is usually refined further by calculating a new and improved set of reference projections that serve for the next cycle of image classification and Euler angle assignment. The paradigm underlying this iterative alignment strategy is that improvements in the class-averages will lead to better defined Euler angles, which in turn will result in an even better model. This is most certainly true for the initial rounds of refinements. However, as the nominal resolution of the model increases, an increasing fraction of data will be weak and very close to the noise level. This can result in noise bias, an effect that has recently been investigated in depth by Stewart and Grigorieff.[20]

4.2   Electron Crystallography

The study of membrane protein structure has been a stronghold of electron crystallography ever since the structure of bacteriorhodopsin, a light-driven archebacterial proton pump, was solved at near atomic resolution in 1990.[4] In fact, most electron crystallographic image processing programs were developed in the course of the structure determination of bacteriorhodopsin,[4,11,21] and provided many insights into how to deal with electron microscopic images in general. In contrast to single-particle approaches, electron crystallography requires crystalline samples. Yet different from their 3D relatives that are used in X-ray crystallography, crystals used in electron crystallography are usually only a single layer thick. For this reason, these crystals are referred to as 2D crystals.

4.2.1 2D Crystallization and Advantages of Crystalline Samples over Single Particles

Most frequently, 2D crystals of membrane proteins are obtained by reconstitution. The principle of this approach, which relies on the insertion of purified protein into lipid bilayers upon removal of detergents from a protein–lipid-containing solution, is sketched in Figure 9 (for more detail see Refs. 22 and 23). Notably and in contrast to 3D crystals of membrane proteins, packing contacts between molecules are almost exclusively formed between the membrane-embedded parts of the protein. This property makes 2D crystallization of membrane proteins somewhat easier than their 3D crystallization because the likelihood that suitable contacts can be formed between transmembrane helices of neighboring molecules is higher than the chance of encountering matched surfaces between the soluble parts of the molecules. Nevertheless, obtaining 2D crystals that are ordered to near atomic resolution is just as challenging as obtaining 3D crystals, which is the reason why the majority of structures solved by cryoEM to date are limited to about 5–10 Å resolution. Still, these resolutions are better than what can routinely be obtained by single-particle approaches and the improvement in resolution is based on the fact that working with a crystalline sample has tremendous advantages over structure determination from single-particle projections: (1) there is no lower limit to the size of the molecules because one does not need to identify individual-particle projections, (2) the crystal lattice overcomes the alignment/classification problem, (3) the coherent diffraction of thousands of molecules greatly enhances the signal over the noise, and (4) the

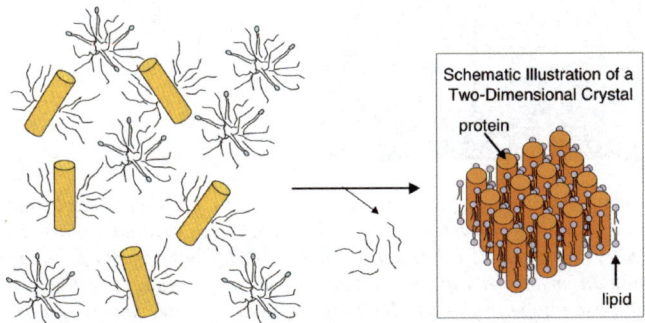

Figure 9 *Principle of 2D crystallization by reconstitution. This figure illustrates the principle of the most commonly used approach toward 2D crystallization of membrane proteins. Briefly, a preparation of the protein of interest, purified in detergents, is mixed with lipids that have also been solubilized in detergent. After equilibration of the two components, the detergent molecules can be removed by either dialysis or incubation with hydrophobic BioBeads.[23] This will cause the lipids to spontaneously form a bilayer once the detergent concentration drops below the critical micelle concentration. At that point, the protein also enters the newly formed bilayer, and may form a regular 2D array if favorable interactions between transmembrane helices from neighboring molecules allow them to pack tightly. To be useful, these arrays should contain a few thousand molecules/complexes, which usually is the case by the time the crystals reach a size of ~0.5 × 0.5 μm*

periodic nature of the crystals causes the information about the molecules to be encoded in discrete diffraction maxima, which does not apply for single particles whose transforms are continuous. This latter point is illustrated in Figure 10, which shows part of a low-dose image of a gap-junction 2D crystal and the corresponding, calculated transform.

4.2.2   Image Filtering and Lattice Straightening

The discrete nature of the calculated diffraction pattern of 2D crystals provides access to powerful tools for processing of image data (see Refs. 4, 11, 21, 22). For instance, the discrete nature of the calculated diffraction pattern allows images to be digitally filtered, which in a single step removes most of the image information that is not related to the crystal (see Figure 11). This step is possible because in contrast to traditional X-ray diffraction patterns, calculated diffraction patterns contain both amplitude and phase information, which allows images to be calculated by simply inverting a masked transform. Moreover, the calculated diffraction patterns also contain information about distortions of the lattice. This information can be extracted by

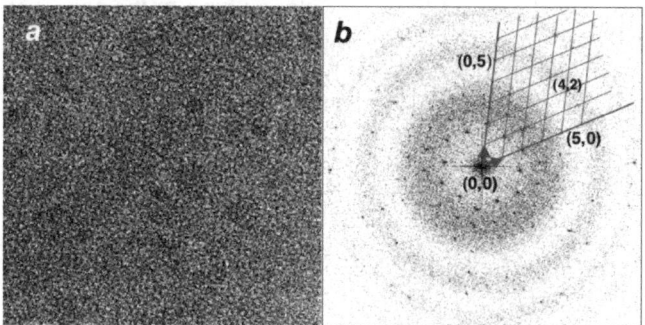

Figure 10 *Fourier transform of 2D crystals. The regular packing of molecules on a lattice causes the molecular transform to be sampled at discrete points. This property of 2D crystals is easily observed in calculated diffraction patterns and is shown here, using an unstained gap-junction 2D crystal as an example. Panel a shows part of a low-dose image. Although the image appears to have some regularity, no individual molecules can be identified. The crystallinity of the sample is very obvious in the calculated diffraction pattern (panel b), which shows a set of discrete reflections superimposed on an alternating pattern of dark and bright, ring-shaped regions that are known as Thon rings. The arrows drawn into the transform indicate the direction of the unit cell vectors (h,k). Discrete diffraction maxima occur at the intersections of lines that run parallel to the unit cell vectors. In this example, three reflections – (0,5), (4,2), and (0,5) – have been explicitly identified. The (0,0) term at the center of the transform represents the undiffracted electron beam. The Thon ring pattern is not related to the crystal but represents how the CTF of the objective lens (Section 3.2) modulates the continuous transform of the amorphous carbon support film. Although the information about the crystal is contained in the discrete diffraction maxima, their positioning in the transform makes them subject to modulation by the CTF as well, which needs to be considered in the image processing, data merging, and structure calculation (see Section 4.2.3 and Figure 12)*

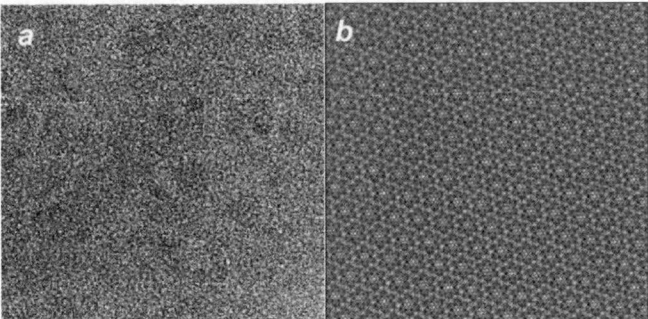

Figure 11 *Digital image filtering. The discrete nature of the diffraction pattern of a 2D crystal allows much of the noise in the image to be removed by a process known as digital filtering. In this procedure the raw image (panel a) is Fourier transformed, and the calculated diffraction pattern is modified to only contain the information present in the strong diffraction maxima, while all remaining transform points are made zero. Inversion of this masked transform generates a digitally filtered image (panel b) that shows the repeating motif of the crystal much more clearly. In this particular example, a tight mask radius was chosen, limiting the information to only a single pixel around the diffraction maxima. This results in a highly averaged, uniform image. If larger mask radii are used, information about lattice distortions will also be present in the filtered image, which can be exploited to computationally improve crystal quality*

cross-correlation procedures and opens the door to the computational improvement of the lattice by image interpolation. Because of this, even modestly ordered crystals can usually be improved to yield information at better than 10 Å, where the packing arrangement of the transmembrane helices becomes visible. Such intermediate resolution structures are usually the first milestone toward the structure determination of a membrane protein by cryoEM and have been successfully used as templates to constrain modeling of the structure in cases where better crystals were not available.[24,25]

4.2.3 Impact of CTF on Images of 2D crystals and CTF–Correction

As was introduced in Section 3.2, the effects of the CTF are born from the fact that contrast is generated by sampling the interference pattern of the undiffracted and diffracted wavefronts at a certain distance from the specimen plane (=defocus). At any given defocus, the exact pathlength differences and thereby the magnitude of interference will depend on the diffraction angle of the component. Similarly, some components will be sampled at a point where the interference is positive while others will be sampled at a point of destructive interference or where interference results in complete cancellation of the signal. In the calculated diffraction patterns, these effects manifest themselves as an alternating pattern of circular regions of variable intensities known as Thon rings. In the example shown in Figure 10, the pattern is superimposed on the discrete diffraction maxima that originate from the crystal and is caused by the presence of the amorphous carbon support film used to mount the crystals on the electron microscope grid. While the variation in intensities is easy to appreciate, what is less obvious is that positive contrast alternates with negative contrast on neighboring regions of the Thon ring pattern. This alternation in contrast, which in reciprocal space amounts to a 180° phase shift, has a profound influence on the image and is illustrated in Figure 12. From this example it immediately becomes obvious why electron microscopic data need to be corrected for the impact of the CTF. Fortunately, 2D crystals are supported on the electron microscope grid by an amorphous carbon film. This gives rise to a well-defined Thon ring pattern in the calculated transforms of the images (see Figure 10). And consequently, CTF correction can be accomplished by adding 180° to the phases of all diffraction maxima that fall within the regions of the odd numbered Thon rings (first, third, fifth, …), which is easy because the position of each diffraction maximum within the transform can be exactly predicted based on the unit cell dimensions.

4.2.4   *Projection Density Maps and Calculation of 3D Structures*

After CTF correction of the data, a projection density map can be calculated by simple Fourier summation. Although each crystal consists of thousands of molecules, the projection will represent only a single view if the crystal was untilted with respect to the incident electron beam. As was pointed out at the very beginning, views from many different directions are required for the calculation of a 3D structure. In contrast to single particles, the large size of crystals prevents them from adopting random orientations on the carbon support film. Consequently, 3D data need to be generated by tilting the electron microscope grid with respect to the incident beam. Depending on the crystallographic symmetry, each projection of a tilted

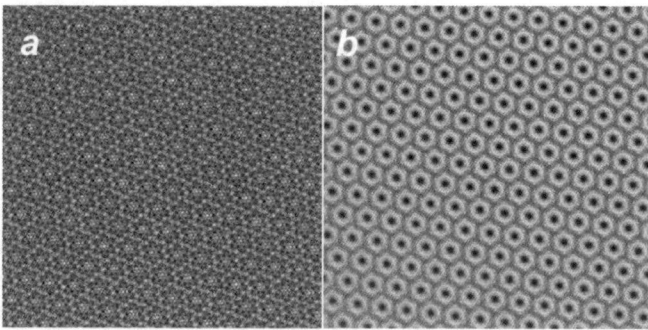

Figure 12 *Impact of CTF. The CTF introduces significant distortions into electron microscopic images. The figure shows digitally filtered images without (panel a) and with (panel b) correction for the CTF. Correction for the CTF is essential to retrieve a faithful image of the molecule. Moreover, zero values in the CTF cause some information to be irreversibly lost from each image. This problem can only be overcome by recording different images at different amounts of underfocus. This will change the distribution of Thon rings (Figure 10) and consequently, the phase values of equivalent reflections in the transform of different images may be out of register by 180°. Therefore, information from different images cannot be merged without careful CTF correction, making this a particularly critical step in the processing of electron crystallographic as well as single-particle data*

crystal adds information for one or more different views of the molecule, which can easily be combined in reciprocal space if the tilt geometry of the crystal with respect to the incident beam is known. Fortunately, this geometry can be determined quite accurately and is less error prone than determination of Euler angles in single-particle approaches. The combination of data from images of tilted crystals results in a collection of so-called "lattice line" data. These data reflect the smooth variation of amplitude and phase values for each reflection along the third axis of reciprocal space. Along this direction, the molecular transform is continuous because, like single particles, 2D crystals do not have a repeat along this direction due to the fact that they consist of only one layer of molecules (or in rare cases two layers that contain molecules in opposite orientations). An example for this type of data is shown in Figure 13. In order to calculate a structure from such data, one needs to extract a set of complete structure factors with indices (h,k,l). While the (h,k) indices are given by the discrete diffraction in the (xy)-plane of the transform, the third index is generated by first fitting curves to the experimental data before sampling each of the curves at points corresponding to an arbitrarily chosen unit cell repeat that is larger than the

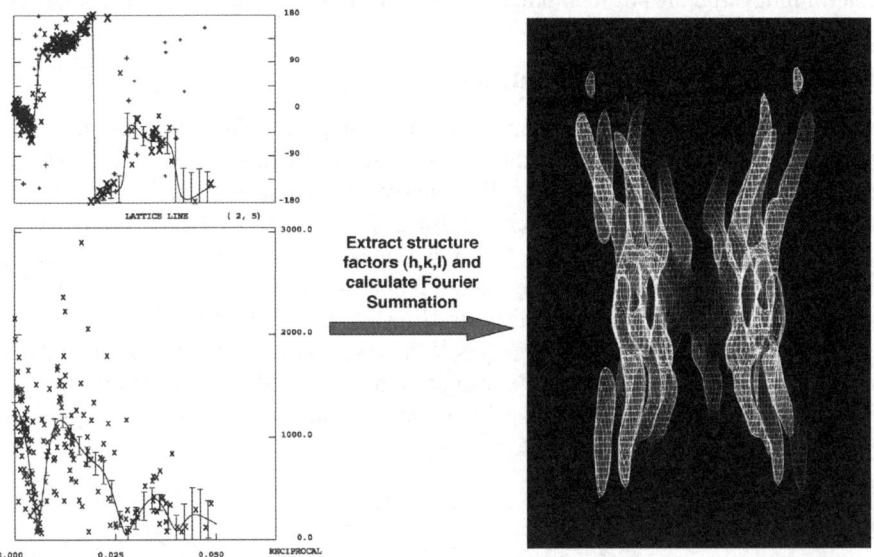

Figure 13 *From lattice line data to structure. While each image of a crystal, tilted or untilted, will yield a discrete diffraction pattern, the data contained in the diffraction maxima fall along continuous lines along the third dimension of the molecular transform. This distribution of data is known as lattice lines and is a direct consequence of the lack of a crystalline repeat along the c-axis of the 2D crystal. An example for the data distribution along the lattice line of a single reflection of the gap-junction 2D crystals is shown to the left. After fitting smooth curves to the experimental data points, complete structure factors (h,k,l) can be read from the curves and are used in the Fourier summation to calculate the structure. A cropped side view of the gap-junction channel is shown to the right. The elongated finger-shaped features represent individual transmembrane regions of this complex intercellular channel*

thickness of the crystal. Once a complete set of structure factors has been obtained, the structure can be calculated using standard crystallographic software.

5   Examples

5.1   Single-Particle Reconstructions

To provide a guide for further reading, this section will list a few examples where cryoEM has been used successfully to determine structures of membrane proteins. Concerning single-particle reconstructions, it was mentioned that the small size of many membrane proteins and the need to maintain them in detergent make structure determination by cryoEM difficult. Nevertheless, there are examples, such as mitochondrial complex I,[26] integrins,[27] ryanodine receptors,[28,29] L-type Ca^{2+}-channel,[30] IP3-receptors,[31,32] pneumolysin,[33] or type III secretion complexes,[34] where structures were obtained from images of unstained samples. Notably, two of these structures, complex I and integrins, were of asymmetric particles, providing hope that this approach may become more fruitful in the future when the emphasis of structure determination will shift from analyzing structures of individual membrane proteins to determining structures of membrane protein-containing macromolecular complexes.

5.2   Electron Crystallography

There are quite a number of examples in the literature describing how electron crystallography has been used to study the structures of membrane proteins. The most prominent are bacteriorhodopsin,[4] light harvesting complex,[35] and aquaporin,[36,37] where near atomic resolutions were obtained. In most other cases, crystal size and order limited resolutions to 5–10 Å. Among those are the G protein-coupled receptor rhodopsin,[38] H^+-ATPase,[39] the bacterial sodium protein antiporter NhaA,[40] the bacterial multidrug resistance transporter EmrE,[41] Sec YEG,[42] the bacterial oxalate transporter,[43] and gap-junction channels.[44] In two of these cases, rhodopsin and gap-junction channels, the intermediate resolution maps served as templates for the generation of a C_α-carbon template, which allowed interpretation of the maps at the level of individual amino acid sidechains.[24,25]

6   Conclusion

Starting from the realization some 40 years ago that the structures of biological macromolecules can be determined from the projection images obtained in an electron microscope, a series of technological and computational developments has transformed molecular microscopy into a vibrant approach poised to make many exciting contributions to the understanding of membrane protein structure and function. The advent of better microscopes and sample stages as well as the development of automated data collection software has eliminated most of the hurdles toward the acquisition of high-quality images. Thus, it seems that the field has come full circle where, once again, progress is limited by the rate at which suitable specimens can be prepared. This applies in particular to membrane-associated signaling and scaffolding complexes, which clearly represent the next challenge in the study of membrane protein structural biology.

References

1. D.J. De Rosier and A. Klug, *Nature*, 1968, **217**, 130.
2. C.C. Blake, R.H. Fenn, A.C. North, D.C. Phillips and R.J. Poljak, *Nature*, 1962, **196**, 1173.
3. H. Muirhead and M.F. Perutz, *Nature*, 1963, **199**, 633.
4. R. Henderson, M. Baldwin, T.A. Ceseka, F. Zemlin, E. Beckmann and K.H. Downing, *J. Mol. Biol.*, 1990, **213**, 899.
5. R. Henderson and P.N.T. Unwin, *Nature*, 1975, **257**, 28.
6. K.A. Taylor and R.M. Glaeser, *Science*, 1974, **186**, 1036.
7. J. Dubochet and A.W. McDowall, *J. Microsc.*, 1981, **124**, RP3.
8. J. Dubochet, J. Lepault, R. Freeman, J.A. Berriman and J.C. Homo, *J. Microsc.*, 1982, **128**, 219.
9. J. Dubochet, M. Adrian, J.J. Chang, J.C. Homo, J. Lepault, A.W. McDowall and P. Schultz, *Q. Rev. Biophys.*, **21**, 129.
10. Y. Fujiyoshi, *J. Elect. Microsc.*, 1989, **38**, S97.
11. L.A. Amos, R. Henderson and P.N.T. Unwin, *Prog. Biophys. Mol. Biol.*, 1982, **39**, 183.
12. O. Skolova, L. Kolmakova-Partensky and N. Grigorieff, *Structure*, 2001, **9**, 215.
13. E.V. Orlova, M. Papakosta, F.P. Booy, M. van Heel and J.O. Dolly, *J. Mol. Biol.*, 2003, **326**, 1005.
14. M.K. Higgins, D. Weitz, T. Warne, G.F. Schertler and U.B. Kaupp, *EMBO J.*, 2002, **21**, 2087.
15. L.A. Kim, J. Furst, D. Gutierrez, M.H. Butler, S. Xu, S.A. Goldstein and N. Grigorieff, *Neuron*, 2004, **41**, 513.
16. M. van Heel, B. Gowen, R. Matadeen, E.V. Orlova, R. Finn, T. Pape, D. Cohen, H. Stark, R. Schmidt, M. Schatz and A. Patwardhan, *Q. Rev. Biophys.*, 2000, **33**, 307.
17. J. Frank, *Three Dimensional Electron Microscopy of Macromolecular Assemblies*, Oxford University Press, New York, NY, 2005.
18. Y. Zhu, B. Carragher, R.M. Glaeser, D. Fellmann, C. Bajaj, M. Bern, F. Mouche, F. de Haas, R.J. Hall, D.J. Kriegman, S.J. Ludtke, S.P. Mallick, P.A. Penczek, A.M. Roseman, F.J. Sigworth, N. Volkmann and C.S. Potter, *J. Struct. Biol.*, 2004, **145**, 3.
19. S.J. Ludtke, P.R. Baldwin and W. Chiu, *J. Struct. Biol.*, 1999, **128**, 82.
20. A. Stewart and N. Grigorieff, *Ultramicroscopy*, 2004, **102**, 67.
21. R. Henderson, J.M. Baldwin, K.H. Downing, J. Lepault and F. Zemlin, *Ultramicroscopy*, 1986, **19**, 147.
22. M. Yeager, V.M. Unger and A.K. Mitra, in *Methods of Enzymology*, Vol. 294, M. Conn (ed), Academic Press, San Diego, CA, 1999, 135.
23. G. Mosser, *Micron*, 2001, **32**, 517.
24. J.M. Baldwin, G.F.X. Schertler and V.M. Unger, *J. Mol. Biol.*, 1997, **272**, 144.
25. S.J. Fleishman, V.M. Unger, M. Yeager and N. Ben-Tal, *Mol. Cell*, 2004, **15**, 879.

26. N. Grigorieff, *J. Mol. Biol.*, 1998, **277**, 1033.

27. B. Adair and M. Yeager, *Proc. Natl. Acad. Sci. USA,* 2002, **99**, 14059.

28. M. Samsó, T. Wagenknecht and P.D. Allen, *Nat. Struct. Mol. Biol.*, 2005, **12**, 539.

29. I.I. Serysheva, S.L. Hamilton, W. Chiu and S.J. Ludtke, *J. Mol. Biol.*, 2004, **345**, 427.

30. M. Wolf, A. Eberhart, H. Glossmann, J. Striessnig and N.Grigorieff, *J. Mol. Biol.*, 2003, **332**, 171.

31. I.I. Serysheva, D.J. Bare, S.J. Ludtke, C.S. Kettlun, W. Chiu and G.A. Mignery, *J. Biol. Chem.*, 2003, **278**, 21319.

32. Q.X. Jiang, E.C. Thrower, D.W. Chester, B.E. Ehrlich and F.J. Sigworth, *EMBO J.*, 2002, **21,** 3575.

33. S.J. Tilley, E.V. Orlova, R.J.C. Gilbert, P.W. Andrew and H.R. Saibil, *Cell*, 2005, **121**, 247.

34. T.C. Marlovits, T. Kubori, A. Sukhan, D.R. Thomas, J.E. Galán and V.M. Unger, *Science*, 2004, **306**, 1040.

35. W. Kühlbrandt, D.N. Wang and Y. Fujiyoshi, *Nature*, 1994, **367**, 614.

36. K. Mitsuoka, K. Murata, T. Walz, T. Hirai, P. Agre, J.B. Heymann, A. Engel and Y. Fujiyoshi, *J. Struct. Biol.*, 1999, **128**, 34.

37. G. Ren, A. Cheng, V. Reddy, P. Melnyk and A.K. Mitra, *J. Mol. Biol.*, 2000, **301**, 369.

38. V.M. Unger, P.A. Hargrave, J.M. Baldwin and G.F. Schertler, *Nature*, 1997, **389**, 203.

39. M. Auer, G.A. Scarborough and W. Kühlbrandt, *Nature*, 1998, **392**, 840.

40. K.A. Williams, *Nature*, 2000, **403**, 112.

41. I. Ubarretxena-Belandia, J.M. Baldwin, S. Schuldiner and C.G. Tate, *EMBO J.*, 2003, **22**, 6175.

42. C. Breyton, W. Haase, T.A. Rapoport, W. Kühlbrandt and I. Collinson, *Nature*, 2002, **418**, 662.

43. T. Hirai, J.A. Heymann, D. Shi, R. Sarker, P.C. Maloney and S. Subramaniam, *Nat. Struct. Biol.*, 2002, **9**, 597.

44. V.M. Unger, N.M. Kumar, N.B. Gilula and M. Yeager, *Science*, 1999, **283**, 1176.

CHAPTER 10

Atomic Resolution Structures of Integral Membrane Proteins Using Cubic Lipid Phase Crystallization

HARTMUT LUECKE

Departments of Molecular Biology & Biochemistry, Physiology & Biophysics, and Information & Computer Sciences, The UCI Center for Drug Discovery, University of California, Irvine, CA 92697-3900, USA

1 Introduction

The determination of high-resolution structural models of macromolecules, obligatory for detailed structure–function studies, requires large amounts of pure, stable target protein. The two principal methods of structure determination are X-ray crystallography, a method that requires growing highly ordered 3D crystals of the target protein, and nuclear magnetic resonance (NMR) spectroscopy, a method that requires the generation of doubly isotopically labeled (^{15}N and ^{13}C) protein.

1.1 Nuclear Magnetic Resonance Techniques

Although isotopic labeling for NMR studies is rather routine, albeit somewhat expensive, the actual determination of 3D structures using multidimensional magnetic resonance techniques faces several limitations. NMR methods applied in solution have to overcome problems such as aggregation due to the high protein concentration, limited stability at relatively high temperatures over extended periods of time, or peak broadening resulting in peak overlap due to slow tumbling rates (which are proportional to the molecular weight, effectively limiting the size of the macromolecular complex that can be studied in detail). Furthermore, the effective resolution of the final structural model tends to be relatively low, largely as a result of the low number of experimentally determined restraints. One advantage of solution NMR over crystallography is the fact that the target protein is studied in a more natural setting as opposed to a target that is part of a crystal lattice. More recently, solid-state magic angle spinning NMR methods have been developed for isotopically labeled proteins, with the distinct advantage that molecular weight is no longer an inherent limitation. Both techniques are described in detail elsewhere in this book.

1.2 Crystallography

Macromolecular crystallography has to contend with its own share of problems. The major bottleneck is the production of well-ordered 3D crystals of the macromolecular target for X-ray diffraction (the related technique of electron diffraction using ordered 2D crystals is discussed briefly in the historical perspective below and elsewhere in this book). Even for soluble proteins, the crystallization step is still largely empirical, where the trial-and-error approach of screening different conditions (hundreds to thousands) typically requires large amounts of stable protein. In screens setup by manual pipetting, the amount of protein per condition screened can be estimated by the product of the volume of protein solution used per condition (typically 1–2 µl) with the concentration of the protein (typically 10–20 mg mL^{-1}, equivalent to 0.2–0.4 mM for a 50,000 Da protein). This amounts to approximately 10–40 µg of protein per condition, or 10–40 mg of pure, stable protein to screen 1000 conditions. This amount can be reduced drastically by employing precision robots that are able to reproducibly pipet volumes as small as 50 nL of varying viscosity, reducing the amount of protein required for large-scale screening by a factor of 20 or more.

1.3 Crystallization Techniques

1.3.1 Vapor Diffusion

Regular (soluble) proteins are usually crystallized using the vapor diffusion technique, whereby a small volume of protein solution (1–2 µL) is mixed with an equal volume of a precipitating reagent, forming the so-called "drop," which is in vapor contact with a much larger volume (500 µL, called the well or reservoir) of the same precipitating reagent.[1] Common precipitating agents are solutions of ammonium sulfate (AS), polyethylene glycol (PEG), or methanol, buffered to varying pH values and containing various inorganic salt additives. Immediately after the protein and the small volume of precipitating agent have been mixed to form the drop, the protein should stay in solution. Then, over time, vapor diffusion will cause equilibration of the precipitating reagent concentration between the drop and the reservoir, causing an increase of the precipitating reagent concentration in the drop. Once the solubility of the protein in the drop has been reduced sufficiently, two things can happen: (1) the individual protein molecules start to aggregate in nonspecific ways, forming an amorphous precipitate, and (2) far more desirable, crystal seed formation and crystal growth commence. This is the standard crystallization approach that led to the vast majority of the currently 35, 144 structures listed in the Protein Data Bank at http://www.rcsb.org.[2]

However, the vapor diffusion crystallization technique works only for proteins that can be kept stable for days or weeks at relatively high concentrations (5–50 mg mL^{-1}) without appreciable inherent aggregation.

1.3.2 Microdialysis Crystallization

A variation of the vapor diffusion technique employs semipermeable dialysis membranes that separate the subcritical protein–reagent mixture from a solution of the reagent at a higher concentration. This process is also called batch crystallization and is usually not employed for screening due to the even larger volume of protein required. Alternatively, the dialysis membrane can be replaced by a layer of oil, which then ensures differential diffusion rates for the various components.

1.4 Special Issues of Membrane Protein Crystallization

Polytopic integral membrane proteins, the star actors of this book, add another layer of complexity to the issue of crystallization. This explains the shocking scarcity of known membrane protein structures. Today, membrane proteins make up only about 0.5% of the Protein Data Bank (99 unique membrane proteins with 188 structural models, see http://blanco.biomol.uci.edu/Membrane_Proteins_xtal.html and http://www.mpibp-frankfurt.mpg.de/michel/public/memprotstruct.html), whereas the various genome-sequencing projects predict 25–30% membrane proteins in most genomes. This 50- to 60-fold underrepresentation of structures of membrane proteins in the Protein Data Bank can be attributed to a few main causes, which will be described below. This underrepresentation is even more stunning when one considers only mammalian membrane proteins: barely a handful of such structures are known today, and only for one of these (number 8 below) was the protein not derived from its natural source but through heterologous expression:

(1) Cytochrome-*c* oxidase from bovine heart.[3]
(2) Cytochrome-bc_1 complex from bovine heart mitochondria[4]
(3) The G-protein coupled light receptor bovine rhodopsin (from the bovine retina)[5]
(4) Aquaporin 1 from human red blood cells[6]
(5) Calcium ATPase from rabbit muscle sarcoplasmic reticulum[7]
(6) ADP/ATP carrier from bovine heart mitochondria[8]
(7) Aquaporin 0 from the bovine eye lens[9]
(8) The Kv1.2 voltage-gated potassium channel from *Rattus norvegicus* (expressed in the yeast *Pichia pastoris*).[10]

Two reasons are chiefly responsible for this scarcity of structural information of membrane proteins:

(1) It is generally very difficult to obtain large amounts of the functional form of most membrane proteins. This is particularly the case for eukaryotic membrane proteins, where folding, trafficking, targeting, and modifications are far

more common and more complex than in prokaryotes during and after protein synthesis. And even in cases where heterologous expression of eukaryotic membrane proteins in prokaryotes would appear feasible, issues such as protein stability, toxicity to the host, and limited insertion volume, namely bilayers, often prove to be virtually insurmountable. The issues surrounding expression and purification of membrane proteins are covered in detail elsewhere in this book.

(2) For common protein purification and concentration methods to be applicable, integral membrane proteins need to be turned into 'soluble' proteins by extracting them from their native phospholipid bilayer. This is usually achieved by a process called detergent extraction, whereby mild detergents are employed to replace the hydrophobic belt formed by the lipid tails of the planar membrane bilayer. Once a membrane protein molecule has been extracted, detergent molecules form a micellar torus around each individual membrane protein molecule that covers the majority of its hydrophobic surface (~30 Å in height). More details about the use of detergents in extracting membrane proteins from their native lipid bilayers and for essentially turning membrane proteins into soluble proteins are presented elsewhere in this book.

1.5   History of Membrane Protein Crystallization

The first membrane protein structure was determined by electron diffraction on ordered 2D crystals. It was the structure of bacteriorhodopsin, a highly efficient light-driven ion pump of the archaeon *Halobacterium salinarum,* where it generates an electrochemical ion gradient across the plasma membrane that is subsequently converted into chemical energy in the form of ATP by a second integral membrane protein, ATP synthase. These two membrane proteins together comprise one of the simplest systems that transduce light energy into chemical energy (photosynthesis). The first structural report presented a very low-resolution model (7 Å) that showed a bundle of seven slightly tilted transmembrane helices.[11] This model was later improved to an atomic model at 3.5 Å resolution.[12] This medium-resolution atomic model was then used to provide the starting phases for the high-resolution bacteriorhodopsin structures obtained from 3D crystals grown using the cubic lipid phase (CLP) approach (the main subject of this chapter),[13,14] detergent-based vapor diffusion,[15,16] and crystallization from bicelles[17] and spherical vesicles.[18] The effect of increasing resolution (from 7 Å for the early electron diffraction studies to 1.55 Å for the CLP-based X-ray diffraction studies) will be discussed further in Section 3.

However, the first structure of any integral membrane protein at a resolution high enough to allow the construction of an atomic model was determined from 3D crystals[19,20] using X-ray diffraction. It is the structure of the photosynthetic reaction center from *Rhodopseudomonas viridis*[20,21] that earned the team the Nobel Prize in Chemistry in 1988.[22] For the first time, the structure of a membrane protein complex could be correlated with its function in detail, in this case a light-driven electron pump across the photosynthetic membrane.

1.6   Aim of this Chapter

This chapter will address issues specific to generating well-diffracting crystals of integral (polytopic) membrane proteins, in particular by use of the CLP crystallization method.

2   Membrane Protein Crystals and Crystallization

Crystals of integral membrane proteins fall into two categories: type I and type II crystals. Type I crystals are stacks of 2D crystals of bilayer-embedded protein. This means that such crystals contain, in addition to the target protein and aqueous buffer, relatively large amounts of bilayer-forming lipids. These membrane stacks have to be highly ordered in all three dimensions to yield useful information upon X-ray diffraction. Well-ordered type I crystals were not observed until the advent of the CLP crystallization technique.[13] Shortly thereafter, type I crystals were grown by crystallization from bicelles[17] and small vesicles[18] as well.

Type II crystals are by definition all nontype I crystals. They are grown from detergent-solubilized membrane proteins using the vapor diffusion or microbatch techniques. Type II crystals are thus similar to crystals of regular soluble proteins, with the notable difference that an appreciable volume of type II crystals is comprised of disordered detergent micelles. This generally results in low relative protein densities (large V_M or Matthews coefficient[23]) with concurrent below-average crystalline order, frequently coupled with anisotropic diffraction. In contrast to type I crystals, type II crystals contain no or very few ordered lipid molecules since they lack bilayers. The first 3D crystals of a membrane protein (bacteriorhodopsin) were reported in 1980.[16]

In functional terms, type I crystals are likely to represent more native conditions since the protein is embedded in a native-like lipid bilayer environment as opposed to being surrounded by detergent molecules (type II crystals). Until recently, all membrane protein X-ray structures had been obtained from type II crystals, *i.e.* crystals containing detergent-solubilized protein molecules.

2.1   Vapor Diffusion of Detergent-Solubilized Membrane Proteins

Traditional vapor diffusion crystallization (see Section 1.3.1) allows for a large number of parameters.[1] Some of the common choices when attempting to crystallize a protein are:

(1) Buffer and its pH value
(2) Temperature (4 °C *vs.* 20 °C)
(3) Initial protein concentration
(4) Precipitant (AS, PEG, *etc.*)
(5) Initial precipitant concentration
(6) Concentration gradient between drop and reservoir
(7) Cations/anions and other additives such as small amphiphiles.

Clearly, an exhaustive screening of all these parameters for each target protein is an impossible task. Furthermore, when confronted with the task of crystallizing an integral membrane protein, additional parameters need to be considered:

(1) Choice of detergent for extraction from lipid bilayer
(2) Choice of detergent for purification and crystallization
(3) Use of antibody fragments to increase hydrophilic surface area[24]
(4) Limited protease digestion to remove highly flexible regions[25]
(5) Homologs from different organisms may be more amenable to crystallization.[26]

The central role of detergents in membrane protein biochemistry has already been discussed in the introduction in Section 1.4. and is discussed in detail elsewhere in this book. Briefly, detergents serve two main purposes:

(1) "Harsh(er)" detergents are generally necessary to *extract* a given membrane protein from its native lipid bilayer (Figure 1). During this step, the protein needs to be dissociated from most bilayer lipids as well as from other proteins it may be interacting with. The result of extraction is ideally a monodisperse solution of the solubilized target protein without contamination by other proteins or nonspecifically bound lipids.
(2) A "mild(er)" detergent might have to replace the "harsh(er)" detergent from the extraction step to achieve a stable protein–detergent complex for *crystallization*

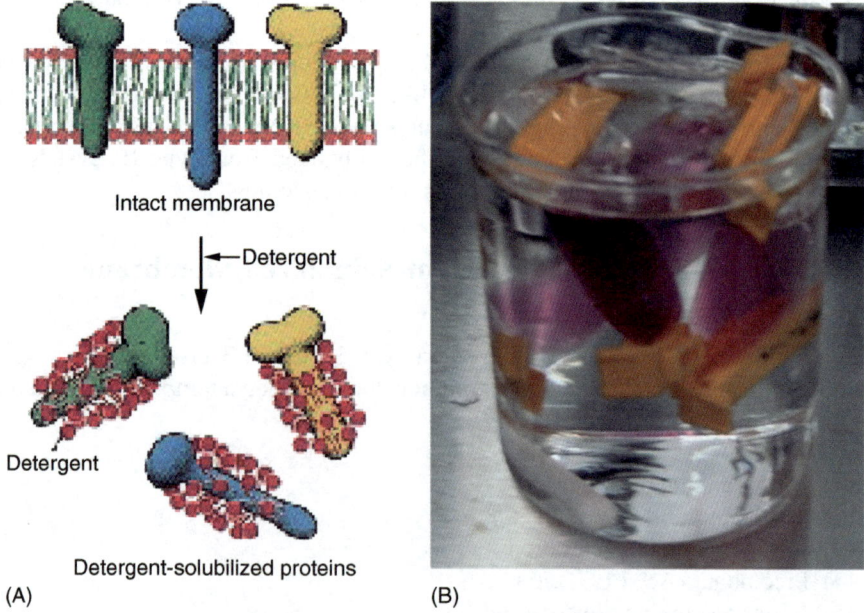

(A)   (B)

Figure 1 *(A) Schematic of the process of detergent-mediated extraction of an integral membrane protein from its native lipid bilayer. (B) Detergent-solubilized sensory rhodopsin in dialysis bags prior to crystallization* *trials*. The acyl chain length and type of each detergent determine how much of the hydrophobic belt of the target protein is covered (to prevent nonspecific aggregation), and also, how much of the hydrophilic surfaces remain accessible (to form crystal contacts).

Over 25 different detergents have been employed to grow well-diffracting crystals of membrane proteins. The most commonly used detergent is octyl-β-D-glucoside. The choice of detergent is generally dictated by the particular protein and typically has to be optimized by trial-and-error.

Once each target protein molecule is stably inserted into a detergent micelle, crystallization trials by regular vapor diffusion can be carried out. It is quite common for such detergent-solubilized membrane proteins to yield crystals using the vapor diffusion technique; sometimes these crystals are nice-looking with sharp edges and, if they contain a chromophore, with deep color – but unfortunately, such crystals often only diffract to rather poor resolution, somewhere in the 7–25 Å range. The poor diffraction and thus order on the atomic scale tend to be due to a combination of the following facts: (1) The detergent micelle forms a large, structurally ill-defined belt around the hydrophobic section of the membrane protein, leaving only a relatively small area for productive crystals contacts. (2) Heterogeneity of the protein due to different conformations, different post-translational modifications such as different degrees of glycosylation or phosphorylation, *etc.* may not prevent crystal formation, but is likely to prevent order on the 2–3 Å scale.

2.2 The Cubic Lipid Phase (CLP) Crystallization Method

Since the report of the first membrane protein 3D crystals of the light-driven ion pump bacteriorhodopsin in 1980,[16] it took almost 20 years for that structure to be solved at high resolution.[14,15,27]

The first membrane protein to have its structure elucidated was a different protein, the photosynthetic reaction center.[20] Since then, a small trickle of membrane protein structures, all based on crystals grown using vapor diffusion of detergent-solubilized proteins, have laid the foundation of our knowledge of this crucially important class of proteins.

A major breakthrough was achieved in 1996, when Landau and Rosenbusch[13] reported a fundamentally new way of growing 3D crystals of membrane proteins. This method exploits a phase of certain lipid–water mixtures called the Pn3m CLP (Figure 2) to serve as a reservoir for bilayer-embedded protein. For its formation, this phase requires lipids that are different from typical bilayer-forming lipids. CLP-forming lipids tend to have a narrower hydrophobic portion; in general they are mono-acyl lipids, *i.e.* lipids with only one acyl chain as compared to the common bilayer-forming lipids with two acyl chains. When such lipids (monoolein, monovaccenin) are mixed vigorously with the appropriate proportion of aqueous solvent (20–50% w/w) at a temperature above ~20 °C, the Pn3m CLP is formed (Figure 3).[28,29] This phase consists of a bicontinuous matrix, in which one compartment is predominantly hydrophilic (a 3D network of aqueous channels lined by the lipid headgroups), and the other compartment is hydrophobic (a spatially offset 3D network formed by highly curved bilayers of the mono-acyl lipid tails).

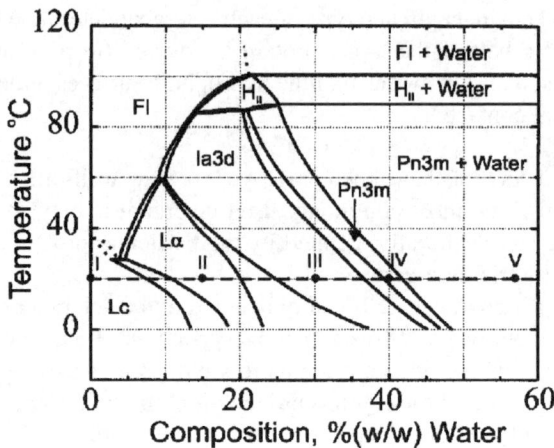

Figure 2 *Phase diagram of lipid–water mixtures. The Pn3m phase is used for CLP crystallization*

Thus, a hydrophilic solute (salt, buffer, soluble protein) in this matrix will partition into the aqueous channels, whereas a hydrophobic solute (membrane protein, cholesterol, other lipids) will partition into the hydrophobic portion of the bicontinuous matrix. Since each of the two compartments extends in all three dimensions, solutes in either compartment are free to diffuse within their respective hydrophilic and hydrophobic environments. When a detergent-solubilized membrane protein was part of the aqueous phase during formation of the CLP, it is thought that the protein – as the curved CLP bilayers form – will shed its detergent micelle and enter into the more native-like hydrophobic compartment (Figure 4). In this state, the protein is generally stable in the CLP for months. Only upon addition of a precipitating agent, such as high salt, PEG, or 2-methyl-2,4-pentanediol, do crystal seed formation and growth commence. The CLP then serves as a reservoir for membrane protein molecules as the crystals grow. During the process, the growing crystals, which are comprised of bilayer-forming lipids (*i.e.* neither monoolein nor monovaccenin) and protein, will slowly destroy the nearby CLP.

Once crystallization screens are set up using the CLP approach, several issues arise, such as:

(1) How does one visualize small crystals of a colorless protein in the often heterogeneous CLP? The majority of the crystals grown using CLP (see Table 1) are of proteins with chromophores. Adding covalently linked dyes to proteins without a chromophore may be one approach to see small crystals. Using a UV-sensitive microscope would be another.

(2) If crystals are obtained, they need to be removed from the rather viscous CLP without undue mechanical stress. Two methods have been developed. (A) A mother liquor solution with an additional suitable detergent is added to a 'blob' of CLP-containing crystals to be harvested. After several hours, the CLP typically begins to soften, and crystals can be removed with cryo-loops.[27]

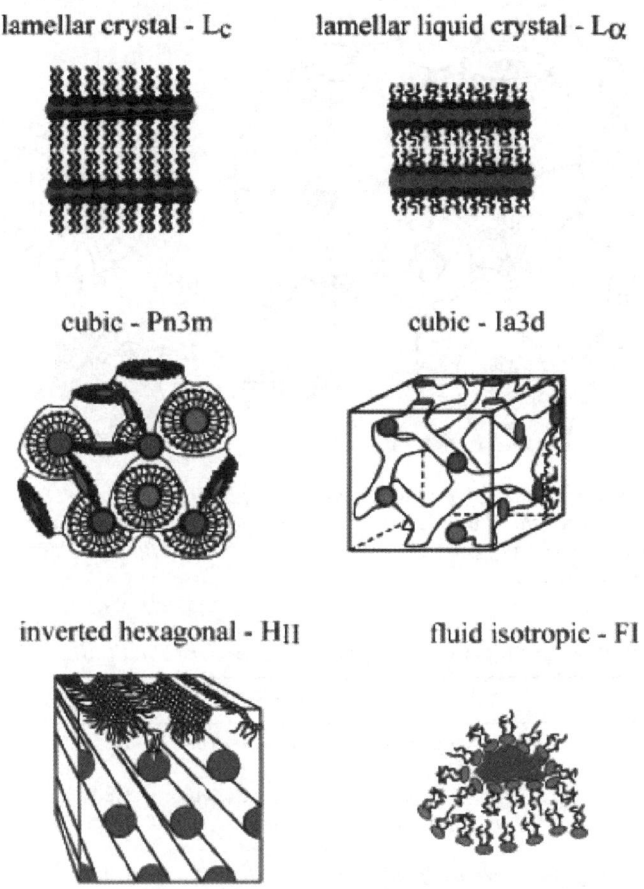

lamellar crystal - L$_c$

lamellar liquid crystal - L$_\alpha$

cubic - Pn3m

cubic - Ia3d

inverted hexagonal - H$_{II}$

fluid isotropic - FI

Figure 3 *Schematic of the various phases of lipid–water mixtures. Pn3m phase is the CLP employed for membrane protein crystallization*

(B) The CLP is incubated with lipases that slowly degrade the CLP enzymatically, generating free crystals.[32]

Since CLP crystals are type I membrane protein crystals, which are comprised of stacked, planar bilayers with embedded protein; bilayer-forming lipids have to be either carried along by the target protein or added to the crystallization mix. In the case of bacteriorhodopsin, the protein possesses a very high affinity for native archaeal lipids (diether lipids); these lipids, which are essential for crystal formation, are carried along by the protein all the way from the detergent extraction step through the various purification steps into the CLP setup. These lipids are an integral part of the crystals formed, with the majority of them visible in the electron density in a bilayer arrangement (Figure 5). For bacteriorhodopsin, during refinement of the structure, numerous well-defined difference density features were noted in the form of long narrow cylinders oriented parallel to the crystallographic *c*-axis.[14] From

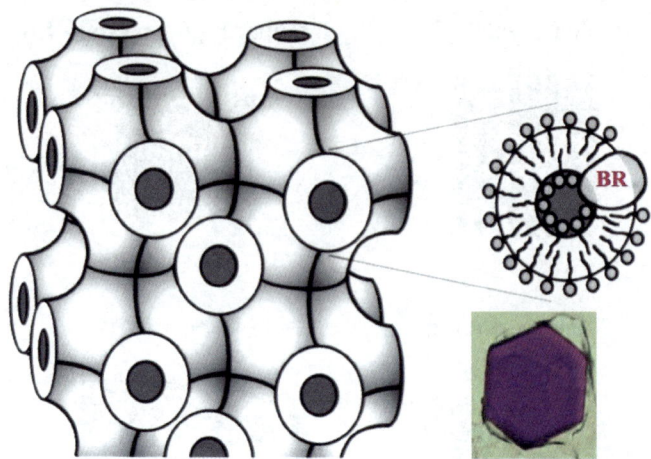

Figure 4 *Schematic of hydrophilic (dark gray) and hydrophobic (white) compartments of the CLP on the left. The lower right hand corner is an image of a bacteriorhodopsin crystal grown from CLP. These crystals are ~50–200 μm in diameter and 10–20 μm in thickness*

Table 1 *Membrane proteins crystallized using the CLP method*

Membrane protein	Source organism	Expressed in	Resolution [Å]	Reference
Bacteriorhodopsin	H. salinarum	H. salinarum	1.55	14
				34
Halorhodopsin	N. pharaonis	N. pharaonis	1.80	35
Sensory rhodopsin	N. pharaonis	E. coli	2.4	30,32
			2.1	
Complex of sensory rhodopsin and a 2TM transducer fragment	N. pharaonis	E. coli	1.94	36
Sensory rhodopsin	Anabaena sp.	E. coli	2.0	33
Reaction center	R. sphaeroides	R. sphaeroides	2.35	37
Outer membrane transporter, BtuB			No structure	38

these features, 18 lipid chains were identified (Figure 6). Four of these are modeled shorter than full length because of lack of contiguous electron density. Four individual pairs of full-length chains could be linked with a glycerol backbone, thereby identifying four archaeal (native) diether lipids. Lipid head-group densities were observed in several instances but were not modeled because of their lower quality and ambiguity. Mass spectrometry of dissolved crystals confirmed the presence of archaeal lipids.

For CLP crystals of other proteins, bilayer-forming lipids had to be added to the crystallization setup in order to obtain crystals.[30–32] Furthermore, in the case of the first eubacterial sensory rhodopsin from *Anabaena* sp., the lipids form undulating bilayers rather than planar ones (Figure 7).[33]

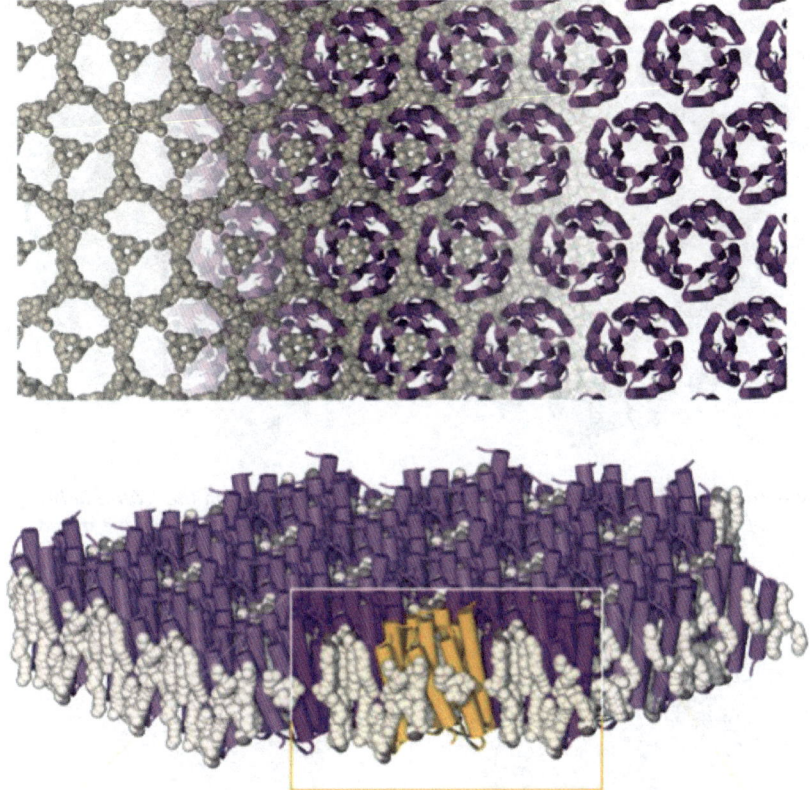

Figure 5 *Two views of the arrangement of one layer of lipids (gray) and bacteriorhodopsin (many purple and one yellow molecule showing seven transmembrane helices) in the CLP-derived crystals that diffract to better than 1.4 Å. The top panel shows a top view of one such layer, which has the same packing and lipid content as the naturally occurring 2D crystals of purple membrane. Note the dense packing of protein trimers with intervening layers of ordered lipid molecules*

Membrane proteins crystallized using the CLP method are shown in Table 1.

As can be seen from the resolution values in this table (between 1.55 and 2.4 Å), CLP crystals, once they formed, tend to be highly ordered. This may be counterintuitive when one considers that a large volume of these crystals is occupied by lipids. However, the lipids usually pack one or two acyl chains wide between protein molecules, ensuring high order within the bilayer (Figure 5). Most of the lipids make very specific contacts with grooves on the surface of the membrane protein (Figure 6).

Because most membrane proteins exhibit an appreciable electrical dipole moment perpendicular to the bilayer, type I crystals tend to have either crystallographic[30,33,39] or noncrystallographic[17] twofold symmetry perpendicular to the stacking direction, or they may show strong merohedral twinning perpendicular to the stacking direction (http://bass.bio.uci.edu/~hudel/br/twinning/).[27]

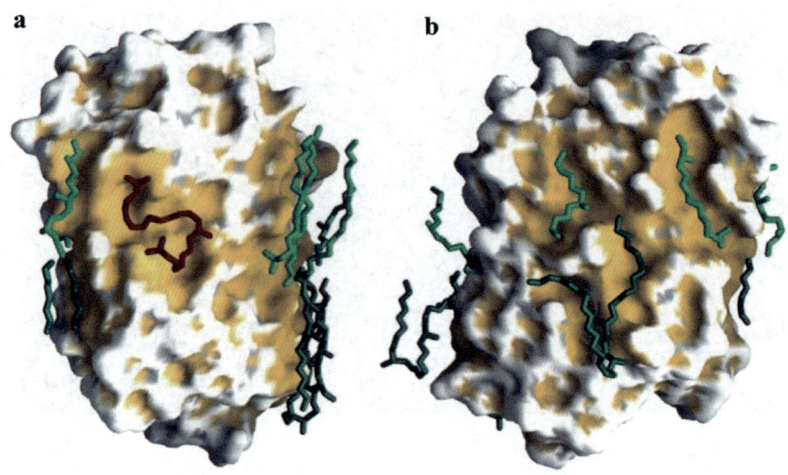

Figure 6 *(a,b) Lipid molecules interact specifically with groves on the hydrophobic surface of bacteriorhodopsin (archaeal diether lipids in green and squalene in red)*

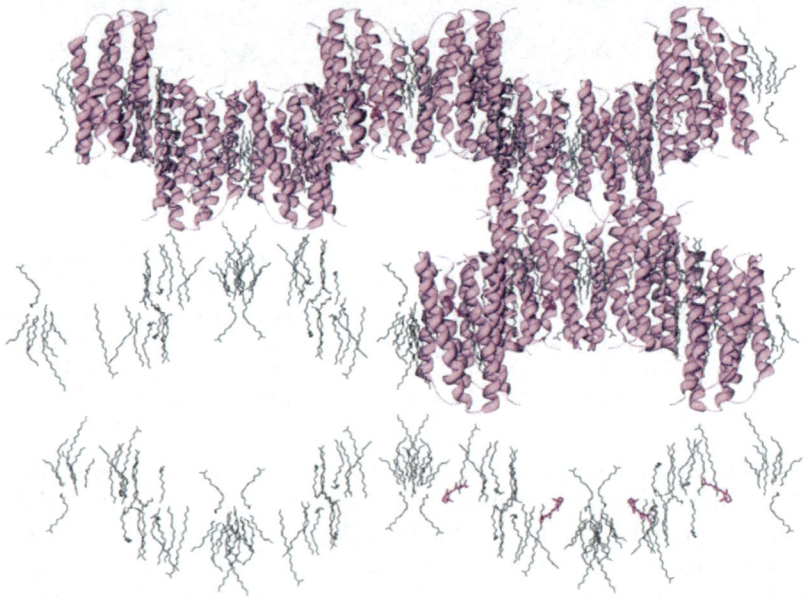

Figure 7 *Side view of lipid (gray) and sensory rhodopsin (purple) for a CLP-derived crystal of Anabaena sp. sensory rhodopsin. The bilayer undulation may be caused by protein–protein contacts in the vertical dimension*

While the number of high-resolution X-ray structures of membrane proteins using the CLP crystallization approach has been lower than most people had hoped for when the method was first reported in 1996 (six different proteins in 8 years, not counting numerous bacteriorhodopsin photocycle intermediates and mutants), there is still hope that eventually we will be able to devise more comprehensive screens for the multitude of parameters involved. The most important parameter of all is probably the requirement for bilayer-forming lipids that at the same time form specific contacts with the target membrane protein. In this regard, one can exploit the knowledge about conditions for the formation of 2D crystals since in at least two cases,[14,30] the CLP-grown 3D crystals were simply stacks of 2D crystals previously studied by electron diffraction.

Effective screening for lipids may be achieved by a more systematic approach of adding lipids from various lipid extracts to the crystallization mixture, or by the design of specific lipids.[40]

2.3 Crystallization from Bicelles

Shortly after the description of the CLP method, Bowie and co-workers published another novel method for crystallizing membrane proteins in type I crystals, this time from a bicelle-forming lipid–detergent mixture.[17] Bicelles are small disks whose core is composed of a nearly planar lipid bilayer, whereas the edges are covered by short, detergent-like lipids. Membrane proteins can enter into the central bilayer-like region (Figure 8A).

Bacteriorhodopsin was crystallized from such a solution of bicelle-embedded bacteriorhodopsin (Figure 8B). The structure was solved by molecular replacement and refined to 2.0 Å resolution. Despite very different packing (head-to-tail dimers *vs.* parallel trimers), the root-mean-square deviation between the bicelle-derived structure and CLP-derived structure is only 0.72 Å on C_α atoms. Thus, somewhat surprisingly, the structure of bacteriorhodopsin is not greatly influenced by the formation of trimers (as observed in native purple membrane and in some of the CLP crystal forms), bound lipids, or the packing between stacked bilayers.

2.4 Crystallization from Spherical Micelles

Two years after publication of the CLP method, Kouyama and co-workers[18] reported yet another protocol for growing type I crystals of bacteriorhodopsin: they incubated purple membrane, which is comprised of naturally occurring 2D crystals of bacteriorhodopsin, at 32 °C in the presence of a small amount of the neutral detergent octylthioglucoside together with the precipitant AS. Over time, a large fraction of the membrane fragments was converted into spherical vesicles with a diameter of approximately 50 nm, which subsequently assembled into hexagonal crystals with an increase in the precipitant concentration. These 3D crystals diffract X-rays to 2.5 Å resolution, and like crystals grown using the CLP method, are made up of stacked planar bilayers. These stacked membranes have been suggested to be produced by successive fusion of the spherical vesicles. This implies that in this case crystallization is achieved without prior solubilization of the protein. Crystallization without prior extraction from the native lipid bilayer has also been demonstrated using the CLP method.[41]

Bicelle Crystallization Method

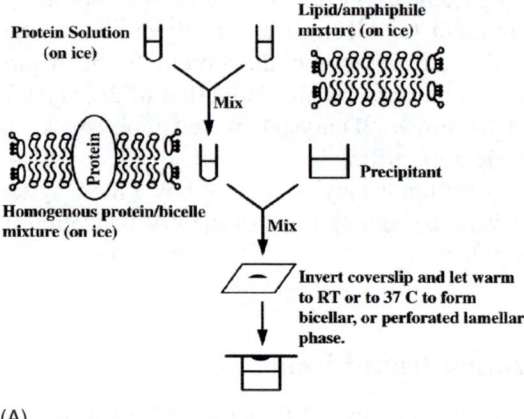

(A)

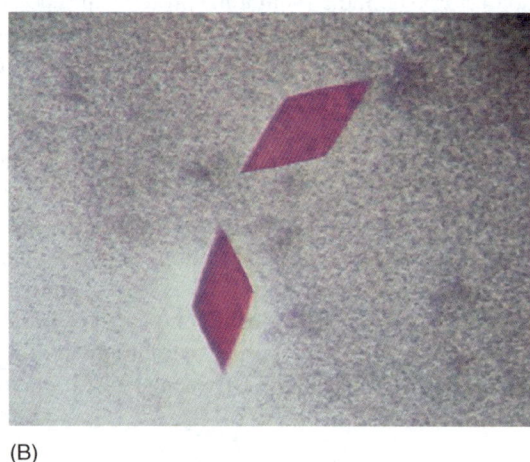

(B)

Figure 8 *Crystallization from bicelles: (A) schematic of the process. (B) Crystals of bacteriorhodopsin grown using the method*

3 Advantages of Structures in a Native Setting at High Resolution

The advantages of increased resolution can be demonstrated with the case of bacteriorhodopsin. The first structure, reported in 1975 and determined by electron diffraction on 2D crystals, was carried out at a nominal resolution of 7 Å (Figure 9A).[11] Using the same method, the resolution was improved to 3.5 Å in 1996 (Figure 9B).[12] 3D crystals from the CLP approach resulted in a refined 1.55 Å structure,[14] a significant improvement (Figure 9C).

The 1.55 Å ground-state structure of bacteriorhodopsin unambiguously identifies the configuration of the retinal and its binding site, specifies all residue and backbone interactions, locates all hydrogen-bonded water molecules in the protein interior, and describes the arrangement of lipids in the surrounding bilayer.[42]

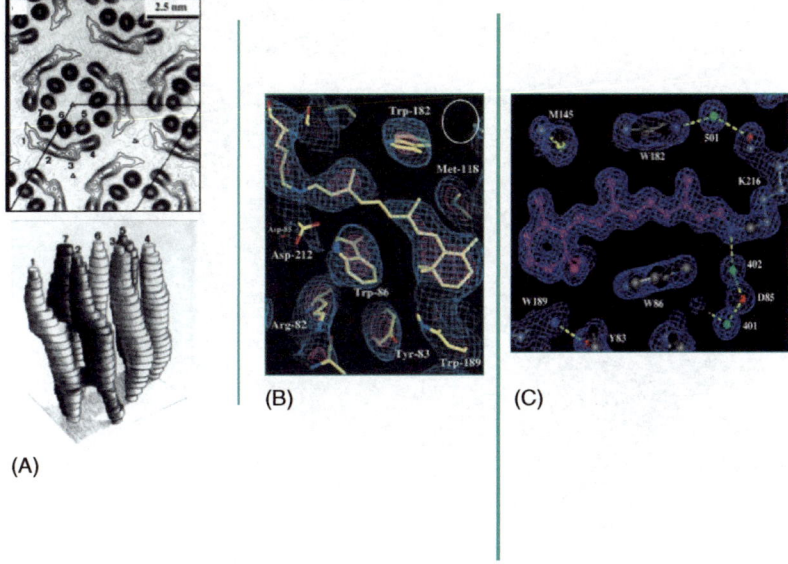

(B)   (C)

(A)

Figure 9 *Comparison of the effect of resolution for various bacteriorhodopsin structures. Densities of (A) a 7 Å and (B) a 3.5 Å resolution structure by electron diffraction, and (C) a 1.55 Å resolution structure by X-ray diffraction from crystals grown in CLP*

A further advantage of type I crystals is the fact that the target protein is embedded in native-like bilayers. Thus, artifacts due to binding of detergent, lack of native lipid contacts, or due to crystal packing are less likely to arise than in crystals containing detergent. In the case of bacteriorhodopsin, the photocycle, whose kinetics are very sensitive to even the slightest changes in the lipid environment of the protein, proceed normally in CLP crystals (Figure 10).

High-resolution structural studies of white- and yellow-illuminated crystals, representing the ground state and the M intermediate of the bacteriorhodopsin photocycle (Figure 10), respectively, have advanced our understanding of the conformational changes that bring about light-driven ion transport.[43,44]

With numerous high-resolution structures of the ground and intermediate states of bacteriorhodopsin available, a new level of understanding of the prototypical ion pump bacteriorhodopsin has been achieved.[45] Based on the exact positions of ordered (but nevertheless mobile) water molecules in combination with the elucidation of detailed conformational changes of the retinal and protein at various stages of the photocycle, a very detailed mechanism for the molecular events during the photocycle has been derived.

4   Conclusions

The CLP crystallization method, devised by Landau and Rosenbusch, provides a sorely needed novel approach to the crystallization of integral membrane proteins.

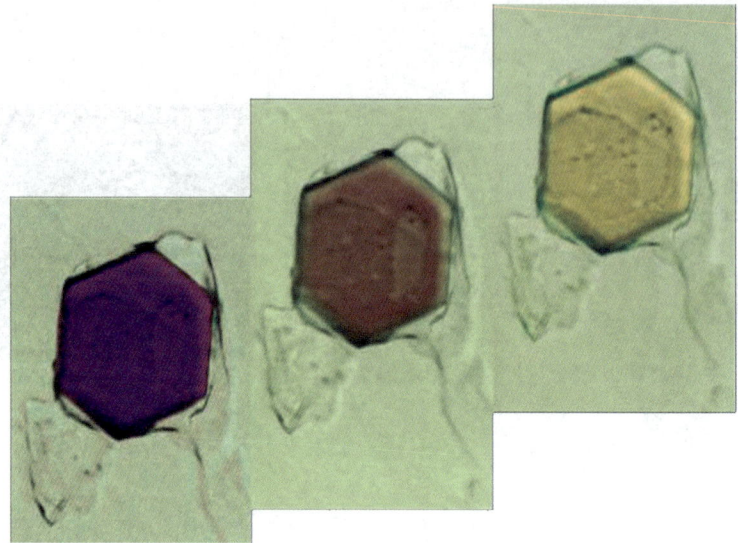

Figure 10 *Crystal of the bacteriorhodopsin D96N mutant grown in CLP. The crystals form typically thin hexagonal plates of approximately 80 × 80 × 15 μm and are strongly merohedrally twinned along the thin dimension (crystallographic c-axis). The same crystal is shown in all three panels: in the left panel fully light-adapted during white-light illumination, in ground state; in the center panel during dim yellow-light (λ > 520 nm) illumination; in the right panel fully bleached in the M state during bright yellow-light illumination*

However, the initial optimism after the first report of CLP-based membrane protein crystallization[13] and the 1.55 Å resolution X-ray structure[14] has given way to the realization that the CLP crystallization method, in order to become more widely applicable, requires a significant amount of additional research, especially with regard to the types of lipids used in the formation of the CLP as well as with regard to the types of lipids added to allow the formation of planar bilayers required for the growth of type I crystals.

There is concern that the CLP method might not be suitable for membrane proteins with large intra- or extracellular domains, because there may not be sufficient room in the CLP arrangement for these large, hydrophilic domains to be accommodated. Another issue is the wide spacing between planar bilayers in the type I crystals required in such a scenario. The CLP crystallization of a photosynthetic reaction center,[37] which contains a large cytoplasmic domain, showed that the CLP method can be used for these types of membrane proteins as well.

Other issues with respect to widespread application of the CLP method are the automation of setup preparation (which requires mechanical mixing), the amount of membrane protein needed per condition tested, and the screening for successful crystal formation. Improvements in these areas are currently being pursued vigorously in several research groups.

So while the jury is still out as to whether the CLP crystallization method will one day become a truly general way of growing highly ordered 3D crystals of integral membrane proteins, the impressive achievements of the past 9 years clearly justify continued efforts in improving this method.

References

1. A. McPherson, Introduction to protein crystallization, *Methods*, 2004, **34**(3), 254–265.
2. N. Deshpande *et al.*, The RCSB Protein Data Bank: a redesigned query system and relational database based on the mmCIF schema, *Nucleic Acids Res.*, 2005, **33**(Database issue), D233–237.
3. T. Tsukihara *et al.*, The whole structure of the 13-subunit oxidized cytochrome c oxidase at 2.8 A, *Science*, 1996, **272**(5265), 1136–1144.
4. D. Xia *et al.*, Crystal structure of the cytochrome bc1 complex from bovine heart mitochondria, *Science*, 1997, **277**(5322), 60–66.
5. K. Palczewski *et al.*, Crystal structure of rhodopsin: A G protein-coupled receptor, *Science*, 2000, **289**(5480), 739–745.
6. K. Murata *et al.*, Structural determinants of water permeation through aquaporin-1, *Nature*, 2000, **407**(6804), 599–605.
7. C. Toyoshima *et al.*, Crystal structure of the calcium pump of sarcoplasmic reticulum at 2.6 A resolution, *Nature*, 2000, **405**(6787), 647–555.
8. E. Pebay-Peyroula *et al.*, Structure of mitochondrial ADP/ATP carrier in complex with carboxyatractyloside, *Nature*, 2003, **426**(6962), 39–44.
9. S.B. Long, E.B. Campbell and R. Mackinnon, Crystal structure of a mammalian voltage-dependent Shaker family K+ channel, *Science*, 2005, **309**(5736), 897–903.
10. W.E. Harries *et al.*, The channel architecture of aquaporin 0 at a 2.2-A resolution, *Proc. Natl. Acad. Sci. USA*, 2004, **101**(39), 14045–14050.
11. R. Henderson, The structure of the purple membrane from Halobacterium hallobium: analysis of the X-ray diffraction pattern. *J. Mol. Biol.*, 1975, **93**(2), 123–138.
12. N. Grigorieff,, E. Beckmann and F. Zemlin, Lipid location in deoxycholate-treated purple membrane at 2.6 A, *J. Mol. Biol.*, 1995, **254**(3), 404–415.
13. E.M. Landau and J.P. Rosenbusch, Lipidic cubic phases: a novel concept for the crystallization of membrane proteins, *Proc. Natl. Acad. Sci. USA*, 1996, **93**(25), 14532–14535.
14. H. Luecke *et al.*, Structure of bacteriorhodopsin at 1.55 A resolution, *J. Mol. Biol.*, 1999, **291**(4), 899–911.
15. L. Essen *et al.*, Lipid patches in membrane protein oligomers: crystal structure of the bacteriorhodopsin-lipid complex, *Proc. Natl. Acad. Sci. USA*, 1998, **95**(20), 11673–11678.
16. H. Michel and D. Oesterhelt, Three-dimensional crystals of membrane proteins: bacteriorhodopsin, *Proc. Natl. Acad. Sci. USA*, 1980, **77**(3), 1283–1285.

17. S. Faham and J.U. Bowie, Bicelle crystallization: a new method for crystallizing membrane proteins yields a monomeric bacteriorhodopsin structure, *J. Mol. Biol.*, 2002, **316**(1), 1–6.

18. K. Takeda *et al.*, A novel three-dimensional crystal of bacteriorhodopsin obtained by successive fusion of the vesicular assemblies, *J. Mol. Biol.*, 1998, **283**(2), 463–474.

19. H. Michel, Three-dimensional crystals of a membrane protein complex. The photosynthetic reaction centre from Rhodopseudomonas viridis, *J. Mol. Biol.*, 1982, **158**(3), 567–572.

20. J. Deisenhofer *et al.*, X-ray structure analysis of a membrane protein complex. Electron density map at 3 A resolution and a model of the chromophores of the photosynthetic reaction center from Rhodopseudomonas viridis, *J. Mol. Biol.*, 1984, **180**(2), 385–398.

21. J. Deisenhofer and H. Michel, The photosynthetic reaction centre from the purple bacterium Rhodopseudomonas viridis, *Biosci. Rep.*, 2004, **24**(4–5), 323–361.

22. J. Deisenhofer and H. Michel, Nobel lecture. The photosynthetic reaction centre from the purple bacterium Rhodopseudomonas viridis. *Embo. J.*, 1989, **8**(8), 2149–2170.

23. B.W. Matthews, Solvent content of protein crystals, *J. Mol. Biol.*, 1968, **33**(2), 491–497.

24. C. Hunte and H. Michel, Crystallisation of membrane proteins mediated by antibody fragments, *Curr. Opin. Struct. Biol.*, 2002, **12**(4), 503–508.

25. D.A. Doyle *et al.*, The structure of the potassium channel: molecular basis of K+ conduction and selectivity, *Science*, 1998, **280**(5360), 69–77.

26. G. Chang *et al.*, Structure of the MscL homolog from Mycobacterium tuberculosis: a gated mechanosensitive ion channel, *Science*, 1998, **282**(5397), 2220–2226.

27. H. Luecke, H.T. Richter and J.K. Lanyi, Proton transfer pathways in bacteriorhodopsin at 2.3 angstrom resolution, *Science*, 1998, **280**(5371), 1934–1937.

28. M. Lindstrom, *et al.*, Aqueous lipid phases of relevance to intestinal fat digestion and absorption, *Lipids*, 1981, **16**(10), 749–754.

29. H. Qiu and M. Caffrey, Phase behavior of the monoerucin/water system, *Chem. Phys. Lipids*, 1999, **100**(1–2), 55–79.

30. H. Luecke *et al.*, Crystal structure of sensory rhodopsin II at 2.4 angstroms: insights into color tuning and transducer interaction, *Science*, 2001, **293**(5534), 1499–1503.

31. J.L. Spudich and H. Luecke, Sensory rhodopsin II: functional insights from structure, *Curr. Opin. Struct. Biol.*, 2002, **12**(4), 540–546.

32. A. Royant *et al.*, X-ray structure of sensory rhodopsin II at 2.1-A resolution, *Proc. Natl. Acad. Sci. USA*, 2001, **98**(18), 10131–10136.

33. L. Vogeley *et al.*, Anabaena sensory rhodopsin: a photochromic color sensor at 2.0 A, *Science*, 2004, **306**(5700), 1390–1393.

34. H. Belrhali *et al.*, Protein, lipid and water organization in bacteriorhodopsin crystals: a molecular view of the purple membrane at 1.9 A resolution, *Struct. Fold Des.*, 1999, **7**(8), 909–917.

35. M. Kolbe *et al.*, Structure of the light-driven chloride pump halorhodopsin at 1.8 A resolution, *Science*, 2000, **288**(5470), 1390–1396.

36. V.I. Gordeliy *et al.*, Molecular basis of transmembrane signalling by sensory rhodopsin II-transducer complex, *Nature*, 2002, **419**(6906), 484–487.

37. G. Katona *et al.*, Lipidic cubic phase crystal structure of the photosynthetic reaction centre from Rhodobacter sphaeroides at 2.35A resolution, *J. Mol. Biol.*, 2003, **331**(3), 681–692.

38. L.V. Misquitta *et al.*, Membrane protein crystallization in lipidic mesophases with tailored bilayers, *Struct. (Camb.)*, 2004, **12**(12), 2113–2124.

39. S. Rouhani *et al.*, Crystal structure of the D85S mutant of bacteriorhodopsin: model of an O-like photocycle intermediate, *J. Mol. Biol.*, 2001, **313**(3), 615–628.

40. Y. Misquitta *et al.*, Rational design of lipid for membrane protein crystallization, *J. Struct. Biol.*, 2004, **148**(2), 169–175.

41. P. Nollert *et al.*, Detergent-free membrane protein crystallization, *FEBS Lett.*, 1999, **457**(2), 205–208.

42. J.P. Cartailler and H. Luecke, X-ray crystallographic analysis of lipid-protein interactions in the bacteriorhodopsin purple membrane, *Annu. Rev. Biophys. Biomol. Struct.*, 2003, **32**, 285–310.

43. H. Luecke *et al.*, Structural changes in bacteriorhodopsin during ion transport at 2 angstrom resolution, *Science*, 1999, **286**(5438), 255–261.

44. H. Luecke *et al.*, Coupling photoisomerization of retinal to directional transport in bacteriorhodopsin, *J. Mol. Biol.*, 2000, **300**(5), 1237–1255.

45. H. Luecke, Atomic resolution structures of bacteriorhodopsin photocycle intermediates: the role of discrete water molecules in the function of this light-driven ion pump, *Biochim. Biophys. Acta*, 2000, **1460**(1), 133–156.

Section 3

New Membrane Protein Structures

Section 7

CHAPTER 11

Aquaporins: Integral Membrane Channel Proteins

ROBERT M. STROUD, WILLIAM E.C. HARRIES, JOHN LEE, SHAHRAM KHADEMI AND DAVID SAVAGE

Department of Biochemistry and Biophysics, University of California–San Francisco, S-412 Genentech Hall, San Francisco, CA 94143-2440, USA

1 Introduction

The first of what today are called 'aquaporins' (AQPs) were bacterial glycerol channels. Alfred Fischer recognized the phenotype over 100 years ago. He described pathogenic bacteria that, when placed in hyperosmotic glycerol solutions, failed to undergo lysis.[1] He concluded that the membranes had to be highly permeable to glycerol in these organisms, but not in the organisms that shriveled up and died. This stimulated interest in glycerol channels, but since the substrates are uncharged the field lagged behind that of ion channels where electrical phenomena were evident. The glycerol channel from *Escherichia coli* has a special place in AQP history. Genetic analysis pioneered by E.C.C. Lin and co-workers[2] identified the 'glycerol facilitator' GlpF. This work led to the cloning and sequencing of the gene,[3] followed by characterization of the 281 amino acid (29,780 Dalton) GlpF protein.[4] The crystal structure of GlpF became the first AQP structure to be determined at near atomic resolution.[5]

Early functional studies of GlpF characterized[6] it as a highly selective transmembrane channel that conducts water, glycerol, and other small uncharged organic molecules such as urea, glycine, and D,L-glyceraldehyde. Inside the cell, glycerol is rapidly phosphorylated by glycerol kinase to produce glycerol-3-phosphate, which is no longer a substrate for any back flow. This phosphorylation retains the substrate in the cell and maintains the gradient of the substrate glycerol from outside to inside. Glycerol-3-phosphate proceeds by dehydrogenation to dihydroxyacetone phosphate (DHAP), or onward to phospholipid synthesis, where glycerol provides the attachment base for fatty acid chains and phosphatidyl headgroups in about two-thirds of the cellular phospholipids. Stimulating *E. coli* growth on glycerol is the glp regulon, which is inducible by glycerol-3-phosphate.

GlpF is also stereo- and enantio-selective in conductance of linear carbohydrates (alditols).[2,5] Aldoses or sugars – the cyclized alditols – are not conducted through the GlpF channel. The structure of the GlpF channel shows that it is indeed too small to conduct cyclic molecules, illustrating the basis of stereo- and enantio-selectivity.[5,7]

GlpF conducts water at about one-sixth of the rate of its *E. coli* homolog, AQPZ, which conducts water but not glycerol. The presence of two AQPs even in *E. coli* illustrates the necessity for differentiation of function to cover the spectrum of activities within the AQP family. In mammals, 13 AQPs have been identified, termed AQP0 up to AQP12.[8] MIP26 from the eye lens was one of the first mammalian proteins to be recognized as a water channel in 1974.[9,10] Water channels were recognized in plants in the early 1980s. In 1990, the genetic similarity between GlpF, the major intrinsic protein (MIP) of the eye lens that was well known to act as a water channel, and a soybean nodulin-26 had been described by Saier and Baker.[11] The red blood cell contains a water channel protein, that was called CHIP28 (abbreviated from channel-like integral membrane protein, a 28 kDa protein). These all now fall into the 'Aquaporin family' so named by Peter Agre in 1992. Peter Agre shared the Nobel Prize in 2003 for his pioneering work in the ensuing decade that elaborated properties of the 'AQP' family.[12]

Aquaporins comprise three functionally distinct subgroups that include transmembrane water conducting channels (aquaporins), channels that conduct glycerol (perhaps the most relevant physiological substrate in humans) called aquaglyceroporins (*i.e.*, GlpF), and most recently, aquaammonioporins.[13] These channels conduct water, but also variously conduct urea, D,L-glyceraldehyde, linear polyalcohols (called alditols), other small organic molecules,[14–16] and ammonia.[8] In humans, AQP3, AQP7, AQP9, and AQP10 are in the aquaglyceroporin subclass.[17] The AQPs can conduct their substrates at close to the diffusion-limited maximum rate through a pore of this cross section. Many eukaryotic AQPs are regulated by phosphorylation, pH, osmolarity, or the binding of other proteins or ligands.[18,19]

The amino acid sequence similarity between any two AQPs typically shows conservation in the range of 28–32%. They are all constructed around a highly conserved structural backbone fold of six transmembrane and two half membrane-spanning helices numbered M1 to M8 that surround a central water filled channel (Figure 1). The family arose by tandem intragenic duplication[20] such that the N-terminal segment displays ~20% conservation with the C-terminal segment.[21] This duplication occurred early in evolution since bacteria contain both an aquaglyceroporin (GlpF) and an aquaporin (AQPZ). Near the center of each segment is a conserved -Asn-Pro-Ala- signature sequence (-NPA-).

A key question is how AQPs exclude conduction of all charged molecules and ions including hydroxide, hydronium ions,[22] and protons. The recent structures of 5 AQPs instruct as to the determinants of AQP selectivity.[5,23–25] Mutational analysis and molecular mechanics seek to evaluate the contributions of each factor to this property.[26] There is only one exception to the absolute insulation against ions in human AQPs: AQP6. AQP6 conducts ions at low pH. This adaptation most probably controls its action in membranes of intracellular organelles.[27]

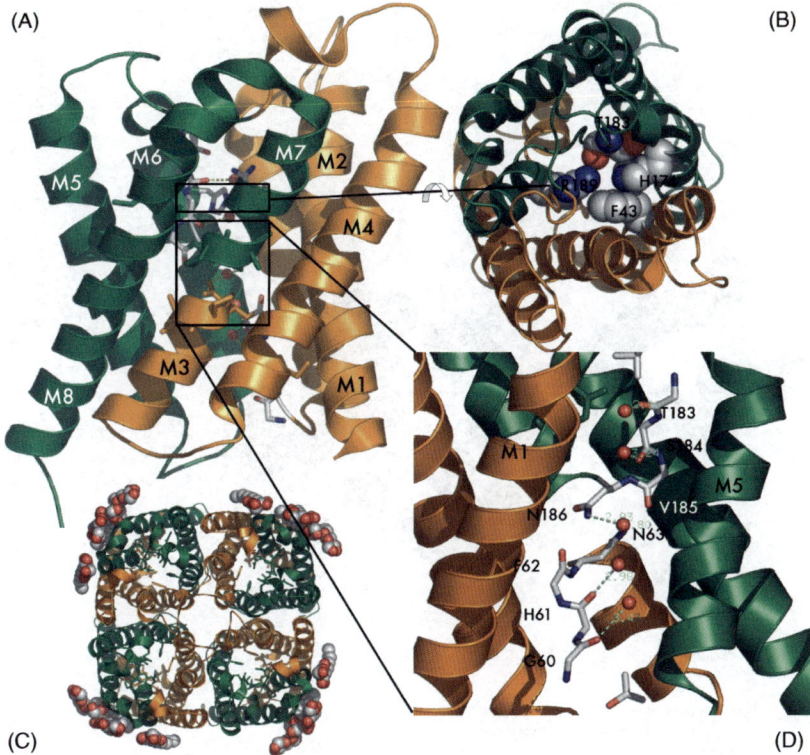

Figure 1 *(A) Each aquaporin monomer is composed of six transmembrane helices and two half-length helices (M3, M7) that meet in the center of the bilayer. They are numbered M1 to M8. (B) The expansion of the upper rectangle shows the 'selectivity' filter in AQPs in which conserved R206 is a key player. (C) Aquaporin monomers associate in the plane of the membrane to form tetramers. (D) In the expansion from the lower rectangle in (A) focus is on the region where the two NPA regions meet. Residues that contribute to the orientation of the central water molecules include N63 and N186 of the NPA regions*

In this chapter we focus on the information, which derives from the five structures of AQPs at near atomic resolution. The amino acid sequences of all five AQPs are aligned in Figure 2A. The residues that line the pore are colored and represented by spheres in Figure 2B to indicate where in the structure they lie. Each AQP has different conductivity, substrate specificity, and role in the biology of the organism. Thus, they serve to extract features of the molecular structures that encode these properties. The common elements of structure serve to illustrate the elements of function that are preserved in the family. We begin by focusing on the mechanisms by which the AQPs, which are normally filled with a continuous line of hydrogen-bonded water molecules, remain insulating to protons and ions. This is followed by what we have learned about the selectivity of AQPs for what they do conduct. Finally, we focus on functional associations between AQPs and their role in higher levels of cellular organization in the eye lens.

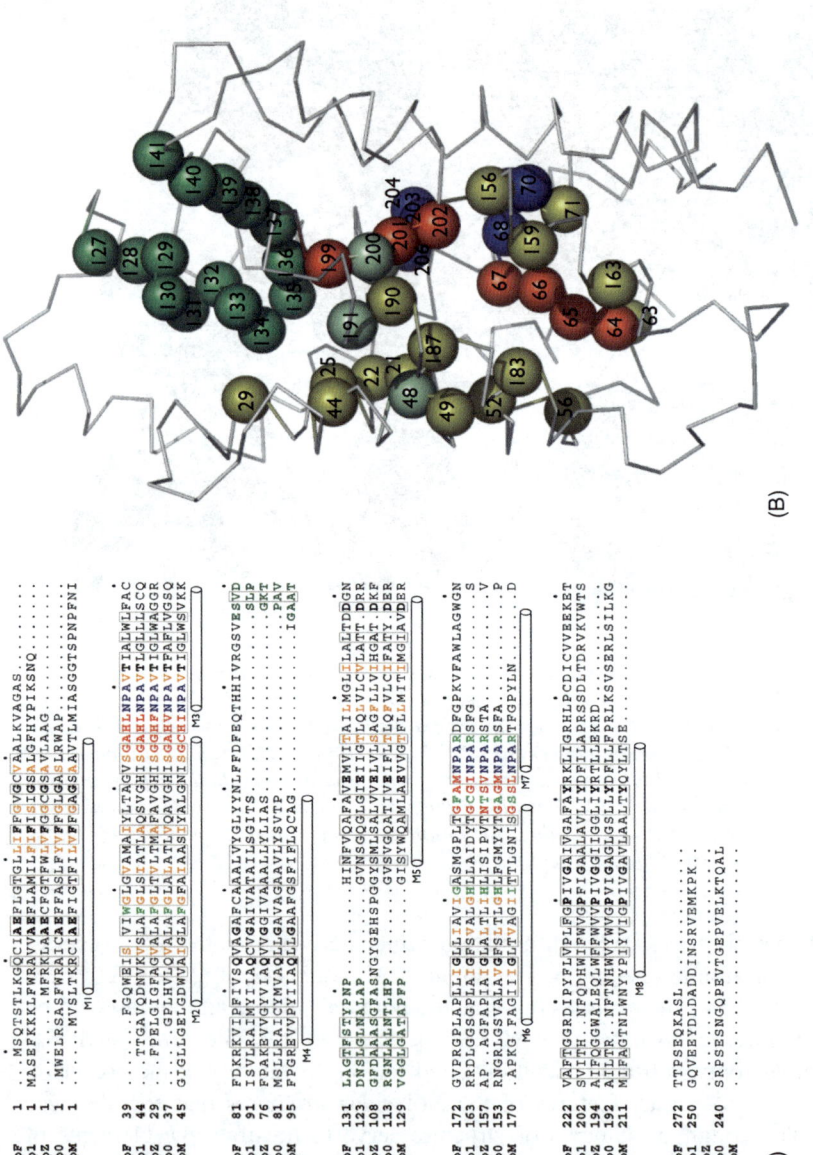

(B)

(A)

GlpF	1	..MSQTSTLKQCIAEFLGTGLLIFPGVCVAALKVAGAS.........
AqP1	1	MASEFKKKLFWRAVVAEFLAMILFIFISIGSALGFHYPIKSNQ.....
AqpZ	1	....MFRKLAAECFGTFWLVFGGCGSAVLAAG................
AqP0	1	.MWELRSASFWRAICAEFFASLFYVFFGLGASLRWAP...........
aqpM	1	..MVSLTKRCIAEFIGTFLVFFGAGSAAVTLMIASGGTSPNPFNI...

GlpF	39	....FGQWEIS.VIWGIGVAMAILTAGVSGAHLNPAVTIALWLFAC
AqP1	44	....TTGAVQDNVKVSLAFGSIATLAQSVGHISGAHLNPAVLGLLLSCQ
AqpZ	29	FPELGIGFAGVALAFGLTVLTMAFAVGHISGGHFNPAVTIGLWAGGR
AqP0	37	.GPLHVLQVALAFGLALALGVQAVGHISGAHVNPAVTFAFLVGSQ
aqpM	45	GIGLGELGDWVAIGLAFGFAIAASIYALGNISGCHINPAVTIGLWSVKK

GlpF	81	FDKRKVIPFIVSQVAGAFCAAALVYGLYYNLFFDFEQTHHIVRGSVSVVD
AqP1	91	ISVLRAIMYIIAQCVGAIVATAILSGITS................SLJP
AqpZ	76	FPAKKVVGVVIAQVVGGIVAAAALYVIAS.................PAV
AqP0	81	MSLLERICYMVAQLLGAVAGAAVLYSVTP.................IGAAT
aqpM	95	FPGREVVPIIIAQLLGAAFGSFILPLQCAG.................

GlpF	131	LAGTFSTYPNP..........HINFVQAFAVEMVITAILMGLILALTDDGN
AqP1	103	QFSLGNAAAR..........DFSNGQGLGIEIIGTLQLVLCVHLANT.DKR
AqpZ	108	RGNPAAANGYGEHSPGGYSMLQALVVELVLTAILCLVIFATDEF
AqP0	113	...........GISWQAMLAEVVVTFLQILIIFATDER
aqpM	129	VGGLGATAFPP..........FLMIGHGIAVDBR

GlpF	172	GVPRGPLAPLLIGLLIAVIGASMGPLTGFAMNPARDFGPKVFAWLAGWGN
AqP1	163	RRDLGGSAPLAIGFSVALGHLLAIDYTGCGINPARSFG........V
AqpZ	157	AP.AGFAGRIAIAIGFSLTLHLIISISPVNTSVNPARSTA........V
AqP0	153	RNGRLGSVALAVGFSLTLGHLFGMYYTGAGMNPARSFA.........P
aqpM	170	APKG.FAGIILETVAGIITTLGNISGSLNPARTFGPYLN.........D

GlpF	222	VAFTGRDIPYFLVPLFGPIVGAIVGAFAYRKLIGRHLPCDICVVBEKT
AqP1	202	SVITH..NFPQDHWIFWVGPFIGAALAVLIYDFILAPSSDLTDVKVWTS
AqpZ	194	AIFPGGWALEQLWFPWVPIVGGLIIGLLYRTLLEKRD
AqP0	192	AILTR..NPTNHHWVWWVPIVGAGLGSLLYDFLLFPRLKSVSERLSILKG
aqpM	211	MIPAGTNLWNYPIYVIGPIVGAVLAALTCYLTSE..

GlpF	272	TTPSEQKASL.....
AqP1	250	GQVEEYDLDADDINSRVEMKPK....
AqpZ		
AqP0	240	SRPSESNGQPEVTGEPVELKTQAL
aqpM		

Figure 2 *Alignment of aquaporin sequences mapped to pore-lining residues. (A) Alignment of the five aquaporins whose structures are known. The ruler on the top is GlpF numbering at 10 residue intervals, as is the secondary structure information. Color scheme is: grey boxes, conserved residues; orange: pore residues; red: residues whose mainchain carbonyls project into the pore; blue: NPA residues; light green: selectivity filter residues; dark green: inter-repeat loop. These colors also map on figure 2B. (B) Structure of GlpF with residues mapped from alignment as detailed above*

2 The Exclusion Barrier to Ions and Protons in Aquaporins

The electrochemical gradient across biological membranes, established by active transport, is the basis of energy generation and storage in most organisms. In bacteria the gradients are usually protonic. In mammals they are generally Na^+/K^+ ion gradients. These gradients must be maintained and any disruption can induce cellular distress. The cell, however, also requires the passive and active concurrent transport of many compounds. Thus, the membrane and the proteins therein must balance both selectivity and permeability. Aquaporins demonstrate this balance by selectively conducting water and small amphipathic molecules, with the complete exclusion of ions (particularly protons).[28,29] It has long been known that bulk water and channels able to conduct water, such as gramicidin, can also conduct protons by the Grotthuss "hop and turn" mechanism where hydrogen bonds among a chain of waters are rapidly realigned, resulting in proton transfer with no net water movement.[30] Though aquaporins contain a similar single-file chain of water, they display no such conductance of protons.[29] The explanation of this phenomenon has interested both theoretical and experimental scientists alike and there are three current explanations, all of which are probable components in varying magnitudes, of the proton exclusion barrier.[5]

2.1 Global Orientational Tuning by the NPA Motif

The first concept for how this might be encoded is 'global orientational tuning', that centers on the conserved asparagine-proline-alanine (NPA) region near the middle of the pore.[26] This conserved sequence, present in both the N-termini and C-termini of all aquaporins, is seen as the signature motif of aquaporins and high-resolution X-ray structures suggest a possible role in proton exclusion (see Figure 3). The two NPA motifs cap the N-terminal end of helices M3 and M7, which meet at the center of the membrane along the quasi-twofold axis. At the twofold, the proline rings are in van der Waals contact with each other, while the asparagine side chain is constrained by two hydrogen bonds in a way that orients the amide chemical moiety into the pore, toward hydrogen bond acceptors on the permeant substrate.[5]

In the case of water, the central water molecule's two lone electron pairs accept two hydrogen bonds from the two asparagine amide groups and are therefore aligned to donate two hydrogen bonds to the neighboring water molecules on either side of the central water molecule. Each of these neighboring water molecules is a hydrogen-bond donor first to one of the line of equally spaced carbonyl oxygen atoms that run outward from the NPA region to the external surfaces of the pore. In turn, it is therefore oriented to present its second hydrogen outward toward the next water molecule, and so on throughout the length of the conduction pathway. While this 'global orientation' is deduced from the crystal structure, the resolution of the X-ray structures is not quite adequate to 'see' the hydrogen orientation. This alignment has been reiterated multiple times in molecular mechanics simulations and supports the conclusion from the structure. This ordering of the line of water molecules implies that as a water molecule enters the pore its dipole will be oriented generally toward

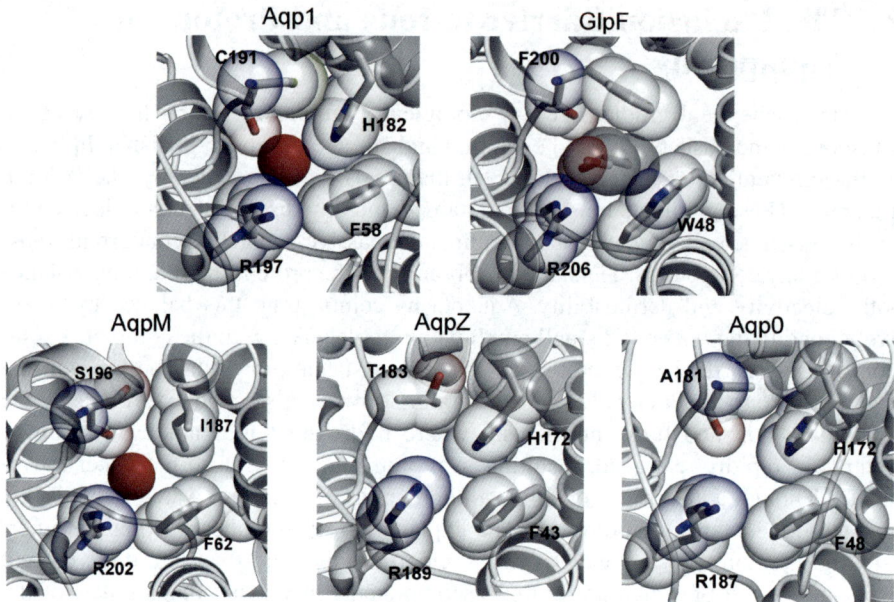

Figure 3 *The selectivity filters are compared. The view is down the channel in the glycerol and water channel GlpF from E. coli, in the water channel AQP1 from bovine red blood cells, AQPM from the archaebacterium Methanothermobacter marburgensis which may serve to conduct H₂S as well as water, AQPZ the bacterial water conducting AQP from E. coli, and the AQP from the bovine eye lens AQP0. The conserved Arg 206 is shown at the left side of each one. While water and glycerol in GlpF are seen throughout the channels, only those substrate molecules seen within the slab shown are shown*

the entrance of the channel, it moves through the pore until it reaches the center where it rotates to interact with the NPA motifs. It then continues to rotate until its dipole points toward the exit, and it progresses through the rest of the pore.[26,31]

The hypothesis is that this ordering makes it difficult for the central water molecule to receive a proton from its nearest neighbor to become a hydronium ion (H_3O^+) since its hydrogen points toward the source of the proton. Viewed in a different way, if a proton were to leave the central water molecule toward the inside, that water would donate a proton to its neighbor. However, the water molecule would then need to rotate in its position to regenerate the hydrogen-bonded alignment. The covariant alignment of the line of water molecules might make this difficult and so create a barrier. This mechanism is bolstered by multiple molecular dynamics simulations that reiterate the co-alignment of the entire line of water molecules. However, what is needed is a simulation where the energetic cost of moving a proton or a hydronium ion along the axis of the channel is calculated. That kind of simulation is much harder to accomplish since it involves the movement of an ion, and the energetic costs of partial dehydration of the ionic species must be built in correctly. There is no good estimate of this effect. One simulation described the barrier as centered at the NPA region and was able to assign a value of nearly 2 kcal mol^{-1} to the barrier, by selectively 'turning off' specific coulombic interactions *in silico*.[32]

2.2 Helix Dipole

Recent structural work has elucidated the chemical toolkit of transmembrane channels and how these proteins function with the chemical implications of their location in a low dielectric medium. One such tool is the use of helix dipoles to interact, both favorably and unfavorably, with charged species.[33] In aquaporins, the half-membrane spanning helices M3 and M7 meet at the quasi-twofold axis and project their N-termini into the pore.[5] These two helix dipoles work in synergy, creating an electrostatic field that opposes entry of positive charge into the channel. Electrostatic calculations in which one can switch the dipole effect off, have set the barrier height at roughly 3 kcal mol^{-1}. Furthermore, by calculating the static field of proton conductance, Roux and colleagues predicted that mutating one of the two NPA motifs to DPA would be enough to negate the dipole effect and therefore the channel would conduct protons.[32] In our hands, the N68D point mutant, while functional for water conductance, still does not conduct protons in proteoliposome assays (unpublished results). This argues against the role of the helix dipoles in maintaining the ordering of water, insofar as the alignment is responsible for proton exclusion.

2.3 Electrostatic Desolvation Penalty

A third hypothesis to explain the insulation to proton conductance is based on the electrostatic free energy of transferring charge from a high dielectric medium such as bulk water to the low dielectric protein channel. The enthalpy of hydration for most ions is on the order of 100 kcal mol^{-1}, and thus, dehydration is an extremely unfavorable process. Ion conducting channels, like potassium channels, have evolved an elaborate fourfold axis of mainchain carbonyls that can favorably "solvate" potassium, yet do not act as binding sites to slow conduction all together.[34] The exclusion of ions from channels that do not compensate for the water of hydration is the rule and conduction is the carefully constructed exception.

In the aquaporins, electrostatic simulations compare the channel with a simple macroscopic hydrophobic channel and find very little difference, suggesting that it is simply the electrostatic nature of the pore and an inability to solvate charge that is the barrier to proton conduction.[35] This is the mechanism we suggested based on the structure that would prevent leakage of ions such as Na$^+$ through the channel.[5] It clearly must apply to the bulk transfer of a hydronium or a Zundel ion ($H_5O_2^+$) through a channel. This mechanism differs in principle from the hypothetical 'hopping' of protons between water molecules along the line, which one might imagine could even occur in a concerted fashion such that there would never be a formal charge within the line or in the center of the channel. It is this latter mechanism that the orientation dependent proposals seek to address.

3   Selectivity in the Aquaporin Family

The extraordinary permeation rate of more than one billion molecules per second makes AQP one of the fastest membrane channels. The high rate of permeation along with the strict selectivity for water and polyols raises fascinating questions concerning the structural basis for substrate selectivity in the AQP family.

GlpF from *E. coli* was the first member of the AQP family whose structure was determined by X-ray crystallography to a high resolution of 2.2 Å.[5] GlpF facilitates passive and selective permeation of water and small-uncharged organic molecules, such as glycerol, across plasma membranes of the cell. The next water channel whose crystal structure was determined was AQP1 from bovine red blood cells.[23] To date, the atomic resolution crystal structures of five members of the AQP superfamily have been determined, namely GlpF, AQP1, AQPZ,[24] AQP0,[25] and AQPM (unpublished). Comparisons of the structures of aquaporins with aquaglyceroporins may explain the basis for their different profiles of selectivity. All AQPs form homotetramers of four AQP monomers that contain single channels arranged around a fourfold symmetry axis. The tetramers are generally stable even in the presence of detergents, as shown by ultracentrifugation, size exclusion chromatography, and mass spectrometry.[5,36]

The AQP channel pathway, defined by bound water or glycerol, is bounded by extracellular and intracellular vestibules and connected by a ~28 Å long amphipathic channel.[7] The channel contains two highly conserved regions, the selectivity filter, which is the narrowest point in the entire channel, and the NPA region.[5,36] The selectivity filter in aquaporins is generally narrower than that in the aquaglyceroporins, matching the difference between a water molecule, and an alcohol containing a carbon backbone (Figure 3).

The selectivity filters for AQP1 and AQPZ, water channels from cow and *E. coli*, respectively, have 'diameters' (defined in spherical terms) of ~2.0 Å.[23,24] while GlpF, an aquaglyceroporin from *E. coli*, has the widest selectivity filter of 3.4 Å.[5] To a first approximation, the preference of aquaporins for water and of aquaglyceroporins for glycerol can be explained by the size of the selectivity filter. The larger glycerol molecule, compared to the size of a water molecule, needs a wider selectivity filter in GlpF. AQP0, a water channel from eye lens, has the narrowest selectivity filter (~1.4 Å) of all the AQPs with high resolution crystal structures, which is consistent with AQP0 being a very poor water channel.[25]

The selectivity filter in GlpF is strongly amphipathic, with the planes of two perpendicular aromatic rings (W48 and F200) forming a hydrophobic corner (Figure 3). The alkyl backbone of glycerol is tightly packed against this corner, leaving no space for any substitution at the C-H hydrogen positions. All the three hydroxyl groups of glycerol are hydrogen bond acceptors from successive NHs of the guanidinium group of R206 and hydrogen bond donors to the carbonyl oxygens of G199 and F200, respectively. The buried carboxyl group of conserved E152, orienting the three adjacent carbonyls of G199, F200, and A201 in the extracellular/periplasmic vestibule, may also increase the negative charge on the carbonyl oxygens of F200 and A201.[5] The binding of permeant molecules in the selectivity filter makes it possible for the negative charge of E152, acting through the amides of 199–201, to form an electrostatic interaction with positively charged R206, through substrate. The amide carbonyls of F200 and A201 act as hydrogen-bond acceptors from successive hydroxyl OHs of the substrate, which in turn accept hydrogen bonds from each of two NH groups of the positively charged R206. This implies that permeant molecules should be polarizable in cross-section, such as glycerol OHs and water.

In AQPZ, the selectivity filter is formed by three hydrophilic residues of H174 (191, the residue abbreviations and numbers in the parentheses refer to the GlpF amino acid sequence), and R189 (206) and T183 (200), and one hydrophobic residue of F43 (48). The AQP1 selectivity filter is almost identical, with cysteine substituted for threonine T183. This cysteine explains the inhibition of water transport in AQP1 by mercury.

The selectivity filters in AQPZ and AQP1 contain side chains from three polar residues and one non-polar residue, whereas in GlpF, the ratio is inverted; the selectivity filter contains only one polar residue. This might partially explain the higher glycerol conductivity and lower water conductivity in GlpF.

A microbial MIP from *Lactococcus lactis*, named Gla$_{Llac}$, has been demonstrated to be permeable to glycerol (by expression in oocytes), at the same rate as *E. coli* GlpF, and permeable to water at the same rate as in *E. coli* AQPZ. [37] In the aquaglyceroporin Gla$_{Llac}$, as compared to AQP1, small non-polar residues, V223 and P232, replace H182 and C191, respectively. Interestingly, F58 is replaced by Y49, a polar residue. Consequently, the selective filter of Gla$_{Llac}$ is made of two polar and two non-polar residues. These substitutions with small residues would result in an enlargement of the constriction region with a potential aperture larger than the aquaporin one, a size compatible with a glycerol channel, and the switch of F58 to Y49 provides the necessary polar environment for an efficient water channel.[38]

4 Permeation of Substances Other than Water and Glycerol

4.1 Conductance of Other Molecules

Despite the similarity in the general architecture across the aquaporin family, *in vivo* and *in vitro* conductance assays show the selectivity spectra of aquaporins extending to the permeation of small molecules such as CO_2 (AQP1,[39] tobacco aquaporin NtAQP1),[40,41] nitrate (AQP6),[42] and urea (AQP3, AQP7,[43] AQP9,[44] AQP10).[45] For example, cells such as red blood cells or plant leaves have situational demands for the specific and efficient conductance of a gas such as CO_2, and the characterization of aquaporins with this functional capacity supports the observations that certain aquaporins may be gas channels.

The archaeal aquaporin AQPM probably conducts CO_2 or H_2S, and the structure of AQPM, especially at the selectivity filter, suggests an adaptation for the conductance of a permeant molecule that is larger and less polar than water (Figure 4). The hydrophobicity and size of the channel, especially at the selectivity filter, lead to the speculation that the differences in AQPM compared to that of AQP1 may be an adaptation that enables it to function as a multifunctional channel that conducts either H_2S or CO_2 in addition to water. *Methanothermobacter marburgensis* relies on H_2S as the terminal electron acceptor in its energy production pathway, and due to the structural similarity of H_2S to H_2O, the mechanism of selectivity for these molecules must necessarily be very similar and points to H_2S as a likely candidate for conductance by AQPM.

The wider and more hydrophobic selectivity filter of the AQPM channel (2.54 Å), in comparison to that of AQP1 (1.86 Å), is well structured to accommodate the passage of H_2S (Figure 3), which is larger and less polar than water. In the selectivity filter of AQP1, a passing water molecule would be coordinated by the hydrogen bond donor Nε of R197 (206) and the carbonyl oxygen of C191 (200). The hydrogen-bond acceptor

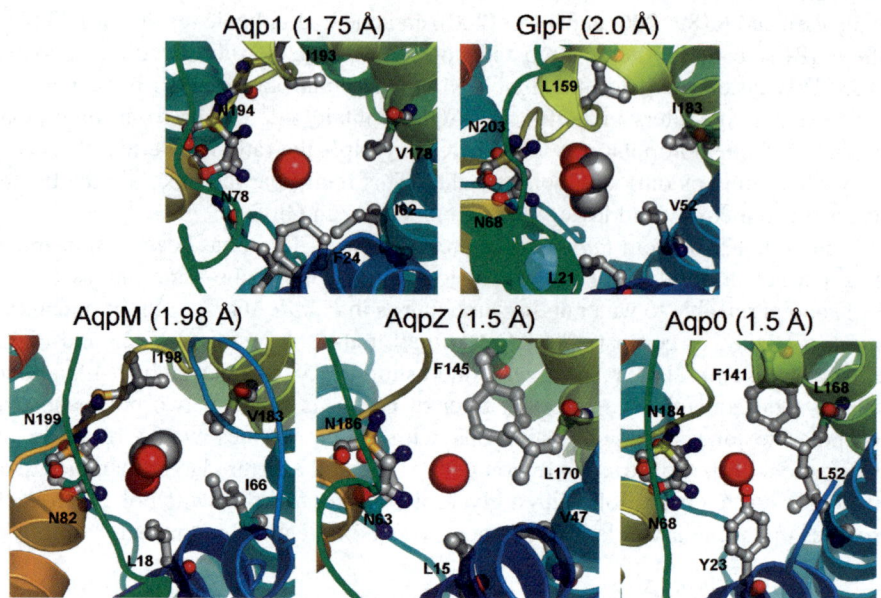

Figure 4 *The center of the channel in AQPs is formed by two NPA motifs that focus hydro-gen bond donors onto the central water or substrate molecule. The nitrogens of the asparagine side chains are shown at the left side in the same orientation as in Figure 3, but further down the channel*

H182 (191) facilitates the passage of the polar water molecule by providing a secondary hydrogen-bond partner for the passing H_2O molecule. H_2S, which is significantly less polar and larger (dipole moment (μ) = 0.97 Debye, van der Waals diameter (d) = 3.1 Å) than water (μ = 1.85 Debye, d = 2.8 Å), requires a channel that is larger and less charged than that required for the efficient selection and passage of water. The presence of a hydrogen-bond partner, such as the histidine residue found opposite (across the channel) to the arginine residue at the selectivity filter in AQP1, facilitates the hydrogen bonding requirements of a polar water molecule. But it may be repulsive to the effec-tively non-polar hydrogen atoms of an H_2S molecule. Therefore, an aliphatic residue such as I187 (191), instead of a histidine residue is favorable for the passage of H_2S, although inhibitory for that of H_2O. Initial permeation experiments using AQPM prote-oliposomes indeed indicate the conductance of H_2S by AQPM (unpublished data) and also indicate that AQPM and possibly other aquaporins are gas-conducting channels.

5   Aquaporin Monomer Associations and their Functional Implications

5.1   The Eye Lens: A Brief History of Aquaporin 0 Research

AQP0 was an early member of the aquaporin family of proteins to be isolated and to have channel properties ascribed to it. AQP0, also known as major intrinsic protein of 26,000 daltons (MIP26), was first isolated from vertebrate ocular lens fiber cells.[9,46] The ocular lens is devoid of blood vessels and innervation and thus, is remarkably transparent. The lack of vascular supply structures in the fiber cell means that diffusion pathways are of paramount importance to the establishment and maintenance of lens homeostasis and transparency. The transparency of the lens, together with its ability to undergo subtle shape changes during the dynamic focusing of accommodation, provides for a clear and accurate image of the world to be projected onto the retina. The transparent nature of the lens is contingent on several crucial features that permit light to pass through with minimum of scattering: (1) the maintenance of a highly ordered molecular structure of the crystallin proteins; (2) terminally differentiated fiber cells containing very few, if any organelles; (3) intracellular and intercellular spaces are kept smaller than the wavelength of ambient light.[47–49] The highly ordered structure of the intracellular crystallin proteins and the maintenance of very small intercellular spaces are both correlated with movement of water into and out of the lens fiber cells. Facilitation of water diffusion throughout the lens is therefore thought to be the primary role of AQP0. In animals, AQP0 makes up approximately 60% of the lens fiber cell plasma membrane protein complement[9,46] and is therefore abundantly present in locations where it can effectively regulate the overall movement of water throughout the lens.

Due to morphological and physiological evidence, AQP0 was initially thought to be a gap junction protein.[50,51] The gap junction-like appearance of highly ordered arrays of AQP0 molecules traversing the entire thickness of the plasma membrane of the lens fiber cell gave credence to the gap junction identification. In addition, a pronounced quaternary association of AQP0 molecules arranged in multimeric structures with apparent transmembrane pores also appeared to be gap junction-like. Transmission electron microscopy images of fiber cells sectioned perpendicular to the plasma membrane also revealed gap junction-like structures formed between AQP0 structures present in adjacent fiber cell plasma membranes. The morphology of these structures revealed in these cross-sections is almost identical to that of proper connexons, differing only in the "gap" dimensions, with AQP0 structures having a significantly thinner (11–15 Å) association.

We find that the AQP0 monomer channel is exclusively a water channel, albeit a rather poor one when compared with all the other AQPs. Early functional studies ascribed a wide range of transport properties to AQP0, very similar to that of known gap junctions. This was probably caused by contamination with endogenous lens fiber cell connexin proteins that also exist in significant amounts and co-purify with AQP0 using the methods employed during the early days of discovery. Having optimized the purification protocols, we obtain uncontaminated AQP0 and have also more precisely characterized permeation. The permeation rates still exhibit a rather wide range, with recent published functional studies measuring the water permeation rates at between $0 - 43$ times the background membrane permeation rate for water.[52–58]

5.2 Aquaporin 0 Monomer Structure and Organization

Bovine AQP0 (bAQP0) is a 263 amino acid protein with a mass of 28,223 Daltons. Each monomer has in its structure, a patent transmembrane channel with an average channel diameter of 2.1 Å. During purification and crystallization, bAQP0 remains a homotetramer. Individual monomers assemble together as tetramers, with the tetrameric fourfold axis in the crystal aligned with the c-axis. In the plane of the membrane the tetramer is ~60 Å wide (~74 Å corner to corner) and ~53 Å tall. Each bAQP0 monomer is ~35 Å in diameter and contains one channel at its center that is oriented parallel to the fourfold axis of the tetramer as in other AQPs and thus each tetramer contains 4 individual monomer channels. One of the structural hallmarks of the aquaporin family of proteins is the association of the monomer subunits into tetrameric quaternary structures. The tetramer structure is the overwhelmingly favored *in vivo* orientation and is also maintained with care throughout the purification and crystallization protocols required for 3-dimensional crystal formation. The tetramer association of aquaporin monomers is seen in all of the high-resolution X-ray crystallographic structures, even though crystal space groups are quite variable. In addition, 2-dimensional crystals can also be readily formed with tetramer organization. It is important to note that the functional unit for water permeation is the monomer channel. There is currently no evidence that the tetramer form is required for monomer channel permeation, nor that the central void that forms along the tetrameric fourfold axis is a functional channel.

All of the aquaporins that have been crystallized to date have exhibited typical integral membrane charge distributions; a hydrophobic zone in the area closely associated with the inside of the membrane and relatively hydrophilic extracellular and cytoplasmic domains facing the extramembraneous milieu. All of the aquaporin monomers mentioned above orient themselves in very specific and reproducible arrangements to bury their hydrophobic surfaces to the maximum extent possible. This orientation tendency assures that the channel spans the membrane. AQP0 tetramers also assemble into larger scale structures of many closely associated tetramers. Hundreds and sometimes thousands of tetramers associate to form large rafts of AQP0 in the ocular lens fiber cell plasma membrane.

In AQP0, the total surface area (excluding the channel surface) of each monomer[2] is 10,788 Å and of each tetramer[2] is 29,632 Å. When assembled as a tetramer, the buried lateral interface surface area between monomers[2] is 13,519 Å, for a ratio of total/buried surface area of approximately 2.2:1. The association of individual monomers into a tetramer results in coaxial monomer channels with a central void along the fourfold symmetry axis. The profile of each aquaporin monomer is roughly triangular when viewed down the z-axis (from extracellular or cytoplasmic aspect perpendicular to the membrane) with relatively straight sides radiating from the fourfold axis and a curved outer aspect that touches the membrane lipid. The constraints that hold the monomers in the same orientation are not obvious or identical in every aquaporin but are strong enough to lock the monomers consistently in a preferred orientation. In general, the outer surface of the monomer that faces the membrane environment is the most hydrophobic region of the monomer and this might be the most influential factor for consistent orientation of the monomers.

In the case of AQP0, there are two "lock and key" interactions between adjacent monomers on the cytoplasmic sides of the monomers, where positively charged residues fit into negatively charged pockets. These interactions contribute to the tetramer stability seen in the aquaporin family of proteins. The "lock and key" arrangement is seen most dramatically in AQP0 but is also present in the other aquaporins mentioned above (Figure 5).

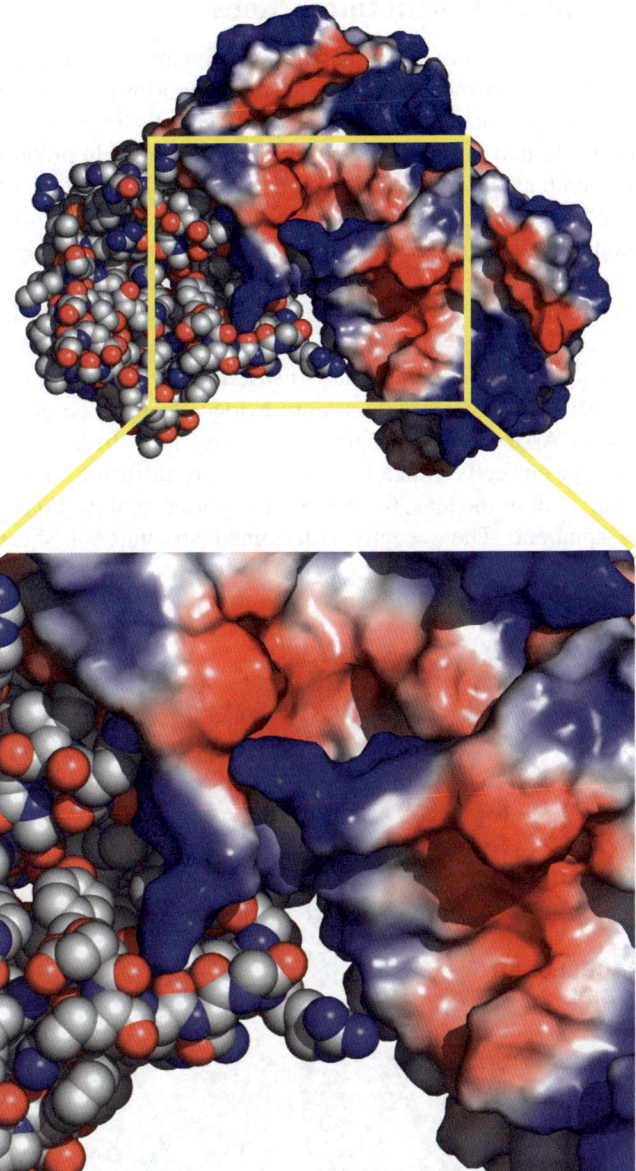

Figure 5 *Three monomers arranged in the typical tetramer orientation. The surface of two monomers are colored according to its electrostatic charges, red is negative, blue is positive, and gray is neutral. The coloration demonstrates the charges involved in the monomer association into the stable tetramer quaternary structures.*

5.3   Extracellular Domain Interactions

Another key interaction between aquaporin monomers/tetramers is the potential close association of the extracellular domains of aquaporins in one fiber cell with the extracellular domains of aquaporins in adjacent fiber cells, in an arrangement that is closely analogous to that of the gap junction forming connexin proteins. The potential exists for the channels in one cell to line up coaxially with the channels of an adjacent cell and thus form a patent contiguous path that directly links the cytoplasm of adjacent cells (Figure 6). The physiological ramifications of this association in single cells are probably minimal when compared to multicellular organisms, where the establishment of a fundamental level of intercellular communication is possible. Tissues and organs usually work as unitary structures where fundamental necessities such as water and nutritional balance and distribution are essential to the establishment of homeostasis. Nowhere is this more evident than in the ocular lens where AQP0 is present. Assemblies of AQP0 in one cell have been seen aligned with assemblies in adjacent cells. Since the lens has only diffusion to rely on to move materials into and out of the lens, the potential importance of AQP0 as a water channel becomes apparent. The recently determined structure of sheep AQP0 was

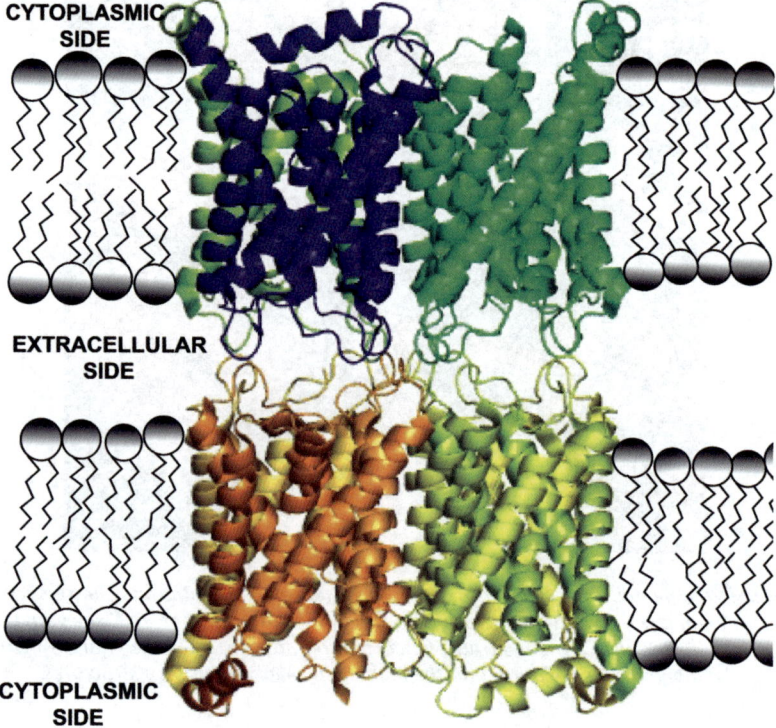

Figure 6 *Diagram of the association of AQP0 tetramers in the extracellular space, where the extracellular domains from each monomer come into close proximity to the extracellular domains of the AQP0 in the adjacent neighboring cell*

obtained using electron diffraction of 2D crystals that happened to crystallize with their extracellular domains oriented so that the monomer channels are almost perfectly coaxial.[59] If this arrangement does indeed mimic the *in vivo* orientation then the formation of patent channels linking the cytoplasm of adjacent cells is possible. Close examination of the opposed extracellular domains shows that these domains do not tightly associate in a way that would close off the channel from the extracellular space. It appears that water from the cytoplasm of each adjacent cell and from the extracellular space is relatively free to move down osmotic and pressure gradients as lens physiology dictates for optimal lens transparency.

The ocular lens and thus the individual fiber cells that makeup the majority of the lens is also subject to considerable physical stress during the process of accommodation, when the muscle fibers of the ciliary body put tension on the lens. This tension significantly distorts the overall shape of the lens, as well as the individual fiber cells, in order to change the refractive power of the lens. The fiber cells must have a mechanism to recover from the shape changes induced by these forces so that the lens returns to its nominal dimensions. In addition, a mechanism to prevent delamination or separation of adjacent fiber cells is also of paramount importance in order to maintain lens integrity and optical clarity. The remarkably high concentration of AQP0 in the fiber cell plasma membrane and its significant extracellular domain region give rise to the suggestion that AQP0 might also have a cell adhesion role. This supposition requires that some sort of attractive force or binding arrangement is present. The two high-resolution structures of AQP0 provide no indication or evidence that such a force is present. The extracellular surface of the AQP0 monomer has a slightly positive charge on an average, and thus opposed monomers would tend to repel each other and the extracellular loops of AQP0 do not seem to have a "lock and key" mechanism that would attractively interact with the extracellular domains of AQP0 monomers in an adjacent cell.

In conclusion, the AQP0 monomers of one tetramer are tightly associated with each other through a "lock and key" structural mechanism with strong charge attractions. There is currently no evidence of a cell adhesion role mediated through extracellular domain interactions, although this remains an area of great interest to structural biologists and vision researchers.

Acknowledgment

Work in the Stroud laboratory in this field is supported by a grant from the National Institutes of Health grant number GM24485.

References

1. A. Fischer, *Vorlesungen über Bakterien*, 1903.
2. K.B. Heller, E.C. Lin and T.H. Wilson, *J. Bacteriol.*, 1980, **144**, 274.
3. G. Sweet, C. Gandor, R. Voegele, N. Wittekindt, J. Beuerle, V. Truniger, E.C. Lin and W. Boos, *J. Bacteriol.*, 1990, **172**, 424.
4. D.L. Weissenborn, N. Wittekindt and T.J. Larson, *J. Biol. Chem.*, 1992, **267**, 6122.

5. D. Fu, A. Libson, L.J. Miercke, C. Weitzman, P. Nollert, J. Krucinski and R.M. Stroud, *Science*, 2000, **290**, 481.
6. M.O. Eze and R.N. McElhaney, *J. Gen. Microbiol.*, 1978, **105**, 233.
7. P. Nollert, W.E. Harries, D. Fu, L.J. Miercke and R.M. Stroud, *FEBS Lett.*, 2001, **504**, 112.
8. K. Liu, H. Nagase, C.G. Huang, G. Calamita and P. Agre, *Biol. Cell*, 2005.
9. J. Alcala, N. Lieska and H. Maisel, *Exp. Eye Res.*, 1975, **21**, 581.
10. R.M. Broekhuyse and E.D. Kuhlmann, *Exp. Eye Res.*, 1974, **19**, 297.
11. M.E. Baker and M.H. Saier Jr., *Cell*, 1990, **60**, 185.
12. P. Agre, *Angew Chem. Int. Ed. Engl.*, 2004, **43**, 4278.
13. L.M. Holm, T.P. Jahn, A.L. Moller, J.K. Schjoerring, D. Ferri, D.A. Klaerke and T. Zeuthen, *Pflugers Arch.*, 2005, **450**, 415.
14. G.M. Preston, T.P. Carroll, W.B. Guggino and P. Agre, *Science*, 1992, **256**, 385.
15. C. Maurel, J. Reizer, J.I. Schroeder, M.J. Chrispeels and M.H. Saier Jr., *J. Biol. Chem.*, 1994, **269**, 11869.
16. J.H. Park and M.H. Saier Jr., *J. Membr. Biol.*, 1996, **153**, 171.
17. A.S. Verkman, L.B. Shi, A. Frigeri, H. Hasegawa, J. Farinas, A. Mitra, W. Skach, D. Brown, A.N. Van Hoek and T. Ma, *Kidney Int.*, 1995, **48**, 1069.
18. T.L. Anthony, H.L. Brooks, D. Boassa, S. Leonov, G.M. Yanochko, J.W. Regan and A.J. Yool, *Mol. Pharmacol.*, 2000, **57**, 576.
19. A. Engel, Y. Fujiyoshi and P. Agre, *EMBO J.*, 2000, **19**, 800.
20. G.M. Pao, L.F. Wu, K.D. Johnson, H. Hofte, M.J. Chrispeels, G. Sweet, N.N. Sandal and M.H. Saier Jr., *Mol. Microbiol.*, 1991, **5**, 33.
21. G.J. Wistow, M.M. Pisano and A.B. Chepelinsky, *Trends Biochem. Sci.*, 1991, **16**, 170.
22. A. Finkelstein, *Water movement through lipid bilayers, pores and plasma membranes, theory and reality*. Wiley, New York, NY, 1987.
23. H. Sui, B.G. Han, J.K. Lee, P. Walian and B.K. Jap, *Nature*, 2001, **414**, 872.
24. D.F. Savage, P.F. Egea, Y.C. Robles, J.D. O'Connell and R.M. Stroud, *PLoS Biology*, 2003, **1**, 334.
25. W.E. Harries, D. Akhavan, L.J. Miercke, S. Khademi and R.M. Stroud, *Proc. Natl. Acad. Sci. USA*, 2004, **101**, 14045.
26. E. Tajkhorshid, P. Nollert, M.O. Jensen, L.J. Miercke, J. O'Connell, R.M. Stroud and K. Schulten, *Science*, 2002, **296**, 525.
27. M. Yasui, A. Hazama, T.H. Kwon, S. Nielsen, W.B. Guggino and P. Agre, *Nature*, 1999, **402**, 184.
28. M.L. Zeidel, S.V. Ambudkar, B.L. Smith and P. Agre, *Biochemistry*, 1992, **31**, 7436.
29. M.L. Zeidel, S. Nielsen, B.L. Smith, S.V. Ambudkar, A.B. Maunsbach and P. Agre, *Biochemistry*, 1994, **33**, 1606.
30. R. Pomes and B. Roux, *Biophys. J.*, 2002, **82**, 2304.
31. B.L. de Groot and H. Grubmuller, *Science*, 2001, **294**, 2353.
32. N. Chakrabarti, B. Roux and R. Pomes, *J. Mol. Biol.*, 2004, **343**, 493.
33. B. Roux and R. MacKinnon, *Science*, 1999, **285**, 100.
34. D.A. Doyle, J. Morais Cabral, R.A. Pfuetzner, A. Kuo, J.M. Gulbis, S.L. Cohen, B.T. Chait and R. MacKinnon, *Science*, 1998, **280**, 69.

35. A. Burykin and A. Warshel, *Biophys. J.*, 2003, **85**, 3696.
36. B.L. Smith and P. Agre, *J. Biol. Chem.*, 1991, **266**, 6407.
37. A. Froger, J.P. Rolland, P. Bron, V. Lagree, F. Le Caherec, S. Deschamps, J.F. Hubert, I. Pellerin, D. Thomas and C. Delamarche, *Microbiology*, 2001, **147**, 1129.
38. D. Thomas, P. Bron, G. Ranchy, L. Duchesne, A. Cavalier, J.P. Rolland, C. Raguenes-Nicol, J.F. Hubert, W. Haase and C. Delamarche, *Biochim. Biophys. Acta*, 2002, **1555**, 181.
39. B. Yang, N. Fukuda, A. van Hoek, M.A. Matthay, T. Ma and A.S. Verkman, *J. Biol. Chem.*, 2000, **275**, 2686.
40. A.S. Verkman, *J. Physiol.*, 2002, **542**, 31.
41. G.V. Prasad, L.A. Coury, F. Finn and M.L. Zeidel, *J. Biol. Chem.*, 1998, **273**, 33123.
42. M. Ikeda, E. Beitz, D. Kozono, W.B. Guggino, P. Agre and M. Yasui, *J. Biol. Chem.*, 2002, **277**, 39873.
43. K. Ishibashi, K. Yamauchi, Y. Kageyama, F. Saito-Ohara, T. Ikeuchi, F. Marumo and S. Sasaki, *Biochim. Biophys. Acta*, 1998, **1399**, 62.
44. K. Ishibashi, M. Kuwahara, Y. Gu, Y. Tanaka, F. Marumo and S. Sasaki, *Biochem. Biophys. Res. Commun.*, 1998, **244**, 268.
45. K. Ishibashi, T. Morinaga, M. Kuwahara, S. Sasaki and M. Imai, *Biochim. Biophys. Acta*, 2002, **1576**, 335.
46. R.M. Broekhuyse, E.D. Kuhlmann and A.L. Stols, *Exp. Eye Res.*, 1976, **23**, 365.
47. J.H. Kinoshita, P. Kador and M. Catiles, *JAMA*, 1981, **246**, 257.
48. S. Trokel, *Invest. Ophthalmol.*, 1962, **1**, 493.
49. J.H. Kinoshita and L.O. Merola, *Invest. Ophthalmol.*, 1964, **47**, 577.
50. E.L. Benedetti, I. Dunia and H. Bloemendal, *Proc. Natl. Acad. Sci. USA*, 1974, **71**, 5073.
51. D.A. Goodenough, J.S. Dick II and J.E. Lyons, *J. Cell. Biol.*, 1980, **86**, 576.
52. G. Chandy, G.A. Zampighi, M. Kreman and J.E. Hall, *J. Membr. Biol.*, 1997, **159**, 29.
53. S.M. Mulders, G.M. Preston, P.M. Deen, W.B. Guggino, C.H. van Os and P. Agre, *J. Biol. Chem.*, 1995, **270**, 9010.
54. B. Yang and A.S. Verkman, *J. Biol. Chem.*, 1997, **272**, 16140.
55. K.L. Nemeth-Cahalan and J.E. Hall, *J. Biol. Chem.*, 2000, **275**, 6777.
56. K. Varadaraj, C. Kushmerick, G.J. Baldo, S. Bassnett, A. Shiels and R.T. Mathias, *J. Membr. Biol.*, 1999, **170**, 191.
57. C. Kushmerick, S.J. Rice, G.J. Baldo, H.C. Haspel and R.T. Mathias, *Exp. Eye Res.*, 1995, **61**, 351.
58. C. Kushmerick, K. Varadaraj and R.T. Mathias, *J. Membr. Biol.*, 1998, **161**, 9.
59. T. Gonen, P. Sliz, J. Kistler, Y. Cheng and T. Walz, *Nature*, 2004, **429**, 193.

CHAPTER 12

Gas Channels for Ammonia

SHAHRAM KHADEMI AND ROBERT M. STROUD

Department of Biochemistry and Biophysics, UCSF S412C Genentech Hall, San Francisco, CA 94143, USA

1 Introduction

Ammonium transport across biological cell membranes is a critical physiological process in all domains of life.[1–4] [In this chapter, Am refers to (NH_3 and NH_4^+) and MA to (CH_3NH_2 and $CH_3NH_3^+$).] Although Am is the preferred source of nitrogen for most microorganisms, it is highly toxic in high concentrations. Simple diffusion also occurs through the lipids of cell membranes and can support growth; however, at low concentrations of Am, other regulatable mechanisms of Am transport are usually required. Functional complementation of a yeast mutant, defective in Am transport, led to the isolation of Am transporters from yeast (which were called *M*ethylammonium/ammonium *P*ermeases or MEPs) and from plants *Arabidopsis* (which were called *Am*monium *T*ransporters or AMTs)[5,6] simultaneously. Since then, related transporters have been found in bacteria[7–11] (AmtB), yeast,[12,13] invertebrates[14] (*e.g.*, the *Caenorhabditis elegans* genome contains four AMT homologs), and several species of plants, including *Arabidopsis*,[15] rice,[16] and tomato.[17]

The AMT/MEP genes have not been identified in mammalian species. However, Am transport in mammals is well documented in the physiology of the kidney.[18] Both the marginal homology between AMTs and red blood cell (RBC) Rh proteins[19] and the functional complementation of an *Saccharomyces cerevisiae* Δ(*mep1,2,3*) mutant in which all three MEPs are deleted, led Marini *et al.*[20] to argue that the Rh proteins could be orthologs of AMT in mammalian erythrocytes. Members of the MEP/AMT family vary in length from ~400 to 500 amino acids, although some members have long C-terminal regions increasing their length to ~600 amino acids.[21]

The transport rates of most AMT/MEP proteins can be conveniently measured by the transport of radioactive ^{14}C-labeled MA.[22] Plant, bacteria, and yeast grow on very low concentrations (<5 μM) of ammonium salts as the sole nitrogen source, reflecting the "high affinity" of AMT/MEP for ammonium. For example, *S. cerevisiae* has three MEP proteins: MEP1, MEP2, and MEP3. MEP1 and MEP2 show the highest affinity for Am with "response K_m" (the concentration that evokes half maximal conductance) of ~5–10 and 1–2 μM, respectively. The lowest affinity belongs to MEP3 with a "response K_m" of ~2 mM.[12]

As the intracellular concentration of Am rises, it becomes toxic to the cell and therefore its uptake is under tight regulation. In bacteria and archaea, genes encoding AmtB are linked to and potentially co-transcribed with *glnK*, which encodes a member of the P_{II} signal transduction protein family. Members of this family are small homotrimeric proteins that regulate activities of metabolic enzymes, permeases, and signal transduction enzymes. This protein family shows high conservation, with family members in eukaryotes (plants and eukaryotic algae), archaea, and bacteria.[23] GlnK serves to regulate the enzyme activity and gene expression in response to the intracellular nitrogen status.[24] GlnK has been shown to associate with AmtB,[25] and this is consistent with the fact that both are found on the same operon.[26,27] Both proteins are trimers,[28] which reflect the symmetry required for optimal interaction between two proteins. Under nitrogen-limiting conditions, GlnK is predominantly in its fully uridylylated state (uridylylation occurs at Tyr51) and it does not bind to AmtB, which is then active and conducts ammonia from the surrounding milieu. When the Am concentration gets too high inside the cell, GlnK is rapidly deuridylylated and the unmodified form of GlnK associates tightly with AmtB. This association blocks transport by AmtB and therefore ammonia is no longer transported into the cell.[29]

There is a fascinating "detective story" as to the molecular species (NH_4^+ *vs.* NH_3) that is transported by AMT/MEP/Rh proteins, and whether it is driven by a proton gradient. The X-ray structure of AmtB from *Escherichia coli* at 1.35 Å resolution made it possible to specify the most probable mechanism of conduction by this family of membrane proteins.[30] In this chapter we review the pathway to understand the mechanism of conduction by this family of proteins that we originally derived based on the structure and measurement of the transported species by AmtB.

2 The Structure of Ammonia Channel

2.1 Overall Structure of AmtB: A New Family of 11-Crossing Proteins

Based on amino acid sequence analysis, members of AMT/MEP family were predicted to contain between 9 and 12 putative transmembrane (TM) helices.[4–7,31,32] This variation in the predicted number of TM helices seems to be in part because of differences between the various algorithms used for helix prediction, but may also reflect some real differences within the family.[29] The ambiguity in predicted topology was resolved by purification and characterization of AmtB proteins[29,30] and ultimately by the structure determination.[30] Matrix-assisted laser desorption/ionization mass spectroscopy (MALDI-MS) and N-terminal amino acid sequencing revealed that 20 amino acids from the N-terminus of *Aquifex aeolicus* AmtB (AmtB_AQFX) and 22 residues from *E. coli* AmtB (AmtB_Ecoli) were excised in the mature proteins.[30] Mass spectrometry and N-terminal sequencing results, coupled with prediction of signal peptide cleavage sites[30] using neural network approaches,[33] led to the identification of these regions as signal sequences, which are removed upon insertion of proteins into the cell membrane (Figure 1). Thus the membrane topology for AmtB proteins with 11 TM helices has an extracellular N-terminus and a cytoplasmic C-terminus.[30]

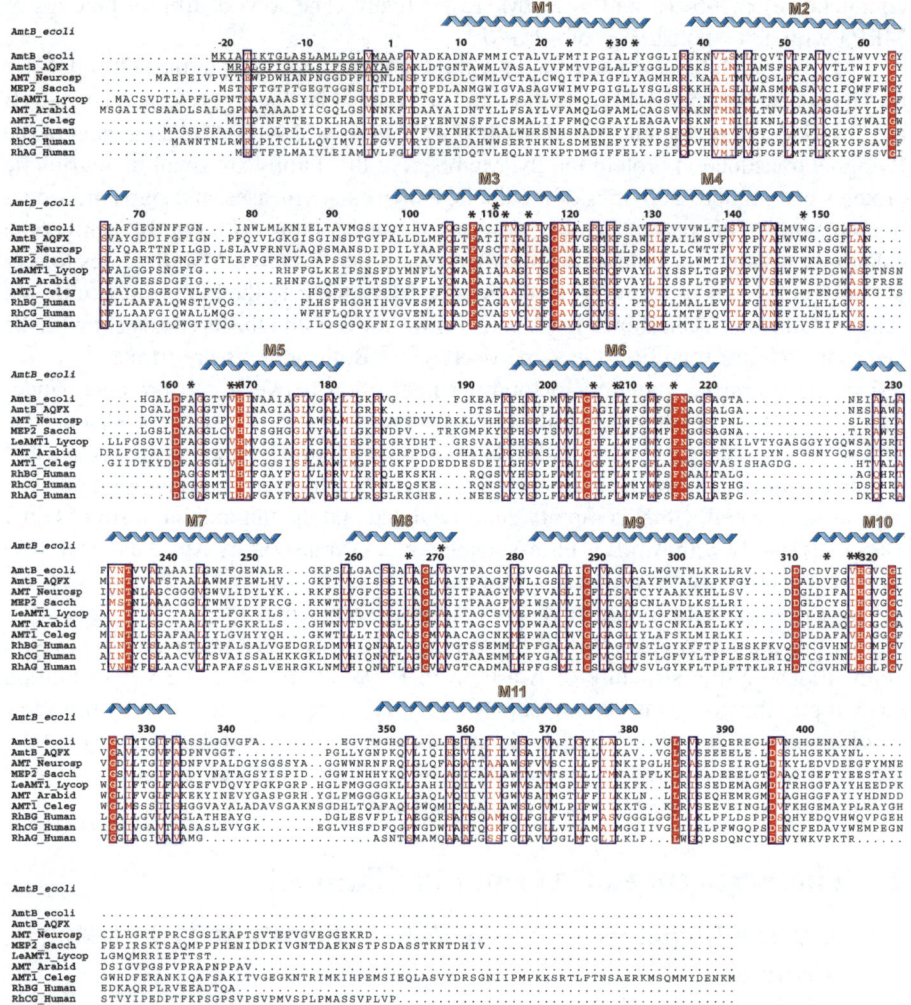

Figure 1 *The amino acid sequence alignments of AmtB/MEP/Rh homologs from E. coli (AmtB_Ecoli), A. aeolicus (AmtB_AQFX), Neurospora (AMT_Neurosp), Saccharomyces cerevisiae (MEP2_Sacch), Lycopersicon esculentum (LeAMT1_Lycop), Arabidopsis thaliana (AMT_Arabid), Caenorhabditis elegans (AMT1_Celeg), and human Rh factors (RhBG, RhCG, RhAG). Transmembrane helices M1–M11 are indicated above the sequence. The numbering is that of AmtB_E. coli. Conserved amino acids are in white and red filled rectangles. Similar residues are in red surrounded by blue lines. The signal sequences in E. coli and A. aeolicus are underlined. Residues that line the lumen of the channel are labeled with an asterisk (*) above. Eleven transmembrane helices are identified by helical motifs above the sequence, labeled by transmembrane helix numbers M1–M11*

Crystals of AmtB_Ecoli diffracted to a resolution of 1.35 Å (P6$_3$, a = 96.5 Å, c = 94.6 Å),[30] which is the highest resolution for a membrane protein to date, reflecting repeated attempts to improve the resolution by finding optimal detergent and lipid conditions. AmtB_Ecoli crystallizes as a homotrimer arranged around a threefold symmetry axis that lies perpendicular to the membrane plane where each monomer contains a single pathway for Am (Figure 2A,B). AmtB was shown to be a homotrimer in the membrane fraction when expressed in *E. coli*, which reflects the physiological form of the protein, and when reconstituted into lipid bilayers.[34] It also exists as a trimer in detergent micelles.[28,30]

The only channel-like pathway from one side to the other lies within each monomer and is surrounded by 11 membrane-spanning α-helices (M1–M11) arranged in a right-handed helical bundle. The interfaces between subunits are almost as hydrophobic as the lipid-contacting exterior, suggesting that a monomer could be transiently stable in the membrane upon synthesis, prior to forming trimers.

The trimer of AmtB_Ecoli has a net negative charge of −7.5 (13.5 positive + 21 negative charges) on the periplasmic surface and a net positive charge of +9 (42 positive + 33 negative charges) on the cytoplasmic side, which is consistent with the "positive inside" trend in membrane proteins. As described for other membrane proteins, the polar aromatic side chains of Tyr and Trp, in this case residues Tyr62 (at the extracellular side) and Tyr180, Trp250, and Trp297 (at the cytoplasmic side), would lie at the hydrocarbon/headgroup interface.

At the extracellular side, the threefold axis is surrounded by three tightly packed copies of just helix M1 of each monomer, which seem to seal the central axis against passage. Toward the cytoplasmic side, the three M1s (+16° to each other) veer away from the threefold axis to leave an open pocket ~10 Å across, formed by three copies of M1 and M6. M1 has a kink (22°) in the helix secured by the only *cis*-proline (Pro26) in AmtB, which is a residue not conserved in the superfamily. It is often the case that kinks in transmembrane helices are conserved among homologous structures when there is helix-distorting proline in just one member, even when this proline is not conserved in homologous structures. M1 and M6 are not long enough to span the bilayer, consistent with the trimer being the stable physiological quaternary structure and preserving a dimpled region at the central symmetry axis.

The 11-crossing AmtB structure is the first fold of its kind for a membrane protein. Within each AmtB channel, M1–M10 diverge outward from the central plane in a right-handed helical bundle to generate a vestibule on each side of the cell membrane. The interhelix angles in each monomer are ~11°–58°. At 1.35 Å resolution, the AmtB structure provides a metric of distances and helix packing in membrane proteins. Packing of α-helices in membrane proteins of known structures shows that transmembrane helices are often packed at distances close enough for Cα–H ... O=C hydrogen bond formation.[35–37] This type of interaction is facilitated by glycine, which is overrepresented in transmembrane segments and allows closer interhelical contact for purely steric reasons, there being no side chain.[36] The structure of AmtB reveals four intramonomer glycine Cα–H ... O=C hydrogen bonds within the *trans*-bilayer region. Three of them are between M1 and M6 (Gly204, Gly211, Gly34), and one is between M8 and M10 (Gly325). Located next to the electron withdrawing O=C group, Cα–H$_2$ has a residual slightly positive charge. The electrostatic

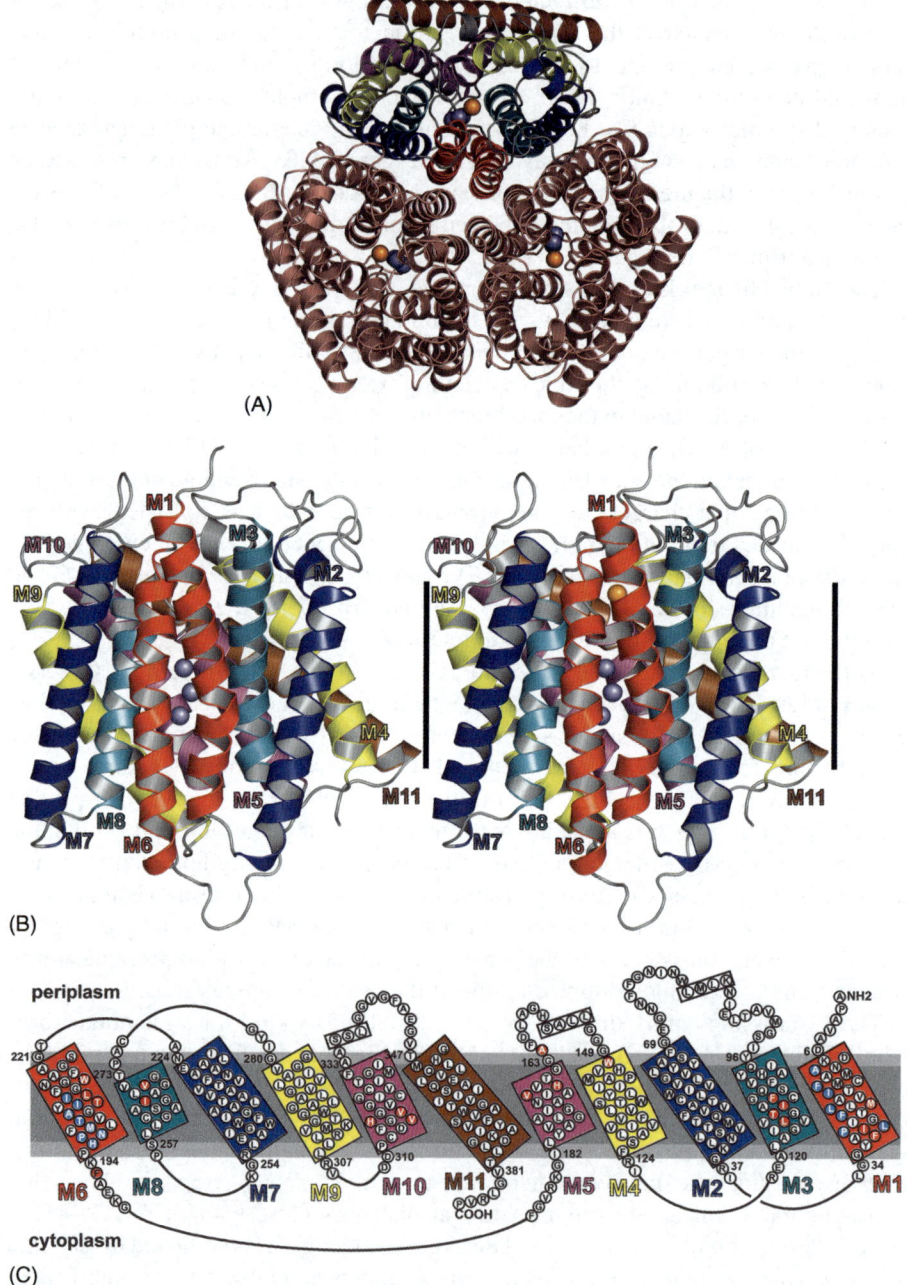

Figure 2 *Three-dimensional (3D) fold of AmtB. (A) Ribbon representation of the AmtB trimer is viewed from the extracellular side. A threefold axis perpendicular to the page relates the three monomers. In the upper monomer, the quasi twofold axis is vertical and parallel to the plane of the page, and relates five helices on the left side to five on the right. Corresponding quasi twofold related helices are shown in the*

component of these interactions, although weak, may be accentuated owing to the low dielectric within the bilayer and so can be many times stronger than they would be in a solvent-accessible situation. This is estimated to be 2.5–3.0 kcal mol^{-1} per bond *in vacuo*, or about half the energy of a conventional N–H … O hydrogen bond *in vacuo*,[36] and so may stabilize the AmtB structure in the context of the hydrophobic lipid bilayer. Cys109 and Cys56, found only in AmtB_Ecoli, are close enough to form a disulfide bond within the transmembrane region, which also may increase the stability of the protein in the membrane.

2.2 A Membrane Protein with In-Plane Quasi Twofold Symmetry

There is no readily detectable evidence of gene duplication in the sequence of AMT/MEP/Rh family examined without a structure. However, the structure of AmtB shows a quasi twofold axis in the mid-plane of the membrane that intersects the trimer threefold axis. As such it shows how to look for any residue of duplication. The quasi twofold symmetry relates M1–M5 to M6–M10. This clearly reflects an origin of the first ten crossings in a primordial gene duplication event. M11 is an additional ~50 Å long-straight helix inclined –50° perpendicular to the membrane plane that surrounds the lipid-accessible side of each monomer (Figure 2B,C).

The quasi twofold symmetry and the structural duplication with opposite polarity with respect to the membrane plane observed in AmtB are seen in a number of other membrane proteins including GlpF[38] and all other aquaporins (AQPs),[39–41] the SecY protein of the translocon,[42] the ClC chloride channel[43,44] (recently shown to be an H$^+$–Cl$^-$ exchange transporter[45]), the bacterial homolog of the Na$^+$/Cl$^-$-dependent neurotransmitter transporters for biogenic amines,[46] and lactose permease LacY[47] and its "major facilitator super-family" (MFS) homologs. However, of these

---

same color. The quasi twofold axis intersects the threefold axis. The blue spheres are potential ammonia molecules. The orange sphere represents an ammonium ion. (B) A stereo view of the monomeric ammonia channel viewed down the quasi twofold axis. The vertical bar (35 Å) represents the inferred position of the hydrophobic portion of the bilayer. In this and all subsequent figures the extracellular side is up. The N and C termini of the structure are labeled N and C, respectively. Three NH$_3$ molecules, reflecting 20%, 15%, 20% occupancy peaks seen only when crystallized in presence of ammonium sulfate, are shown as blue spheres. (C) The amino acid sequence of AmtB is arranged topologically as in the structure, with helices viewed as if from inside the channel looking away from the threefold axis. The quasi twofold axis is perpendicular to the center of the figure. Five helices comprise each segment, labeled M1–M5 and M6–M10. Related helices M1 and M6, M2 and M7, etc. are boxed in similar colors. Sidechains of residues in red circles contribute to the substrate-contacting walls of the channel. Residues in blue circles contribute sidechains to the inter monomer contacts that immediately surround just the threefold axis of the trimer, since many oligomeric membrane proteins either need to insulate against passage of alternate molecules there, or utilize this special location for stability as in the aquaporins (AQP), or for conductance as in K$^+$ channels. The deduced location of the cell membrane is illustrated in gray (35 Å) with light gray for the head group region (to 40 Å thickness)

proteins, only in the AQP family were the vestiges of structural duplication recognized in the gene sequences[48] prior to structure determination. Owing to the opposite polarity of the duplicated segments relative to the membrane, the primordial gene duplication event of such membrane proteins must have occurred prior to generation of enough functional surrounding to support any transport of molecules from one side of the membrane to the other.

This duplication with opposite polarity is seen in several other membrane proteins, yet in almost all cases it was not identified in the gene sequence prior to structure determination. Thus one can ask whether in future, algorithms may be designed that could detect or predict duplication in the structure based on very low overall conservation, for conserved residues that duplicate functional roles at key regions. One can ask if there are residues that might reflect twofold relatedness consistent with the remnants of gene duplication. Thus, Asp160 – (seven residues) – His168 is found at the N-terminus of M5, while Asp310 – (seven residues) – His318 is found 150 residues downstream at the N-terminal end of M10. One could foresee that for membrane proteins, prior to structure determination, it might be possible to identify regions of helical membrane crossings, and then to search for such repeats of potentially functional motifs as shown in this example. Remarkably, there are no other duplications of conserved residues. Even helices such as M6 that have three highly conserved residues and four more that are closely similar across Amt/MEP/Rh family members, do not have any matches in M1 which would be the correlated region of the first repeated segment. Therefore, it seems that vestiges of the duplication implied by the structure can be present even if observed only in two pairs of residues with a common length of seven residues between the two in each repeat. An indication that these residues are significant is their conservation, and their congruent roles from opposite sides of the structure.

2.3   The Ammonia Pathway

To identify any sites for Am or MA binding and to determine the mechanism of conductance, the structure of AmtB in the presence of either 25 mM ammonium sulfate at pH 6.5 or 100 mM methyl ammonium sulfate at pH 6.5 was determined and compared with the structure without any Am by difference methods. The comparison of the structure of AmtB in the presence of Am/MA showed no significant conformational changes *versus* AmtB with no Am, which is at least consistent with AmtB acting as a channel, rather than as an Am transporter that would harness alternating conformational states. Although the original name AmtB is suggestive of a transporter, the evidence coupled with our study (see further below) suggests that it acts as a channel that serves to conduct Am in the direction of an Am gradient.

The channel pathway in AmtB begins with a wide vestibule on the outer surface reaching into one of the two most constricted hydrophobic regions (Figure 3A). At the mid-membrane center of the pathway, there are two in-line almost coplanar histidine side chains of residues 168 and 318, which are conserved throughout the family of AMT/MEP/Rh proteins, and these are followed by the second constricted hydrophobic region and the intracellular vestibule. Between the two hydrophobic constrictions the channel wall is narrow and contains mostly aliphatic non-polar residues throughout its ~20 Å length, consistent with the conduction of uncharged NH_3.

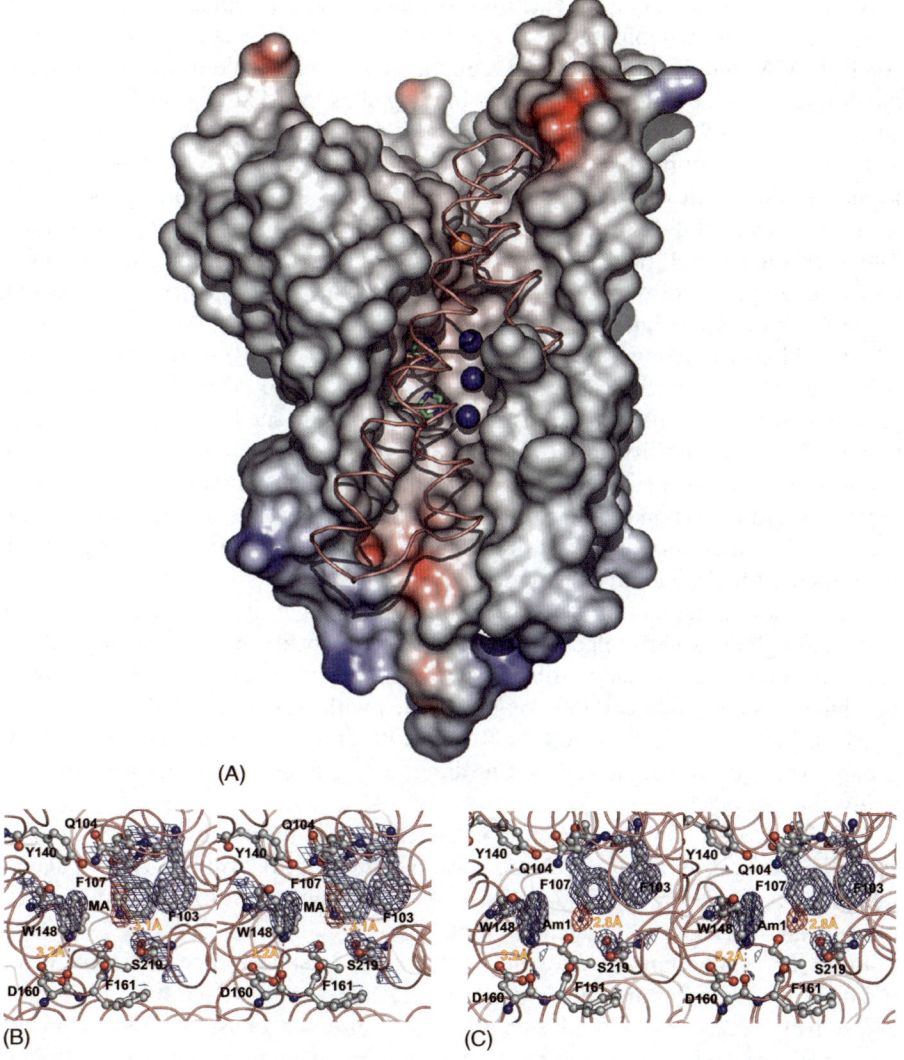

(A)

(B)   (C)

Figure 3 *The ammonia conducting channel in AmtB. (A) The surface of the lumen of ammonia channel is colored according to electrostatic potential, after removal of surfaces of helices M9, and parts of M8 and M10 (whose helical backbones are indicated by pink lines). The positions of two histidines near the three NH_3 sites (blue spheres) are shown in green and blue stick representation, and were not included in the surface electrostatic calculation. (B) Stereo views of the $CH_3NH_3^+/NH_4^+$ binding site Am1. The electron-density map (2Fo–Fc) is contoured at 2σ for the protein (blue), and (Fo–Fc) $CH_3NH_3^+$ omit map contoured at 4.5 σ (red) for AmtB in 100 mM $MASO_4$ at pH 6.5. The MA order is 67% occupancy for each (CH_3 and NH_3) group. Hydrogen bonds between the NH of $CH_3NH_3^+$ and $O\gamma$ of Ser219, and between $N\varepsilon1H$ of Trp148 and O of Asp160 are indicated by yellow dashed lines, with bond distance in yellow. (C) As in part (B), for AmtB in 25 mM $AmSO_4$ at pH 6.5. Am order is 67% occupancy (6.7 electrons)*

Am has the same number of electrons as water, thus, if it replaces H_2O, it is difficult to identify it unambiguously as Am from an electron density map. AmtB also conducts MA, and the density for MA might be distinguishable from H_2O/Am, and the difference could uniquely mark any binding sites for MA. One such MA difference map site (60% ordered, where this "order parameter" represents the integrated electron content, normalized based on the highest occupied water molecule) is located against both aromatic rings of Trp148 and Phe107 and corresponds to the lowest water site of the extracellular vestibule (Figure 3B). This provides a favorable 2π-cation site for NH_4^+ and for $CH_3NH_3^+$, stabilized by ring currents. MA displaces two separate peaks (60% ordered) in the H_2O structure. One is Am1 (67% ordered) in the Am structure (Figure 3C). Although the concentrations of Am and MA are in ~1000-fold excess relative to the K_m for transport (K_m [Am]) ~ 10 µM, K_m [MA] ~ 50 µM), their "order" parameters are relatively low (~60–67%). The low "order" parameter primarily reflects either high movement in the site or low statistical occupation. The high mobility and low occupancy of the Am1 site is consistent with it acting as a "recruitment site" for attracting Am, rather than a binding site. The $-NH_3^+$ moiety is hydrogen-bond donor to the $O\gamma$ of Ser219. Perhaps in a special way, the 2π-cation site may serve to assist in the deprotonation of the Am ion as it proceeds on to the first hydrophobic constriction.

Ammonium at the recruitment site is connected by hydrogen bonding to the bulk water through 30 water molecules seen in the outer vestibule. At physiological pH, Am at the outer recruitment vestibule will be in the form of charged NH_4^+. The outer vestibule provides eight carbonyl oxygens (each with δ –0.4e charge) from Ala158, Asp160, Phe161, Ser219, Ser68, Val91, Trp148, and Ala162 (the lowest carbonyl group). Thus, the vestibules can act as an attractive funnel for water and for positively charged NH_4^+.

The first hydrophobic constriction in the channel is formed by Trp148, Phe103, Phe161, and Tyr140, all of which are conserved in the AMT subfamily. The diameter is 1.2 Å, therefore, the side chains of Phe107 and Phe215 must move apart dynamically during any conduction event.

In 25 mM ammonium sulfate, the crystal structure shows three additional peaks not present without ammonium sulfate (Am2, Am3, Am4 of order 20%, 15%, 20%, respectively) adjacent to imidazole rings of conserved quasi twofold symmetry related His168 and His318, at the center of the narrow hydrophobic channel (Figure 4A,B). Their partial order, even when crystallized at 1000-fold excess relative to the "K_m" for conductance, indicates that these Am-specific peaks are poorly ordered, and may be occupied alternately with each other. The π-cation stabilization is possible at Am2 from the side chains of Phe215 and Trp212. Am2 and Am3 are surrounded by hydrophobic hydrocarbon sidechains.

At 1.35 Å resolution, the orientation and hydrogen bonding of the imidazole rings are unambiguously determined. The $N\epsilon2H$ of His168 is hydrogen-bond donor to the oxygen of OH in conserved Thr273. The unprotonated His168 $N\delta1$ and His318 $N\delta1H$ are fixed by hydrogen bonding to each other, and $N\epsilon2$ of His318 accepts a hydrogen bond from Am4. $C\epsilon1$–H is the closest atom from the imidazole ring of His168 to Am2 (3.2 Å) as is also from His318 to Am3 (3.4 Å). This is almost unprecedented to find the C–H seemingly playing a functional role.

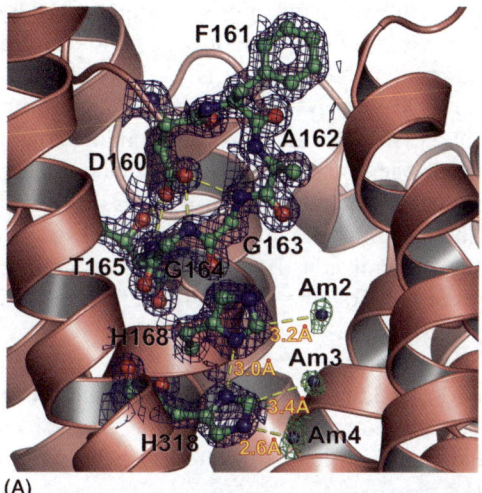

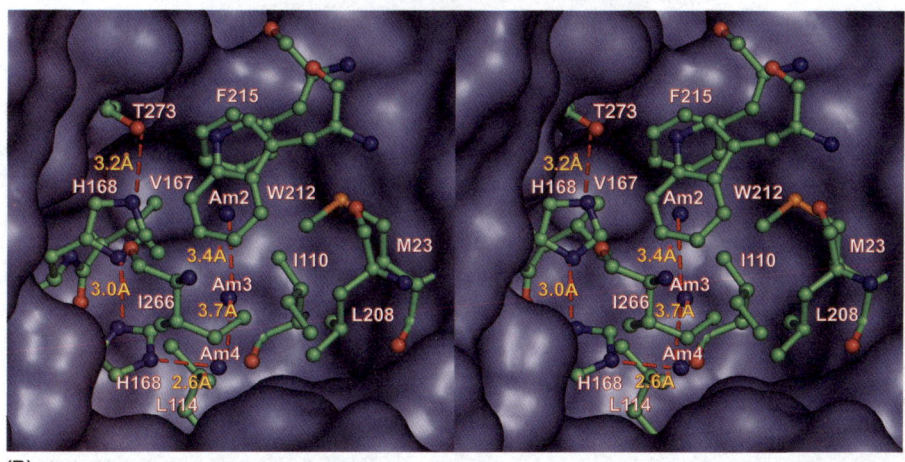

Figure 4 *(A) Electron density (2Fo–Fc) contoured at 1.5σ (blue) for the two-histidine region and surrounding structure including conserved Asp160 which accepts four short hydrogen bonds (dashed yellow). Additional peaks Am2, Am3, Am4 are seen when crystallized with 25 mM ammonium sulfate and are defined in the Fo–Fc omit map at 1.5σ (in red), indicating putative NH3 molecule positions (blue spheres). The hydrogen bonding network shows interactions between the histidines 168, 318 and NH3 peaks in yellow (distances in red). (B) Stereo view of the two-histidine center of the channel and surrounding hydrophobic residues are shown in ball and stick representation. The surface representation covers other surrounding amino acids. Three ammonia-dependent sites are shown (blue spheres) with associated distances (dashed yellow line, red labels)*

Cε1–Hs are capable of donating a hydrogen bond to Am2 and Am3. Cε1–H hydrogen-bond donors have been seen in other proteins.[49–51] In this hydrophobic environment, the imidazole Cε1–H will appear more acidic than that of imidazole in aqueous solution. The low effective dielectric constant (ε) increases the coulombic attraction forces as $1/\varepsilon$. While imidazole nitrogen might act as hydrogen-bond donor N–H or acceptor :N, the Cε1–H bonds can only be donors, easing passage of a molecule that is an acceptor, as is :NH$_3$ but not NH$_4^+$. This and the aliphatic hydrophobic channel are compatible with NH$_3$ as substrate for AmtB. Thus this signature structure beautifully harnesses all of its hydrogen-bonding potential to accommodate the passage of small hydrogen bond accepting molecules. The pK_a of NH$_3$ must be lowered to below 6 at sites Am3 and Am4 since peaks clearly reflect NH$_3$ at a pH of 6.5 in the crystals. Therefore, its pK_a must be reduced as it enters the pathway. At some point close to the aromatic constriction, NH$_4^+$ gives up its proton on the side of entry and NH$_3$ is transported. Together the cation-attractive vestibule possibly down to the Am2 site can recruit NH$_4^+$.

The cytoplasmic side incorporates similar features as related by the quasisymmetric structure. Phe31, Tyr32, and Val314 surround the second cytoplasmic constriction. The first exit-peak of density within the channel pathway is hydrogen bonded to Asp313 and could be close to where NH$_3$ would reacquire a proton on the inside to become NH$_4^+$.

Mutation of conserved Asp160 to Ala160 completely destroys transport of MA, implying that it plays a key role; and Javelle *et al.*[52] proposed that it might provide a primary binding site for NH$_4^+$. However, the crystal structure of AmtB reveals that there is no peak of density nearby even in the presence of 25 mM ammonium sulfate or 100 mM methyl ammonium sulfate. Cys326 shields one planar face of Asp160 and the indole ring of Trp148 shields the other face of Asp160,[30] and Asp160 itself is not accessible to bulk solvent and is conserved across the superfamily, underscoring its key role that appears to be primarily structural. Conserved Asp160 is a helix-capping residue for M5 (Figure 4A), forming a hydrogen bond with Thr165 and the carbonyls of Gly164 and Gly163 at the N-terminal end of M5. Thus, Asp160 carboxyl orients the carbonyls of Asp160, Phe161, Ala162 that, along with carbonyls of Ser68, Ser219, Val147, Trp148 from the outside ends of helices M2, M4, M6, line the vestibule and make it cation attracting. Asp310, the conserved quasi twofold relative of Asp160, similarly acts as a helix cap for M10.

3   Reconstituted into Liposomes AmtB Acts as a Channel that Conducts NH$_3$

To determine the substrate specificity and rates of conductance of AmtB, a fluorescence-based assay was adapted to measure the influx of Am into vesicles by monitoring the pH-sensitive fluorescence of 5-carboxy fluorescein (CF).[30,53,54] Rapid mixing of CF-loaded vesicles in the buffer with ammonium chloride (0.5 or 5 mM) was initiated at pH 6.8. In Am-containing solution, NH$_4^+$ is in equilibrium with NH$_3$ with a pK_a of 9.25. Any NH$_3$ influx through the lipids and through AmtB would reacquire a proton inside from H$_2$O, forming NH$_4^+$ and OH$^-$ and so raise intravesicular pH (Figure 5A). The rate constant of ammonia influx indicated by the rate of pH change, 115.6 ($n = 6$) $\pm$ 13.2 (typical curve-fitting error in a single curve) sec^{-1} for 5 mM NH$_4$Cl outside, is tenfold faster than for protein-free liposomes (12.8 [$n = 6$] $\pm$0.7 sec^{-1}), which indicates a 10:1 signal-to-noise ratio for conductance by the protein *versus* lipids at the lipid/protein ratio of 200 (by weight).

In such methods it is important to eliminate the possibility of artifacts because of changes in osmolarity that could alter the concentration of CF, and hence its degree of self-quenching.[30] Reconstitution of other non-conducting membrane proteins tested does not lead to leakage because of the reconstitution procedure *per se*.

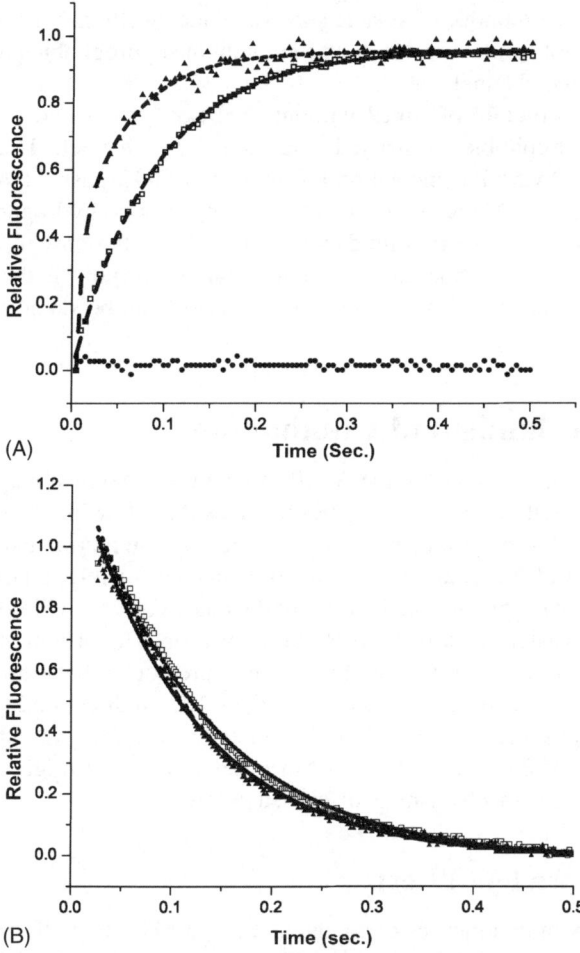

Figure 5 *Channel conductance in proteoliposomes. (A) The time course of change in pH inside of vesicles containing CF buffered by 20 mM HEPES is detected by fluorescence change. The initial pH was 6.8 both inside and outside the vesicles. To initiate flux, 5 mM NH₄Cl was added externally to protein-free liposomes (open squares), and to AmtB-containing proteoliposomes (solid triangles) (protein to lipid ratio 1:200 by weight). To control for possible osmotic effects, 5 mM NaCl was added to proteoliposomes (solid circles) instead of 5 mM NH₄Cl. The dashed lines are exponential fits to the data. (B) Water conductance was assessed by the concentration-dependent self-quenching of CF-containing liposomes (open squares) and proteoliposomes (solid triangles) upon addition of 500 mM sucrose at t = 0. The osmotic change leads to conductance of water through the lipids or through proteins. The dashed lines are exponential fits to the data*

Likewise, when the Am concentration was lowered tenfold to 0.5 mM, the rate dropped only 40%, which is generally consistent with the fact that the K_m for transport is below 0.5 mM Am.

To see if there is water conductivity by AmtB, osmotic permeability of water was measured by the concentration-dependent self-quenching of vesicular CF, using an aliquot of the same batch of AmtB proteoliposomes[30,53] (Figure 5B), and also by light scattering as a monitor of vesicle shrinkage and swelling.[38] Thus it was shown that AmtB does not conduct water, consistent with the hydrophobic pathway through the structure of the channel.

Likewise, the structure of AmtB without $(NH_4)_2SO_4$ shows no ordered water in between the hydrophobic-constricted regions of the channel. Thus the channel removes water of hydration and a proton from NH_4^+ as NH_3 passes through the channel. In contrast to the AQPs, the AmtB channel provides no hydrogen-bonding sites along its length except for two imidazole C–HS. Thus, the cost of dehydration of H_2O itself cannot be compensated by coordinating oxygen in the lumen of the channel, and neither line of water nor water transport can be supported through the channel.

4   The Mechanism of Conduction

The novel mechanism based on the AmtB structures, difference maps for Am/MA, and NH_3 conductivity in proteoliposomes involves the following: a vestibular attraction of total Am (but predominantly NH_4^+), a site that can recruit and possibly assist in deprotonation of NH_4^+, and a hydrophobic channel for NH_3 that lowers the pK_a to <6[30] (Figure 6). This mechanism of conductance can reconcile many otherwise seemingly inconsistent data that led to different proposals of how members of the AMT/MEP family might function. Thus, it had previously been proposed that the AMT/MEP family members variously conduct NH_3 bidirectionally,[11,55] transport NH_4^+,[56–58] cotransport NH_3/H^+,[59] and exchange NH_4^+/H^+ in Rh proteins.[60,61] However, some of this apparently inconsistent data is interestingly consistent with the proton-stripping mechanism as discussed below.

4.1   pH-Dependent Effects

There are two common metrics of conductance by AMTs. First, the rate of transport can be measured using radioactive $^{13}CH_3NH_3^+$, and second, the so-called K_m (the concentration of Am that supports half maximal transport) can be determined. In general, there is no observed dependence of these properties on pH that corresponds to change in the expected ratios of NH_3 and NH_4^+ in bulk solution. NH_4^+ (or $CH_3NH_3^+$) becomes less basic in the channel as it becomes progressively desolvated until it is deprotonated and becomes uncharged ($pK_a < 6$). Entry of the uncharged NH_3 or CH_3NH_2 species into the hydrophobic channel at pH values above the new pK_a could eliminate all dependence of the transport rate on pH, and hence, any dependence on the concentration of the uncharged species in the bulk solution, even though it conducts NH_3 whose concentration in solution would change dramatically

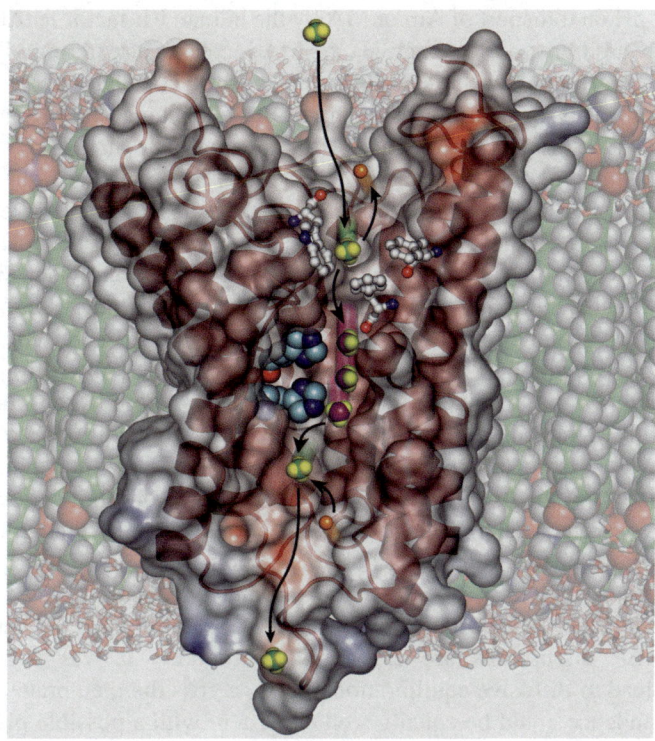

Figure 6 *The deduced mechanism of conductance is summarized. The green and purple spheres are nitrogen atoms of NH_4^+ and NH_3, respectively. The mechanism involves an extracellular vestibule attracting Am, a recruitment site for NH_4^+, and a hydrophobic channel that strips the proton (orange spheres) off the NH_4^+ and conducts NH_3. NH_3 is reprotonated at the cytoplasmic site at physiological pH and forms NH_4^+*

as the pH is changed. Correspondingly, the solution equilibrium between NH_3 and NH_4^+ has little to do with the mechanism because both species can enter the vestibule to become NH_3. If at all, lower pH might contribute to reduce the rate of Am conductance, not by reducing the level of NH_3 in bulk solution but by opposing proton release from NH_4^+ in the vestibule. The vestibule may also have some preference for NH_4^+ or NH_3 that could be reflected as an indirect effect of pH in either direction; a tendency to recruit NH_4^+ would favor a small increase in rate at lower pH.

In *C. glutamicum*, two ammonia channels are present, Amt and AmtB. The K_m for MA is 53 ± 11 μM at pH 6.0, and remains unchanged at pH 8.5, leading to the conclusion that NH_4^+ is the transported species because the NH_3 concentration in bulk solution would change 300-fold over this pH range, while the NH_4^+ concentration is little changed.[7,58] Likewise, the K_m for transport by *Arabidopsis* AMT2 is not pH dependent between pH 5.0 and 7.5.[62]

Effects of pH on transport of Am or MA by the human Rh-factor RhAG show that the rate of inward transport of (external) ^{14}C MA increases as external pH rises from 5.5 to 8.5,[60] suggesting that the neutral species may be conducted. However, a lack of effect of changes in membrane potential on conductance of MA and a small decrease in conductance as the internal pH rises led to the conclusion that transport was electrically net neutral. Arguing that an increase in the outward-directed proton gradient might drive transport of the suggested Am ion, presumed to be of the predominant species in solution, it was suggested that RhAG might be an $H^+/CH_3NH_3^+$ antiporter. However, the increase in conductance was very strongly dependent on the external pH than on internal pH, and can be reconciled if the deprotonation of $CH_3NH_3^+$ in the vestibule might well become easier as the pH is raised.

The rate of MA uptake in RhAG was saturable and well fitted by a Michaelis Menten equation with the Hill coefficient of $n = 1$, $K_m = 1.6$ mM for total Am. NH_4^+ competed with MA,[60] suggesting a competition for a single site. We presume that this site is Am1. However, Rh factors may have other roles, their K_m is 300-fold higher for NH_4^+ and the Am1 site differs in lacking some of the π-cation stabilizing rings. Trp148 is Leu or Val in Rh factors; Phe/Tyr103 is Ile in Rh; and Phe107 is conserved. This is consistent with low affinity for cations.

Based on the growth of yeast or *E. coli* in minimal nitrogen-limiting conditions of Am (≤ 1 mM), which is especially deleterious at pH values below 7 where the effect of AmtB/MEP disruptions are especially profound,[11] it was suggested that Amt/MEPs lead to diffusive equilibration of NH_3 across the membrane.[55] However, any pH dependence could be equally well consistent with a possible pH-dependent effect on H^+ release from NH_4^+ in the vestibule. Consistent with this conclusion, the increase in rates of transport of MA *versus* pH does not increase by a factor of 10 per unit increase in pH, as would be expected for a bulk pH effect on the concentration of NH_3, but increases by a factor of ~2 per pH unit.[60]

4.2   Competitive Inhibition

Ammonia channels conduct Am and MA though no larger secondary or tertiary amines. However, dimethylamine and ethylamine inhibit the uptake of CH_3NH_2 by Amt or by AmtB from *C. glutamicum*, while trimethyl and tetramethyl amine do not.[58] Although the channel is too small to accommodate these molecules, a competition for the smaller cationic forms can take place at the Am1 site in the vestibule. This inhibition of MA conductance by Am in RhAG is pH independent , yet the values of pK_a of MA and Am are different.[60] Moreover, this pH independence is expected because the competition is between total MA and Am, and so would again be independent of pH of bulk solution.

4.3   Transmembrane Potential

Transmembrane potential accentuated conductance by the channel has been noted because the K_m for MA decreases as transmembrane voltage (negative in the direction of conductance) increases, interpreted as indicating that NH_4^+ is conducted.[59] However, the electric field could concentrate NH_4^+ on the NH_4^+ rich side of the membrane and increase its local concentration in the vestibule. Thus, the K_m (app) might appear to be lower because the local concentration of $NH_4^+/CH_3NH_3^+$ is higher in the vestibule than in the bulk solution.

By expression of human RhAG in oocytes, Westhoff *et al.*[60] show that the rate of conductance of MA remains constant even when the transmembrane potential is modulated from –35 to –9 mV (inside the oocyte), concluding that transport is not electrogenic. The NH_4^+/H^+ antiport mechanism suggested by these authors is also consistent with the net transport of uncharged CH_3NH_2 as shown by our results. In an interesting way, our proton-stripping mechanism makes it *seem* as if a proton is transferred from inside to outside since NH_4^+ leaves a proton initially on the outside, to pick one up on the other side. Unlike an antiporter, however, there would be no net transfer of a proton; it could also be viewed rather as the cotransfer of a "proton hole."

All the observations noted so far are reconciled by the recruitment of NH_4^+, reduction of its pK_a, and conductance of NH_3 as the primary mechanism. However, observations of Am-dependent currents through ammonia channels cannot be explained, and demand further examination. A two-electrode voltage clamp experiment was used to vary the transmembrane potential in oocytes, transiently expressing the tomato paralog LAMT1. Inward currents increased with voltage, and with external Am ion from ~3 μM Am upward. The currents show a Hill coefficient of 1, implying that there is no cooperative effect of one conducted molecule of Am on another.[59] This electrogenic behavior suggests that the channel transports ammonium ions (NH_4^+) (or NH_3 plus a H^+) rather than uncharged NH_3. The measured currents are the same from pH 5.5 to 8.5[59] consistent with the conductance of $NH4^+$ ions, rather than NH_3 and H^+ which might have showed some pH dependence. Although we have shown that AmtB conducts predominantly NH_3 (assayed in proteoliposomes), we have not shown that it does not conduct any occasional non-stoichiometric NH_4^+/H^+ ions that could give rise to these currents. A stoichiometric measure of the conductance of each species NH_3/NH_4^+ is required to establish whether this takes place at some low level. We do not see any hydrogen-bonded pathway that could act as a conductor for H^+, nor any change in conformation of AmtB determined in the presence of Am/MA. Faced with these observations one possibility is therefore that an occasional NH_4^+ ion can reach Am2, stabilized by ring currents of the rich aromatic environment at the constriction, and pass through the two-histidine region, possibly using the acid/base properties of the imidazole nitrogen to assist in proton transfer. This would require a transient conformational change that would need to be induced. The structure can clearly guide experimental measures of the ratio of conductances. Alternatively, the currents could be carried by another ion or another Am-dependent pathway in the oocytes.

5   The Rh Proteins

Identification of the first Am transporters from yeast and plants prompted investigations of whether mammals have any Am transporters. Sequence analysis revealed that mammalian Rh proteins are the closest gene family to AMT/MEP. Subsequently, Marini *et al.*[20] demonstrated that the expression of human RhAG can specifically mediate Am transport in Am-uptake deficient yeast by testing the growth of triple-*mepΔ* strains, expressing the RhAG protein, on media containing less than 5 mM ammonium. Furthermore, the expression of RhAG enhances resistance to a toxic concentration of MA (250 mM) in triple-*mepΔ* yeast, consistent with RhAG promoting export of the Am analogs. Expression of RhAG in the cells growing on arginine – a nitrogen source, whose catabolism leads to Am production – increases Am excretion by the yeast cell, and this indicates Am export by RhAG. While bacteria, plants, and fungi can acquire nitrogen from their environment in the form of Am,[63] animals and humans prefer assimilated forms of nitrogen, amino acids, for nutrition, and they use Am for excretion of nitrogen through Rh proteins.

Since the first time Rh glycoproteins were shown to be Am transporters, multiple Rh proteins were identified in various organs.[60,64] RhAG is expressed in erythrocytes and erythroid precursors[65,66] where it may function to extrude ammonia.[67] Nonerythroid RhBG and RhCG are expressed in important sites for ammonia metabolism, including kidney, liver, skin, testis, the central nervous system, and the intestinal tract.[18,64,68–72]

Rh proteins most probably have a similar membrane topology to AmtB. Current models for Rh proteins are oligomers of Rh subunits (D, CCEe) associated noncovalently with RhAG subunits.[73–75] Functional studies in *Xenopus oocytes* injected with cRNA encoding RhAG, RhBG, and RhCG indicated that the uptake was independent of the membrane potential and the Na^+ gradient. The RBC can be used to investigate the function of RhAG in its natural context. Ripoche *et al.*[76] conducted a stopped-flow analysis of intracellular pH changes of resealed RBC ghosts, from human and mouse genetic variants with defects of proteins that comprise the Rh complex, submitted to inward-directed MA or Am gradients (in iso-osmotic conditions). The permeability of a single Rh complex unit has been estimated to be about $1–2 \times 10^6$ molecules of NH_3/sec, suggesting that RhAG behaves as a channel rather than as a transporter.[75,76]

The majority of data on Rh proteins is consistent with their action as an NH_3 channel as in AmtB. However, it has been speculated that Rh factors – prominent in mammals and found alongside Amt proteins in, for example, green algae, which contain two AMTs and two Rh proteins – may function physiologically as channels for CO_2.[77,78]

Although AmtB and Rh proteins have only marginal sequence conservation or similarity, the AmtB structure makes it possible to map the sequence of Rh factors onto the 3D structure of AmtB in attempt to further address questions of Rh function. Most particularly the two [aspartic acid–(seven residues)–histidine] motifs (residues D160, H168 and D310, H318), which are conserved in AMT/MEP proteins and play key roles with the histidines inside the hydrophobic channel of AmtB_Ecoli, are conserved in Rh proteins RhAG, RhBG, RhCG, but not in RhD and RhCCEe proteins.[75] The structure allows docking of models of the various Rh factors against each other to help design experiments to define the stoichiometry of this assembly.

6   Comparison with Aquaporins

H_2O and NH_3 have similar sizes and molecular dipole moments (1.85 and 1.47 D, respectively), and they adopt the same molecular orbital hybridization, although

H_2O has one more lone pair and one less bonding pair of electrons. In spite of the high molecular similarity between H_2O and NH_3, AmtB did not show a significant permeability for water. This can be explained by the hydrophobic nature of ammonia channel.

Although the existence of the membrane water channel was predicted in the 1950s,[79] before the family of water channels was named and defined by Agre and co-workers in 1992,[80] some believed that water moved relatively freely by diffusion through the lipid bilayer of a cell membrane and that cells would not need water-specific transmembrane channels. Water channels have been classified into orthodox AQPs that transport H_2O with high specificity and aquaglyceroporins that in addition to H_2O also transport glycerol.[40,81]

As membranes in several tissues and cell types have permeabilities for CO_2[82] and NH_3,[83] diffusion through the lipid bilayer does not seem to be the only pathway for gaseous compounds. Therefore, in addition to glycerol, AQPs are suggested to conduct other small solutes such as CO_2[84–87] (even though disputed[88]) and NH_3.[89,90] Since AMT/MEP family and some members of the AQP family conduct NH_3, we compare the structure and function of these two families.

Tonoplast intrinsic proteins (TIPs) are a group of AQP homologs in plants, which are expressed in tonoplast vacuolar membranes.[91] TaTIP2;1 expression in *Arabidopsis* roots is upregulated in response to nitrogen starvation,[92] suggesting that TIP2 functions in remobilization of Am from vacuoles under starvation conditions by transporting Am across the vacuole membrane.[89] Using the functional complementation of a yeast mutant deficient in all three Am transporters (Δ*mep1–3*) by wheat (*Triticum aestivum*) TIP2 aquaporin homologs and also using the expression in *Xenopus oocytes*, TaTIP2;1 and mammalian AQP3, AQP8, and AQP9 were shown to conduct NH_3,[89,90,93] and this group of AQPs are called aquaammoniaporins. These three mammalian AQPs are found primarily in nitrogen-handling organs such as the liver and the kidney.

Aquaporins are a family of membrane channels with a common primary structure characterized by a tandem repeat of two similar halves. High resolution crystal structures of several AQP homologs are available, including the glycerol facilitator GlpF,[38] AQP1,[39] AQPZ,[40] AQP0,[41] and AQPM. The channel has the following three regions: an extracellular vestibule, an extended narrow pore of ~28 Å, and a cyclo-plasmic vestibule.[40] The amphipathic nature of the channel is important for rapid transport of water and glycerol. The most constricted region of the water or glycerol pathway, the so-called selectivity filter, is located between the extracellular vestibule and the central plane of the bilayer. The channel in GlpF is slightly wider and key hydrophobic residues allow the passage of the carbon backbone of glycerol. Homology modeling was used to examine the differences in the selectivity filter of aquaammoniaporins and AQP1 (an orthodox water channel). The selectivity filter in AQP1 is surrounded by R197/F58/H182/C191. TIPs incorporate isoleucine or valine in the positions of H182 and glycine or alanine at C191 in AQP1, respectively. The F58 of bovine AQP1 is substituted by histidine, tryptophan, leucine, and alanine in aquaammoniaporins. These substitutions make the selectivity filter in aquaammoniaporins wider and more hydrophobic. Both the dipole moment (1.49 D for NH_3 *vs.* 1.85 D for H_2O) and the dielectric constant of the liquid (22 for NH_3 *vs.* 80 for H_2O)

of NH_3 are considerably lower than that of H_2O. A more hydrophobic selectivity filter in the AQP family will be in favor of NH_3 transport, and this is consistent with a hydrophobic channel in the AMT/MEP family. The site-directed mutation of the selectivity filter residues in TIP2;1 to the corresponding residues from AQP1 resulted in the loss of a NH_3 conduction in TIP2;1.[89] In general, a hydrophilic selectivity filter makes AQPs good water channels and poor NH_3 channels. A hydrophobic channel in AMT/MEP, or a hydrophobic selectivity filter in aquaammoniaporins, facilitates NH_3 conduction.

7   Comparison with K^+ Channel

Both NH_4^+ and K^+ have similar ionic radii (0.148 nm *vs.* 0.149 nm) and their hydration shells also have a similar size of 0.331 nm.[14] This explains NH_4^+ permeability through most K^+ channels.[94–97] In contrast, members of the high-affinity Am transporters superfamily AMT/MEP/Rh do not transport K^+.[6,59] The crystal structure of AmtB reveals the molecular basis for the discrimination in AMT/MEP/Rh.[30] The hydrophobic channel of AmtB conducts only NH_3, not NH_4^+ nor any other ion that would require replacement of its hydration shell while in the narrow portion of the channel. The molecular basis of K^+ channel conduction and selectivity are elucidated in the structure of KcsA K^+ channel from *Streptomyces lividans*.[98] In K^+ channels, 16 carbonyl oxygens (charged δ $-0.4e$ per oxygen), disposed in four layers of four backbone carbonyl groups, mimic the hydration water and stabilize the passage of K^+ or NH_4^+.[99,100] The KcsA channel also provides a water-filled cavity in the most energetically costly position, at the center of the bilayer, to stabilize ions during conduction. The energy penalty for passing charged NH_4^+ or K^+ through the hydrophobic channel of AMT/MEP is prohibitive, and thereby AMT/MEP/Rh selectively conducts neutral CH_3NH_2/NH_3 and blocks ions such as NH_4^+/K^+.

Acknowledgment

The research on AmtB in the author's laboratory was supported by the National Institutes of Health (NIH), Grant number GM 24485.

References

1. J. Broach, C. Neumann and S. Kustu, *J. Bacteriol.*, 1976, **128**, 86.
2. M.A. Knepper, R. Packer and D.W. Good, *Physiol. Rev.*, 1989, **69**, 179.
3. M.H. Saier Jr., B.H. Eng, S. Fard, J. Garg, D.A. Haggerty, W.J. Hutchinson, D.L. Jack, E.C. Lai, H.J. Liu, D.P. Nusinew, A.M. Omar, S.S. Pao, I.T. Paulsen, J.A. Quan, M. Sliwinski, T.-T. Tseng, S. Wachi and G.B. Young, *Biochim. Biophys. Acta*, 1999, **1422**, 1.
4. G.H. Thomas, J.G. Mullins and M. Merrick, *Mol. Microbiol.*, Field Publication Date: Jul 2000, 2000, **37**, 331.
5. A.M. Marini, S. Vissers, A. Urrestarazu and B. Andre, *EMBO J.*, 1994, **13**, 3456.

6. O. Ninnemann, J.C. Jauniaux and W.B. Frommer, *EMBO J.*, 1994, **13**, 3464.
7. R.M. Siewe, B. Weil, A. Burkovski, B.J. Eikmanns, M. Eikmanns and R. Kramer, *J. Biol. Chem.*, 1996, **271**, 5398.
8. A. Van Dommelen, V. Keijers, J. Vanderleyden and M. de Zamaroczy, *J. Bacteriol.*, 1998, **180**, 2652.
9. N. Michel-Reydellet, N. Desnoues, M. de Zamaroczy, C. Elmerich and P.A. Kaminski, *Mol. Gen. Genet.*, 1998, **258**, 671.
10. M.L. Montesinos, A.M. Muro-Pastor, A. Herrero and E. Flores, *J. Biol. Chem.*, 1998, **273**, 31463.
11. E. Soupene, L. He, D. Yan and S. Kustu, *Proc. Natl. Acad. Sci. USA*, 1998, **95**, 7030.
12. A.M. Marini, S. Soussi-Boudekou, S. Vissers and B. Andre, *Mol. Cell. Biol.*, 1997, **17**, 4282.
13. M.C. Lorenz and J. Heitman, *EMBO J.*, 1998, **17**, 1236.
14. S.M. Howitt and M.K. Udvardi, *Biochim. Biophys. Acta*, 2000, **1465**, 152.
15. S. Gazzarrini, L. Lejay, A. Gojon, O. Ninnemann, W.B. Frommer and N. von Wiren, *Plant Cell*, 1999, **11**, 937.
16. N. von Wiren, A. Bergfeld, O. Ninneman and W.B. Frommer, *Plant Mol. Biol.*, 1997, **35**, 681.
17. F.R. Lauter, O. Ninnemann, M. Bucher, J.W. Riesmeier and W.B. Frommer, *Proc. Natl. Acad. Sci. USA*, 1996, **93**, 8139.
18. M.A. Knepper, *Kidney Int. Suppl.*, Field Publication Date: Jul 1991, **33**, S95.
19. A.M. Marini, A. Urrestarazu, R. Beauwens and B. Andre, *Trends Biochem. Sci.*, 1997, **22**, 460.
20. A.M. Marini, G. Matassi, V. Raynal, B. Andre, J.P. Cartron and B. Cherif-Zahar, *Nat. Genet.*, 2000, **26**, 341.
21. G.H. Thomas, J.G. Mullins and M. Merrick, *Mol. Microbiol.*, 2000, **37**, 331.
22. S.L. Hackette, G.E. Skye, C. Burton and I.H. Segel, *J. Biol. Chem.*, 1970, **245**, 4241.
23. A.J. Ninfa and P. Jiang, *Curr. Opin. Microbiol.*, 2005, **8**, 168.
24. G. Thomas, G. Coutts and M. Merrick, *Trends Genet.*, Field Publication Date: Jan 2000, **16**, 11.
25. G. Coutts, G. Thomas, D. Blakey and M. Merrick, *EMBO J.*, 2002, **21**, 536.
26. G. Thomas, G. Coutts and M. Merrick, *Trends Genet.*, 2000, **16**, 11.
27. T. Dandekar, B. Snel, M. Huynen and P. Bork, *Trends Biochem. Sci.*, 1998, **23**, 324.
28. D. Blakey, A. Leech, G.H. Thomas, G. Coutts, K. Findlay and M. Merrick, *Biochem. J.*, 2002, **364**, 527.
29. N. von Wiren and M. Merrick, *Regulation and Function of Ammonium Carriers in Bacteria, Fungi, and Plants*, R.K. Eckhard Boles (ed), Springer GmbH, 2004.
30. S. Khademi, J. O'Connell III, J. Remis, Y. Robles-Colmenares, L.J. Miercke and R.M. Stroud, *Science*, 2004, **305**, 1587.
31. A. Javelle, B. Andre, A.M. Marini and M. Chalot, *Trends Microbiol.*, 2003, **11**, 53.
32. H. Vermeiren, V. Keijers and J. Vanderleyden, *DNA Seq.*, 2002, **13**, 67.

33. H. Nielsen, J. Engelbrecht, S. Brunak and G. von Heijne, *Protein Eng.*, 1997, **10**, 1.

34. M.J. Conroy, S.J. Jamieson, D. Blakey, T. Kaufmann, A. Engel, D. Fotiadis, M. Merrick and P.A. Bullough, *EMBO Rep.*, 2004, **5**, 1153.

35. M.C. Wahl and M. Sundaralingam, *Trends Biochem. Sci.*, 1997, **22**, 97.

36. A. Senes, I. Ubarretxena-Belandia and D.M. Engelman, *Proc. Natl. Acad. Sci. USA*, 2001, **98**, 9056.

37. A. Senes, M. Gerstein and D.M. Engelman, *J. Mol. Biol.*, 2000, **296**, 921.

38. D. Fu, A. Libson, L.J. Miercke, C. Weitzman, P. Nollert, J. Krucinski and R.M. Stroud, *Science*, 2000, **290**, 481.

39. H. Sui, B.G. Han, J.K. Lee, P. Walian and B.K. Jap, *Nature*, 2001, **414**, 872.

40. D.F. Savage, P.F. Egea, Y. Robles-Colmenares, J.D. O'Connell III and R.M. Stroud, *PLoS Biol.*, 2003, **1**, E72.

41. W.E. Harries, D. Akhavan, L.J. Miercke, S. Khademi and R.M. Stroud, *Proc. Natl. Acad. Sci. USA*, 2004, **101**, 14045.

42. B. Van den Berg, W.M. Clemons Jr., I. Collinson, Y. Modis, E. Hartmann, S.C. Harrison and T.A. Rapoport, *Nature*, 2004, **427**, 36.

43. R. Dutzler, E.B. Campbell, M. Cadene, B.T. Chait and R. MacKinnon, *Nature*, 2002, **415**, 287.

44. R. Dutzler, E.B. Campbell and R. MacKinnon, *Science*, 2003, **300**, 108.

45. A. Accardi and C. Miller, *Nature*, 2004, **427**, 803.

46. A. Yamashita, S.K. Singh, T. Kawate, Y. Jin and E. Gouaux, *Nature*, 2005, **437**, 215.

47. J. Abramson, I. Smirnova, V. Kasho, G. Verner, H.R. Kaback and S. Iwata, *Science*, 2003, **301**, 610.

48. J.H. Park and M.H. Saier Jr., *J. Membr. Biol.*, 1996, **153**, 171.

49. Z.S. Derewenda, U. Derewenda and P.M. Kobos, *J. Mol. Biol.*, 1994, **241**, 83.

50. Z.S. Derewenda, L. Lee and U. Derewenda, *J. Mol. Biol.*, 1995, **252**, 248.

51. E.L. Ash, J.L. Sudmeier, R.M. Day, M. Vincent, E.V. Torchilin, K.C. Haddad, E.M. Bradshaw, D.G. Sanford and W.W. Bachovchin, *Proc. Natl. Acad. Sci. USA*, 2000, **97**, 10371.

52. A. Javelle, E. Severi, J. Thornton and M. Merrick, *J. Biol. Chem.*, 2004, **279**, 8530.

53. A. Roos and W.F. Boron, *Physiol. Rev.*, 1981, **61**, 296.

54. N.A. Priver, E.C. Rabon and M.L. Zeidel, *Biochemistry*, 1993, **32**, 2459.

55. E. Soupene, H. Lee and S. Kustu, *Proc. Natl. Acad. Sci. USA*, 2002, **99**, 3926.

56. M.Y. Wang, M.Y. Siddiqi, T.J. Ruth and A. Glass, *Plant Physiol.*, 1993, **103**, 1259.

57. R. Tate, A. Riccio, M. Merrick and E.J. Patriarca, *Mol. Plant Microb. Interact.*, 1998, **11**, 188.

58. J. Meier-Wagner, L. Nolden, M. Jakoby, R. Siewe, R. Kramer and A. Burkovski, *Microbiology*, 2001, **147**, 135.

59. U. Ludewig, N. von Wiren and W.B. Frommer, *J. Biol. Chem.*, 2002, **277**, 13548.

60. C.M. Westhoff, M. Ferreri-Jacobia, D.O. Mak and J.K. Foskett, *J. Biol. Chem.*, 2002, **277**, 12499.

61. C.M. Westhoff, D.L. Siegel, C.G. Burd and J.K. Foskett, *J. Biol. Chem.*, 2004, **279**, 17443.
62. C. Sohlenkamp, C.C. Wood, G.W. Roeb and M.K. Udvardi, *Plant Physiol.*, 2002, **130**, 1788.
63. N. von Wiren, S. Gazzarrini, A. Gojon and W.B. Frommer, *Curr. Opin. Plant Biol.*, 2000, **3**, 254.
64. I.D. Weiner and J.W. Verlander, *Acta Physiol. Scand.*, 2003, **179**, 331.
65. Z. Liu and C.H. Huang, *Biochem. Genet.*, 1999, **37**, 119.
66. N.D. Avent, *J. Pediatr. Hematol. Oncol.*, 2001, **23**, 394.
67. M.B. Hemker, G. Cheroutre, R. van Zwieten, P.A. Maaskant-van Wijk, D. Roos, J.A. Loos, C.E. van der Schoot and A.E. von dem Borne, *Br. J. Haematol.*, 2003, **122**, 333.
68. Z. Liu, Y. Chen, R. Mo, C. Hui, J.F. Cheng, N. Mohandas and C.H. Huang, *J. Biol. Chem.*, 2000, **275**, 25641.
69. Z. Liu, J. Peng, R. Mo, C. Hui and C.H. Huang, *J. Biol. Chem.*, 2001, **276**, 1424.
70. N.D. Avent, *Trends Mol. Med.*, 2001, **7**, 94.
71. F. Quentin, D. Eladari, L. Cheval, C. Lopez, D. Goossens, Y. Colin, J.P. Cartron, M. Paillard and R. Chambrey, *J. Am. Soc. Nephrol.*, 2003, **14**, 545.
72. N.L. Nakhoul and L.L. Hamm, *Pflugers Arch.*, 2004, **447**, 807.
73. S.A. Eyers, K. Ridgwell, W.J. Mawby and M.J. Tanner, *J. Biol. Chem.*, 1994, **269**, 6417.
74. K. Ridgwell, S.A. Eyers, W.J. Mawby, D.J. Anstee and M.J. Tanner, *J. Biol. Chem.*, 1994, **269**, 6410.
75. C.L. Van Kim, Y. Colin and J.P. Cartron, *Blood Rev.*, 2005, June 13 (Epub ahead of print).
76. P. Ripoche, O. Bertrand, P. Gane, C. Birkenmeier, Y. Colin and J.P. Cartron, *Proc. Natl. Acad. Sci. USA*, 2004, **101**, 17222.
77. E. Soupene, N. King, E. Feild, P. Liu, K.K. Niyogi, C.H. Huang and S. Kustu, *Proc. Natl. Acad. Sci. USA*, 2002, **99**, 7769.
78. E. Soupene, W. Inwood and S. Kustu, *Proc. Natl. Acad. Sci. USA*, 2004.
79. V.W. Sidel and A.K. Solomon, *J. Gen. Physiol.*, 1957, **41**, 243.
80. G.M. Preston, T.P. Carroll, W.B. Guggino and P. Agre, *Science*, 1992, **256**, 385.
81. M.S. Jang, Y.M. Lee, Y.L. Choi, Y.S. Cho and Y.C. Lee, *J. Microbiol.*, 2004, **42**, 139.
82. G.J. Cooper, Y. Zhou, P. Bouyer, Grichtchenko, II and W.F. Boron, *J. Physiol.*, 2002, **542**, 17.
83. D. Kikeri, A. Sun, M.L. Zeidel and S.C. Hebert, *Nature*, 1989, **339**, 478.
84. N.L. Nakhoul, B.A. Davis, M.F. Romero and W.F. Boron, *Am. J. Physiol.*, 1998, **274**, C543.
85. G.V. Prasad, L.A. Coury, F. Finn and M.L. Zeidel, *J. Biol. Chem.*, 1998, **273**, 33123.
86. X.C. Sun, K.T. Allen, Q. Xie, W.D. Stamer and J.A. Bonanno, *Invest. Ophthalmol. Vis. Sci.*, 2001, **42**, 417.
87. N. Uehlein, C. Lovisolo, F. Siefritz and R. Kaldenhoff, *Nature*, 2003, **425**, 734.
88. B. Yang, N. Fukuda, A. van Hoek, M.A. Matthay, T. Ma and A.S. Verkman, *J. Biol. Chem.*, 2000, **275**, 2686.

89. T.P. Jahn, A.L. Moller, T. Zeuthen, L.M. Holm, D.A. Klaerke, B. Mohsin, W. Kuhlbrandt and J.K. Schjoerring, *FEBS Lett.*, 2004, **574**, 31.

90. L.M. Holm, T.P. Jahn, A.L. Moller, J.K. Schjoerring, D. Ferri, D.A. Klaerke and T. Zeuthen, *Pflugers Arch.*, 2005, **450**, 415.

91. N. Frangne, M. Maeshima, A.R. Schäffner, T. Mandel, E. Martinoia and J.-L. Bonnemain, *Planta*, 2001, **212**, 270.

92. L.H. Liu, U. Ludewig, B. Gassert, W.B. Frommer and N. von Wiren, *Plant Physiol.*, 2003, **133**, 1220.

93. K. Liu, H. Nagase, C.G. Huang, G. Calamita and P. Agre, *Biol. Cell*, 2005.

94. M.H. Jacobs, *Cold Spring Harbor Symp. Quant. Biol.*, 1940, **8**, 30.

95. M. Haas and B. ForbushIII, *J. Bioenerg. Biomembr.*, 1998, **30**, 161.

96. T.N. Nagaraja and N. Brookes, *Am. J. Physiol.*, 1998, **274**, C883.

97. D.P. Schachtman, J.I. Schroeder, W.J. Lucas, J.A. Anderson and R.F. Gaber, *Science*, 1992, **258**, 1654.

98. D.A. Doyle, J. Morais Cabral, R.A. Pfuetzner, A. Kuo, J.M. Gulbis, S.L. Cohen, B.T. Chait and R. MacKinnon, *Science*, 1998, **280**, 69.

99. J.H. Morais-Cabral, Y. Zhou and R. MacKinnon, *Nature*, 2001, **414**, 37.

100. Y. Jiang, A. Lee, J. Chen, M. Cadene, B.T. Chait and R. MacKinnon, *Nature*, 2002, **417**, 523.

CHAPTER 13

Channels in the Outer Membrane of Mycobacter

GEORG E. SCHULZ

Institut für Organische Chemie und Biochemie, Albert-Ludwigs-Universität, Albertstrasse 21, D-79104 Freiburg im Breisgau, Germany

1   Introduction

Bacteria need external walls as a shield against all kinds of stress from their environment. For instance, Gram-positive bacteria such as *Staphylococcus aureus* possess an exceptionally thick peptidoglycan cell wall. In contrast, Gram-negative bacteria such as *Escherichia coli* have a comparatively thin peptidoglycan layer that is covered by a second outer membrane. In this Gram scheme, the mycobacteria were originally assigned to the group of Gram-positive bacteria. Also, their ribosomal RNA sequences were most closely related to this group.[1] Recently, however, comparisons based on many more genes indicate that the mycobacteria should be considered a third group equidistant from the Gram-positive and Gram-negative bacteria.[2]

The outer membranes of the Gram-negative bacteria and the mycobacteria are essentially impermeable for most nutrients; therefore, channels are needed to support cell metabolism. For the Gram-negative bacteria these channels were named porins. They are passive, in general not very selective channels, discriminating against nonpolar solutes, but allowing the unhindered passage of all kinds of polar molecules with molecular masses below about 500 Da.[3] These porins were detected a couple of decades ago and characterized by their electrical conductance. For this purpose they were incorporated into artificial lipid bilayers separating two chambers filled with a salt solution. Their conductance was then determined by applying a voltage between the two chambers and measuring the resulting electric current.[4] Later on, these porins were crystallized and structurally elucidated.[5,6]

Until a decade ago, mycobacteria were not expected to possess porins because they were considered similar to Gram-positive bacteria. A report on a mycobacterial porin detected in electrical conductance experiments came therefore as a surprise.[7] Porin

```
MspA:    mkaisrvliamv----aaiaalftstgtsha
MspB:    mtafkrvliamisallagttgmfvsagaaha
MspC:    mkaisrvliamisalaaavaglfvsagtsha
MspD:    ----mrylvmmf---allvsvtlvsprpana
```

```
          .            .               .                  .
A:   GLDNELSLVDGQDRTLTVQQWDTFLNGVFPLDRNRLTREWFHSGRAKYIVAGPGADEFEGTL  62
B:   ............................................................
C:   ............................................................
D:   -V..Q..V....G.......AE.........................T.H...........
```

```
          .              .                  .               .
A:   ELGYQIGFPWSLGVGINFSYTTPNILIDDGDITAPPFGLNSVITPNLFPGVSISADLGNGP 123
B:   ............................................................
C:   .......................................G....E...............
D:   .....V.................G....Q....DTI.........................
```

```
       .              .               .                 .
A:   GIQEVATFSVDVSGAEGGVAVSNAHGTVTGAAGGVLLRPFARLIASTGDSVTTYGEPWNMN 184
B:   ...............PA...........................................
C:   ...............PA...........................................
D:   ............K..K.A..........................................
```

Figure 1 *Sequences of the porins MspA, MspB, MspC, and MspD from M. smegmatis.[13] The sequences of the signal peptides are given in lower case letters and those of the mature proteins in upper case letters. Dots stand for the MspA sequence. Strands β1 through β12 of MspA are underlined*

conductance data are rather conclusive but require only a small number of single molecules because the current through a single channel can be readily measured.[8] It took a further number of years until these porins were produced in amounts large enough for chemical characterization. This was first achieved for the soil bacterium *Mycobacter smegmatis*.[9–11] The porin MspA of this organism was characterized with respect to amino acid sequence and electrical conductance,[9] overall structure as derived from electron micrographs of porin aggregates,[10] as well as high-resolution structure.[12] Three more porins of *M. smegmatis* showing quite different signal peptides but very similar mature protein sequences are known (Figure 1).[13] Among the mycobacteria, *M. leprae* and *M. tuberculosis* are of high medical relevance. It is therefore hoped that the data on MspA are applicable for these organisms.

2 Structure Determination

2.1 Protein Production

When the sequence of porin MspA from *M. smegmatis* was known,[9] a crystal structure analysis became a reasonable goal. For this purpose the porin had to be produced in milligram amounts. *E. coli* was selected as the expression host because this system is very well established, allowing site-directed mutagenesis with commercial kits and protein labeling with selenium in order to solve the phase problem arising during an X-ray diffraction analysis.[14] Using a pET vector, MspA was expressed in sufficient amounts as a soluble protein, which was then purified by ion-exchange chromatography and ammonium sulfate fractionation.[12] In a standard sodium dodecylsulfate-polyacrylamide gel electrophoresis (SDS-PAGE) procedure (5 min boiling), the protein had an apparent mass of about 21 kDa which agreed well with the

19,404 Da calculated mass of the 184 amino acid residue polypeptide. In size-exclusion chromatography (Figure 2) the protein ran in a single peak at an apparent mass of 46 kDa, which is closer to a dimer than to a monomer. However, because of the inherent inaccuracy of the size-exclusion mass determination, it seemed appropriate to name this peak "monomer," distinguishing it from the channel-forming high-numbered oligomers.

Since MspA consists of merely 184 amino acid residues,[12,13] and since conductance measurements revealed that MspA forms a larger channel than Gram-negative bacterial porins,[9] which need 280 residues for a minimum-sized β-barrel channel,[6] it was expected that MspA forms oligomers grouped around one or more channels. This was confirmed by electron micrographs showing an MspA oligomer rather than a monomer or a dimer.[10] The oligomerization was accomplished by incubating soluble MspA overnight at the highest possible protein concentration available, which was in the centrifuged pellet of an ammonium sulfate precipitate.[12] The assembly worked only in the presence of a detergent, for example, the applied octyltetraoxyethylene. The detergent was used in all subsequent steps of the analysis.

The resulting material was analyzed by size-exclusion chromatography in conjunction with SDS-PAGE.[12] The size-exclusion run shown in Figure 2 revealed large aggregates at the exclusion volume, oligomers of at least six subunits in a double peak, and residual monomers (or dimers) in a single peak. In all the runs, the oligomer peak contained MspA exclusively. This was demonstrated by boiling the material in 80% dimethylsulfoxide which gave rise to a single band at the monomer

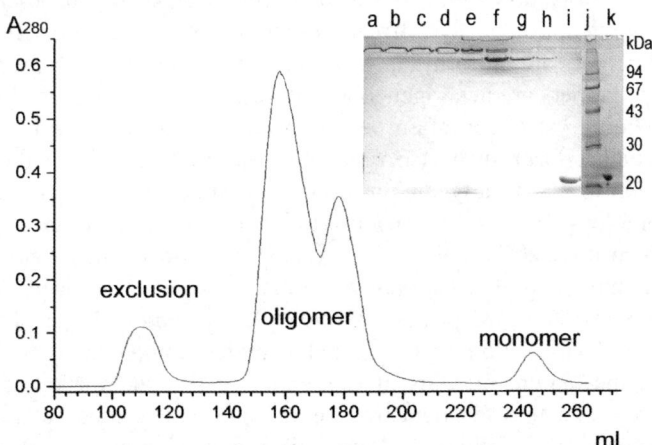

Figure 2 *Size-exclusion chromatography of the assembled recombinant MspA mutant A96R.[12] The inserted SDS-PAGE shows lanes a through j prepared under standard conditions (5 min boiling). The lanes correspond to fractions a (starting at 140 mL), b (144), c (148), d (152), e (156), f (160), g (168), h (180), and i (244). Lane j presents the molecular mass markers and lane k shows a pool of the whole double peak after 5 min boiling in 80% dimethylsulfoxide demonstrating that it contains only MspA. Only fractions a and b yielded ordered crystals suitable for an X-ray analysis. Corresponding runs with wildtype MspA and with all other mutants showed a single peak at about 178 mL instead of the depicted two peaks at 158 and 178 mL[16]*

mass in lane *k* of the SDS gel. In contrast, the standard SDS-PAGE analysis (5 min boiling in sample buffer) showed an apparent mass of more than 100 kDa for all oligomer peak fractions (Figure 2). This behavior agreed with that of native MspA oligomers, which survive on boiling in sample buffer but dissociate on boiling in 80% dimethylsulfoxide.[9]

2.2 Crystallization

The soluble protein was produced in relatively large amounts of about 5 mg L^{-1} culture medium.[12] In order to establish the monomer structure as well as the oligomer structure, the purified protein was subjected to extensive crystallization screenings. However, crystals could not be detected in any screen, indicating that soluble MspA is unlikely to assume a uniform and rigid conformation. In retrospect, the channel structure showed clearly that the oligomerization is crucial to the structure of segment 70–120, forming the composite β-barrels. In a soluble monomer (or dimer), this large 50-residue loop is most likely mobile thereby prohibiting crystallization. Crystallization would be equally unlikely if the soluble protein were a dimer associated via a nonpolar, not well-defined interface because such a dimer fails to assume the defined and uniform shape essential for crystallization.

In order to establish the channel structure, it was necessary to crystallize material from the oligomer peak of the size-exclusion chromatography (Figure 2). The oligomer peak fractions were therefore subjected to numerous crystallization screenings which eventually gave rise to crystals that appeared to be suitable for X-ray analysis.[12] In all tests, however, these crystals were intrinsically disordered yielding only very low-resolution X-ray diffraction patterns. Such a disorder indicates that either the oligomers are uniform but form undefined contacts with the other molecules or the oligomers are heterogeneous with respect to the number of subunits.

As disordered crystals can often be improved by mutations on the protein surface,[15] more than a dozen such MspA mutants were produced, first at the DNA level and then as proteins.[16] Usually, the positions to be mutated are determined by structures of homologs from other species or sequence-based surface predictions. However, since no such homologs were known and since surface predictions for oligomeric membrane proteins are not very reliable, the positions were derived from the MspA variants MspB, MspC, and MspD of *M. smegmatis* (Figure 1). In view of poor sequence conservation in the signal peptides compared to high sequence homology in the mature proteins, it was expected that the variants assumed the MspA structure and that the nonconserved amino acid residues marked solvent-accessible residues of the oligomer. Among the produced mutants, only Ala96→Arg (A96R) showed some improvement in the crystalline order.[12] An arginine was introduced because of its intermediate sidechain mobility and its multiple hydrogen bond donor capability, which is desirable for a defined contact. Moreover, the arginine lifted the isoelectric point of MspA somewhat above the low 4.2 value of the wildtype.

The size-exclusion chromatogram of mutant A96R depicted in Figure 2 turned out to be the only one with two oligomer peaks at 158 and 178 mL. Wildtype MspA and other mutants showed merely a single peak at 178 mL.[16] The column calibration assigned a mass of 170 kDa to the additional peak at 158 mL, which is in the range of an octamer. The SDS-PAGE analysis along the two oligomer peaks revealed a number of bands above 100 kDa in lanes *a* to *h* of Figure 2, which changed with the analyzed fraction but exclusively contained MspA as derived from the completely dissociated material in lane *k*. Multiple bands of membrane proteins are rather common and likely to report conformational heterogeneities introduced by boiling in an SDS solution. Such heterogeneity cannot be explained in detail but indicates that the size-exclusion oligomer peak fractions should not be pooled. Rather, each fraction should be subjected to separate crystallization trials.

Since mutant A96R improved the crystalline order somewhat, crystallization attempts with this mutant were largely extended and refined.[12] Finally, it turned out that the very first two fractions *a* and *b* of the 158 mL peak (Figure 2) yielded crystals which diffracted X-rays somewhat beyond 2.5 Å resolution. However, these crystals still displayed a relatively large mosaic spread indicating that the crystal was composed of micrometer-sized crystallites with varying orientations. It should be noted that the selected fractions comprised merely 0.5% of the total protein. Presumably, the protein formed hetero-numbered oligomers and the purity required for crystallization was only available at the very rise of the additional size-exclusion peak at 158 mL, which was only obtained with mutant A96R.

2.3   X-Ray Analysis

For data collection, the crystals have to be shock-frozen to 100 K.[17] For this purpose, the crystals are usually soaked in a cryo-protectant that prevents ice-crystal formation. Unfortunately, the freezing procedure destroyed almost all MspA crystals as it dramatically amplified the existing mosaic spread.[12] Eventually, some crystals survived shock-freezing in the presence of polyoxyethylene with an average mass of 600 Da. An initial X-ray data set showed that the crystal belonged to the tetragonal space group I422. The observed cell axes corresponded to a packing parameter V_M of 9.6 $\text{Å}^3\,\text{Da}^{-1}$ under the assumption that the crystallographic asymmetric unit contained a single MspA subunit.

A self-rotation analysis of the data set resulted in a large peak at a 45° rotation corresponding to an eightfold rotation axis. This peak coincided with the direction of the crystallographic fourfold axis suggesting that the oligomer is a regular C_8-symmetric octamer with two subunits per asymmetric unit. Any alternative packing with four subunits per asymmetric unit was very unlikely because the corresponding packing parameter of 2.15 $\text{Å}^3\,\text{Da}^{-1}$ is too dense for such a large molecule.[18] Further trials succeeded in collecting a native (actually mutant A96R) X-ray diffraction data set of reasonable quality at 2.5 Å resolution, opening the door for structure analysis at high resolution.

In order to solve the common phase problem of X-ray analyses, the protein was labeled with selenium replacing the sulfur atoms of methionines. Since MspA contains only one methionine (Met183), which cannot be expected to suffice for accurate phasing, a second methionine was introduced by converting Leu88. Accordingly, seleno-labeling was performed with the double mutant L88M-A96R. The labeled protein was produced and crystallized, and the phases were established using the multiwavelength anomalous diffraction method.[14] The resulting structure (Figure 3) demonstrated that the porin was indeed the expected C_8-symmetric octamer and thus contradicted the proposed tetrameric structure derived from electron micrographs.[10] As expected, the eightfold rotation axis coincided with the crystallographic fourfold axis and the asymmetric unit contained two subunits. The corresponding packing parameter of 4.3 $\text{Å}^3\,\text{Da}^{-1}$ is in the usual range for membrane proteins.[19]

3   Structure Description

3.1   The Channel

In Figure 3 the structure of the mycobacterial porin is illustrated as a ribbon model emphasizing one of the subunits. The construction looks like a goblet with a thick rim at the top, a stem, and a base.[12] The protein consists exclusively of β-strands

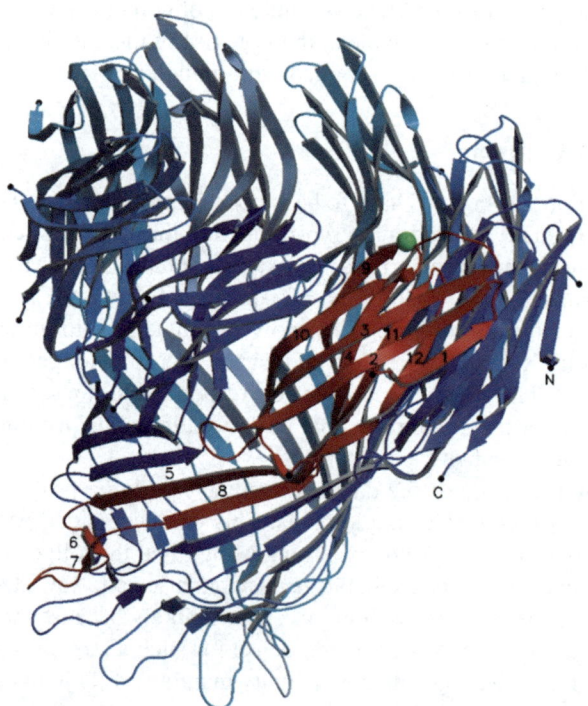

Figure 3 *Ribbon model of porin MspA from M. smegmatis.[12] The protein is a homo-octamer consisting of 8 × 184 amino acid residues forming a goblet-like structure around a single central channel. One of the subunits is emphasized in red. The upper part of the goblet consists of eight rim domains. The lower part contains two consecutive β-barrels with 8 × 2 = 16 strands forming the stem and the base regions of the goblet. A peptide used for raising an antibody is labeled with a green ball. Since this antibody reacted with intact M. smegmatis cells, the ball marks the external surface of the porin*

forming several sheets. This corresponds to the predominant β-structure of the Gram-negative bacterial porins and also to all other outer membrane proteins. The mycobacterial porin, however, follows a different construction principle. While the Gram-negative porins consist of three parallel channels running through three parallel β-barrels,[20] the mycobacterial porin forms only a single central channel running through two concentric β-barrels composed of eight identical parts. A view from the external medium into the eightfold symmetric channel is provided in Figure 4.

In order to prevent permeation of large solutes, both Gram-negative and mycobacterial porins contain constriction zones which, however, show completely different architecture. The Gram-negative porins draw one of the loops from the external β-barrel to the inside where it forms a constriction eyelet with an open diameter of around 8 Å.[6,20] In contrast, the mycobacteria narrow the major β-barrel to a second β-barrel, which also forms a bulge toward the interior, constricting the open channel diameter to about 10 Å. The resulting eyelet is defined by the carboxylates of Asp90 and Asp91 (Figure 4), which cause a rather strong electric field around them. Similar electric fields are known from the constriction eyelets of Gram-negative porins.[21] With such a field, the eyelets function as polarity filters: polar solutes can pass through the channel whereas nonpolar solutes meet a diffusion barrier because it is energetically unfavorable if they replace the oriented dipoles of water molecules within the constriction.

If the MspA channel were a macroscopic entity, its electrical conductance would be determined by the distribution of the available cross-section along the channel axis and by the specific conductance of the applied salt solution.[20] In a standard conductance measurement, one or more porins are incorporated in an artificial bilayer separating two chambers filled with $1M$ KCl (specific conductance of 1.13 S cm^{-1}).[4] KCl has the advantage that the mobilities of K$^+$ and Cl$^-$ are almost identical, which avoids

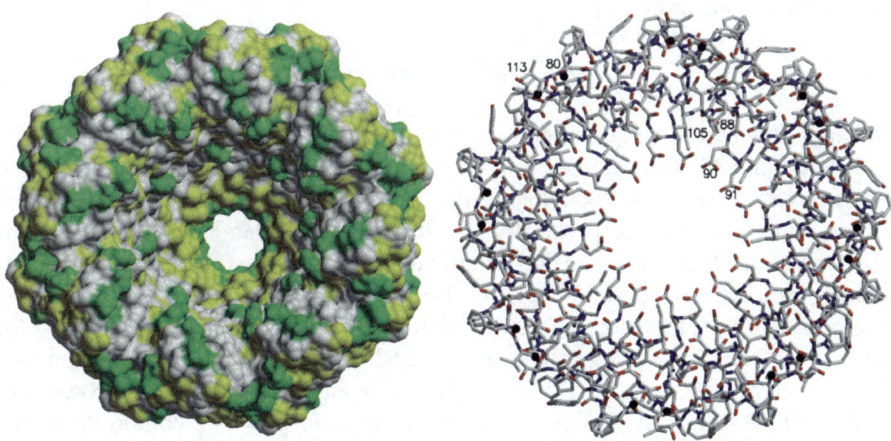

Figure 4 *View from the external medium into the channel of MspA.[12] On the left side the full channel is depicted using a surface representation with polar residues in green and nonpolars in yellow. The pore diameter is 10 Å. The right-hand side shows the channel constriction at high resolution. The diffusion eyelet is lined by 8 × 2 = 16 aspartates causing a surrounding electric field that prevents nonpolar solutes from passing*

interference from a Donnan potential. At the microscopic level, however, the calculation has to take surface effects into account. If the bulk solvent is considered to begin 3.0 Å outside the surface non-hydrogen atoms and if all interactions with charges on the protein surface are neglected, the shape and size of the MspA channel give rise to a calculated conductance of 5.1 nS (nano-Siemens). Because the calculated conductance is only 10% higher than the measured value of 4.6 nS,[9] the MspA channel behaves essentially like a macroscopic rod of conducting bulk solvent. In this respect, MspA differs greatly from Gram-negative porins. The major porin from *Rhodobacter capsulatus*, for example, contains three channels each with a narrow crooked geometry around the eyelet. If the three cross-sectional areas of the trimer were added up and converted to a hypothetical single circular cross-sectional distribution along the channel axes, the conductance calculated with the 3.0 Å criterion would result in 13 nS. This is four times the experimental value of 3.3 nS,[4] indicating that small eyelets with bent access conduits diminish the ionic conductance appreciably below that of a straight channel with an equivalent cross-sectional distribution.

3.2   β-Barrels

Polypeptides tend to form secondary structures in order to remove their amide dipoles from the nonpolar protein interior.[22] In contrast to a β-sheet with exposed dipoles along the sheet edges, a circularly closed sheet called a β-barrel is particularly stable. Such barrels are known for soluble proteins but they are especially abundant in outer membrane proteins.[6] With their amide dipoles compensated within the wall, β-barrels with nonpolar sidechains pointing to the outside fit well to the nonpolar environment in a membrane. The number of strands n in a β-barrel ranges from 6 to 22. The barrel diameter D is determined by n and the angle α between β-strands and barrel axis according to

$$D = (4.4 \text{ Å}) \cdot n/\pi \cdot \cos \alpha \tag{1}$$

The angle α can also be described by the so-called shear number.[6]

The common Gram-negative bacterial porins possess 16- and 18-stranded barrels with average angles of 43° and 40°, respectively.[6] TolC of these organisms contains a 12-stranded β-barrel composed of three subunits with a large angle of 51°.[23] The mycobacterial MspA has a major 16-stranded barrel composed of eight subunits with an even larger angle of 54° giving rise to a diameter of 38 Å. The large barrel is appended by a minor narrow barrel which keeps the 16 β-strands but reduces the angle to 37°, resulting in a diameter of merely 28 Å which allows an open constriction diameter of about 10 Å (Figure 4). The same angle is observed in the composite β-barrel of α-hemolysin.[24] All β-strands involved in these two composite barrels are sketched in Figure 5. While the stem and base of the goblet are formed by regular β-barrels, the thick rim contains β-sandwiches, the construction of which is shown in Figure 5. Each sandwich consists of two 4-stranded β-sheets. These sheets are hydrogen bonded to each other according to the scheme depicted in Figures 5 and 6. This construction combines eight overlapping 8-stranded β-sheets in a novel type of β-barrel.

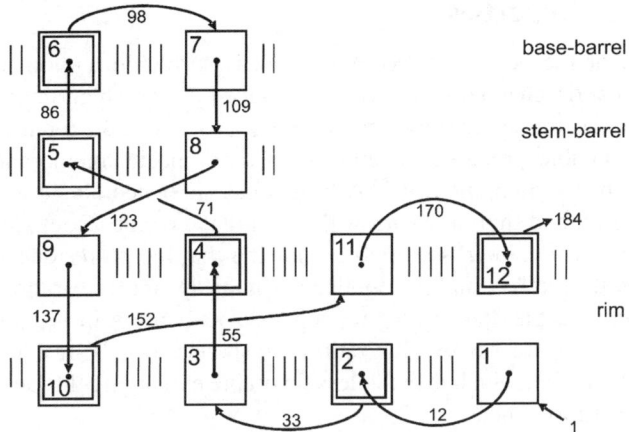

Figure 5 *β-sheet topology of MspA.[12] Strands β6 and β7 form the composite 16-stranded β-barrel at the base of the goblet and strands β5 and β8 the composite barrel of the stem (Figure 3). Strands β1–β4 and β9–β12 assemble in a sandwich of two 4-stranded β-sheets forming a domain at the rim of the goblet. The contact between the rim domains extends the β-sheets to neighboring subunits, as indicated by the hydrogen bond markings at strands β10 and β12. Some residue numbers in loops are given*

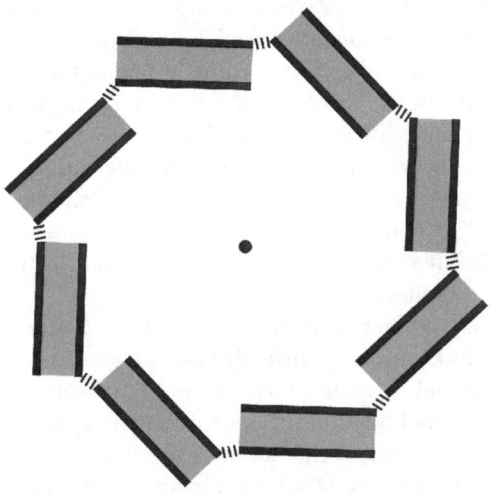

Figure 6 *Structural organization of the goblet rim viewed from the external side. Each rim domain consists of a β-sheet sandwich shown in more detail in Figure 5. The 4-stranded sheets are prolonged over two subunits to form 8-stranded sheets. The resulting construction is a barrel assembled from overlapping 8-stranded β-sheets*

3.3   Protein Properties

The precursor of the octameric MspA channel is a soluble monomer (or dimer). Since the octameric channel can only be produced at very high concentrations in detergent micelles, the *in vivo* assembly is likely to occur in the membrane. A similar case of a soluble precursor assembled to a composite membrane channel is known for the bee venom, melittin.[25] A further case is α-hemolysin, which forms a composite 14-stranded short and narrow β-barrel in the membrane.[24] Here, the structure of a precursor homolog is available, showing that the 36-residue loop forming the composite β-barrel is highly mobile.[26] Obviously, the size ratio between the 36-residue loop and the 264-residue remainder of α-hemolysin is small enough to allow crystallization. In MspA, this ratio corresponds to a 50-residue loop and a 134-residue remainder, which is much less favorable explaining the lack of any crystal of the soluble monomer.

It should be noted that examples of soluble proteins, which fail to crystallize without a detergent micelle, are also known for monotopic membrane proteins.[27] This type of protein harbors a relatively large nonpolar surface which, *in vivo*, dips into a membrane. *In vitro*, however, it causes undefined associations that prevent crystal formation. The associated proteins are soluble because they bury their nonpolar faces in transient dimers or trimers. Since these associations are transient, the proteins tend to run close to the monomer mass on a size-exclusion column. Similar behavior is to be expected and was actually observed for soluble MspA (Figure 2).

The crystal packing of MspA showed a strong base-to-base contact that is crucial for the crystalline order. This contact is depicted in Figure 7. It derives its strength and its exact definition from the eight hydrogen bonds formed by the introduced Arg96, shown in detail in the inset of Figure 7. Arginine was certainly the right choice because its guanidinium group mediates a hydrogen-bonded connection between Asp93 and a backbone carbonyl group. Interestingly, the size-exclusion peak at 158 mL in Figure 2 occurred only for mutant A96R and not for the wildtype or for other mutants.[16] It is therefore conceivable that this peak contained a hydrogen-bonded channel dimer in a detergent micelle, which in fractions *a* and *b* was pure enough to yield ordered crystals.

The assembly of any protein in an oligomer with a specific high-numbered rotational symmetry is an intrinsically difficult task because one subunit more or less changes the interfaces only slightly.[22] The assembly of soluble MspA to octamers may therefore contain small fractions of nonamers and heptamers. Since such impurities do not matter *in vivo*, MspA did not need to acquire a highly precise interface during evolution. For ordered crystals, however, an extremely homogenous oligomer ensemble is required. This explains why suitable crystals could only be obtained with the contact-defining mutant A96R that facilitated separation because it discriminated against nonoctamer contact partners.

The fact that only a tiny 0.5% fraction of the protein yielded analyzable crystals raises the question whether the crystals contain an impurity rather than the natural channel. In view of the preparation procedure, it is conceivable that the native MspA channel may be a heptamer, like that of α-hemolysin.[24] However, this possibility is excluded by the electron micrographs of linear aggregates of the natural MspA

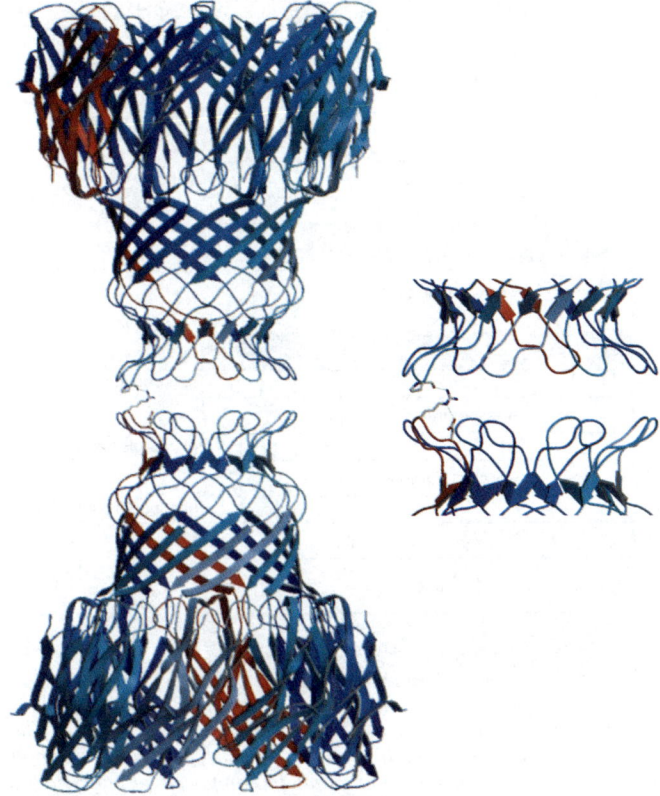

Figure 7 *The crucial base-to-base contact in the crystals of MspA mutant A96R.[12] This contact involves eight connecting hydrogen bonds (inset) and assembles hexadecamers that associate laterally to form the three-dimensional packing array. The hydrogen bonds introduced by the mutation A96R explain why the wildtype and all other mutants yielded only disordered crystals unsuitable for an X-ray analysis*

channels, which showed clearly a fourfold rotational symmetry.[10] This fourfold axis is consistent with an eightfold axis but excludes sevenfold (or ninefold) symmetry. The crystallized recombinant and (re)natured octamer is therefore most likely the native species.

4 The Outer Membrane

4.1 Membrane Structure

The mycobacterial outer membrane is of high medical interest because it excludes drugs very effectively.[28–35] The present model of this membrane is sketched in Figure 8. The inner, cytoplasmic membrane is covered by a thin peptidoglycan cell wall that is covalently linked through phosphodiester bonds to a galactofuranose layer.[36] This layer carries covalently bound arabinofuranose branches connected to

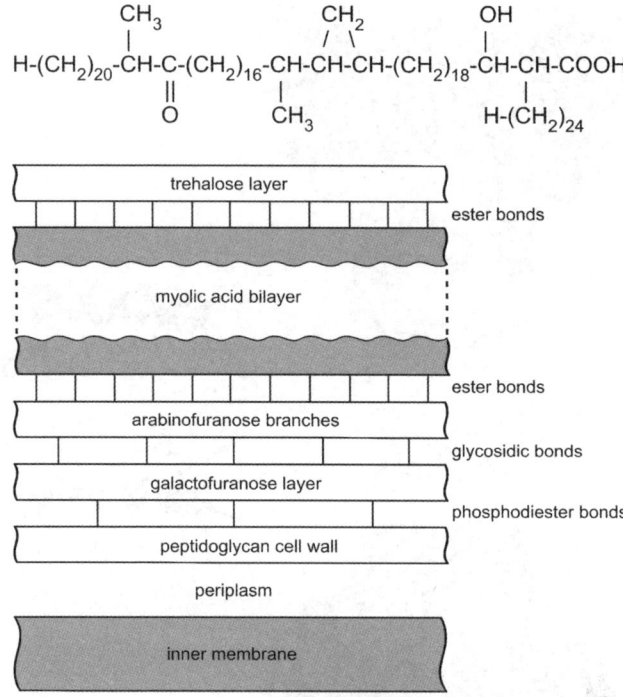

Figure 8 *The chemical structure of the outer membrane of mycobacteria. The mycolic acid example shown at the top is among the largest found in M. tuberculosis.[38] It contains a total of 89 carbon atoms and is composed of four pieces of saturated aliphatic chains of around 20 carbons each. The outer membrane consists of a mycolic acid bilayer that is covalently attached via sugar layers to the outer side of the thin peptidoglycan cell wall. Based on electron microscopic data the present model of the outer membrane assumes extended mycolic acids resulting in a thickness of about 90 Å for the nonpolar bilayer*

mycolic acids, which are β-hydroxy fatty acids with up to 90 carbon atoms.[37,38] The structural formula of a large mycolic acid is given at the top of Figure 8. It seems to be constructed from four pieces of linear saturated aliphatic chains with around 20 carbon atoms each. Three of these pieces form the main chain and the fourth piece is added as a branch at the C_{α} carbon.[39] The inner leaf of the outer membrane is covered by an outer leaf consisting of trehalose molecules carrying a short mycolic acid at each glucose unit. The thickness of the outer membrane amounts to about 90 Å, as derived from electron micrographs.[39,40] This value is very high but corresponds to the extremely low permeability of this membrane and to the length of the mycolic acids that may extend to more than 80 Å. Furthermore, it appears to be supported by electron micrographs of MspA showing a channel length of around 90 Å.[10] In accordance with these data, the present model assumes a dense lateral packing of stretched mycolic acids forming a 90 Å thick nonpolar bilayer.[11]

The crystal structure reveals the actual shape and the surface properties of MspA (Figure 9). The outer dimensions agree rather well with the electron microscopic data, but the outer surface is incompatible with a 90 Å thick nonpolar membrane structure. Actually, the outer surface of MspA is only nonpolar in the base and stem regions. This nonpolar girdle is merely 37 Å wide and its upper border is clearly marked by Arg35 and Arg38. These arginines form a ring of 16 guanidinium groups around the molecule, which are always charged ($pK = 12.5$) and therefore certainly outside the nonpolar bilayer. Moreover, the outer surface above this ring of arginines is generally polar (Figure 9). Consequently, the outer surface of MspA suggests that

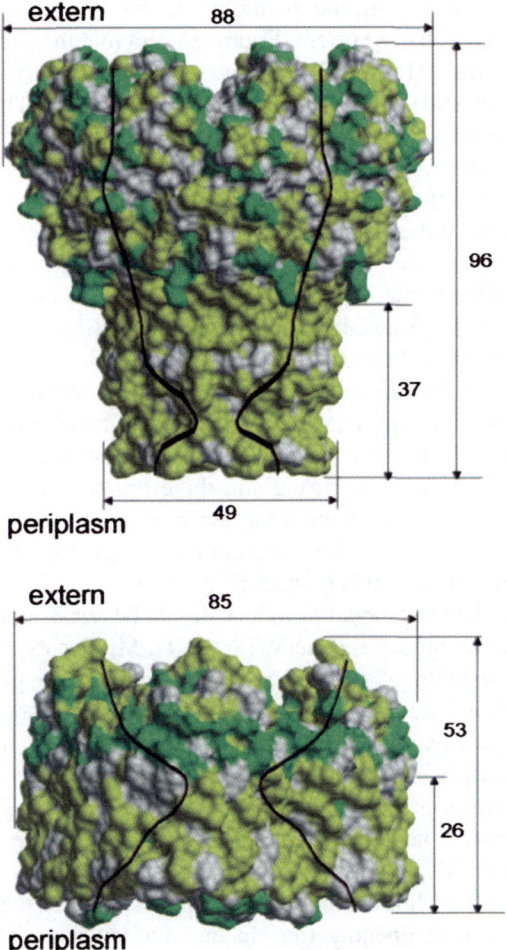

Figure 9 *Dimensions and outer surfaces of MspA from M. smegmatis (top) and from the Gram-negative bacterium Rhodobacter capsulatus (bottom).[12] The nonpolar parts are yellow and the polar parts are green. The 37 Å nonpolar girdle of MspA and the respective 26 Å girdle of the Gram-negative porin are indicated. The central channel of MspA is outlined. It shows the constriction near the periplasmic end. For the Gram-negative porin the cross-sections of the three parallel channels are added and converted to a hypothetical single circular cross-section, the diameter of which is outlined*

the nonpolar bilayer of the outer membrane is merely 37 Å thick rather than 90 Å thick. This conclusion agrees with the construction of the Gram-negative bacterial porins (Figure 9), which carry a nonpolar girdle that is about 25 Å wide and therefore matches the nonpolar aliphatic part of the outer membrane.[20]

4.2 Porin Localization

In the light of the Gram-negative bacterial porins fitting into their outer membrane, the surface structure of MspA suggests that the stem and base regions of MspA sit in the mycolic acid bilayer with the guanidinium ring (Arg35 and Arg38) forming hydrogen bonds to the trehalose esters (Figure 8), and that the rim domain protrudes into the external medium. This implies that the nonpolar bilayer is merely about 37 Å thick and the mycolic acids are not extended as suggested.[36,41] This proposal clearly contradicts the interpretation of the electron micrographs of the mycobacterial outer membrane[39,40] but is compatible with the low membrane permeability, because 37 Å is still 50% thicker than the 25 Å of Gram-negative bacteria. Moreover, the mycolic acids are laterally crosslinked through the galactofuranose layer (Figure 8) and they are much longer and therefore much more tightly entangled than their counterparts.

The protrusion of the goblet rim into the external medium is supported by experiments with an antibody produced against a peptide at the top of the rim domain (Figure 3). This antibody is bound to the intact *M. smegmatis* bacterium, confirming that the top of the rim domain is exposed to the environment.[12] Furthermore, it is most unlikely that *in vivo* the channel is formed by two base-to-base associated octamers (Figure 7), which would double the derived membrane thickness. *In vivo*, such a contact would have to be strong and therefore conserved during evolution. This disagrees with the crystallization behavior of wildtype MspA pointing to a weak contact as well as with the observed amino acid exchanges in the MspA variants around position 100 near this contact (Figure 1).

The proposed localization model of MspA should be consistent with the observed exchanges in the homologous sequences (Figure 1). MspB carries only the substitutions Ala138→Pro and Glu139→Ala in a loop at the top of the goblet. According to the model this modifies the antigenic properties, which is reasonable and therefore supportive. MspC carries the exchanges of MspB and in addition mutations Ala96→Gly and Asn102→Glu at the base of the goblet. Gly96 renders the loop pointing into the periplasm more mobile and Glu102 at the periplasmic side of the constriction adds another negative charge near the 16-aspartate ring (Figure 4), both supporting the model. MspD contains 18 amino acid exchanges, the most dramatic of which is Asp91→Gly, diminishing the 16-aspartate ring of the constriction to an 8-aspartate ring and thus opening the channel for larger and less polar solutes. Mutations Leu8→Val, Ile68→Val, and Gly141→Ala are in the core and at the interface between the rim domains and may therefore modify the oligomerization. Thus one may speculate that MspD formed a nonamer, which, in conjunction with Asp91→Gly, would open the channel drastically. The remaining exchanges at positions 1, 2, 5, 13, 21, 22, 47, 49, 136, and 139 are all at the top and at the outer surface of the rim domain and should therefore mostly affect the antigenicity. Further exchanges at positions 96, 102, 103, and 104 are at the base and point into the periplasm. In summary, all amino acid exchanges are consistent with the proposed location of MspA in the outer membrane.

5 Conclusion

The mycobacterial porin follows the generally accepted thesis that outer membrane proteins consist of β-structure and inner membrane proteins of α-helices. No generally accepted explanation for this difference is yet available. MspA contains two β-barrels of the common type for outer membrane proteins that have the same strand number but different diameters as well as a novel type with overlapping β-sheets (Figure 6). A comparison of the outer surface of MspA with those of Gram-negative bacterial porins indicates that the outer mycobacterial membrane has a thickness of about 40 Å rather than the 90 Å of the present model. Antigenicity, crystallization properties, and the exchanges observed in three closely related variants suggest how MspA is located in the membrane. The electron micrographs of linearly associated native MspA indicate that the crystallized octamers reflect the native oligomerization state. Moreover, the channel structure corresponds to the electrical conductance of native MspA.

Mycobacteria of high medical relevance are *M. leprae* and *M. tuberculosis*, causing the diseases named after them. Unfortunately, these species grow very slowly rendering the direct identification of their proteins and an analysis of their outer membranes a tedious task.[41] These pathogens contain porins that have been identified by their electrical conductance but have not yet been chemically characterized.[42,43] The genome of *M. tuberculosis* contains no sequence-related homolog to MspA. Still, structural homologs are expected, because it is well known from the porins of Gram-negative bacteria that their amino acid sequences change rapidly during evolution without much effect on the structure or function.[44] Since the death toll caused by these mycobacteria is on the rise,[45] the intensive search for these porins will continue.

Acknowledgments

I thank Michael Niederweis and Michael Faller for the cooperation in the structure analysis project, and Linda Böhm and Dirk Grüninger for their help in the preparation of the manuscript.

References

1. C. Pitulle, M. Dorsch, J. Kazda, J. Wolters and E. Stackebrandt, *Int. J. Syst. Bacteriol.*, 1992, **42**, 337.
2. L.M. Fu and C.S. Fu-Liu, *Tuberculosis*, 2002, **82**, 85.
3. H. Nikaido, *Science*, 1994, **264**, 372.
4. R. Benz and K. Bauer, *Eur. J. Biochem.*, 1988, **176**, 1.
5. M.S. Weiss, T. Wacker, J. Weckesser, W. Welte and G.E. Schulz, *FEBS Lett.*, 1990, **267**, 268.
6. G.E. Schulz, *Biochim. Biophys. Acta*, 2002, **1565**, 308.

7. J. Trias, V. Jarlier and R. Benz, *Science*, 1992, **258**, 1479.

8. J. Trias and R. Benz, *Mol. Microbiol.*, 1994, **14**, 283.

9. M. Niederweis, S. Ehrt, C. Heinz, U. Klöcker, S. Karosi, K.M. Swiderek, L.W. Riley and R. Benz, *Mol. Microbiol.*, 1999, **33**, 933.

10. H. Engelhardt, C. Heinz and M. Niederweis, *J. Biol. Chem.*, 2002, **277**, 37567.

11. C. Heinz, S. Karosi and M. Niederweis, *J. Chromatogr. Sect. B*, 2003, **790**, 337.

12. M. Faller, M. Niederweis and G.E. Schulz, *Science,* 2004, **303**, 1189.

13. C. Stahl, S. Kubetzko, I. Kaps, S. Seeber, H. Engelhardt and M. Niederweis, *Mol. Microbiol.*, 2001, **40**, 451.

14. W.A. Hendrickson, A. Pähler, J.L. Smith, Y. Satow, E.A. Merritt and R.P. Phizackerley, *Proc. Natl. Acad. Sci. USA*, 1989, **86**, 2190.

15. A. Pautsch, J. Vogt, K. Model, C. Siebold and G.E. Schulz, *Proteins: Struct. Funct. Genet.*, 1999, **34**, 167.

16. M. Faller, Kristallstruktur eines mycobacteriellen Porins, Ph.D. Thesis, Albert-Ludwigs-Universität, Freiburg im Breisgau, 2004.

17. E. Garman, *Acta Crystallog. Sect. D*, 1999, **55**, 1641.

18. K.A. Kantardjieff and B. Rupp, *Protein Sci.*, 2003, **12**, 1865.

19. A. Pautsch and G.E. Schulz, *J. Mol. Biol.*, 2000, **298**, 273.

20. G.E. Schulz, in *Bacterial and Eukaryotic Porins*, R. Benz (ed)., Wiley-VCH, Weinheim, 2004, 25.

21. M.S. Weiss, U. Abele, J. Weckesser, W. Welte, E. Schiltz and G.E. Schulz, *Science*, 1991, **254**, 1627.

22. G.E. Schulz and R.H. Schirmer, *Principles of Protein Structure,* Springer Verlag, New York, 1979.

23. V. Koronakis, A. Scharff, E. Koronakis, B. Luisi and C. Hughes, *Nature*, 2000, **405**, 914.

24. L. Song, M.R. Hobaugh, C. Shustak, S. Cheley, H. Bayley and J.E. Gouaux, *Science*, 1996, **274**, 1859.

25. T.C. Terwilliger and D. Eisenberg, *J. Biol. Chem.*, 1982, **257**, 6016.

26. R. Olson, H. Mariya, K. Yokota, Y. Kamino and E. Gouaux, *Nat. Struct. Biol.*, 1999, **6**, 134.

27. D.P. Kloer, S. Ruch, S. Al-Babili, P. Beyer and G.E. Schulz, *Science*, 2005, **308**, 267.

28. D.E. Minnikin, *Res. Microbiol.*, 1991, **142**, 423.

29. C.E. Barry III, R.E. Lee, K. Mdluli, A.E. Sampson, B.G. Schroeder, R.A. Slayden and Y. Yuan, *Prog. Lipid Res.*, 1998, **37**, 143.

30. M. Daffé and P. Draper, *Adv. Microb. Physiol.*, 1998, **39**, 131.

31. P.A. Lambert, *J. Appl. Microbiol.*, 2002, **92**, 46S.

32. B.A. Dmitriev, S. Ehlers, E.T. Rietschel and P.J. Brennan, *Int. J. Med. Microbiol.*, 2000, **290**, 251.

33. M. Jackson, D.C. Crick and P.J. Brennan, *J. Biol. Chem.*, 2000, **275**, 30092.

34. H. Nikaido, S.H. Kim and E.Y. Rosenberg, *Mol. Microbiol.*, 1993, **8**, 1025.

35. J. Liu, E.Y. Rosenberg and H. Nikaido, *Proc. Natl. Acad. Sci. USA*, 1995, **92**, 11254.

36. P.J. Tonge, *Nat. Struct. Biol.*, 2000, **7**, 94.

37. N. Rastogi, E. Legrand and C. Sola, *Rev. Sci. Tech.*, 2001, **20**, 21.

38. M. Watanabe, Y. Aoyagi, M. Ridell and D.E. Minnikin, *Microbiology*, 2001, **147**, 1825.

39. P.J. Brennan and H. Nikaido, *Annu. Rev. Biochem.*, 1995, **64**, 29.

40. T.R. Paul and T.J. Beveridge, *J. Bacteriol.*, 1992, **174**, 6508.

41. M. Niederweis, *Mol. Microbiol.*, 2003, **49**, 1167.

42. T. Lichtinger, B. Heym, E. Maier, H. Eichner, S.T. Cole and R. Benz, *FEBS Lett.*, 1999, **454**, 349.

43. B. Kartmann, S. Stenger and M. Niederweis, *J. Bacteriol.*, 1999, **181**, 6543, *Corrigendum* p. 7650.

44. E. Schiltz, A. Kreusch, U. Nestel and G.E. Schulz, *Eur. J. Biochem.*, 1991, **199**, 587.

45. D. Bleed, C. Watt and C. Dye, "Global tuberculosis control", *WHO Report*, Geneva, 2001.

CHAPTER 14

The Structure of the SecY Protein Translocation Channel

BERT VAN DEN BERG[1] AND IAN COLLINSON[2]

[1]University of Massachusetts Medical School, Program in Molecular Medicine, 373 Plantation Street, Worcester, MA 01605, USA
[2]University of Bristol, Department of Biochemistry, University Walk, Bristol BS8 1TD, UK

1 Introduction

Protein transport across the endoplasmic reticulum (ER) membrane in eukaryotes is an early and decisive step in the biosynthesis of many proteins. These proteins can be divided into two groups: soluble proteins, such as those secreted from the cell, and membrane proteins, such as those in the plasma membrane or in membranes of other organelles of the secretory pathway. In eubacteria and archaea, protein transport occurs through the plasma membrane, and is an important step in the biosynthesis of secreted and membrane proteins. Soluble proteins cross the membrane completely and usually have N-terminal cleavable signal sequences, characterized by a short hydrophobic segment (typically 7–12 amino acid residues). Membrane proteins have different topologies with one or more trans-membrane (TM) segments that contain about 20 hydrophobic residues, and soluble domains that are either translocated through the membrane or stay in the cytosol. Both types of proteins use the same machinery for translocation across the membrane: a protein-conducting channel with a hydrophilic interior.[1,2] This channel, in contrast to those that transport ions and small molecules, has the unusual property that it can open in two directions: perpendicular to the plane of the membrane to let a polypeptide segment across, and within the membrane to let a hydrophobic TM segment of a membrane protein exit laterally into the lipid phase. The protein-conducting channel is formed by an evolutionarily conserved heterotrimeric membrane protein complex called the Sec61 complex in eukaryotes and the SecY complex in eubacteria and archaea. In this review, we will discuss the recently determined electron cryo-microscopy[3] and X-ray structure[4] of bacterial SecY complexes, and summarize our current understanding of how the channel functions in bacterial protein translocation and membrane protein integration.

1.1 The Sec61/SecY Complex

The largest subunit of the heterotrimeric Sec61/SecY complex is the α-subunit, called Sec61α in mammals, Sec61p in *Saccharomyces cerevisiae* and SecY in eubacteria and archaea (for review, see Ref. 5). This subunit spans the membrane ten times, with both the N- and C-termini in the cytosol. The β-subunit is called Sec61β in mammals, Sbh1p in *S. cerevisiae*, SecG in eubacteria and Secβ in archaea. In eukaryotes and archaea, it spans the membrane once with the N-terminus in the cytosol. The eubacterial β-subunit (SecG) spans the membrane twice. The γ-subunit is called Sec61γ in mammals, Sss1p in *S. cerevisiae* and SecE in eubacteria and archaea. In most species, it is a single-spanning protein with its N-terminus in the cytosol. In some eubacteria, *e.g. Escherichia coli*, it has two additional N-terminal TM segments that are not essential for its function. The α- and γ-subunits of the Sec61/SecY complex are found in all organisms, showing low, but significant sequence conservation. The β-subunits are homologous among eukaryotes and archaea, but have no obvious sequential similarity with the eubacterial SecG. The α- and γ-subunits are essential for viability of yeast and eubacteria, whereas the β-subunits are not. Together, these observations indicate that the α- and γ-subunits constitute the core of the channel-forming complex.

The initial evidence that the Sec61/SecY complex forms a protein-conducting channel came from systematic crosslinking experiments, in which photoreactive probes were placed at different positions of a polypeptide substrate.[6] Substrates with probes at positions predicted to be within the membrane could be crosslinked to the α-subunit of the Sec61 complex, but not to other membrane proteins. These data indicated that the α-subunit surrounds the polypeptide chain as it passes through the membrane. Strong support for the notion that the Sec61/SecY complex forms a channel came from experiments in which the purified complex was reconstituted into proteoliposomes and was shown to be the essential membrane component for protein translocation.[7–9]

1.2 The Three Different Translocation Modes

The protein-conducting channel formed by the Sec61/SecY complex is a passive pore that allows a polypeptide chain to slide back and forth. The channel therefore needs to associate with partners that provide a driving force for translocation. Depending on the partner, the channel can function in three different translocation modes.

The first mode, co-translational translocation, involves the ribosome as the major channel partner (Figure 1). This translocation mechanism is found in all organisms and cells, and it is responsible for the integration of most membrane proteins. Co-translational translocation begins with a targeting phase, during which a ribosome-nascent chain complex is directed to the membrane by the signal recognition particle (SRP) and its membrane receptor (SRP-receptor). Once the ribosome is bound to the

CO-TRANSLATIONAL   **POST-TRANSLATIONAL**

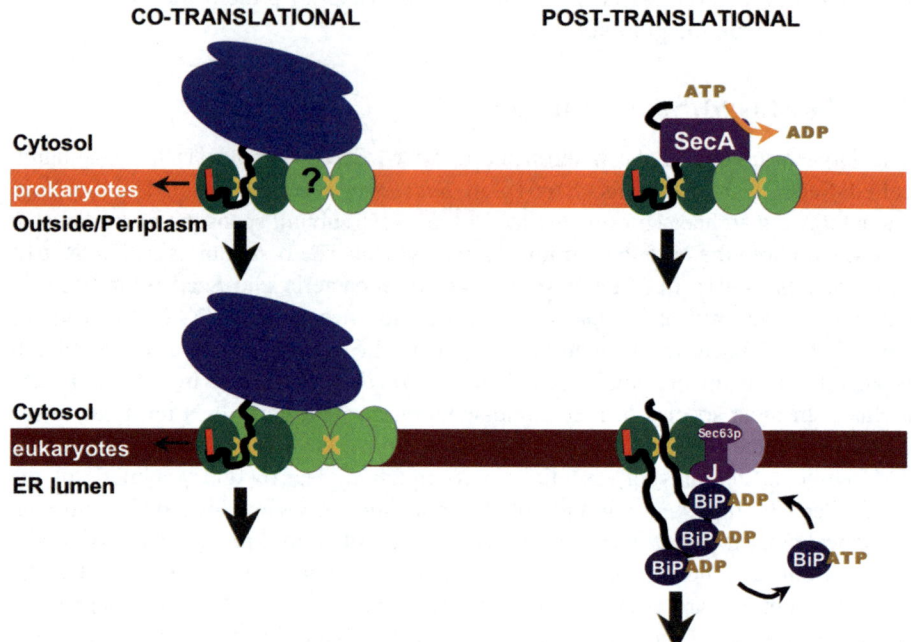

Figure 1 *Schematic overview of the different translocation modes. On the left and right the pathways for co- and post-translational translocation are shown respectively. At the top of the figure, translocation across the cytosolic membrane of bacteria is shown, whereas the bottom shows translocation across the ER membrane in eukaryotes. The olive green arrows indicate the direction in which the translocating chain would move. The active SecY complex (i.e. the one forming the actual channel) is shown in dark green and the passive complex(es) in light green. The gold central semicircles represent the hydrophobic seal ring in the centre of the protein channel formed by a single SecY complex. The ribosome is shown pushing the polypeptide (black) through the channel, with the N-terminal signal sequence (red) of the substrate bound to the SecY complex. In eukaryotes the co-translational complex is thought to be a tetramer. In prokaryotes this is not yet clear, hence the question mark. The post-translational pathway in bacteria involves SecA, which hydrolyzes ATP to push the protein across the membrane. A dimeric SecYEG complex likely plays a key role in this reaction. In eukaryotes, the Sec complex associates with additional membrane proteins during post-translational translocation. One of these is Sec63p (in yeast), which contains a J-domain that interacts with and activates the luminal Hsp70 chaperone homologue BiP to drive a molecular ratchet (see text)*

Sec61/SecY channel, the elongating polypeptide chain is moved from the ribosome into the membrane channel, with the energy for translocation provided by GTP hydrolysis during translation.

The second mode is post-translational translocation, a process by which proteins are transported after completion of their synthesis in the cytosol. Proteins that are translocated post-translationally have a less hydrophobic signal sequence and may therefore escape the interaction with SRP during their synthesis.[10] The mechanism of post-translational translocation has been elucidated in *S. cerevisiae*,[11] and it is likely to be the same in higher eukaryotes (Figure 1). Here, the channel partners are the Sec62/63 complex, a membrane protein complex, and the lumenal protein BiP, a member of the Hsp70 family of ATPases. In yeast, the Sec62/63 complex is a tetramer which together with the Sec61 complex forms a seven-component Sec complex.[12,13] In addition to the essential proteins Sec62p and Sec63p, yeast contains the non-essential components Sec71p and Sec72p. Mammalian cells have Sec62p and Sec63p, but lack the other two proteins.[14,15]

The driving force for post-translational translocation in eukaryotes is generated by a ratcheting mechanism.[11] A polypeptide in the channel can slide in either direction, but its binding to BiP inside the ER lumen prevents movement back into the cytosol, resulting in net forward translocation. BiP with bound ATP has an open peptide-binding pocket, which interacts with a lumenal domain of Sec63p called the J-domain. This interaction stimulates rapid ATP hydrolysis and closure of the peptide-binding pocket around the incoming polypeptide chain. When the polypeptide has moved a sufficient distance in the forward direction, another BiP molecule can bind to it, and this process is repeated until the polypeptide chain has completely traversed the channel. When ADP is exchanged for ATP, the BiP peptide-binding pocket opens and the polypeptide chain is released.

The third mode of translocation, found only in eubacteria, also occurs post-translationally, and is used by most secretory proteins (for review, see Ref. 16). In this case, the channel partner is a cytosolic ATPase called SecA (Figure 1). SecA likely undergoes conformational changes coupled to its ATPase cycle and pushes polypeptides through the SecY channel in a stepwise manner.[17] We will discuss the mechanism of SecA-mediated translocation in more detail later.

Archaea probably have both co- and post-translational translocation,[18,19] but it is unclear how they perform the latter, because they lack both SecA and the Sec62/63 complex.

2 Structure Determination of the SecY Complex by Electron Cryo-Microscopy

A wealth of genetic and biochemical data on the Sec-dependent secretory pathway has been chronicled in the literature. This has laid the foundations for subsequent work aimed at properly understanding how proteins are able to pass through or insert into a hydrophobic membrane barrier. Clearly, the description of the mechanism of the protein translocation reaction requires, as a first step, information on the architecture of the SecY channel. In order to obtain structural information in sufficient detail for this purpose, crystallographic approaches are required.

Two such approaches have been adopted for the SecY complex. The first of these, two-dimensional (2D) crystallization in combination with electron cryo-microscopy, relies on the efficient insertion of membrane proteins into a reconstituted lipid bilayer. Conditions are sought whereby the amount of proteins incorporated into the bilayer reaches a high enough density to initiate lattice formation, and hence 2D crystallization. This is usually achieved by adding together detergent-solubilized pure protein and lipids in specific ratios. Reconstitution and crystallization of proteins within the bilayer follows detergent removal, by dialysis or adsorption.

Subsequently, the crystals can then be visualized with electron microscopy, and the images can be used to extract structure factors and reconstruct a 3D structure.[20]

Potentially, the calculated maps can be used to build atomic models;[21–24] however, more usually the resolution attained is between 6 and 8 Å, which is enough to resolve TM α-helices, but not the amino acid side chains.[25–27] In spite of this obvious disadvantage, there are several good reasons for the application of this strategy. The amount of material required is usually about ten times less than that needed to initiate classical 3D-crystallization trials (see below). Moreover, since electron cryomicroscopy structures are determined in a lipid bilayer, such structures are likely to represent the native conformation of the membrane protein. A further advantage offered by the membranous environment of the crystal is that the protein is much more stable and therefore more likely to crystallize.

For the first structure determination of a protein translocation channel, the SecYEG complex from *E. coli* was chosen. The pBAD22 vector was used to produce the complex in a homologous overexpression system,[28,29] which is under control of the arabinose promoter.[29,30] The plasmids contained each of the three subunits (in the order E, Y and G), each with a separate ribosome initiation site. The SecE subunit contained an N-terminal hexa-histidine tag. Relatively higher levels of protein expression (~1 mg L^{-1} cell culture) could be obtained. The complex was subsequently purified in the detergent nonaethylene glycol monododecyl ether ($C_{12}E_9$) to yields sufficient for a 2D-crystallization screen.[29] Crystals were grown when the lipid phosphatidyl-ethanolamine was used at a protein to lipid ratio of about 0.2.[29] Images of the best crystals were recorded in vitreous ice in a range of tilt angles (0–55°).[3,29] Those containing detail to high resolution were then selected and the information combined to reconstruct a 3D map of 8 Å resolution.[3]

The crystals were unusual in that they had formed a double membrane, with the larger cytosolic loops facing inwards to form crystal contacts between the two layers.[3] Within each layer, pairwise associations of the heterotrimeric SecYEG complexes are observed, related by a twofold screw axis in the plane of the membrane. SecYEG dimers are known to form once the detergent is diluted.[31] The reconstitution of the SecY complex for post-translational protein translocation experiments *in vitro* and the 2D crystallization are performed via similar procedures involving the removal of detergent;[29] therefore, it seems likely that the dimers we observe in the EM structure and those that participate in post-translational translocation in eubacteria are the same (see below).

The resolution of the EM structure was sufficient to visualize all 30 predicted TM α-helices in the dimeric complex (Figure 2). Each SecY consisted of a core of 13 TM segments with an additional two peripherally located TM segments. One interesting feature was a highly tilted helix at the dimer interface corresponding to the essential third TM helix of SecE (homologous to the single TM segment present in Sec61γ). As is the case in most EM structures, most of the loops connecting the helices could not be seen, but some weak density corresponding to domains on the surface of the bilayer could be identified (see below). Due to the limited resolution neither the amino acid side chains nor the α-helical segments could be assigned; further interpretation of the structure required a more detailed electron density map. This was achieved by employing X-ray crystallography.

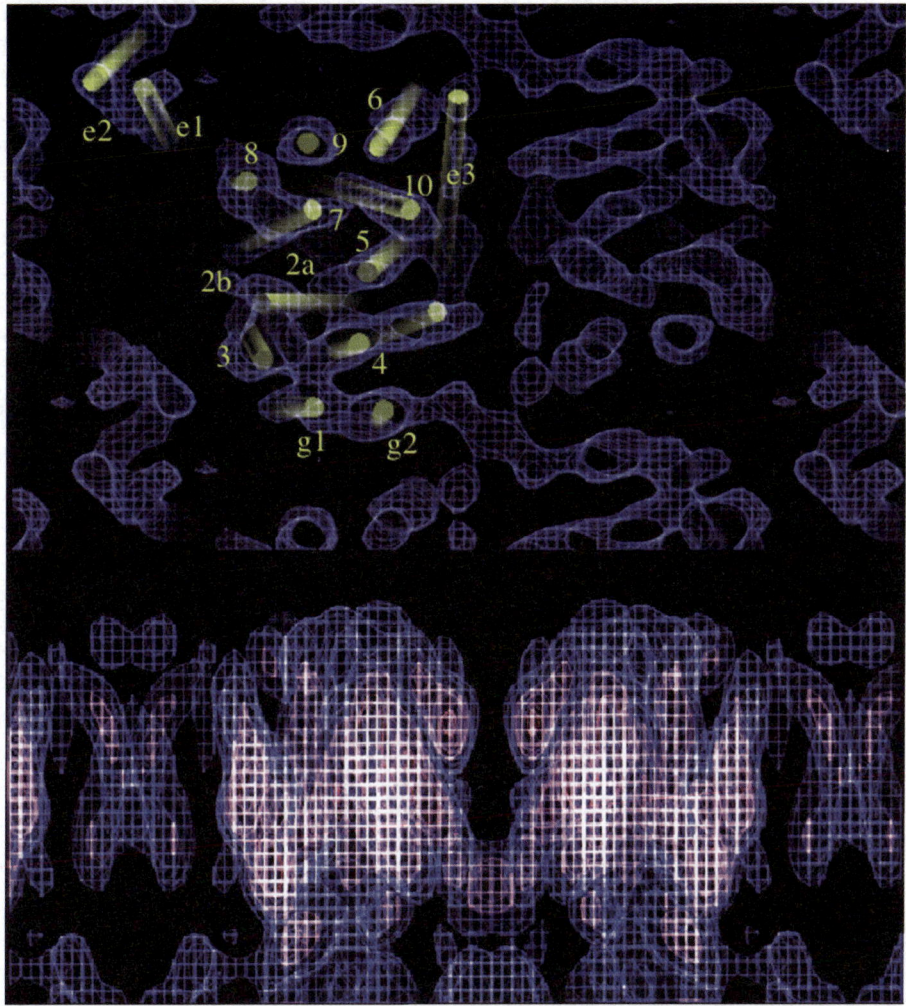

Figure 2 *The structure of the dimeric SecYEG complex from E. coli determined by electron cryo-microscopy. The upper panel shows the view from the cytosol. One of the complexes has its 15 TM domains labelled, which are also overlaid with gold tubes. SecY, 1–10, with 2a being the plug; SecE, e1–e3; SecG, g1 and g2. The side view on the bottom of the figure has the cytosolic surface facing downwards. The two monomers can be clearly distinguished*

3 Determination of the X-ray Crystal Structure of the SecY Complex

In order to determine the high-resolution structure of the protein translocation channel, the SecY complex was subjected to 3D-crystallization trials. The SecY complexes were purified via the procedures used to produce material for the 2D approach (see above); crystals made in this way are grown by the classical 3D procedure, but in the presence of specific detergents. The choice of detergent for the purification and crystallization steps in these experiments is critical and needs to be screened empirically, in order to obtain crystals that diffract to high enough resolution for structure determination.

The *E. coli* complex was purified and subjected to large crystallization screens. After a thorough screening only very small and fragile protein-containing crystals could be grown. They diffracted X-rays very poorly and could not be improved. It became clear that, in the absence of lipids, the complex was not stable enough to withstand the long periods of time required to grow 3D crystals at room temperature. A search was then initiated to identify alternative samples that might be more stable and amenable to crystallization.

One approach to this problem is to focus on a series of homologues from alternative species, in particular those from thermophilic organisms. We employed this strategy to clone and overexpress nine different bacterial SecY complexes and the SecY complex from *S. cerevisiae* in the same way as for the *E. coli* complex. The levels of protein expression for *Thermotoga maritima* and the archaeon *M. jannaschii* were comparable to that of the *E. coli* complex. For the purification of these SecY complexes we modified the protocol utilized for the *E. coli* preparation.[29]

At the time the project was initiated only the *secE* and the *secY* genes were annotated in the *M. jannaschii* genome database; therefore, our initial efforts focused on the archaebacterial SecYE complex. However, despite extensive efforts to optimize our screens, the crystals of the SecYE complex did not diffract beyond 6.5 Å.

Often, the key and most difficult step in a crystallographic project is to push the resolution beyond 4 Å, so that the solution of the phase problem can be approached; this is often achieved empirically. Following the discovery in the database of the *sec61β* gene, it was cloned into the context of the pBAD22 vector to overexpress the heterotrimeric SecYEβ complex. Importantly, the addition of this subunit increased the stability of the complex, allowing a larger range of detergents to be screened; some of which were too harsh for the dimeric complex (most notably dodecyl-*N*, *N*-dimethylamine-*N*-oxide [LDAO], nonyl-glucoside and the short-chain phospholipid diheptanoyl-phosphatidylcholine [DHPC]). Crystals that appeared as very thin plates were rapidly obtained in DHPC, but were difficult to manipulate. To improve the quality and size of the crystals we introduced a thrombin cleavage site to remove the histidine tag after the purification step. This strategy has the additional advantage that an additional purification step is easily achieved by passing the cleaved protein over a second nickel column to remove impurities as well as uncleaved protein.[4] In addition to cleavage of the hexa-histidine tag, it was essential to streamline the cation exchange step in order to remove contaminating SecYE from the SecYEβ complex (apparently the β-subunit was made in substoichiometric amounts relative to the other subunits). The crystals from the best preparations grew in 50–55% PEG400 (~45% PEG400 for selenomethionine-substituted protein), 50 mM glycine pH 9.0–9.5, and they diffracted anisotropically to a maximum resolution of 3.4–3.5 Å (space group $P2_12_12$).

The initial model for the wild-type protein proved very difficult to refine. Therefore, we inspected the structure and made several mutants with the intention to strengthen the crystal contacts between the C-terminus of SecY and the cytosolic loop between TM 5 and 6. These domains were chosen as in the crystals each of them was close to the same domain in adjacent monomers. One of these mutants (Lys422Arg/Val423Thr), called Y1, crystallized in thin plates with 35% PEG400, and diffracted to 3.2 Å. Interestingly, although the space group of the crystals remained the same, the crystal packing and unit cell dimensions were completely different for the Y1 mutant. The inclusion of the Y1 dataset in the cross-crystal averaging procedures greatly facilitated the refinement process, and was essential to the refinement and determination of a reliable model.

4 Description of the Structure of the SecY Complex

Given the evolutionary conservation and the sequence similarities of SecY complexes mentioned above, it is likely that the structure of the SecY complex described below is representative of all species. Indeed, the X-ray crystal structure of the archaeal complex[4] demonstrates that all TM segments are essentially super-imposable on those of the lower resolution electron cryo-microscopy structure derived from 2D crystals of the *E. coli* SecY complex in a lipid bilayer (above[32]). This observation also means that the structure of the SecY monomer complex in detergent is very similar to that in a lipid bilayer, except for the extra three TM helices found in the eubacterial complex (two in SecE and one in SecG). This comparison with the *E. coli* structure reveals the location of these extra domains; the peripheral helices noted in the EM structure belong to the non-essential TM 1 and 2 of SecE. The most important difference between the two structures is the oligomeric state of the SecY complex; the membrane-bound structure is a dimer of SecYEG and the X-ray structure in detergent is a monomer. This is not too surprising as high concentrations of detergent are known to dissociate oligomeric associations of membrane proteins, which is also true of the SecY complex.[31]

In the X-ray crystal structure, the SecY complex contains one copy of each of the three subunits (Figure 3). Viewed from the cytosol, the complex has an approximately square shape. The two small subunits (SecE and Secβ) are located at the periphery of the complex. The SecE subunit contacts SecY extensively, occupying two sides of the square. As predicted, the SecY subunit contains ten TM segments. These are organized into N- and C-terminal domains, comprising TM 1–5 and 6–10 respectively (Figure 3). The two domains are connected at the back of the complex by the loop between TM 5 and TM 6. SecY displays pseudo-symmetry, such that its C-terminal domain is essentially an inverted version of its N-terminal domain. The domain organization of SecY and the locations of the two small subunits at the periphery leave one side of the complex (referred to as the front, Figure 3) as the only site that could open laterally towards lipid. Such a lateral gate is essential for the function of the SecY complex. Therefore, the complex can be likened to a clamshell that can open at the front towards lipid, with the hinge located at the back of the complex between TM 5 and TM 6. The SecE subunit could serve as a brace that prevents the two domains from separating completely.

The X-ray structure suggests that the channel pore is located at the centre of a single copy of the SecY complex,[4] rather than at the interface of three or four complexes.[32–35] Disulfide bridge formation between cysteines in a translocation substrate

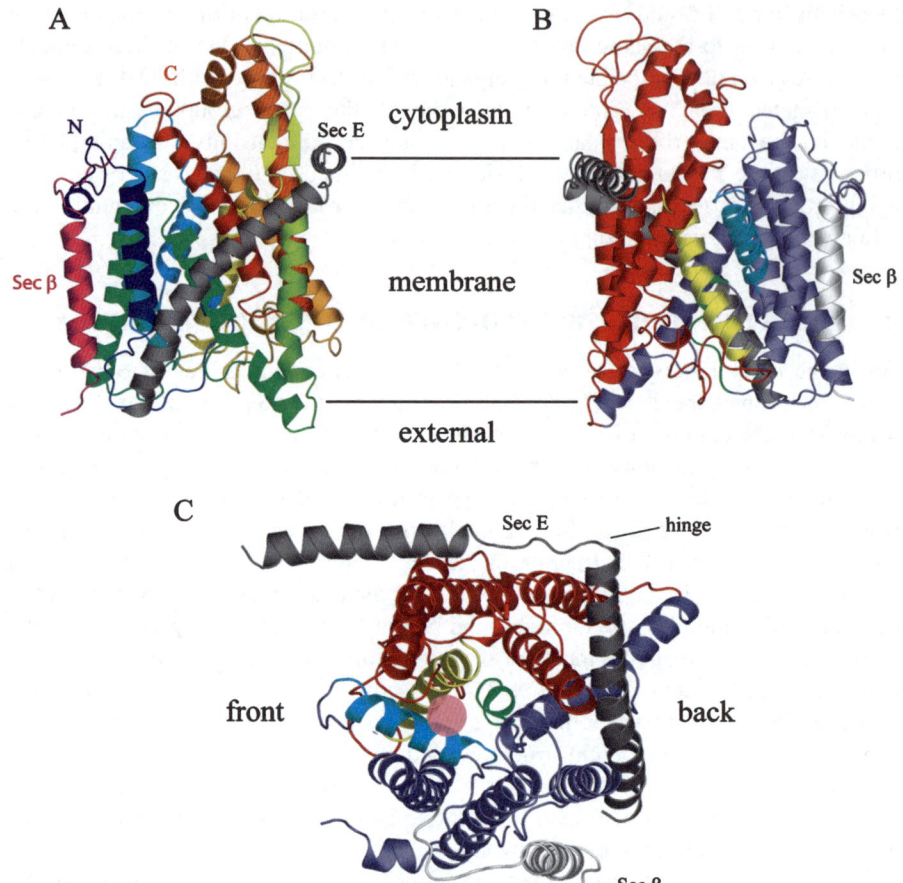

Figure 3 *(A) The structure of the M. jannaschii SecY complex viewed from the back. The individual helices from SecY are coloured from blue to red going from the N-terminus to the C-terminus. SecE is coloured grey, Secβ is coloured magenta. (B, C) SecY consists of two halves. (B) Front view of the SecY complex, with the N-terminal domain of SecY (TM 1–5) shown in blue, with the exception of TM 2b (cyan). The C-terminal domain (TM 6–10) is shown in red, with the exception of TM 7 (yellow). (C) The signal sequence of a precursor protein (shown as a magenta sphere) intercalates at the front of the SecY complex, between TM 2b and TM 7. The plug (TM 2a) in the centre of SecY, blocking the pore of the closed channel, is shown in green. The proposed hinge region between TM segments 5 and 6 is indicated*

and cysteines in SecY supports the notion that the polypeptide chain moves through the centre of a single SecY molecule.[36] In addition, almost all of the conserved residues in the SecY complex are located in the centre of the complex, and not at the periphery.[4] Mutations that allow proteins with defective or missing signal sequences to be transported (*prl* mutations[37,38]) are also located in the centre of the SecY complex.[4] Moreover, the lateral surfaces of the complex are entirely hydrophobic, and therefore would not be able to form an interfacial hydrophilic pore upon association in the membrane. Together, these observations suggest that the pore is contained within a single SecY complex. Indeed, the structure exhibits a partially hydrophilic cytoplasmic funnel that could mark the channel entrance. The funnel tapers to a close in the middle of the membrane and is blocked on the extra-cytoplasmic side by the presence of a small helical segment (TM 2a), coined the 'plug' (Figure 4A).[4] The crystal structure of the archaebacterial SecY complex therefore represents a closed channel, as expected from the fact that it was crystallized in the absence of translocation partners and substrate. Interestingly, the 'plug' domain and the short loop ('hinge') connecting TM 1–5 and 6–10 could also be identified as weak densities in the EM map.

Opening of the channel appears to require movement of the plug (Figure 4). Cysteines introduced into the plug and the TM segment of SecE of the *E. coli* SecY complex form a disulfide bridge *in vivo*,[39] suggesting that the plug moves towards the back of the complex into a cavity located on the extracellular side. This disulfide bridge formation can only be explained by assuming a considerable (more than 20 Å) movement of the plug. Not surprisingly, locking the channel into a permanently open state by inducing disulfide bridge formation is lethal to cells. The channel is probably in a dynamic equilibrium, with the plug moving between its position in the closed and open states. The unoccupied channel would be predominantly in the closed state, but the equilibrium could be shifted towards the open state; for example, by the binding of a signal sequence or, in the case of a membrane protein, by binding of a TM segment (see below).

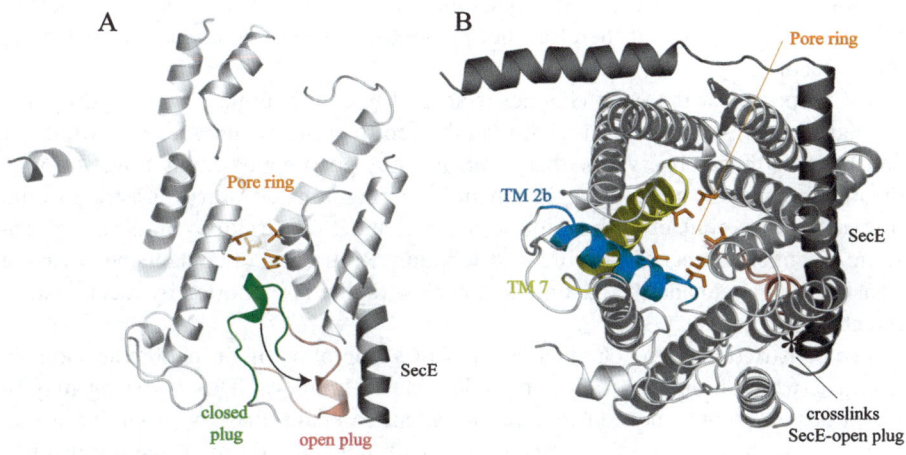

Figure 4 *Plug movement leads to opening of the SecY channel. (A) View from the side with the front half of the model cut away. The modelled movement of the plug towards the SecE subunit is indicated by an arrow, with the observed position of the closed plug shown in green and the hypothetical position of the open plug shown in pink. The side chains of residues in the pore ring are coloured in orange. (B) Cytosolic view with the plug modelled in its open position. TM 2b and TM 7 located at the front of the complex are shown in cyan and yellow respectively. The star indicates the region where introduced cysteines result in crosslinks between the plug and the TM segment of SecE*[39]

Crosslinking experiments have shown that the hydrophobic core of a signal sequence forms a short helix containing about two turns, which intercalates between TM 2 and TM 7 of Sec61/SecY at the front of the molecule and contacts phospholipids.[40] The translocation substrate is inserted as a loop, with the signal sequence intercalated into a 'lateral gate' in the channel wall, and the following polypeptide segment located in the pore proper. The X-ray structure is in excellent agreement with the crosslinking data, since the front of the complex between TM 2b and TM 7 is the only side that could open laterally to intercalate a signal sequence (Figure 3). In the case of a TM segment, which can be regarded as a very long and hydrophobic signal sequence, the lateral gate would span the total width of the membrane, composed of TM 2b and TM 3 on one side and TM 7 and TM 8 on the other. The intercalation would require a hinge motion at the back of Sec61/SecY to open the 'mouth' of the 'clamshell'. The separation of the two halves of the molecule might destabilize interactions that keep the plug in the centre of the molecule, thus promoting channel opening. In support of this model, many signal sequence suppressor mutations in SecY appear to destabilize the structure of the closed channel.[4] Once the signal sequence is inserted into the walls of the channel, the polypeptide segment distal to the signal sequence would be moving through the pore and prevent the plug from returning to its closed-state position.

The binding of a channel partner (SecA or the ribosome) may also regulate channel opening. Support for the notion that ribosomes destabilize the closed state of the channel comes from electrophysiological experiments, in which increased ion conductance is observed when a non-translating ribosome is bound to the channel.[1] The ribosome binds exclusively to the cytosolic loops located in the C-terminal half of Sec61/SecY,[41] and would therefore not prevent the separation of the two halves of the molecule.

We propose that the open channel is shaped like an hourglass, with hydrophilic funnels on both sides of a constriction in the centre of the membrane. This would be consistent with the observation that a translocating polypeptide chain moves through the membrane in an aqueous environment.[1,2] During translocation, a substrate would primarily make contact with residues at the channel constriction, minimizing substrate–channel interactions during translocation. Restriction of contacts between the translocating chain and the channel to a narrow region is supported by recent experiments.[36]

The constriction of the channel consists of a ring of six hydrophobic amino acid residues, which in many species are isoleucines (Figure 4).[4] This pore ring may fit like a gasket around the translocating polypeptide chain, thereby providing a seal that restricts the passage of ions and other small molecules during protein translocation. In this model, the membrane barrier would be provided by the channel itself, sensibly making it valid for all modes of translocation.

In addition to plug movement, widening of the pore is required to allow polypeptide chain translocation. The diameter of the pore ring as observed in the crystal structure is too small even to allow passage of an unfolded, extended polypeptide chain. Widening of the channel presumably occurs upon the intercalation of a signal sequence at the front of Sec61/SecY (opening of the 'clamshell'). In addition, further widening of the pore might occur by rearrangements of the helices which line the pore. Flexible glycine-rich sequences in the cytosolic loops between TM 4 and TM 5 and between TM 9 and TM 10 could allow the channel to accommodate movement of these helices. Pore widening would be required to explain the experimentally observed translocation of α-helices, a 13-residue disulfide-bonded polypeptide loop or domains with amino acid side chains modified to include bulky groups.[42,43] The flexibility of the pore region is supported by molecular dynamics simulations, which show that a ball of 10–12 Å or a helix with a diameter of 10 Å could move through the pore (P. Tian and I. Andricioaei; J. Gumbart and K. Schulten, personal communications).

Considerable controversy exists regarding the size of the open translocation channel. The estimated maximum dimensions the pore could attain based on the X-ray structure (~15 × 20 Å) would be much smaller than the estimate of a pore diameter of 40–60 Å, based on the observation that large cytoplasmic reagents can pass through the membrane channel to quench fluorescent probes in a nascent polypeptide chain.[44] The reason for the discrepancy between the structural data and the fluorescence quenching data is not clear. It is uncertain as to why a very large channel would be required, since (1) proteins are transported through the pore in a largely unfolded state as extended chains or α-helices and (2) the TM segments of membrane proteins are likely integrated into the lipid bilayer sequentially (*i.e.* there is no need to 'store' multiple TM segments in the channel). A large hydrophilic channel could only be generated if several Sec61/SecY molecules were associated with their fronts and fused their individual pores. However, at least for post-translational translocation in eubacteria, the functional translocation complex is thought to be a dimer with a back-to-back orientation of the individual SecY complexes (Figure 5), as observed in the 2D crystals of the *E. coli* SecY complex.[3] Additional support for this notion comes from crosslinking experiments[45] and from experiments in which a tandem SecY molecule was shown to be active in post-translational translocation.[46]

Thus, based on the structural data we favour a model in which the functional SecY complex forms an oligomer with several independent translocation pores. This immediately raises the question why oligomerization occurs. The answer is not yet known, but one possibility is that it serves to create binding sites for the recruitment of other components. These could include signal peptidase, which cleaves signal sequences from translocating polypeptides, oligosaccharyl transferase in eukaryotes and TRAM/YidC, which are multi-spanning membrane proteins that may serve as membrane chaperones (see below). All these proteins are close to the channel, but have no strong affinity for the isolated Sec61 complex. Oligomerization of the Sec61 complex might also regulate translocation partner binding (SecA and the ribosome). Oligomers could provide a larger number of binding interactions, resulting in stronger partner binding during translocation, while dissociation of the oligomers could weaken the interaction and facilitate partner release upon termination of translocation.[4]

5   Post-Translational Translocation in Bacteria

The mechanism by which the cytoplasmic ATPase SecA moves polypeptide chains through the SecY channel is still poorly understood, but some new insights have

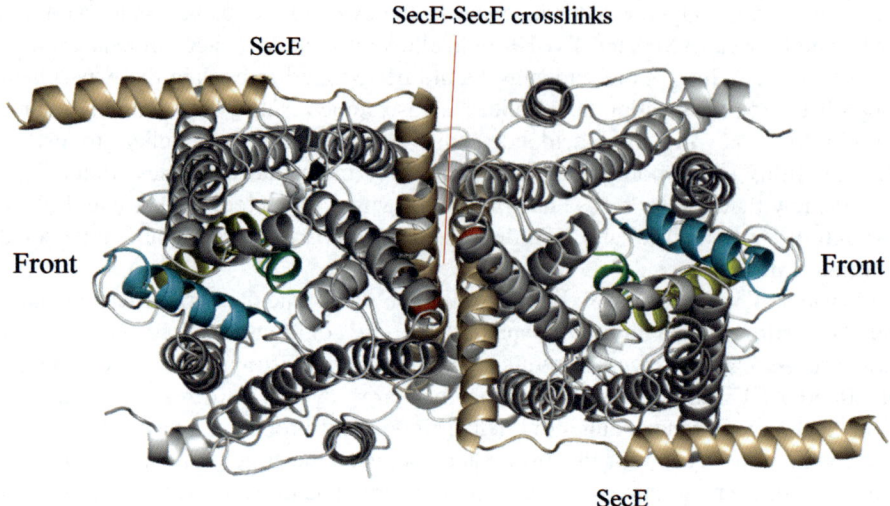

Figure 5 *Schematic model of the dimeric SecY complex thought to be the functional species in post-translational translocation, with a back-to-back arrangement of the individual SecY channels.[3,4] TM 2b and TM 7 at the front of the complexes are coloured in blue and yellow respectively. The plugs are shown in dark green. Cysteines introduced at the residues indicated in red result in efficient disulfide formation between two SecE subunits (light brown)[45]*

been provided by recent structural and biochemical studies. SecA exists in equilibrium between monomeric and dimeric states.[47–50] When isolated in solution, SecA is predominantly a dimer. Dissociation into monomers is stimulated upon interaction with ligands, such as lipids[48,49,51] or synthetic signal peptides, suggesting that the monomer is the active species in translocation. This is supported by the observations that crosslinked products corresponding to SecA dimers are lost upon interaction with the SecY complex, and that monomeric mutants of SecA retain significant translocation activity *in vitro* and *in vivo*.[48,52] Furthermore, a single copy of SecA is found in a complex containing an arrested translocation substrate and a SecY dimer.[53] Thus, although the exact nature of the complex during protein translocation is unclear, it is most likely that the active complex consists of one SecA associated with two SecY complexes or possibly two SecAs associated with four SecY complexes.[34,53,54] SecA-induced tetramerization of SecY complexes has indeed been observed by EM.[55]

Compared to the structure of the SecA dimer, the recently determined X-ray structure of monomeric SecA shows an open conformation, in which several domains have undergone dramatic movements, while the nucleotide-binding fold domains remain at the same position.[56] In this open conformation, SecA displays a large groove that is the likely polypeptide-binding site; this idea has support from crosslinking and mutagenesis studies.[57,58] The groove is similar in dimensions to that seen in other peptide-binding proteins such as OppA, DnaK and SecB.[59–61] It is still unclear whether SecA binds signal sequences in a more specific way.

It is likely that SecA pushes the polypeptide substrate through the SecY channel, as originally proposed,[17] but it is not clear how this happens. A pushing mechanism implies that there are two polypeptide-binding sites that alternate in their affinities for the polypeptide substrate and that can move relative to each other. One possibility is that both sites are located in SecA, similar to helicases. However, since only one peptide-binding groove is apparent from the SecA structures, it is possible that SecY provides the second binding site.

It has been proposed that SecA inserts deeply into the SecY channel, reaching the other side of the membrane.[17,62-65] This mechanism has been inferred from the fact that SecA is accessible to proteases and labelling reagents added from the outside of the cell. However, the insertion model is unlikely since (1) the structural data indicate that SecA is too large to insert into the channel and (2) the modification sites are spread out over the entire SecA molecule.[66] Thus, the previous data are probably better explained by assuming that SecA adopts a protease resistant conformation upon SecY binding[67] and is accessible to labelling reagents through the open SecY channel.

5.1 A Model for Post-Translational Protein Translocation in Bacteria

The X-ray and EM structures allow us to propose the following model for the translocation of secretory proteins. Initially, the SecY channel (orange) is closed because the plug (green) blocks the pore (Figure 6, stage 1). Next, a channel partner binds, in this case SecA (Figure 6, stage 2). Although part of an oligomer (here a dimer is shown), only one copy of the SecY complex, defined by the orientation of the bound partner, forms the active pore. The other copy of the SecY complex may form additional binding surfaces for SecA. The closed state of the 'active' channel may be destabilized by the interactions with the partner. In the next step, the substrate inserts as a loop into the channel, with its signal sequence (red) intercalated at the front of the SecY complex between TM 2b and TM 7, and with its mature region in the pore (Figure 6, stage 3). Insertion requires a hinge motion to separate TM 2b and TM 7, and leads to pore widening and displacement of the plug to its open position close to the γ-subunit. The mature region of the polypeptide chain is then transported through the pore, and the signal sequence is cleaved at some point during translocation (Figure 6, stage 4). In this model SecA pushes the polypeptide into the SecY channel by a mechanism that is unclear, but without inserting deeply into the channel. During polypeptide movement, the pore ring (purple) forms a seal around the chain, hindering the permeation of other molecules. Finally, when the polypeptide has passed through, SecA dissociates and the plug returns to its closed-state position (Figure 6, stage 5).

6 Conclusions and Outlook

Structural studies of ribosome/Sec61 complexes, of SecA and particularly of the SecY channel have significantly advanced our understanding of the mechanism of

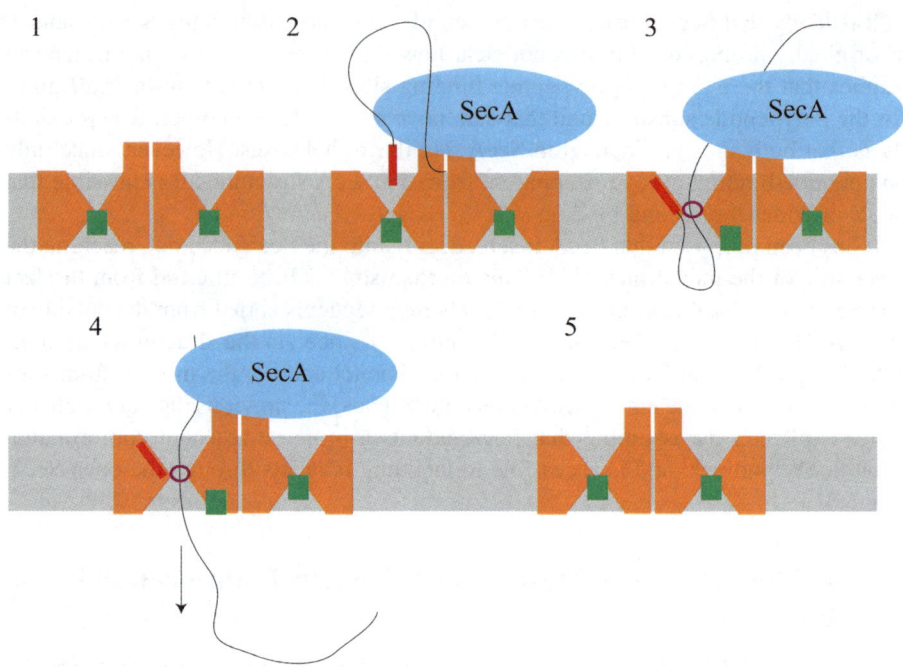

Figure 6 *Scheme showing SecA-mediated post-translational translocation in bacteria. See text for details*

protein translocation. Interpretation of these structures has been made possible by genetic and biochemical data accumulated in many laboratories over the years. The recent structural data have led to a number of new hypotheses that need to be tested experimentally. These include: (1) How does SecA move polypeptides through the SecY channel? (2) How are the individual SecY complexes organized within the oligomeric assembly in co- and post-translational translocation? (3) What is the role of the oligomerization of the Sec61/SecY channel? (4) How do interacting partners of the Sec61/SecY channel regulate its function? (5) How are membrane proteins integrated and folded? Answering these questions will require further structural studies, with the architecture of an active channel (bound to a translocation partner and with a polypeptide substrate in the pore) determined at high resolution being the most important goal for the future.

References

1. S.M. Simon and G. Blobel, *Cell*, 1991, **65**, 371.
2. K.S. Crowley, G.D. Reinhart and A.E. Johnson, *Cell*, 1993, **73**, 1101.
3. C. Breyton, W. Haase, T.A. Rapoport, W. Kühlbrandt and I. Collinson, *Nature*, 2002, **418**, 662.

4. B. van den Berg, W.M. Clemons Jr., I. Collinson, Y. Modis, E. Hartmann, S.C. Harrison and T.A. Rapoport, *Nature*, 2004, **427**, 36.

5. T.A. Rapoport, B. Jungnickel and U. Kutay, *Annu. Rev. Biochem.*, 1996, **65**, 271.

6. W. Mothes, S. Prehn and T.A. Rapoport, *EMBO J.*, 1994, **13**, 3937.

7. L. Brundage, J.P. Hendrick, E. Schiebel, A.J.M. Driessen and W. Wickner, *Cell*, 1990, **62**, 649.

8. J. Akimaru, S.I. Matsuyama, H. Tokuda and S. Mizushima, *Proc. Natl. Acad. Sci. USA*, 1991, **88**, 6545.

9. D. Gorlich and T.A. Rapoport, *Cell*, 1993, **75**, 615.

10. D.T. Ng, J.D. Brown and P. Walter, *J. Cell Biol.*, 1996, **134**, 269.

11. K.E. Matlack, B. Misselwitz, K. Plath and T.A. Rapoport, *Cell*, 1999, **97**, 553.

12. R.J. Deshaies, S.L. Sanders, D.A. Feldheim and R. Schekman, *Nature*, 1991, **349**, 806.

13. S. Panzner, L. Dreier, E. Hartmann, S. Kostka and T.A. Rapoport, *Cell*, 1995, **81**, 561.

14. H.A. Meyer, H. Grau, R. Kraft, S. Kostka, S. Prehn, K.U. Kalies and E. Hartmann, *J. Biol. Chem.*, 2000, **275**, 14550.

15. J. Tyedmers, M. Lerner, C. Bies, J. Dudek, M.H. Skowronek, I.G. Haas, N. Heim, W. Nastainczyk, J. Volkmer and R. Zimmermann, *Proc. Natl. Acad. Sci. USA*, 2000, **97**, 7214.

16. H. Mori and K. Ito, *Trends Microbiol.*, 2001, **9**, 494.

17. A. Economou and W. Wickner, *Cell*, 1994, **78**, 835.

18. R. Ortenberg and M. Mevarech, *J. Biol. Chem.*, 2000, **275**, 22839.

19. V. Irihimovitch and J. Eichler, *J. Biol. Chem.*, 2003, **278**, 12881.

20. W. Kühlbrandt, *Q. Rev. Biophys.*, 1992, **25**, 1.

21. W. Kühlbrandt, D. Wang and Y. Fujiyoshi, *Nature,* 1994, **367**, 614.

22. R. Henderson, J.M. Baldwin, T.A. Ceska, F. Zemlin, E. Beckmann and K.H. Downing, *J. Mol. Biol.,* 1990, **213**, 899.

23. K. Murata, K. Mitsuoka, T. Hirai, T. Walz, P. Agre, J. Heymann, A. Engel and Y. Fujiyoshi, *Nature*, 2000, **407**, 599.

24. T. Gonen, P. Sliz, J. Kistler, Y. Cheng and T. Walz, *Nature*, 2004, **429**, 193.

25. K. Rhee, E. Morris, J. Barber and W. Kühlbrandt, *Nature*, 1998, **396**, 283.

26. M. Auer, G. Scarborough and W. Kühlbrandt, *Nature*, 1998, **392**, 840.

27. K. Williams, *Nature*, 2000, **403**, 112.

28. L.M. Guzman, D. Belin, M.J. Carson and J. Beckwith, *J. Bacteriol.*, 1995, **177**, 4121.

29. I. Collinson, C. Breyton, F. Duong, C. Tziatzios, D. Schubert, E. Or, T.A. Rapoport and W. Kühlbrandt, *EMBO J.*, 2001, **20**, 2462.

30. L. van den Berg, W.M.J. Clemons, I. Collinson, Y. Modis, E. Hartmann, S.C. Harrison and T.A. Rapoport, *Nature*, 2004, **427**, 36.

31. P. Bessonneau, V. Besson, I. Collinson and F. Duong, *EMBO J.*, 2002, **21**, 995.

32. C. Breyton, W. Haase, T.A. Rapoport, W. Kühlbrandt and I. Collinson, *Nature*, 2002, **418**, 662.

33. R. Beckmann, D. Bubeck, R. Grassucci, P. Penczek, A. Verschoor, G. Blobel and J. Frank, *Science*, 1997, **19**, 2123.

34. E.H. Manting, C. van der Does, H. Remigy, A. Engel and A.J. Driessen, *EMBO J.*, 2000, **19**, 852.

35. D.G. Morgan, J.F. Menetret, A. Neuhof, T.A. Rapoport and C.W. Akey, *J. Mol. Biol.*, 2002, **324**, 871.

36. K.S. Cannon, E. Or, W.M. Clemons Jr., Y. Shibata and T.A. Rapoport, *J. Cell Biol.*, 2005, **169**, 219.

37. K.L. Bieker, G.J. Phillips and T.J. Silhavy, *J. Bioenerg. Biomembr.*, 1990, **22**, 291.

38. A.I. Derman, J.W. Puziss, P.J. Bassford and J. Beckwith, *EMBO J.*, 1993, **12**, 879.

39. C.R. Harris and T.J. Silhavy, *J. Bacteriol.*, 1999, **181**, 3438.

40. K. Plath, W. Mothes, B.M. Wilkinson, C.J. Stirling and T.A. Rapoport, *Cell*, 1998, **94**, 795.

41. D. Raden, W. Song and R. Gilmore, *J. Cell Biol.*, 2000, **150**, 53.

42. K. Tani, H. Tokuda and S. Mizushima, *J. Biol. Chem.*, 1990, **265**, 17341.

43. J. de Keyzer, C. van der Does and A.J. Driessen, *J. Biol. Chem.*, 2002, **277**, 46059.

44. B.D. Hamman, J.C. Chen, E.E. Johnson and A.E. Johnson, *Cell*, 1997, **89**, 535.

45. A. Kaufmann, E.H. Manting, A.K. Veenendaal, A.J. Driessen and C. van der Does, *Biochemistry*, 1999, **38**, 9115.

46. F. Duong, *EMBO J.*, 2003, **22**, 4375.

47. R.L. Woodbury, S.J. Hardy and L.L. Randall, *Protein Sci.*, 2002, **11**, 875–882.

48. E. Or, A. Navon and T. Rapoport, *EMBO J.*, 2002, **21**, 4470.

49. J. Benach, Y.T. Chou, J.J. Fak, A. Itkin, D.D. Nicolae, P.C. Smith, G. Wittrock, D.L. Floyd, C.M. Golsaz, L.M. Gierasch and J.F. Hunt, *J. Biol. Chem.*, 2003, **278**, 3628.

50. H. Ding, J.F. Hunt, I. Mukerji and D. Oliver, *Biochemistry*, 2003, **42**, 8729.

51. Z. Bu, L. Wang and D.A. Kendall, *J. Mol. Biol.*, 2003, **332**, 23.

52. E. Or, D. Boyd, S. Gon, J. Beckwith and T. Rapoport, *J. Biol. Chem.*, 2005, **280**, 9097.

53. F. Duong, *EMBO J.*, 2003, **22**, 4375.

54. C. Tziatzios, D. Schubert, M. Lotz, D. Gundogan, H. Betz, H. Schagger, W. Haase, F. Duong and I. Collinson, *J. Mol. Biol.*, 2004, **340**, 513.

55. A.K. Veenendaal, C. van der Does and A.J. Driessen, *Biochim. Biophys. Acta*, 2004, **1694**, 81.

56. A.R. Osborne, W.M. Clemons Jr. and T.A. Rapoport, *Proc. Natl. Acad. Sci. USA*, 2004, **101**, 10937.

57. E. Kimura, M. Akita, S. Matsuyama and S. Mizushima, *J. Biol. Chem.*, 1991, **266**, 6600.

58. L. Kourtz and D. Oliver, *Mol. Microbiol.*, 2000, **37**, 1342.

59. S.H. Sleigh, P.R. Seavers, A.J. Wilkinson, J.E. Ladbury and J.R. Tame, *J. Mol. Biol.*, 1999, **291**, 393.

60. X. Zhu, X. Zhao, W.F. Burkholder, A. Gragerov, C.M. Ogata, M.E. Gottesman and W.A. Hendrickson, *Science*, 1996, **272**, 1606.

61. Z. Xu, J.D. Knafels and K. Yoshino, *Nat. Struct. Biol.*, 2000, **7**, 1172.

62. Y.J. Kim, T. Rajapandi and D. Oliver, *Cell*, 1994, **78**, 845.

63. J. Eichler and W. Wickner, *Proc. Natl. Acad. Sci. USA*, 1997, **94**, 5574.

64. C. van der Does, T. den Blaauwen, J.G. de Wit, E.H. Manting, N.A. Groot, P. Fekkes and A.J. Driessen, *Mol. Microbiol.*, 1996, **22**, 619.
65. V. Ramamurthy and D. Oliver, *J. Biol. Chem.*, 1997, **272**, 23239.
66. J.F. Hunt, S. Weinkauf, L. Henry, J.J. Fak, P. McNicholas, D.B. Oliver and J. Deisenhofer, *Science*, 2002, **297**, 2018.
67. C. van der Does, E.H. Manting, A. Kaufmann, M. Lutz and A.J. Driessen, *Biochemistry*, 1998, **37**, 201.

Structure and Function of the Translocator Domain of Bacterial Autotransporters

PETER VAN ULSEN[1], PIET GROS[2] AND JAN TOMMASSEN[1]

[1]Department of Molecular Microbiology and Institute of Biomembranes, Utrecht University, Padualaan 8, 3584 CH Utrecht, The Netherlands
[2]Department of Crystal and Structural Chemistry, Bijvoet Center for Biomolecular Research, Utrecht University, Padualaan 8, 3584 CH Utrecht, The Netherlands

1 Introduction

The cell envelope of Gram-negative bacteria consists of two membranes, the inner and outer membranes (IM and OM, respectively). The OM not only protects the bacterial cell from the environment, but also presents a barrier for proteins that are to be secreted into the external milieu.[1] To overcome this barrier, many different transport machineries, often consisting of multiple components, have evolved. In contrast, the autotransporter secretion mechanism appears, at first sight, to be a simple secretion system with the secreted protein mediating its own secretion across the OM.[2]

Autotransporters are modular proteins that consist of an N-terminal signal sequence, a secreted passenger domain and a C-terminal translocator domain (Figure 1). The signal sequence directs the autotransporter to the Secretion machinery which executes transport across the IM;[3,4] whereas, the translocator domain mediates transport of the passenger across the OM. The mechanism of OM passage is under debate[5] and will be discussed later in this chapter. After OM passage, the passenger either remains attached to the cell surface or is released into the extracellular milieu, often through autoproteolytic cleavage facilitated by a serine-protease domain in the N-terminal half of the passenger (Figure 1). The name autotransporter was coined because of the apparent absence of a dedicated secretion machine.

On the basis of sequence homology, autotransporters have been identified in the genome sequences of virtually all pathogenic Gram-negatives available to date. The functions of the secreted passengers are very often unknown, but functions that have

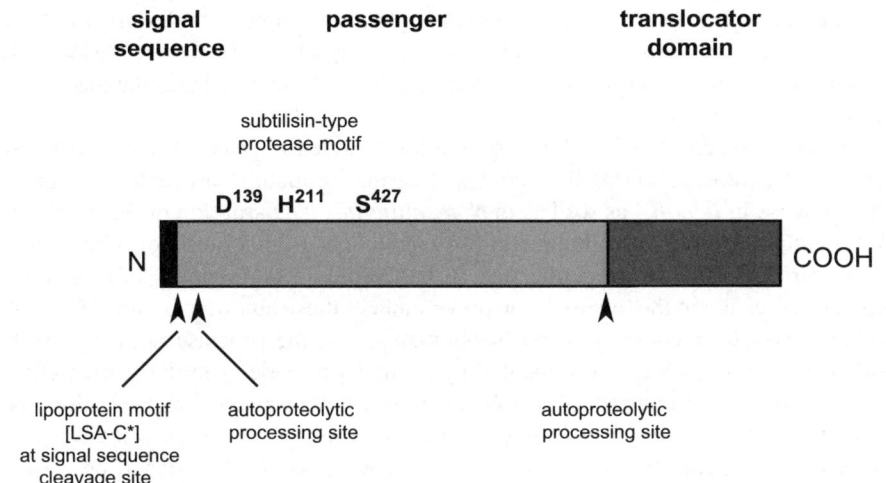

Figure 1 *Domain organization of the NalP autotransporter of N. meningitidis. With a signal sequence, a passenger and a translocator domain present, NalP shows the classical autotransporter domain organization. At the end of the NalP signal sequence, a lipoprotein box is present. The passenger domain contains a subtilisin-type protease motif, the active-site residues of which are indicated. Autoproteolytic cleavage at the indicated sites releases the passenger domain from the lipid moiety and the translocator domain, resulting in its secretion into the medium*

been elucidated clearly suggest that the secreted proteins contribute to the virulence of the organisms.[6] The very first autotransporter discovered was the IgA protease of *Neisseria gonorrhoeae*,[7] which proteolytically cleaves human secretory IgA. Other examples are the adhesins AIDA-I of *Escherichia coli* and Pertactin of *Bordetella pertussis*, the cytotoxin VacA of *Helicobacter pylori,* and the IcsA protein of *Shigella flexneri*, which polymerizes actin.[6]

2   The NalP Autotransporter

NalP is an autotransporter of the Gram-negative pathogen *N. meningitidis*, the causative agent of meningitis and sepsis – diseases with high mortality and morbidity – which particularly affect young children.[8] The NalP autotransporter was identified in a homology-based screen of the sequenced genomes of *N. meningitidis* for genes putatively encoding autotransporters.[9] It has the classical signal sequence, passenger and translocator domain architecture of an autotransporter (Figure 1). A subtilisin-type serine-protease domain is present in the passenger with the active-site Ser at position 427 (Figure 1). At the C-terminal end of the signal sequence, the protein contains a lipoprotein motif Leu-Ser-Ala-Cys, where Cys is the first residue after the signal sequence. Lipidation at the Cys of this motif was indeed demonstrated in *E. coli* using a chimaeric protein consisting of an N-terminal fragment of NalP, including the signal sequence, fused to β-lactamase[10] and in *N. meningitidis* after overproduction of NalP (P. van Ulsen, unpublished results).

Lipidation is exceptional among autotransporters; it prompted the designation NalP, which stands for neisserial autotransporter lipoprotein. However, despite the presence of this hydrophobic moiety, NalP is efficiently secreted into the medium of *N. meningitidis*.[10]

The protease activity of NalP is required for the release of the passenger from the cell surface. Substitution of the active-site serine by alanine prevented autoproteolytic release in *E. coli*[11] as well as in *N. meningitidis*.[10] Disruption of the *nalP* gene in *N. meningitidis* revealed that the protein modulates the secretion of other neisserial autotransporters, in particular that of IgA protease and App.[10] NalP-mediated proteolysis results in the release of novel variants of these autotransporters. The IgA-protease passenger consists of two subdomains, *i.e.*, the protease domain and the 40-kDa α-peptide, which are separated by a small γ-peptide.[7] Similarly, the passenger domain of App consists of a domain with a protease motif and an additional α-peptide.[10] In *N. meningitidis*, autocatalytic processing results in the release of only the protease domain of App and IgA protease, whereas NalP-mediated processing results in the release of the protease domains extended with the α-peptides.[10] Both α-peptides contain nuclear localization signals[10,12] that have the potential to redirect the attached protease domain to a different target. For example, it has been shown that the IgA protease α-peptide targets fused proteins to the nucleus of cultured epithelial cells.[12] Therefore, NalP-mediated processing of neisserial autotransporters may have important implications for infection.

3   The Translocator Domain of Autotransporters

The autotransporter translocator domain is a 25–30-kDa domain that is inserted into the OM.[13,14] In some autotransporters, however, the translocator domain is extended with a 10–15-kDa domain that is sensitive to extracellularly added proteases and, therefore, cell-surface exposed.[13,15] Secondary structure predictions suggested that translocator domains adopt a β-barrel conformation common to OM proteins (OMPs).[13,14] Outer membrane proteins consist of antiparallel amphipathic β-strands that form a β-barrel with the hydrophobic residues facing the membrane lipids.[16,17] These proteins are further characterized by a high content of aromatic amino acid residues, located at positions in the β-strands that are close to the head groups of the membrane lipids. Although the amino acid sequences of the autotransporter translocator domains are not very similar, they all have a high number of aromatic amino acid residues, and their secondary structure predictions suggest that they contain 14 amphipathic β-strands.[14] Furthermore, the three C-terminal residues are conserved in both autotransporters and OMPs[2,18] and were shown to be essential for OM insertion of the OMP PhoE[18] and the translocator domain of the autotransporter Hap of *Haemophilus influenzae*.[19] Autotransporter secretion is also dependent on the Omp85 complex, the machinery for the assembly of integral OMPs.[20] *N. meningitidis* depleted of this essential protein accumulated full-length autotransporters and did not secrete the passengers.[20] However, in contrast to the assembly of OMPs, the secretion of autotransporters, and thus the folding and OM insertion of the translocator domains, seems to occur independent of the periplasmic chaperones Skp and SurA.[21]

The translocator domain mediates the transport of the passenger across the OM to the cell surface.[13] In analogy to the OMPs, the β-barrel configuration of the translocator domain could result in an aqueous channel inside the barrel. This presumption led to a model in which the passenger is transported to the cell surface through this channel.[7,13] The model, however, poses several questions. First, all elucidated OMP structures show an even number of β-strands with the N and C termini located in the periplasm: How then is the N-terminally located passenger directed to the cell surface? Second, does the process require energy and, if so, what is the source of energy? ATP is absent in the periplasm and there is also no proton gradient across the OM. Finally, is the translocation channel wide enough to allow for the passage of even folded protein subdomains? In an attempt to address these questions, we have recently solved the crystal structure of the translocator domain of the autotransporter NalP.[5]

4 Purification and *In Vitro* Folding of the NalP Translocator Domain

Solving the structure of membrane proteins is notoriously difficult because of the amphiphilic nature of the molecules. However, in recent years, considerable progress has been made in solving the structures of OMPs. To date, the structures of 30 β-barrel membrane proteins have been solved (listed at: http://www.mpibp-frankfurt.mpg.de/michel/public/memprotstruct.html; also see Table 1 for an overview of recently solved structures). A major problem for membrane proteins is to obtain sufficient amount of correctly folded protein for either NMR spectroscopy or crystallization experiments. Two strategies can be followed to obtain purified OMPs:[33] they can be purified from solubilized membranes or folded *in vitro* from denatured proteins. In the first method, OMPs are purified in their native conformation from the OM. However, expression levels of OMPs are not very high in most cases. Moderate overexpression in *E. coli* is usually possible, but high overexpression is often lethal. Furthermore, the purification of the protein after extraction can be tedious. Alternatively, the OMP gene can be cloned without the part encoding the signal sequence, and overexpressed in *E. coli* using constructs controlled by strong promoters. In most cases, this strategy results in large amounts of the desired protein (in some cases up to 50% of the total cellular protein) that accumulate in insoluble inclusion bodies in the cytoplasm. After breaking the cells, for example by ultrasonication, the inclusion bodies can easily be collected by centrifugation, resulting in high quantities of rather pure proteins. However, proteins from inclusion bodies must be folded *in vitro*, which can be problematic. The most common approach for *in vitro* folding is to dilute denatured protein, obtained by solubilizing the inclusion bodies in 7–8M urea or 6M guanadinium–HCl, in a buffer containing either lipids or detergents to mimic the membrane environment[34] (Table 1). So far, no crystal structures have been derived from proteins that were folded *in vitro* in lipids, but the use of detergents has resulted in the resolution of several OMP structures (Table 1). Specific experimental details differ considerably depending on the protein and are subject to trial and error.[34] For example, the detergent concentrations used may vary

Table 1 Recently solved structures of OMPs of Gram-negative bacteria*

Protein/Organism	Function	Obtaining the protein		Crystallization conditions†	PDB	References
		Method	Buffer‡			
OmpT E. coli	Protease	Dilution of urea-denatured inclusion bodies	25 mM 3-dimethyl-dodecylammonio-propane-sulfonate, 10 mM glycine (pH 8.3)	d.: Octyl-ß-D-glucopyranoside a.: None p.: 2-Methyl-2,4-pentanediol	1I78	22
FecA E. coli	Ferric citrate receptor	Overproduction in E. coli followed by solubilization in detergent	Triton X-100, buffer unknown	d.: Lauryldimethylamine oxide a.: Heptane-1,2,3-triol p.: Polyethylene glycol 1000	1KMO 1KMP	23
FecA E. coli	Ferric citrate receptor	Overproduction in E. coli followed by solubilization in detergent	5% (v/v) Elugent, buffer unknown	d.: Lauryldimethylamine oxide a.: Heptane-1,2,3-triol p.: Polyethylene glycol 1000	1PNZ 1PO0 1PO3	24
OpcA N. meningitidis	Adhesin	Solubilized from N.meningitidis membranes and dilution of guanidine HCL-denatured inclusion bodies	50 mM bis-tris propane HCl (pH 7.0), 250 mM NaCl, 5% lauryldimethylamine oxide	d.: Decylpentaoxyethylene a.: Heptyl-ß-D-glucoside p.: Polyethylene glycol 4000	1K24	25
BtuB E. coli	Cobalamin receptor	Solubilization of E. coli membranes in detergent and purification	50 mM Tris (pH 6.9), 2mM EDTA, 150 mM n-octyl tetraoxyethylene	d.: Octyltetraoxyethylene a.: None p.: Polyethylene glycol 3350	1NQF 1NQE 1NQG 1NQH	26

Protein / Source	Function	Method	Solubilization buffer	Crystallization (d./a./p.)	PDB	Ref.
NspA *N. meningitidis*	Unknown	Dilution of urea-denatured inclusion bodies	20 mM ethanolamine (pH 11), 1% 3-dimethyldodecyl-ammoniopropane-sulfonate	d.: Decylpentaoxyethylene a.: Isopropanol p.: Polyethylene glycol 3000	1P4T	27
PagP *E. coli*	Acyltransferase	Dilution of guanadine HCl-denatured inclusion bodies	0.5% lauryldimethylamine oxide, 10 mM Tris–HCl (pH 8.0)	d.: Lauryldimethylamine oxide a.: None p.: 2-Methyl-2,4-pentanediol	1THQ	28
FadL *E. coli*	Long-chain fatty acid receptor	Solubilization of *E. coli* membranes in detergent and purification	1% lauryldimethylamine oxide, 1% octyl glucoside, 20 mM Tris–HCl (pH 7.8), 300 mM NaCl, 10% glycerol	d.: Octyltetraoxyethylene plus traces of lauryldimethylamine oxide a.: None p.: Polyethylene glycol 4000 or magnesium formate	1T16 1T1L	29
Tsx *E. coli*	Nucleoside receptor	Solubilization of *E. coli* membranes in detergent and purification	1% lauryldimethylamine oxide, 1% β-octyl glucoside	d.: Octyltetraoxyethylene a.: None p.: Polyethylene glycol monomethylether 550	1TLY 1TLW 1TLZ	30
OprM *P. aeruginosa*	TolC-like channel protein	Solubilization of *P. aeruginosa* membranes in detergent and purification	2.5 % n-octyl-D-glucopyranoside, 20 mM Na-phosphate, 20 mM imidazole, 0.3 M NaCl	d.: Cyclohexylpropyl-ß-D-maltoside, octylpolyoxyethylene a.: None p.: 2-Methyl-2,4-pentanediol	1WP1	31

(continued)

Table 1 (continued)

Protein/Organism	Function	Obtaining the protein		Crystallization conditions†	PDB	References
		Method	Buffer‡			
NalP-TD *N. meningitidis*	Autotransporter translocator domain	Dilution of urea-denatured inclusion bodies	0.5% 3-dimethyldodecyl-ammoniopropane-sulfonate, 20 mM Tris–HCl (pH 8.0), 1M NaCl	d.: Decylpentaoxyethylene a.: Heptyl-ß-D-glucopyranoside p.: Polyethylene glycol 1000 or polyethylene glycol monomethylester 2000 plus 2-methyl-2,4-pentanediol	1UYN 1UYO	5
FpvA *P. aeruginosa*	Pyoverdine receptor	Solubilization of *P. aeruginosa* membranes in detergent and purification	1% Zwittergent 3-14, 50 mM Tris-HCl (pH 8.0)	d.: Octylpentaoxyethylene a.: None p.: Polyethylene glycol 4000	1XKH	32
VceC *V. cholerae*	TolC-like channel protein	Solubilization of *E. coli* membranes in detergent and purification	2% ß-dodecyl-maltoside, 20 mM Na$_2$HPO$_4$ (pH 7.4), 300 mM NaCl, 10% glycerol, 0.5 mM Tris (hydroxypropyl) phosphine	d.: Octyl-ß-glucoside a.: None p.: 2,4-Methyl-pentanediol	1YC9	33

Note: * The Table lists proteins whose structure was solved by X-ray crystallography after the appearance of the review by Buchanan.[34] The reader is also referred to the website: http://www.mpibp-frankfurt.mpg.de/michel/public/memprostruct.html, maintained by Dr H. Michel.
† Listed are the detergent (d.), the additive (a.) and the precipitant (p.).
‡ The buffer conditions listed are used either for *in vitro* folding or for solubilization of the membranes, as indicated by the methods.

from 4 to 500 times the critical micelle concentration, dilution factors may range from 2- to 50-fold and also the type of detergent may be crucial. After successful *in vitro* folding, it may be necessary to exchange the detergent used for folding with the one better suited for crystallization experiments.

The crystals of the NalP translocator domain, used to solve the structure, were obtained using *in vitro* folded protein material.[5] The autocatalytic processing site of NalP is not known, but, based upon sequence alignments of the C-terminal parts of a large set of autotransporters,[14,35] the translocator domain was estimated to consist of residues Asp^{777} to Phe^{1084}.[5] The DNA encoding this fragment was cloned and expressed in *E. coli*, and the resulting inclusion bodies were purified. On SDS-PAGE, the recombinant protein had a similar apparent molecular weight as the native translocator domain detected in cell envelope preparations of *N. meningitidis*.[5] Furthermore, the native translocator domain showed heat-modifiability, *i.e.*, the heat-denatured form of the protein has a different electrophoretic mobility on polyacrylamide gels containing a lower percentage of SDS than the folded protein. This property is common to OMPs[36] and has also been observed for the translocator domain of the *E. coli* autotransporter AIDA-I.[15] Therefore, *in vitro* folding of the denatured recombinant NalP translocator domain could be monitored by assessing heat-modifiability. Successful folding was achieved by diluting the NalP translocator domain, denatured in 7M urea, 20-fold into a buffer containing the detergent 3-dimethyldodecylammoniopropane-sulfonate[5] (Table 1). For the crystallization experiments, the detergent was exchanged for decylpentaoxyethylene during anion-exchange chromatography. Crystals were grown by the hanging-drop vapour diffusion method using 0.06% (w/v) decylpentaoxyethylene, 0.5% (w/v) heptyl-β-glucopyranoside, 9% polyethylene glycol 1000, 200 mM lithium sulfate and 100 mM sodium citrate buffer (pH 4.0) at 28 °C.[5]

5 The Structure of the NalP Translocator Domain

The structure of the NalP translocator domain (Figure 2) shows a β-barrel (residues 819–1084) with its hydrophilic pore occupied by an α-helix. The helix spans the entire pore from the periplasm to the extracellular side.[5] Overall, the protein shares several features with other OMPs.[16,17] The β-barrel consists of 12 antiparallel amphipathic β-strands, with short connecting turns at the periplasmic side and longer connecting loops on the side facing the medium. Aromatic residues are located in two girdles at positions that suggest they align with the lipid head groups (Figure 2). In the absence of the helix, the pore inside the channel has a size of 10 × 12 Å.

The interior of the β-barrel is hydrophilic because of 20 charged residues that point inwards.[5] The α-helix that occupies the aqueous channel is, in the primary structure, located N-terminally from the barrel domain. The α-helix interacts with the barrel wall through seven salt bridges, 16 hydrogen bonds and numerous van der Waals contacts. Its position in the channel is slightly off-centre, and charged residues cluster to one side. Consequently, the helix does not completely block the channel. A narrow water-filled channel remains open connecting the periplasm to the medium.

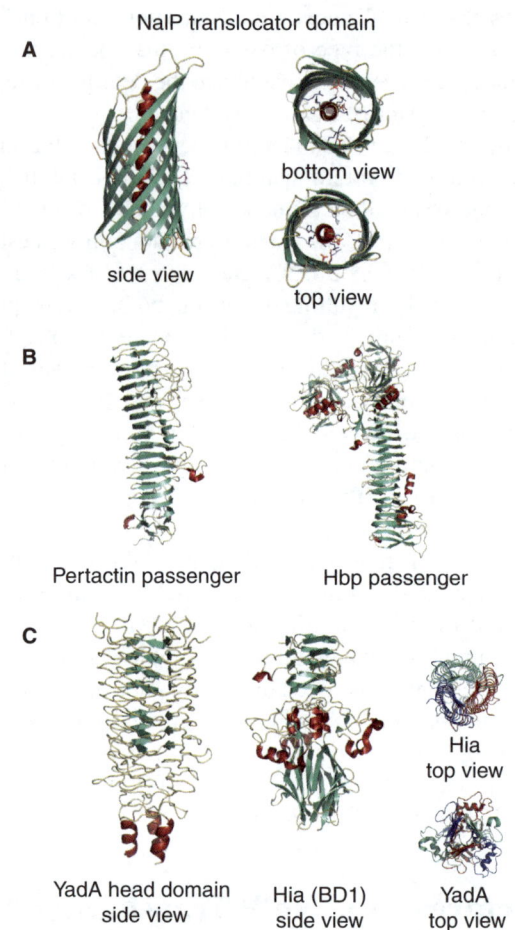

NalP translocator domain

A

side view   bottom view

top view

B

Pertactin passenger   Hbp passenger

C

YadA head domain   Hia (BD1)   YadA
side view   side view   top view

Hia
top view

Figure 2 *Solved crystal structures of autotransporter protein domains. (A) The NalP translocator domain (PDB code 1UYN). In the side view, the aromatic residues and the
hydrophilic patch pointing towards the membrane are indicated in sticks in yellow
and in purple, respectively. The top view represents the view from the medium onto
the membrane; the bottom view represents the view from the periplasm into the
membrane. Here, the charged residues within the barrel are indicated in sticks with
positive residues in blue and negative ones in red. (B) Side views of the structures
of the autotransporter passenger pertactin (1DAB) and Hbp (1WXR). (C) Side and
top views of the subdomains of the OCA autotransporter passengers YadA (1P9H)
and Hia (1S7M). In the top views, the monomers of the trimeric domains are
coloured blue, red and green. All images were made using Pymol*

The channels observed in the crystal structure of the NalP translocator domain
were in complete agreement with the results of biophysical pore-activity measurements.[5] The *in vitro* folded NalP translocator domain was reconstituted in planar
lipid bilayers, and pore activity was monitored by measuring the current after applying a transmembrane potential. In these experiments, the NalP translocator domain

showed openings and closings of pores of two sizes with single-channel conductances of 0.15 nS (frequent) and 1.3 nS (rare). From these measurements, the diameter of the NalP translocator domain pore was estimated to be 2.4 Å for the 0.15 nS channels and 8.4 Å for the 1.3 nS channels. By approximation, these channel sizes match those observed in the crystal structure, with the smaller channels representing the pore when obstructed by the α-helix and the larger ones with the α-helix displaced. Similar pore measurements were performed with the translocator domain of the BrkA autotransporter of *B. pertussis*,[37] which were purified from the OM of *E. coli*. After reconstitution in planar lipid bilayers, larger pores with a single-channel conductance of ~3.0 nS were measured.

The proteins in the crystals formed layers that were organized in ring-like aggregates of monomeric units in alternating up- or downward orientation (C.J. Oomen, unpublished results). This seems to suggest that the protein is monomeric, in contrast to recombinant IgA-protease translocator domain when isolated from the OM of *E. coli*, which showed large multimeric complexes.[38] Furthermore, attempts to show multimers of the NalP translocator domain in the OM of *N. meningitidis* by chemical cross-linking failed,[5] even when the protein was overproduced from a plasmid (P. van Ulsen, unpublished results). Similarly, chemical cross-linking and gel filtration indicated that the purified translocator domain of the *E. coli* autotransporter AIDA-I is monomeric, or dimeric at most.[39]

6 Comparison of the NalP Translocator Domain to Other Translocator Domains and to TolC

The 12-stranded β-barrel found for the NalP translocator domain indicates that two strands were overpredicted in the 14-stranded model that was proposed[14] (Figure 3). In an alternative model, proposed for the translocator domain of the autotransporter EstA of *Pseudomonas aeruginosa*, 11 strands were predicted to be present.[40] The latter model missed only the first β-strand, which indeed, is difficult to predict, as it contains polar residues at the side of the strand that faces the lipids (Figure 3).

The presence of the α-helix was more surprising. An N-terminally located α-helix was modelled only in the case of the translocator domain of the autotransporter Hap of *H. influenzae*,[19] but the authors supposed that it was inserted in the membrane as part of the channel wall. However, a transmembrane helix is generally considered to be incompatible with OM localization, and owing to the different nature of the hydrogen bonding in the two secondary structure motifs, it is also difficult to envisage how the α-helix would interact with the β-strands to constitute a closed barrel.[17] Positioning the α-helix inside the channel elegantly solves this paradox. Secondary structure predictions for a representative subset of autotransporter translocator domains (Figure 3) revealed that an α-helix of considerable length is predicted in all cases at the N-terminal side of the β-barrel sequence.

Strikingly in the SPATE (serine protease autotransporters of Enterobacteriaceae) class of autotransporters, the proteolytic processing site – where the secreted passenger is cleaved from the translocator domain – is located within the predicted α-helix.[41] As a result, part of the helix is secreted with the passenger. The position

processing site Tsh

Block 1

```
                          h  h  h  h
Prn_Bpe     AAANAAVNTGGVGLASTLWYAESNALSKRLGELRLNPDAG-                              GAWGRGFAQRQQLDNRAG-
BrkA_Bpe    ALSGAANAAVNAADLSSIALAAESNALDKRLGELRLRADAG-                             GPWARTFSERQQISNRHA-
App_Nme     PQPQPQRDLISRYANSGLSEFSATLNSVFAVQDELDRVFAEDRRN-                         AVWTSGIRDTKHYRSQDF-
Hap_Hin     EKQRKQKDLISRYSNSALSELSATVNSMLSVQDELDRLFVDQAQS-                         AWTNIAQDKRRYDSDAF-
Tsh_Eco     RNDGQGKAAATFMHISYNNFITEVNNLNKRMGDLRDINGEA-                             GTWVRLLNGSGSADG---
IgAP_Ngo    AVSTNTNSALSDAMASTQSILLDTGAYLTRHIAQKSRADAEKN-                           SVWMSNTGYGRDYASAQY-
NalP_Nme    GVRIFNSLAATVYADSTAAHADMQGRRLKAVSGGLDHG-                                TGLRVIAQTQQDGGTWEQ-
AIDA_Eco    RPENGSYATNMALANSLFLMDLNERKQFRAMSDNTQPESA-                              SVWMKITGGISSGKLNDG-
Ssp-h1_Sma  QAFRQLSGQIHADIASALVNDSRYLREALNGRLRQAEGLASSSAIKADED-                     GAWAQLLGAWDHASGDAN-
EstA_Pae    HPTITGQRLIADYTYSLLSAPWELTLLPEMAHGTLRAYQDELRSQWQADWENWQNVGQWRGFVGGGGQRLDFDSQDS-
```

Block 2

```
              h  h  h  h                  h  h  h  h                          h  h  h  h
Prn_Bpe     RRFD-QKVAGFELGADHAVAVA-   GGRWHLGGLAGYT-   RGDRGFTG-          DGGGHTDSVHVGGYATYIAD----SG
BrkA_Bpe    RAYD-QTVSGLEIGLDRGWSAS-   GGRWYAGGLLGYT-   ADRTYPG-           DGGGKVKGLHVGGYAAYVGD----GG
App_Nme     RAYR-QQTDLRQIGMQKNL-      GSGRVGILFSHN-    RTENTFDD-          GIGNSARLAHGAVFGQYGIDR---
Hap_Hin     RAYQQQKTNLRQIGVQKAL-      ANGRIGAVFSHS-    RSDNTFDE-          QVKNHATLTMSGFAYQWGD---
Tsh_Eco     -GFT-DHYTLLQMGADRKHELG-   SMDLFTGVMATYTDTDASAD-               LYSGKTKSWGGFYASGLFR---SG
IgAP_Ngo    RRFS-SKRTQTQIGIDRSLS-     ENMQIGGVLTYSDSQHTFD-                QAGGKNTFVQANLYGKYYLND---A
NalP_Nme    GGVEGK-MRGSTQTVGIAAKT-    GENTTAATLGMGRSTWS-ENS-              ANAKTDSISLFAGIRHDAGD---I
AIDA_Eco    -QNKTTTNQFINQLGGDIYKFHAEQL-  ADFTLGIMGGYANAKGKTINYTSNKAARNTLDGYSVGVYGTWYQNGENATG
Ssp-h1_Sma  -ATGYQASTYGVLVGLDSAAA-    ADWRLGVATGYTRTSLHG-                 GYGSKADSDNYHLAAYGDKQF---GA
EstA_Pae    AASGDGNGYNLTLGGSYRI-      DEAWRAGVAAGFYRQKLEAGA-              KDSDYRMNSYMASAFVQYQENR---
```

Block 3

```
              h  h  h  h                           h  h  h  h                      h  h  h  h
Prn_Bpe     FYLDATLRASRLENDFKVAGSDGYA-      VKGKYRTHGVGASLEAGRFT-          HADGWFLEPQAELAVFRAG-
BrkA_Bpe    YYLDTVLRLGRYDQQYNIAGTDGGR-      VTADYRTSGAAWSLEGGRFE-          LPNDWFAEPQAEVMLWRTS-
App_Nme     FYIGISAGAFSSGSLSDG-             IGGKIRRRVLIHYGIQARYRAG-        FGWFGIEPHIGATRYFVQ-
Hap_Hin     LQFGVNVGTGISASKMAEEQS-          RKIHRKAINYGVNASYQFR-           LGQLGIQPYFGVNRYFIE-
Tsh_Eco     AYFDVIAKYIHNENKYDLNFAGAG-       KQNFRSHSLYAGAEVGYRYH-          LTDTTFVEPQAELVWGRLQ-
IgAP_Ngo    YVVAGDIGAGSLRSRLQTQQ-           KANFNRTSIQTGLTLGNTLK-          INQFEIVPSAGIRYSRLS-
NalP_Nme    GYLKGLFSYGRYKNSISRSTGADE-       HAEGSVNGTLMQLGALGGVNV-         PFAATGDLTVEGLRYDLLKQD
AIDA_Eco    LFAETMQYNMFNASVKGDLGEEE-        KYMLGLTASAGGYNLNVHTWTSPEGITGEHLQPHLQAVVGMGVT
Ssp-h1_Sma  LALRGGAGYTWHRIDTKRSVNYGMQSDRD-TAKYSARTEQLFAEAGYSVKG-           EWLNLEPFVNLAYVNFE-
EstA_Pae    WWADAALTGGYLDYDDLKRKFALGGGERSEKGDTNGHLWAFSARLGYDIAQQA-DSP-     WHLSPFVSADYARVE-
```

Prn_Bpe
BrkA_Bpe
App_Nme
Hap_Hin
Tsh_Eco
IgAP_Ngo
NalP_Nme
AIDA_Eco
Ssp-h1_Sma
EstA_Pae

```
              h h h h                    h h h h                              h h h h
Prn_Bpe      GGAYRAANG----LRVRDEGGSSVLGRLGLEVGKRIELAGGRQ-----VQPIKASVLQEFDGAGTVHTNGIAHRTELR-
BrkA_Bpe     GKRYRASNG----LRVKVDANTATLGRLGFGRRIELAGGNI-------VQPYARLGWTOEFKSTGDVRTNGIGHAGAGR-
App_Nme      KADYRYEN-----VNIATPGLAFNRYRAGIKADYSFKPAQ--------HISITPYLSLSYTDAASGKVRTRVN-TAVLAQDF-
Hap_Hin      RENYQSEE-----VRVKTPSLAFNRYNAGIRVDYTFTPTD--------NISVKPYFFVNYVDVSNANVQTTVN-LTVLQQPF-
Tsh_Eco      GQTFNWNDSGMDVSMRRNSVNPLVGRTGVVSGKTFSGKD---------WSLTARAGLHYEFDLTDSADVHLKDAAGEHQING-
IgAP_Ngo     SADYKLGDD----SVKVSSMAVKTLTAGLDNGYARAGFAYKVG-----NLTVKPLLSAAYFAAYA-KGGVNVGGKSFAYKA--
NalP_Nme     AFAEKGSALGW--SGNSLTEGTLVGLAGLKLSQPLSDK----------AVLFATAGVERDLNGRDYTVTGGFTGATAATG
AIDA_Eco     PDTHQEDNG----TVVQGAGKNNIQTKAGIRASWKVKSTLDKDTGRRFRPYIEANWIHNTHEFGVKMSDDSOLLSGS--
Ssp-h1_Sma   NNGIAESGGAAALRGDKQHTDATVSTLGLRADTEWQVSPG---GSD----TTVALRSELGWMQHOYGGLERGTGLRFNGGNAPFV
EstA_Pae     VDGYSEKGAS-ATALDYDDQKSKRLGAGLQGKYAF---------------TQLFAEYAHEREYEDDTQDLTMSLNSLPGNRF
                                         *                          :

              h h h h                              h h h h
Prn_Bpe      ------GTRAELGLGMAAALGRGHSLYASYEYSKGPKLAMPWTFHAGYRYSW
BrkA_Bpe     ------HGRVELGAGVDAALGKGHNLYASYEYAAGDRINIPWSFHAGYRYSF
App_Nme      ------GKTRSAEWGVNAEI-KGFTLSLHAAAAKGPQLEAQHSAGIKLGYRW
Hap_Hin      ------GRYWOKEVGLKAEI-LHFOISAFISKSQGSQLGKQQNVGVKLGYRW
Tsh_Eco      RK------DSRMLYGVGLNARFGDNTRLGLEVERSAFGKYNTDDAINANIRYSF
IgAP_Ngo     DNQOYSAGVALLY-RNVTINVNGSITKGKQLEKQKSGOIKIQIRF
NalP_Nme     KTGARNMPHTRLVAGLGADVEFGNWNGLARYSYAGSKQ-YGNHSGRVGVGYRF
AIDA_Eco     RNQGEIKTGIEGVITQNLSVNGGVAYQAGGHGSNAISGALGIKYSF
Ssp-h1_Sma   VDSVPVS-RDGMVLKAGAEVAVNENASLSLGYGLLS-QNHQDNSVNAGFTWRF
EstA_Pae     TLEGYTPQDHLNRVSLGFSQKLAPELSLRGGYNWRKG-EDDTQQSVSLALSLDF
                   *                                    :  : :
```

Figure 3 *Alignment of the translocator domains of 10 representative autotransporters. Aligned are pertactin of Bordetella pertussis (Prn_Bpe), BrkA of B. pertussis (BrkA_Bpe), App of Neisseria meningitidis (App_Nme), Hap of Haemophilus influenzae (Hap_Hin), Tsh of Escherichia coli (Tsh_Eco), IgA protease of N. gonorrhoeae (IgAP_Ngo), NalP of N. meningitidis (NalP_Nme), AIDA-I of E. coli (AIDA_Eco), Ssp-h1 of Serratia marcescens (Ssp-h1_Sma), and EstA of Pseudomonas aeruginosa (EstA_Pae). The alignment was made by adjusting a ClustalW alignment by hand to optimize the positions of the putative ten-residue membrane-spanning segments of the β-strands, which should consist of alternating hydrophilic and hydrophobic residues. The region encompassing the predicted α-helices was not adjusted by hand. The putative membrane-inserted parts of the β-strands are boxed, with the positions of the hydrophobic residues indicated on top (h). The secondary structure elements found in the NalP structure are shaded; light grey, the α-helix; dark grey, the β-strands; black, the mostly hydrophilic residues pointing inward in the channel. The hydrophilic patch in NalP made up of four residues in β-strands one and two, is given in bold italics. The membrane-spanning segments annotated in the alignments of Loveless and Saier[14] and Wilhelm et al.[40] are indicated under the sequences with single and double lines, respectively. The α-helices predicted by Psipred are indicated with a wavy line under the sequences. The level of identical residues is indicated below the alignment; "*", 10 or 9 residues the same; ":", 8 or 7 residues the same; ".", 6 or 5 residues the same. The autocatalytic processing site of the SPATE autotransporters is given in bold and indicated above the alignment*

of the processing site seems to preclude autocatalytic cleavage at the cell surface. Indeed, for several SPATES, it has been shown that substitution of the active-site Ser does not interfere with passenger release from the cell surface,[42-44] suggesting that other proteases may be involved in this process. It marks the SPATEs as a distinct subclass of autotransporters that differs from other serine protease autotransporters, such as IgA protease of the pathogenic *Neisseriaceae*[45] and Hap of *H. influenzae*,[19] for which autocatalytic cleavage has been observed. The actual mechanism of processing of the SPATEs remains to be investigated. However, the position of the processing site suggests that the protease may be active in the periplasm rather than on the extracellular side of the OM. Consistently, secretion of the SPATE Pic of *S. flexneri* was not hampered in an *E. coli* strain that lacked the OM proteases OmpT and OmpP.[43]

The C-terminal region of many autotransporter passengers contains a domain, designated as the linker or folding domain, which is necessary for the correct folding of the passenger at the cell surface.[46,47] In the case of the *E. coli* autotransporter EspP, it was suggested that the α-helix of the translocator domain is a part of this folding domain.[41] In EspP, the folding domain is located directly N-terminally of the predicted α-helix. However in some cases, for example in IgA protease of *N. gonorrhoeae*, the folding domain is located further upstream in the passenger,[47] and thus physically separated from the predicted α-helix in the translocator domains. Only in the SPATEs, part of the α-helix may be secreted and as a result be associated with the folding domain. Missense mutations in the putative α-helix of EspP did not interfere with secretion of the passenger.[41] In contrast, mutations in the domain with homology to the BrkA folding domain did interfere with secretion.[41] Therefore, we conclude that the α-helix is a part of the translocator domain and not of the folding domain. Consistently, mutations in the predicted α-helix of EspP did affect translocator domain localization, but only in the absence of the passenger,[41] suggesting a role for the α-helix in OM insertion or stabilization of the translocator domain. Furthermore, when we expressed the translocator domain of NalP, with or without α-helix fused to a signal sequence in *E. coli*, the amount of protein with the helix produced was much higher than that without the helix,[5] also suggesting a role of the α-helix in the folding or the stability of the β-barrel.

TolC of *E. coli*[48] and its homologues OprM of *P. aeruginosa*[31] and VceC of *V. cholerae*[34] are the only other protein-transporting OMPs whose crystal structure has been solved (Table 1). The OM-embedded domain of the trimeric TolC protein consists of a single 12-stranded β-barrel made up of four β-strands per monomer. Despite the same number of strands, the TolC β-barrel has a pore of 30–35 Å, which is considerably wider than the pore of the NalP translocator domain. This larger pore size is due to an exceptionally high shear number of 20;[17,48] whereas, the shear number for the NalP translocator domain is only 14.[5] The members of the OCA (which stands for oligomeric conserved adhesin family) subfamily of autotransporters, of which the adhesin YadA of *Yersinia enterocolitica* is the prototype,[49] contain a drastically shorter translocator domain. Strikingly these translocator domains are trimeric like TolC and are modelled to form a 12-stranded β-barrel, with each of the monomers donating four strands.[50,51] It will be interesting to determine whether these β-barrels resemble either the NalP translocator domain or the TolC β-barrel, as it will greatly influence the ideas about the secretion mechanism for this subfamily. The solved structures of subdomains of two passengers of the OCA subfamily have revealed that these are also trimeric[52,53] (Figure 2C). In contrast, the passengers of the other classes of autotransporters are considered to be monomers.

7 The Autotransporter Secretion Mechanism

The mechanism of autotransporter secretion across the OM is still far from clear. The NalP translocator domain structure positions the passenger if it is connected to the α-helix at the cell surface. In the through-pore secretion model[7,13] (Figure 4A), the barrel of the translocator domain forms the channel through which the passenger crosses the OM to the cell surface. Artificial passengers that were fused to the N terminus of the translocator domain were also transported,[54,55] which implies that the

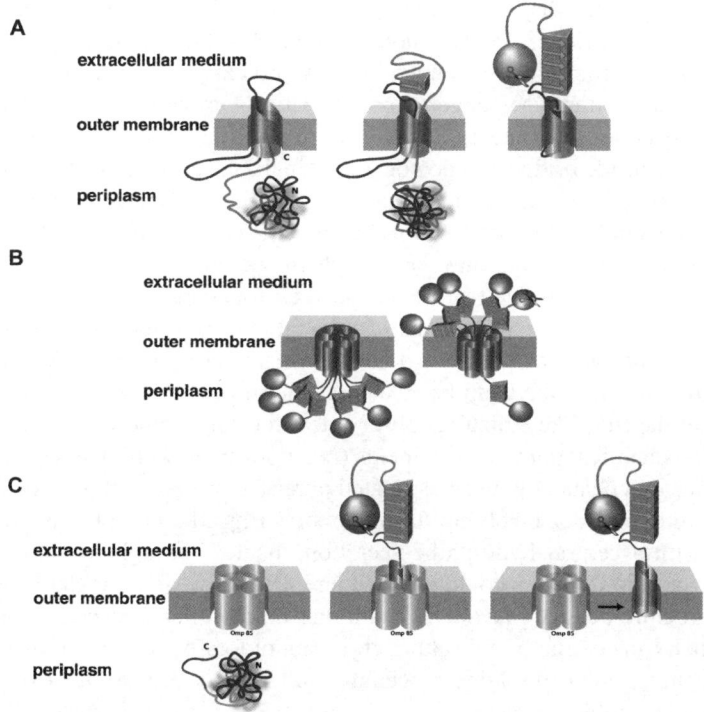

Figure 4 *Proposed mechanisms for passenger translocation across the OM. (A) The passenger is transported through the pore within the β-barrel formed by the translocator domain, whereby folding of the passenger at the cell surface might provide the energy for translocation. (B) The translocator domains form an oligomeric ring-like structure with a central channel through which the passenger domains are transported. (C) The translocator domain, because of its OMP characteristics, targets the autotransporter to the Omp85 complex. While the translocator domain inserts, the passenger is translocated across the OM via the Omp85 complex. The Omp85 complex then opens laterally to release the translocator domain into the OM*

N-terminal domains of the natural passengers have no role in targeting the translocator domain. Therefore, it seems likely that the C-terminal part of the passenger is transported first. It was modelled to dock as an unfolded hairpin within the channel.[7,56] Extracellular folding starting at the tip of the hairpin, could subsequently drive secretion, whereby the ongoing folding would thread the passenger through the channel. The crystallized passengers of pertactin from *B. pertussis*[57] and Hbp from *E. coli*[58] show a large domain with a β-helical conformation that, in the Hbp passenger, is connected to a trypsin-like protease domain at the N terminus[58] (Figure 2B). Such a β-helical conformation has been predicted to be present in the vast majority of the autotransporter passengers.[5] The model predicts that the α-helix is formed in the barrel of the translocator domain only after translocation of the passenger, presumably to close the pore.

The channel in the NalP translocator domain, as shown in the crystal structure, is just wide enough to accommodate two unfolded polypeptides passing each other (Figure 4A). It is too narrow to accommodate bulky structural elements and, therefore, it was suggested that secretion requires an unfolded passenger. Indeed, formation of disulfide bridges in the cholera toxin B subunit, fused as an artificial passenger to the translocator domain of IgA protease, blocked OM passage.[59,60] However, many natural passengers contain pairs of cysteines, separated by 5–10 amino acid residues that potentially could form disulfide bridges, and the actual formation of a disulfide bridge was demonstrated in the case of the autotransporter IcsA of *S. flexneri*.[61] In Gram-negative bacteria, disulfide bonds are formed in the periplasm, thus before OM passage of the passenger; but, apparently, the formation of the disulfide bond did not interfere with IcsA secretion. Furthermore, a disulfide bond also did not prevent cell-surface exposure of a single-chain antibody fused to the IgA-protease translocator domain, although the translocation efficiency was rather low.[62] The same translocator domain, when overproduced in *E. coli*, formed large multimeric ring-like complexes with a central pore or cavity.[38] Measuring the pore size of the ring-like structures observed with electron microscopy, in combination with biochemical pore measurements (*i.e.*, liposome-swelling assays), indicated the presence of a channel with an estimated pore size of ~20 Å. It prompted an alternative secretion model involving an oligomeric ring-like complex of translocator domains with a central hydrophilic secretion channel through which even folded passenger-protein domains could pass (Figure 4B). However, the NalP translocator domain structure does not favour the multimer model; first, because of the position of the α-helix inside the barrel (although it cannot totally be ignored that this is an artefact from the *in vitro* folding procedure) and second, because the exterior of the barrel is hydrophobic with the exception, perhaps, of a polar patch found at the top of the barrel (see Figures 2 and 3), which implies that the central channel of a multimer would also be hydrophobic. Accordingly, the NalP translocator domain was monomeric in the crystal and no evidence for multimerization of either the NalP or the AIDA-I translocator domains was found *in vivo*.[5,39] Additionally, the energy source for the translocation of folded protein domains in the multimeric model remains enigmatic.

Since neither of the two existing models is fully supported by the NalP structure, we have proposed an alternative secretion model.[5] It incorporates the recent observation that autotransporter secretion is dependent on the multimeric Omp85 complex, which constitutes the general OMP-assembly machinery.[20] In this model (Figure 4C), the translocator domain requires the Omp85 complex for its OM insertion. It acts, thereby, as a targeting domain to deliver the passenger to the Omp85 machinery. This machinery then would constitute the actual translocation channel for the passenger. In support of this model it should be noted that integral OMPs, which depend on Omp85 for insertion into the OM, contain large hydrophilic surface-exposed loops – which individually can be up to ~9 kDa in size and sometimes add up to 60 kDa in total (for example in TbpA of *N. gonorrhoeae*[63]). These loops also have to be transported across the OM. Furthermore, proteins that are secreted via the two-partner secretion mechanism[64] have, like the passengers of autotransporters, a predicted β-helical structure.[65,66] The transport of these proteins across the OM is mediated via a dedicated class of OMPs, of which FhaC is the best-studied example.[64] These OMPs show considerable sequence similarity to Omp85[35,20] and have been shown to be multimeric as well.[67] If Omp85 has indeed a direct and active role in the translocation of the passenger, the autotransporter secretion mechanism turns out to be more complex than anticipated, and then even the name 'autotransporter' seems to be inappropriate.

References

1. V.T. Lee and O. Schneewind, *Genes Dev.*, 2001, **15**, 1725.
2. I.R. Henderson, F. Navarro-Garcia and J.P. Nataro, *Trends Microbiol.*, 1998, **6**, 370.
3. R. Sijbrandi, M.L. Urbanus, C.M. ten Hagen-Jongman, H.D. Bernstein, B. Oudega, B.R. Otto and J. Luirink, *J. Biol. Chem.*, 2003, **278**, 4654.
4. L.D. Brandon, N. Goehring, A. Janakiraman, A.W. Yan, T. Wu, J. Beckwith and M.B. Goldberg, *Mol. Microbiol.*, 2003, **50**, 45.
5. C.J. Oomen, P. van Ulsen, P. Van Gelder, M. Feijen, J. Tommassen and P. Gros, *EMBO J.*, 2004, **23**, 1257.
6. I.R. Henderson and J.P. Nataro, *Infect. Immun.*, 2001, **69**, 1231.
7. J. Pohlner, R. Halter, K. Beyreuther and T.F. Meyer, *Nature*, 1987, **325**, 458.
8. Y.L. Tzeng and D.S. Stevens, *Microbes Infect.*, 2000, **2**, 687.
9. P. van Ulsen, L. van Alphen, C.T. Hopman, A. van der Ende and J. Tommassen, *FEMS Immunol. Med. Microbiol.*, 2001, **32**, 53.
10. P. van Ulsen, L. van Alphen, J. ten Hove, F. Fransen, P. van der Ley and J. Tommassen, *Mol. Microbiol.*, 2003, **50**, 1017.
11. D.P. Turner, K.G. Wooldridge and D.A. Ala'Aldeen, *Infect. Immun.*, 2002, **70**, 4447.
12. J. Pohlner, U. Langenberg, U. Wölk, S.C. Beck and T.F. Meyer, *Mol. Microbiol.*, 1995, **17**, 1073.
13. T. Klauser, J. Krämer, K. Otzelberger, J. Pohlner, and T.F. Meyer, *J. Mol. Biol.*, 1993, **234**, 579.
14. B.J. Loveless and M.H.J. Saier Jr., *Mol. Membr. Biol.*, 1997, **14**, 113.
15. M.P.J. Konieczny, I. Benz, B. Hollinderbäumer, C. Beinke, M. Niederweis and M.A. Schmidt, *Antonie van Leeuwenhoek*, 2001, **80**, 19.

16. R. Koebnik, K.P. Locher and P. Van Gelder, *Mol. Microbiol.*, 2000, **37**, 239.
17. G.E. Schulz, *Biochim. Biophys. Acta*, 2002, **1565**, 308.
18. M. Struyvé, M. Moons and J. Tommassen, *J. Mol. Biol.*, 1991, **218**, 141.
19. D.R. Hendrixson, M.L. de la Morena, C. Stathopoulos and J.W. St Geme III, *Mol. Microbiol.*, 1997, **26**, 505.
20. R. Voulhoux, M.P. Bos, J. Geurtsen, M. Mols and J. Tommassen, *Science*, 2003, **299**, 262.
21. J.E. Mogensen and D.E. Otzen, *Mol. Microbiol.*, 2005, **57**, 326.
22. L. Vandeputte-Rutten, R.A. Kramer, J. Kroon, N. Dekker, M.R. Egmond and P. Gros P, *EMBO J.*, 2001, **20**, 5033.
23. A.D. Ferguson, R. Chakraborty, B.S. Smith, L. Esser, D. van der Helm D and J. Deisenhofer, *Science*, 2002, **295**, 1715.
24. W.W. Yue, S. Grizot and S.K. Buchanan, *J. Mol. Biol.*, 2003, **332**, 353.
25. S.M. Prince, M. Achtman and J.P. Derrick, *Proc. Natl. Acad. Sci. USA*, 2002, **99**, 3417.
26. D.P. Chimento, A.K. Mohanty, R.J. Kadner and M.C. Wiener, *Nature Struct. Biol.*, 2003, **10**, 394.
27. L. Vandeputte-Rutten, M.P. Bos, J. Tommassen and P. Gros, *J. Biol. Chem.*, 2003, **278**, 24825.
28. V.E. Ahn, E.I. Lo, C.K. Engel, L. Chen, P.M. Hwang, L.E. Kay, R.E. Bishop and G.G. Privé, *EMBO J.*, 2004, **23**, 2931.
29. B. van den Berg, P.N. Black, W.M. Clemons Jr. and T.A. Rapoport, *Science*, 2004, **304**, 1506.
30. J. Ye and B. van den Berg, *EMBO J.*, 2004, **23**, 3187.
31. H. Akama, M. Kanemaki, M. Yoshimura, T. Tsukihara, T. Kashiwagi, H. Yoneyama, S. Narita, A. Nakagawa and T. Nakae, *J. Biol. Chem.*, 2004, **279**, 52816.
32. D. Cobessi, H. Celia, N. Folschweiller, I.J. Schalk, M.A. Abdallah and F. Pattus, *J. Mol. Biol.*, 2005, **347**, 121.
33. L. Federici, D. Du, F. Walas, H. Matsumura, J. Fernandez-Recio, K.S. McKeegan, M.I. Borges-Walmsley MI, B.F. Luisi and A.R. Walmsley, *J. Biol. Chem.*, 2005, **280**, 15307.
34. S.K. Buchanan, *Curr. Opin. Struct. Biol.*, 1999, **9**, 455.
35. M.R. Yen, C.R. Peabody, S.M. Partovi, Y. Zhai, Y.H. Tseng and M.H. Saier Jr., *Biochim. Biophys. Acta*, 2001, **1562**, 6.
36. N. Dekker, N.K. Merck, J. Tommassen and H.M. Verheij, *Eur. J. Biochem.*, 1995, **232**, 214.
37. J.L. Shannon and R.C. Fernandez, *J. Bacteriol.*, 1999, **181**, 5838.
38. E. Veiga, E. Sugawara, H. Nikaido, V. de Lorenzo and L.A. Fernandez, *EMBO J.*, 2002, **21,** 2122.
39. D. Müller, I. Benz, D. Tapadar, C. Buddenborg, L. Greune and M.A. Schmidt, *Infect. Immun.*, 2005, **73**, 3851.
40. S. Wilhelm, J. Tommassen and K.E. Jaeger, *J. Bacteriol.*, 1999, **181**, 6977.
41. J.J. Velarde and J.P. Nataro, *J. Biol. Chem.*, 2004, **279**, 31495.
42. C. Stathopoulos, D.L. Provence and R. Curtiss III, *Infect. Immun.*, 1999, **67**, 772.

43. I.R. Henderson, J. Czeczulin, C. Eslava, F. Noriega and J.P. Nataro, *Infect. Immun.*, 1999, **67**, 5587.
44. S.K. Patel, J. Dotson, K.P. Allen and J.M. Fleckenstein, *Infect. Immun.*, 2004, **72**, 1786.
45. T. Klauser, J. Pohlner and T.F. Meyer, *Bioessays*, 1993, **15**, 799.
46. Y. Ohnishi, M. Nishiyama, S. Horinouchi and T. Beppu, *J. Biol. Chem.*, 1994, **269**, 32800.
47. D.C. Oliver, G. Huang, E. Nodel, S. Pleasance and R.C. Fernandez, *Mol. Microbiol.*, 2003, **47**, 1367.
48. V. Koronakis, A. Sharff, E. Koronakis, B. Luisi and C. Hughes, *Nature*, 2000, **405**, 914.
49. E. Hoiczyk, A. Roggenkamp, M. Reichenbecher, A. Lupas and J. Heesemann, *EMBO J.*, 2000, **19**, 5989.
50. A. Roggenkamp, N. Ackermann, C.A. Jacobi, K. Truelzsch, H. Hoffmann and J. Heesemann, *J. Bacteriol.*, 2003, **185**, 3735.
51. N.K. Surana, D. Cutter, S.J. Barenkamp and J.W. St. Geme III, *J. Biol. Chem.*, 2004, **279**, 14679.
52. H. Nummelin, M.C. Merckel, J.C. Leo, H. Lankinen, M. Skurnik and A. Goldman, *EMBO J.*, 2004, **23**, 701.
53. H.J. Yeo, S.E. Cotter, S. Laarmann, T. Juehne, J.W. St. Geme III and G. Waksman, *EMBO J.*, 2004, **23**, 1245.
54. J. Jose, J. Kramer, T. Klauser, J. Pohlner and T.F. Meyer, *Gene*, 1996, **178**, 107.
55. J. Maurer, J. Jose and T.F. Meyer, *J. Bacteriol.*, 1996, **179**, 794.
56. J. Jose, F. Jähnig and T.F. Meyer, *Mol. Microbiol.*, 1995, **18**, 378.
57. P. Emsley, I.G. Charles, N.F. Fairweather and N.W. Isaacs, *Nature*, 1996, **381**, 90.
58. B.R. Otto, R. Sijbrandi, J. Luirink, B. Oudega, J.G. Heddle, K. Mizutani, S.Y. Park and J.R. Tame, *J. Biol. Chem.*, 2005, **280**, 17339.
59. T. Klauser, J. Pohlner and T.F. Meyer, *EMBO J.*, 1990, **9**, 1991.
60. T. Klauser, J. Pohlner and T.F. Meyer, *EMBO J.*, 1992, **11**, 2327.
61. L.D. Brandon and M.B. Goldberg, *J. Bacteriol.*, 2001, **183**, 951.
62. E. Veiga, V. de Lorenzo and L.A. Fernandez, *Mol. Microbiol.*, 1999, **33**, 1232.
63. I.C. Boulton, M.K. Yost, J.E. Anderson and C.N. Cornelissen, *Infect. Immun.*, 2000, **68**, 6988.
64. F. Jacob-Dubuisson, C. Locht and R. Antoine, *Mol. Microbiol.*, 2001, **40**, 306.
65. A.V. Kajava, N. Cheng, R. Cleaver, M. Kessel, M.N. Simon, E. Willery, F. Jacob-Dubuisson, C. Locht and A.C. Steven, *Mol. Microbiol.*, 2001, **42**, 279.
66. B. Clantin, H. Hodak, F. Willery, C. Locht, F. Jacob-Dubuisson and V. Villeret, *Proc. Natl. Acad. Sci. USA*, 2004, **101**, 6194.
67. N.K. Surana, S. Grass, G.G. Hardy, H. Li, D.G. Thanassi and J.W. St. Geme III, *Proc. Natl. Acad. Sci. USA*, 2004, **101,** 14497.

CHAPTER 16

X-Ray Crystallographic Structures of Sarcoplasmic Reticulum Ca²⁺-ATPase at the Atomic Level

JESPER VUUST MØLLER[1], POUL NISSEN[2] AND THOMAS LYKKE-MØLLER SØRENSEN[3]

[1]Department of Biophysics, Institute of Physiology and Biophysics, University of Aarhus, Ole Worms Allé 185, DK-8000 Aarhus C, Denmark
[2]Centre for Structural Biology, Department of Molecular Biology, University of Aarhus, Gustav Wieds Vej 10c, DK-8000 Aarhus C, Denmark
[3]Diamond Light Source Ltd., Rutherford Appleton Laboratory, Chilton, Didcot OX11 0QX, UK

1   Introduction

The sarcoplasmic/endoplasmic reticulum (SERCA) and plasma membrane Ca^{2+}-ATPases belong to the family of P-type ATPases whose function is to actively transport cations across intracellular and extracellular membranes by the expenditure of the chemical energy released by ATP hydrolysis.[1–3] P-type ATPases are therefore also known as the cation pumps. Other prominent members in eukaryotes of this primary transporter family are the Na^+,K^+-ATPases present in animal cells and the H^+-ATPases in plant cells. P-type ATPases are also involved in gastric acid secretion (H^+,K^+-ATPase), and in the transport of heavy metals. In addition, P-type ATPases have also been implicated in phospholipid metabolism as "flippases."[3] Together with the proton translocating F_1/F_0- (mitochondrial and chloroplast), V_1/V_0- (vacuolar), and A_1/A_0- (archaeal) ATPases and with the ABC-transporter family, they constitute the stock of known primary transporters in cell membranes.[4] While both V_1/V_0- and A_1/A_0-ATPases are primarily concerned with the establishment of proton gradients, F_1/F_0-ATPases serve in the reverse reaction for synthesis of ATP on the basis of such gradients. The ABC-transporter family couples the ATPase activity of the ABC domain with a broad range of transmembrane modules for the transport of substrates ranging from small metabolites to hydrophobic molecules and proteins. Despite vast differences in the architecture and transport mechanism these ATPases are all characterized by a clear division in the catalytically active center, located in the cytosol, whose activity is coupled to the transport function inside the membraneous domain by transduction of conformational changes in the intervening polypeptide.

Within the P-type family, Ca^{2+}-ATPases primarily serve together with Ca^{2+} channel proteins to maintain the regulatory functions of Ca^{2+} as an intracellular messenger,[5] while Na^+,K^+- and H^+-ATPases are responsible for the creation of cellular energy required for formation of membrane potentials and for driving a number of co-transport processes. Phylogenetically, P-type ATPases belong to the haloacid dehydrogenase (HAD) superfamily of both water-soluble and membrane-bound enzymes.[6] These enzymes use an activated aspartic acid side chain as a nucleophile to form a high energy aspartyl-phosphorylated intermediate after reaction with ATP. The formation of the aspartyl-phosphorylated intermediate during the enzyme cycle has served as a convenient experimental tool to examine the reactions in terms of a number of intermediary steps with well-defined properties. Studies of P-type ATPases relate to a very basic question in biology, namely to understand how cation pumps work. Until recently it has been difficult to explain how P-type ATPases are able to convert the chemical energy, derived from hydrolysis of ATP, into a vectorial mechanism for cation transport. The recent progress, to a large extent, can be attributed to the attainment of 3D crystal structures of Ca^{2+}-ATPase at high resolution, starting with publications by Toyoshima and co-workers[7,8] of the 3D structure of sarcoplasmic reticulum (SR) Ca^{2+}-ATPase from skeletal muscle (isoform 1a) some years ago. These papers have aroused great attention, as they were the first ones to describe the structure of a cation transporting P-type ATPase in both the Ca^{2+}-bound[7] and Ca^{2+}-depleted[8] state. This advance has recently been followed by reports of crystal structures of the Ca^{2+}-ATPase captured at other stages in the enzyme cycle, corresponding to the phosphorylated intermediates.[9–13] Hence, it is now possible to follow the structural changes step by step taking place during transport and thus to gain insight into the transport mechanism as discussed below.

2 The Transport Scheme and Thermodynamics of Ca²⁺ Transport

As shown in Figure 1, the Ca^{2+} transport cycle of sarcoplasmic reticulum Ca^{2+}-ATPase starts with the intramembraneous binding of two Ca^{2+} ions from the cytoplasm and phosphorylation by ATP, leading to the formation of a $Ca_2E1\sim P$ high-energy intermediate that is phosphorylated at Asp351 and accompanied by occlusion of Ca^{2+}. This is followed by the downhill transition of the phosphorylated intermediate to form a low-energy E2P intermediate and translocation of Ca^{2+} to the lumen against an electrochemical gradient. The two Ca^{2+} ions are delivered to the intravesicular lumen where they are exchanged with 2–3 luminal protons.[14–16] In the following step, where E2P is dephosphorylated by hydrolytic cleavage to form

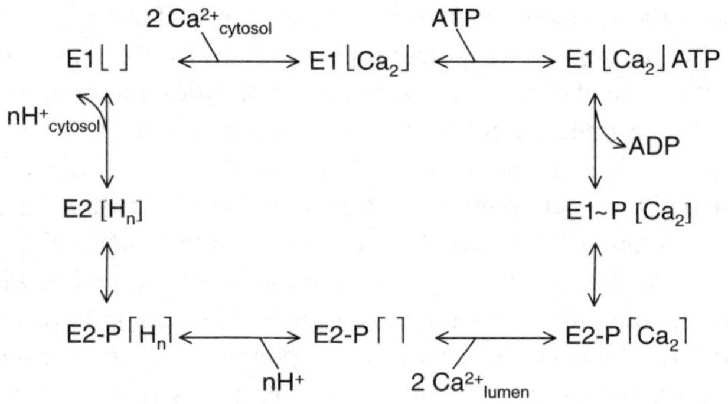

E1⌊ ⌋ ⟵ 2 Ca²⁺cytosol ⟶ E1⌊Ca₂⌋ ⟵ ATP ⟶ E1⌊Ca₂⌋ATP

nH⁺cytosol ↕

E2 [Hₙ]   E1~P [Ca₂]   ⟵ ADP

E2-P⌈Hₙ⌉ ⟷ E2-P⌈ ⌉ ⟷ E2-P⌈Ca₂⌉
       nH⁺   2 Ca²⁺lumen

Figure 1 *Reaction scheme showing the relation of the phosphorylated and nonphosphorylated E1 and E2 states to binding and transport of Ca²⁺ or countertransport of protons. E1~P [Ca₂] denotes Ca²⁺-ATPase with occluded Ca²⁺, while E2[Hₙ] denotes Ca²⁺-ATPase with occluded protons (with or without noncovalently bound phosphate)*

E2, these protons are translocated across the membrane to the cytosolic side of the membrane, from which they are released upon binding of Ca^{2+} to reform Ca_2E1 and start a new transport cycle.

The sarcoplasmic reticulum Ca^{2+}-ATPase thus normally functions as an electrogenic Ca^{2+}/H^+ exchanger where the protons partially compensate for the positive electrostatic charge associated with the intravesicular uptake of Ca^{2+}. At maximal levels of Ca^{2+} accumulation (100–150 nmol Ca^{2+} mg^{-1} protein) isolated SR vesicles can achieve Ca^{2+} concentrations 20,000 – 40,000 times higher than those present in the suspending medium, and under these extreme conditions the Ca^{2+}-ATPase also functions as a Ca^{2+}/Ca^{2+} exchanger.[17,18] In addition, similar evidence for uncoupling of Ca^{2+} transport from ATP hydrolysis in isolated vesicles at 37 °C has been reported.[19] Yet, at lower levels of accumulation, comparable to those present in the muscle cell (30 nmol Ca^{2+} mg^{-1} protein corresponding to accumulation ratios of 2500–5000), Ca^{2+} transport is tightly coupled to ATP hydrolysis. This corresponds to a change of chemical free energy of intravesicular Ca^{2+} as high as 40 – 44 kJ mol^{-1}. At the low ADP/ATP concentration ratios in the cytosol due to the presence of the ADP/ATP translocase in the mitochondria, this will correspond to a pump, working with close to 100% efficiency to convert chemical energy into the establishment of a potent Ca^{2+} gradient.[20,21] Thus the sarcoplasmic reticulum Ca^{2+}-ATPase can be considered as a primarily active pump characterized by little waste of energy and tight coupling of the chemical and vectorial processes under physiological conditions. Mechanisms purporting to describe how Ca^{2+} transport occurs must account for this efficiency.

3   Overall structure of Ca²⁺-ATPase

Figure 2 shows a cavalcade of Ca^{2+}-ATPase structures in various functional states. In the upper left corner is shown the first X-ray crystal structure by Toyoshima *et al.*,[7] obtained at 2.6 Å resolution after solubilization with the non-ionic detergent $C_{12}E_8$ and at a Ca^{2+} concentration of 10 mM, where the Ca^{2+}-ATPase is present in a

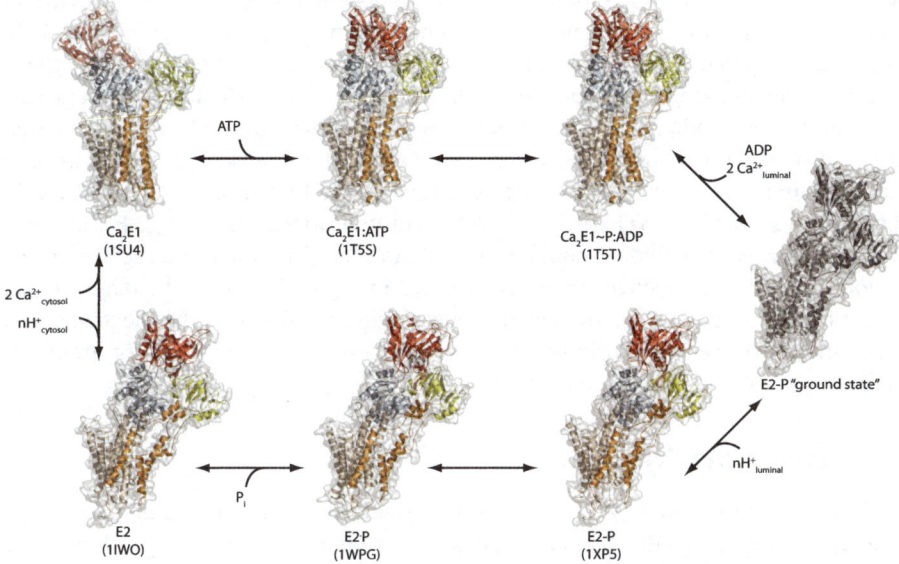

Figure 2 *The known crystal structures of the Ca²⁺-ATPase displayed as they relate to the functional cycle. The crystal structures (with PDB codes indicated) are shown as a semitransparent surface in a cartoon representation with the A-domain in yellow, the N-domain in blue, the P-domain in red, the M1–M3 transmembrane segment in orange, and the M4–M10 segment in wheat. The unknown structure of the E2P ground state is represented by a grey copy of the E2P transition state (1XP5)*

Ca₂E1 conformation. While the data provided new insight into the detailed structure of the ATPase, they also supported many of the features deduced from previous biochemical and biophysical experiments. Thus the existence of ten transmembrane segments[2,22,23] was confirmed, and the presence of three cytosolic domains, comprising the P-(phosphorylation), N-(nucleotide binding), and A-(actuator or β-strand) domain, could clearly be seen. The P-domain was found to be formed by two parts; a central part (comprising amino acid residues 330–360 with the phosphorylation site at Asp351) and a peripheral part formed by the more C-terminal part of the Ca²⁺-ATPase polypeptide chain (residues 605–737), which, like a rosette, surrounds the phosphorylation site with a β,α-Rossman fold as had been predicted by comparative sequence analysis.[24] The intervening part of the peptide chain (residues 360 – 604) formed the N-domain, in agreement with proteolytic dissection.[25] The structure also enabled a detailed description of the two intramembraneous, high-affinity Ca²⁺-binding sites, which were found to be bound alongside each other and coordinating with polar side chains and main chain oxygens in transmembrane helices M4, M5, M6, and M8, in agreement with deductions drawn from site-directed mutagenesis data.[26–28] But surprisingly, the structure was found to be more open and showed clearer separations of the cytosolic domains than had been expected from earlier image reconstructions of 2D crystals at lower (8 Å) resolution.[23] The X-ray data also identified the "nose" seen in those reconstructions as the N-terminal

A-domain instead of being part of the N- or P-domain as had been previously thought. Soon after the first structure, a compact form of the Ca^{2+}-ATPase (1IWO in Figure 2), prepared in the absence of Ca^{2+} and stabilized by the plant sesquiterpene lactone, thapsigargin,[29] was published.[8] In this form there is extensive interaction between the cytosolic A-, P-, and N-domains, accompanied by changes in the disposition of the transmembrane helices, indicative of the absence of high affinity Ca^{2+} binding. These two X-ray structures provided the first structural verification of the existence of Ca^{2+}-ATPase in its two main conformational states, E1 and E2. Representatives of the corresponding phosphorylated forms that are the key to the cation translocation mechanism are also shown in Figure 2 and are the main subject of the following part of this review. However, before addressing the mechanism of Ca^{2+} transport we shall briefly review, in general terms, models for transport processes across biological membranes.

4   Transport Models

Early ideas on mediated transport revolved around the concept of a mobile carrier that, after binding of substrate, moved it across the membrane like a ferry boat. This was hypothesized to account for the phenomena incompatible with diffusion, such as transport saturation at high concentrations, competition, and acceleration of transport (*e.g.*, of radiolabeled substrate) by countertransport of substrate in the opposite direction.[30,31] While the idea of a mobile carrier may be appropriate for transport by a lipid soluble carrier of low molecular mass, biological transporters were later identified as integral membrane proteins, firmly anchored to the membrane, so that any postulated transport, whether by flip-flop or movement of their binding sites over an appreciable distance, would be opposed by extensive and thermodynamically very costly conformational changes.[32–34] Instead, in most contemporary models translocation of substrate is considered to occur by penetration of the membrane through a pore or channel inside the protein, equipped with binding sites for the substrate and entrance gates at both ends which open and close in a concerted manner. Thus unidirectional transport is possible via controlled gating (Figure 3). This mode of translocation, which minimizes the energy barriers associated with transport, is often referred to as the alternating access model. It implies that although the transported substrate can approach the center of the transporter with its binding site by diffusion from either side of the membrane, it can never do so from both sides at the same time, *i.e.*, only one gate is open at any time. In addition, to prevent leakage, the model also incorporates as an important feature, the occluded state, where the bound substrate is trapped inside the pore during the period when one gate closes and the other gate has not yet opened. The model has some features in common with diffusion through ion channel proteins, but generally the flux through the latter are orders of magnitude larger and are considered to be regulated by a single gate.[35–37] Furthermore, although evidence for a very slow passive penetration of Ca^{2+} through Ca^{2+}-ATPase has been obtained,[38,39] this protein like other P-type ATPases requires metabolic input to achieve appreciable translocation rates. As a further difference, it may be noted that while many channel proteins are present as oligomeric complexes that let ions pass through slits or openings between their subunits, the evidence for

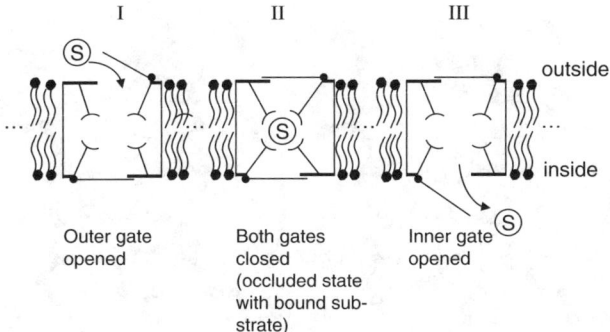

Figure 3 *A schematic presentation of the alternating access transport model with an intramembranous substrate binding site, which regulates the opening of either the outer or inner gate to the binding site*

Ca^{2+}-ATPase and other P-type ATPases point to transport as taking place through a central pathway in the monomeric protein.[2]

5   Initialization of the Cycle: Phosphorylation and Calcium Ion Occlusion

A critical first step of Ca^{2+} translocation in Ca^{2+}-ATPase is the formation of the high-energy intermediate state that fuels subsequent downhill transitions coupled to vectorial transport processes. This state occurs as the result of phosphorylation by ATP which is coupled to the binding and occlusion of Ca^{2+} in the membrane.[2,34] Evidence for the existence of an occluded state of Ca^{2+}-ATPase during Ca^{2+} transport was originally inferred from kinetic experiments where, after phosphorylation by ATP, bound Ca^{2+} was released only towards the luminal side of the membrane, when EGTA had been added to chelate medium Ca^{2+}, to stop further phosphorylation by the pump.[40,41] It is difficult to demonstrate occlusion in this way, since this state is confined to a short period of the transport cycle. However, Ca^{2+} occlusion can be stabilized using transitional state analogues of MgATP such as CrATP[42–44] or ADP:AlF$_4^-$.[45] After complexation with these analogues, Ca^{2+} can remain bound to the ATPase for hours in a calcium-free medium. The crystal structure of Ca^{2+}-ATPase complexed with ADP:AlF$_4^-$ solved at 2.9 Å resolution revealed the basis of phosphoryl transfer coupled to calcium occlusion.[9] As shown in Figure 4A, AlF$_4^-$ closely interacts with both the Asp351 phosphorylatable residue and the β-phosphate of ADP in a nearly linear arrangement with distances between atoms being around 2 Å. These features are characteristic of an associative reaction mechanism in contrast to a dissociative mechanism, where the distance between atoms is larger and a strict linear arrangement is not mandatory. Since calcium at the membraneous-binding sites is required for phosphorylation from ATP to proceed, conformational checkpoints are needed to ensure a tight coupling between Ca^{2+} binding and phosphorylation. An associative mechanism requires external positive charge

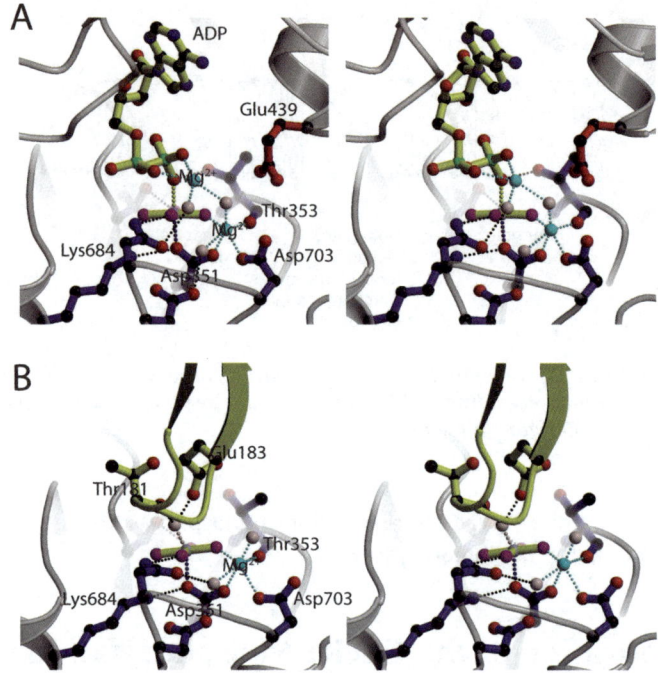

Figure 4 *The phosphorylation and dephosphorylation sites as pinpointed by crystal structures of aluminium fluoride complexes. (A) Stereodiagram of the phosphorylation site as represented by the Ca₂E1:ADP:AlF₄⁻ complex (1T5T). Important residues are shown in a ball-and-stick representation with bond-colours relating to the scheme in Figure 2 and with atoms colored in the following way: carbon in black, oxygen in red, nitrogen in blue, phosphorus in green, aluminium in steel blue, fluoride in magenta, and magnesium in cyan. Water molecules are shown as grey spheres. Polar interactions are indicated by dotted lines with Mg^{2+} coordination in cyan, AlF_4^- coordination in blue and yellow, and hydrogen bonds in black. (B) Similar representation of the dephosphorylation site as observed in the E2:AlF₄⁻ complex (1XP5)*

delivered via metal ions and positively charged side chains, due to the accumulation of negative charge on the non-bonding phosphoryl oxygens, and to ensure a strict linear arrangement.

In the Ca^{2+}-ATPase, positive charge at the phosphorylation site is provided by Mg^{2+} coordinating AlF_4^- with Asp351 and other nearby residues (Asp703, Thr353, and two water molecules). Additional positive charge at the catalytic site is provided by the conserved Lys684 residue interacting via its side chain with Asp351. The adenosine part of ADP is bound to the N-domain by interaction with Phe487 and other conserved amino acid residues in this domain, while three arginine residues (489, 560, and 678) interact with the ribose and the α,β-phosphate groups. Also a second Mg^{2+} ion interacts with the α,β-phosphate groups. The interaction with both the N-domain and the Asp351 phosphorylation site requires conversion of the open structure in Figure 5A into the compact structure shown in Figure 5B which is achieved by tilting the N-domain towards the P-domain. Concurrently, the A-domain moves upwards in the cytosolic direction to make contact with a small crevice now present between the N- and P-domain. The assembly of the cytosolic domains is transmitted to the membrane, because the movement of the A-domain exerts a direct pull on the M1 and M2 transmembrane segments. As a result M2 is lifted up 8–9 Å from the membrane, resulting in a transfer of its C-terminal residues into the cytosolic space and withdrawal of luminal residues between M1 and M2 into the membrane. In M1 the strongly amphipatic (^{54}I E Q F E D L) N-terminal part is also extracted from the hydrophobic membrane phase, but rather than being lifted upwards it rolls over and forms a kink to assume a thermodynamically more relaxed state lying flat with the N-terminal half on the nonpolar membrane interface. As a result of these movements, a putative N-terminal entrance to intramembranous Ca^{2+} binding formed by the hydrophilic amino acid residues in M1–M3 in the Ca_2E1 structure (Figure 5A) is closed and replaced by a hydrophobic plug (Figure 5B). This rearrangement was proposed to prevent leakage of bound Ca^{2+} to the cytosolic space and to result in occlusion, since the conversion of Ca_2E1P to E2P, which is required to open the inner gate, is also blocked under these conditions.

The structure of Ca^{2+}-ATPase in complex with the non-hydrolyzable nucleotide analogue AMPPCP, obtained from crystals prepared under similar conditions as the $Ca_2E1:ADP:AlF_4^-$ complex, has also been solved.[9,10] It is of considerable interest that

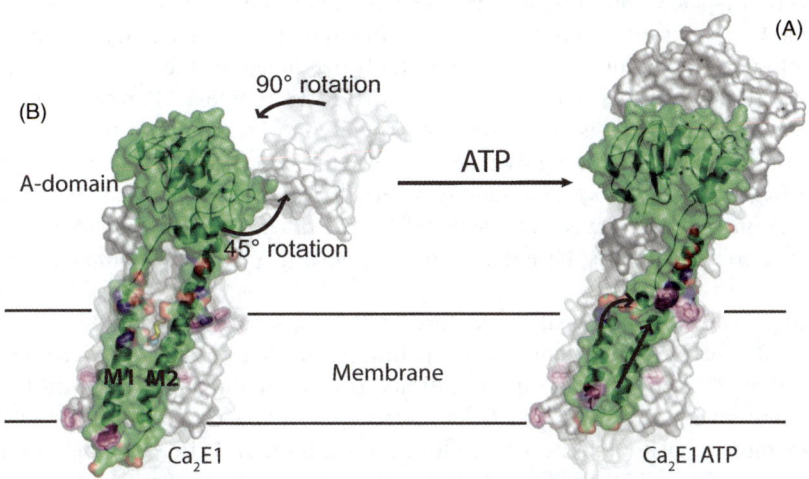

Figure 5 *The Ca^{2+} occlusion mechanism. (A) The intramembraneous Ca^{2+} binding sites as observed in the non-occluding Ca_2E1 state (1SU4). The A-domain and the M1–M2 segment are shown as a semitransparent surface in green with polar nitrogen and oxygen atoms from amino acid side chains indicated by blue and red, respectively. The Glu309 residue is shown as a yellow stick-model and the Ca^{2+} ion of site II is visible as indicated by a cyan sphere. Exposed tryptophan residues are shown in magenta and provide an approximate indication of the position of the membrane surface as represented by horizontal lines. (B) Same representation of the $Ca_2E1:ADP:AlF_4^-$ state (1T5T) showing occlusion by concerted movements of the A-domain and the M1–M2 segment. Note the kink occurring in M1, and the translocation of M1–M2 relative to the membrane. Conformational changes are indicated by arrows*

the structure of the Ca^{2+}-ATPase with nucleotide bound was almost identical with that of the $Ca_2E1:ADP:AlF_4^-$ complex, except that in the AMPPCP bound complex the γ-phosphate group replaced AlF_4^- in an evidently weaker interaction with Asp351 and that the Mg^{2+} ion interacting with the α,β- phosphate was lacking.[9] Thus, as the first conclusion, mere binding of nucleotide is sufficient to induce the compact state, bringing the terminal phosphate group in apposition to Asp351, but without there being any indication for the formation of a transition state complex. This agrees well with kinetic evidence on the ATP phosphorylation reaction which indicates that the conformational change accompanying ATP binding (which must proceed through interaction with the N-domain and its subsequent tilting towards the P-domain) is the rate-limiting step, while the actual phosphoryl transfer to Asp351 is an exceedingly rapid process.[46] The Mg^{2+} ion that we found to be bound to the α,β-phosphate groups of ADP in the $ADP:AlF_4^-$ complex could indicate that after hydrolytic cleavage of ATP and product formation, the leaving group is present as MgADP rather than as ADP, while the Mg^{2+} at the phosphorylation site remains bound to the phosphate and the protein. This idea would be in accordance with functional evidence suggesting a role for a second bound Mg^{2+} ion for Ca^{2+}-ATPase activity.[47,48]

The near identity of the two ATPase forms extends to the membrane sector, and in both cases the N-terminal entrance appears to be blocked as a consequence of the interaction between the N- and P-domains moving the A-domain and the M1–M2 segments in place. One might therefore have expected both forms to be able to occlude Ca^{2+}, but this is not the case. The functional evidence indicates that under physiological conditions Ca^{2+} is only stably occluded in the $Ca_2E1:ADP:AlF_4^-$ complex,[9] and is rapidly released from the AMPPCP complex.[49] Furthermore, Ca^{2+}-ATPase with AMPPCP is susceptible to proteolytic digestion and to reaction with antibodies, and it displays highly reactive sulfhydryl groups, suggestive of a dynamically fluctuating structure that may then permit passage of Ca^{2+} despite the closed entrance indicated by the X-ray structure.[9] On the other hand, the $ADP:AlF_4^-$ transition state complex resists these treatments[9] suggesting a fixed and compact structure of this form.

Why then are the $ADP:AlF_4^-$ and AMPPCP complexes crystallized in near-identical conformations? A recent reinvestigation provides a possible answer and indicates that the crystallization conditions (in particular the 5–10 mM Ca^{2+} concentration used in the crystallization drops) significantly retards the release of intramembraneously bound Ca^{2+}(see Ref. 50) from the AMPPCP complex in membrane bound Ca^{2+}-ATPase. Thus, high levels of Ca^{2+} seem to stabilize the occluded state in the AMPPCP complex. The crystal structure probably adopts this state, which is otherwise only stable with $ADP:AlF_4^-$, at physiological levels of Ca^{2+} in the medium. In fact, conformational stability of transition complexes seems to be a general feature of the ATPase in both its E1 and E2 forms.[2,51] For physiological ligands, the transition states are, on the other hand, of short duration, but by their ability to promote a transient "frozen" conformation, resulting in occlusion, these states are of paramount importance for the strict coupling between membrane translocation and the chemical processes taking place in the cytosolic part.

Toyoshima and Mizutani[10] in their discussion of the AMPPCP complex further emphasize the compatibility of their structure with a gating function of Glu309 in addition to the critical role that this residue plays for intramembraneous binding of Ca^{2+}. This was previously suggested by Lee and East [52] and is supported by site directed mutagenesis data, showing increased leakage by the E309Q mutant[53] and by other nearby residues.[54]

6 The Dephosphorylation Step and Proton Counter Transport

After phosphorylation by ATP and occlusion, the Ca^{2+}-ATPase undergoes a conformational change to the E2P state with opening of the luminal gate, which leads to release of Ca^{2+} to the other side of the membrane. We shall defer discussion of this important event, for which there is so far no representative structure, and jump directly to the subsequent dephosphorylation step where important new information has been gathered.[12,13] In the absence of Ca^{2+} and ADP, AlF_4^- can also bind to the ATPase and induce a transitional state analogue of E2P,[45] the structure of which has now been solved after stabilization of the ATPase with thapsigargin.[13] A striking feature of the $E2:AlF_4^-$(TG) complex formed in this way is that the N-domain, by bending of M4 and M5 with their cytosolic extensions, has moved away from the close association with the P-domain that characterizes the structure with occluded Ca^{2+}. This enables the loop containing the TGES motif of the A-domain, the third cytosolic player, to enter into intimate contact with the phosphorylation site where AlF_4^- is bound to Asp351 and coordinated with Mg^{2+} in the same way as in the $Ca_2E1:AlF_4^-$:ADP complex. To position the TGES motif at the phosphorylation site, the A-domain has to rotate ~110° along an axis normal to the membrane, followed by an upward movement towards the phosphorylation site. As shown in Figure 4B the TGES motif replaces ADP in the $Ca_2E1:AlF_4^-$:ADP binding cavity with Glu183 and Thr181 forming coordinated bonds with AlF_4^- via a bridging water molecule. On the other hand the covalent character of the AlF_4^- bonds with Asp351 is maintained as in the $Ca_2E1:AlF_4^-$:ADP complex. But in this structure the stage is set for hydrolysis at the phosphorylation site by an associative SN2 mechanism where Glu183 functions as a general base catalyst by proton abstraction from the bridging water molecule. The essential catalytic role of Glu183 in the hydrolysis of the physiological intermediate E2P has found striking support in site-directed mutagenesis experiments.[55]

The dephosphorylation reaction is coupled to the transport of protons in the opposite direction of Ca^{2+} translocation, and it should be noted that in the E2P transition state mimicked by $E2:AlF_4^-$ the protonated complexes are in an occluded state inside the membrane.[13,56] The occluded protons cannot be seen directly, but their presence in the $E2:AlF_4^-$ structure can be deduced as all the acidic Ca^{2+} liganding groups (Glu771, Asp800, Glu908, and Glu309) remain inside the nonpolar membrane environment where no Ca^{2+} is present to neutralize their charge. If, for example, a $2Ca^{2+}: 2H^+$ exchange is to occur, two of the charged liganding groups (with Asp800 and Glu771 as likely candidates based on molecular dynamic calculations[57]) must have become neutralized by luminal protons during the preceding luminal Ca^{2+} release. There is ample evidence from kinetic experiments[58–60] that protons, even after dephosphorylation, can remain at least partially bound to the ATPase in an occluded H_nE2 state from which they are released towards the cytoplasm upon addition of Ca^{2+} to induce the $H_nE2 + 2\ Ca^{2+} \rightarrow Ca_2E1 + nH^+$ conversion. In accordance with the presence of occluded protons in the $E2:AlF_4^-$ structure, the Ca^{2+} binding residues have assumed a configuration not compatible with high affinity binding of Ca^{2+}, and furthermore they are separated from the lumen by a dense, 15 Å thick barrier consisting mainly of hydrophobic and some basic amino acid residues (cf. Figure 6). The structure is consistent with the $E2:AlF_4^-$ representing a transitional state, rather than the ground state of E2P where the translocation sites are in contact with the lumen. In further support of the $E2:AlF_4^-$ being a transition state complex with occluded transport sites, luminal and cytosolic Ca^{2+} only is capable of reactivating the $E2:AlF_4^-$ complex very slowly.[13] This is in contrast to the complex of Ca^{2+}-ATPase with BeF_3^- that is considered to stabilize the ATPase in the E2P ground state and that, like E2P, is immediately dissociated by both luminal and cytosolar Ca^{2+} (unpublished observations).[56]

Simultaneously, with our description of the $E2:AlF_4^-$ structure, Toyoshima *et al.*[12] has published the structure of another E2 complex, formed by the interaction of

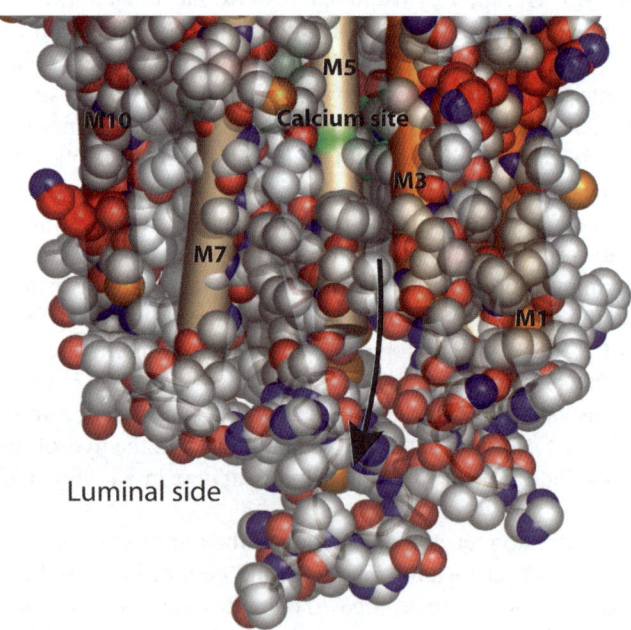

Figure 6 *The lipophilic barrier towards the lumen in the E2P transition state. The membrane region of the $E2:AlF_4^-$ intermediate is depicted as semitransparent spheres in a cartoon representation with the M1–M3 transmembrane segment in orange, and the M4–M10 segment in wheat. Residues constituting the calcium-binding site are shown in green and positively charged residues in red. The arrow indicate the possible release path used by calcium ions in the preceding E2P ground state, though no path exists between the calcium binding residues and the lumen in the depicted conformation*

ATPase with MgF_4^{2-}, which as a structural analogue of phosphate is considered to represent the E2P product state of the dephosphorylation reaction. From a detailed comparison with $Ca_2E1{:}ADP{:}AlF_3$, published alongside with $E2{:}MgF_4^{2-}$ in the same paper, M3 and M4 were concluded to be deeper inserted in the membrane by one helical turn, accompanied by displacements in the M1 and M2 transmembrane segments, but retaining the characteristic kink of the N-terminal part of the M1 helix. In agreement with the E2 status of the MgF_4^{2-} complex the Ca^{2+} ligating groups are not in a proper position for high affinity Ca^{2+} binding; this is in particular caused by the deep insertion of the M4 helix, and concomitant rotation of the M6 helix, removing Asp800 and the other critical residues in M6 from the center of the Ca^{2+} binding cavity.[8] In the cytosolic regions the same movements of the N-, P-, and A-domain were observed as in our study. MgF_4^{2-} was also found to interact with Asp351 and Glu183 of the TGES motif, but the interaction seems to include Gly182, rather than Thr181 as in our $E2{:}AlF_4^-$ structure. A characteristic feature of both structures is that, as a consequence of the movement of the N-domain and rotation of the A-domain, the TGES motif has replaced bound nucleotide at the phosphorylation site in the $Ca_2E1{:}ADP{:}AlF_4^-$ complex. In addition to providing a plausible mechanism for the dephosphorylation mechanism the presence of the TGES motif at the phosphorylation site also gives a direct structural explanation of the ADP insensitivity of the E2P state to dephosphorylation by ADP. It should be noted that insensitivity to ADP is not caused by the absence of nucleotide binding. In fact, Toyoshima *et al.*[12] detected the presence of bound ADP in their structure that either had remained as a contaminant from their affinity purification of Ca^{2+}-ATPase, or subsequently had been added to their MgF_4^{2-} treated ATPase. The adenosine part of the ADP was bound to the N-domain and the β-phosphate interacted with Lys205 at the periphery of the A-domain. Consistent with low affinity ADP binding we have found that the rate of formation of $E2{:}AlF_4^-$ is reduced, but not abolished in the presence of ADP (unpublished observation).

In the E2 state, without bound AlF_4^- and MgF_4^{2-}, the TGES motif is retracted from intimate interaction with the phosphorylation site, permitting the N-domain to regain a closer approach to the P-domain,[8] but otherwise the structure, including the rotation of the A-domain, is remarkably similar to that of both the $E2{:}AlF_4^-$ and $E2{:}MgF_4^{2-}$ forms. One of the consequences of the rotation of the A-domain concerns the A-M3 arm, the polypeptide chain linking the A-domain with M3. In the E1P structure this segment is in a peripheral and loosely attached position. During the rotation of the A-domain this segment is moved towards the center of the protein linked to the C-terminal part of the P-domain, where it forms part of a cluster of five helices that, via a binding site for K^+ (at Glu732 and Lys712), seems to regulate the E2P dephosphorylation rate by an allosteric mechanism.[11] As a result of the changes taking place by a 90° rotation of the M1/M2 segments, an N-terminal opening is formed, lined by a number of polar residues in the M1–M3 residues which leads directly to Glu309, the putative gating residue for Ca^{2+} binding at site II.[8] Thus during the structural changes taking place in the transition from the E1P to the E2 state the originally N-terminal occluded structure has been converted into a more open and loose structure that can be envisaged to act as a cytosolic entrance for Ca^{2+} to allow interaction with the putative gating residue at Glu309.

7   Getting Ca^{2+} in and out of the Membrane

Inside the membrane, the two Ca^{2+} binding sites[58-60] are found located alongside each other at the same level;[7] at Site I the Ca^{2+} binding ligands are formed by side chains from M5 (Glu771), M6 (Thr799, Asp800), and M8 (Glu908). Ca^{2+} binding at site II is mainly located on the M4 helix (liganded by the side chain of Glu309, main chain carbonyls of Val304, Ala305, and Ile307) and with participation of Asn796 and Asp800 from M6. With the exception of Glu908, site-directed mutagenesis of any of the liganding side chains leads to the abolishment of Ca^{2+} occlusion by CrATP.[27] To enable an optimal Ca^{2+} ligand-binding configuration, transmembrane helices M4, M5, and M6 are kinked as a result of the presence of helix-breaking Schellman (bend forming) motifs, which includes the Glu- and Asp-liganding residues, a feature that has been detected by NMR analysis of the conformational dynamics of the isolated transmembrane peptides solubilized in detergent.[61,62] It has been found that conservative mutation of any of the Ca^{2+}-binding sidechains, related to site I, completely destroys high-affinity Ca^{2+} binding by both sites, whereas one Ca^{2+}-binding site can be retained after modification of Glu309 and Asn796, the two side chains specifically involved in binding at site II.[63,64] From this, and from an analysis of the ability of such mutants to become phosphorylated by ATP or P_i,[63,64] it is concluded that binding of the first Ca^{2+} at site I (with medium affinity) sets the stage for binding of Ca^{2+} at site II by a cooperative mechanism that results in high affinity binding of both Ca^{2+} ions. This conformational change involves the cytosolic domains so as to allow ATP to phosphorylate the ATPase at Asp351. However, there is hardly any change in the configuration of the Ca^{2+} binding ligands in the Ca_2E1 and in the occluded $Ca_2E1:ADP:AlF_4^-$ state: as earlier explained, the intramembraneous changes accompanying occlusion primarily comprise M1 and M2 (not involved in intramembraneous binding of Ca^{2+}), with little effect on the disposition of the other transmembrane segments.

The structural evidence for an N-terminal entrance port discussed previously, together with a common Ca^{2+} binding cavity, has been suggested to provide an explanation of the well-established sequential Ca^{2+}-binding mechanism, whereby cytosolic release of bound Ca^{2+} at site I is prevented by Ca^{2+} bound at site II. This phenomenon can be easily demonstrated by comparing the release of bound Ca^{2+} into media containing either EGTA (no Ca^{2+}) or a high (1 mM) concentration of Ca^{2+}.[49,53] In the former case both Ca^{2+} ions are rapidly released, while in the latter situation only half of the Ca^{2+} (see Ref. 45) readily exchanges with unlabeled Ca^{2+} in the medium, because Ca^{2+} bound at site II prevents release of Ca^{2+} at site I. However, it should be mentioned that there is also evidence for the existence of a separate C-terminal entrance of Ca^{2+} that, via a pre-binding site for Ca^{2+} localized at the L6-7 loop, directly leads to intramembraneous Ca^{2+} binding at site I.[65,66] In that case, to account for the reduced Ca^{2+} release at high levels of Ca^{2+} in the medium, the conformational changes taking place during cooperative Ca^{2+} binding, would include closure of the C-terminal gate.

In agreement with the kinetic scheme for Ca^{2+} transport (Figure 1) it is customary to refer to E1P and E2P as "high-energy" and "low-energy" intermediates respectively, that are related to the cation active transport step across the membrane.

In the original formulation of the kinetic scheme for Ca^{2+} transport all E1 and E2 forms, as suggested by de Meis and Vianna,[67] whether phosphorylated or not, were supposed to represent ATPase forms with the same sidedness of the transport sites: cytoplasmic in the case of E1, and intraluminal for the E2 forms. However, it has since then become clear that the nonphosphorylated E2 forms of Ca^{2+}-ATPase (in contrast to E2P) interact with Ca^{2+} towards the cytosolic side only. The present position is to define E2 as those forms of Ca^{2+}-ATPase which harbour intramembranal sites in a protonated and occluded state. In our opinion, all the presently known structures, including the E2·P product and E2-P transition state, fit nicely into this definition of an E2 conformation: in the E2P transition state, stabilized by AlF$_4^-$, the translocation sites are separated from the lumen by a compact hydrophobic barrier and are only momentarily in contact with the cytosolic side insofar so that the occluded protons can be released in exchange with cytosolar Ca^{2+}. In the E1P transition state the intramembraneously bound Ca^{2+} is neither in contact with the cytosol nor with the lumen. We also know the structure of Ca^{2+}-ATPase in the Ca$_2$E1 and E2 forms, but still lack the knowledge of the structure of the important intermediate where the Ca^{2+} binding sites open towards the luminal side. We refer to this intermediate as the E2P ground state to differentiate it from the proton occluded E2P transition state or other E2 conformations of known structure. We can at present only speculate what this state looks like and how energy is transferred from the cytosolic domains to the intramembranous sector to release Ca^{2+} on the luminal side. However, the impetus for these movements seems to come from the A-domain and its attached M3 arm. This is suggested by proteolytic cleavage experiments on Ca^{2+}-ATPase[68] and Na$^+$, K$^+$-ATPase,[69] where it has been shown that cleavage of the polypeptide chain close to the C-terminal end of the A-domain is accompanied by a specific block of the E1P to E2P transition that suggests the involvement of a sequence motif ([232]I G K I) at the C-terminal end of the A-domain.[70] Moreover, recent site-directed mutagenesis experiments have shown that alanine substitutions of any of the residues in the TGES motif and of Ile235 lead to a block of the E1P to E2P transition,[71] in addition to the well-established role of Gly233 in blocking this transition.[72] These residues, by their ability to interact with the P-domain in the E2P state, must play a crucial role for mediating the conformational processes which from the phosphorylation site spread to the membraneous sector to expose the intramembraneously occluded Ca^{2+} to the luminal side.

A somewhat different view on the existence of a unique structure of E2P in the ground state is taken by Toyoshima *et al.*[12] who assume that the E2P product state (as represented by the E2:MgF$_4^{2-}$ complex whose structure they have solved) also can be used as a starting point to represent the E2P ground state with exposure of its low-affinity calcium binding sites to the luminal side. Hence they consider that a close examination of the presently known forms will give important clues about the characteristics of the E2P ground state. They note that in the E2:MgF$_4^{2-}$ structure a distal liganding residue for Ca^{2+} at site II (Val304) is displaced and propose that the distal part M4, as a result of rotation of M1/M2 and increased tilting of the A-domain arising from the interaction with the phosphorylation site, will cause the distal part of M4 to move outwards and thereby create room for passage of Ca^{2+} towards the lumen. In accordance with this view the presence of intraluminal sites has been inferred from kinetic experiments[73,74] and chemical modification of intraluminal carboxyl groups,[75] and from direct binding experiments.[76]

Despite the aforementioned evidence we have some reservations about a simple channel mechanism to account for the translocation step. First of all, it seems to us that the published structures of the E2 and of the product and transition states of E2P are all too much alike for any of them to serve as a representative model of the E2P ground state. In addition the model, as also noted, *e.g.*, by Toyoshima and Inesi,[77] results in ambiguities concerning the sidedness of the E2 and E2P structures. This is especially the case if the E1 to E2 transition is to be taken as an analogue of the E1P to E2P transition, as was done in the original formulation by de Meis and Vianna.[67] The difficulty then arises that we do not understand why the intramembranous binding sites in the E2 state only can interact with Ca^{2+} from the cytosolic side, and conversely that the E2P ground state only interacts with Ca^{2+} via the luminal space. Finally, the barrier of hydrophobic and basic residues separating the intramembranous binding sites from the lumen in the structures representing the E2P transition state and E2P product state is impressive (cf. Figure 6) and does not seem appropriate for the formation of a simple gated channel for Ca^{2+}. While we agree that the pathway for Ca^{2+} probably involves passage between central transmembrane segments such as M6/M4 or M5/M3, translocation probably also entails larger conformational changes than can be envisaged by the channel gated mechanism. The need for more extensive conformational change may arise because of the high energy expenditure required for the Ca^{2+}-ATPase to perform as a primary active transporter with the ability to accumulate Ca^{2+} against large concentration gradients.

8   Compact *vs.* Open Conformations of SERCA

The concentrations of ATP higher than those required to phosphorylate the ATPase exerts a substantial accelerating effect on ATPase turnover over a wide range of nucleotide concentrations, as a result of ATP being bound to the phosphorylated intermediates and E2 states during turnover. A remarkable feature of all the new structures of SERCA is that they are characterized by having compact cytosolic conformations, in agreement with the low resolution structures obtained from 2D crystals,[23,78] but markedly different from the open structure of Ca^{2+}-ATPase with bound Ca^{2+} in the absence of ATP.[7] Since the intracellular concentration of ATP is about four orders of magnitude higher than that of cytosolar free Ca^{2+}, and ATP can be assumed to react with ATPase with about the same rate constant as Ca^{2+}, it is questionable whether the open nucleotide-free conformation is a representative conformational structure during the normal operation of the cycle. However, regardless of this circumstance we believe that it would be a mistake to dismiss the open structure as being of little interest, or even an artefact produced by the crystalline conditions. First of all, we have among our many crystallization attempts, performed under different conditions, been able to reproduce the open structure, but with rather pronounced variations of the position of the A-domain relative to that of the P-domain (T.L. Sørensen, A.M.L. Jensen, J.V. Møller and P. Nissen, unpublished data). Second, the dramatic effect exerted by the binding of Ca^{2+} alone is of major interest, since it is a manifestation of the forces that are decisive in initiating the Ca^{2+}-dependent phosphorylation by ATP, whether or not ATP is initially bound at the start of the reaction. Thus, we can observe that the protein relaxes into an open conformation in the presence of Ca^{2+}. Only in the presence of ATP will the compact Ca_2E1P state (as represented by the $Ca_2E1:ADP:AlF_4^-$ structure) be obtained yielding the high-energy intermediate state. Probably, the effect of orienting the transmembrane segments for high affinity binding of Ca^{2+} is transmitted to the phosphorylation site, via the well-conserved M4 polypeptide chain, connected with Ca^{2+} binding at site II, and possibly also from site I via the L6-7 loop as previously pointed out.[7,79] The significance of the open Ca_2E1 structure is then that it probably reflects some feature which is important to adapt the P-domain for phosphorylation such as the temporary removal of the A-domain from the phosphorylation site to a position which in the absence of ATP is only loosely attached to the P-domain.

9   Conclusions and Perspectives

Recently, a series of structure determinations have nearly completed a structural description of the transport cycle of the sarcoplasmic reticulum Ca^{2+}-ATPase. From these structures Ca^{2+}-ATPase emerges as a molecular machine where globular cytosolic domains and transmembrane helices work in concert like a mechanical pump. This picture can be vividly demonstrated in animated versions of the pump cycle, based on these structures, which, *e.g.*, show how the M3 and M4 membrane segments perform piston like movements in relation to transport and, by reciprocating movements transmitted by the cytosolic extensions of the M1–M5 segments, affect the chemical machinery in the cytosolic domains. The existence of occluded states associated with the Ca^{2+}-dependent phosphorylation by ATP and the protonation-dependent dephosphorylation mechanism allow the tight coupling between transport in the membrane and the chemical reactions taking place at the cytosolic phosphorylation site. However, important elements are still missing in this picture: we do not know any detailed structures of the E2P ground state and do not have a detailed understanding of the reaction cycle such as the modulatory (accelerating) effect of ATP, which includes the effect on steps not directly involved in phosphorylation by ATP, and also the understanding of the entrance and exit pathways for binding and release of Ca^{2+} is still lacking. There are some additional unclarified issues, as the way Ca^{2+}-ATPase are regulated by protein modifiers like phospholamban and sarcolipin, and how pharmacological agents such as thapsigargin affect the functional cycle of the ATPase. Further, functional/structural studies using substrate analogues, new inhibitors and mutated forms of SERCA will probably help to clarify these matters and complete the description of the cycle.

References

1. S. Lutsenko and J.H. Kaplan, *Biochemistry*, 1995, **34**, 15607.
2. J.V. Møller, B. Juul and M.le Maire, *Biochim. Biophys. Acta*, 1996, **1286**, 1.
3. K.B. Axelsen and M.G. Palmgren, *J. Mol. Evol.*, 1998, **46**, 84.
4. P.L. Pedersen, *J. Bioenerg. Biomembr.*, 2002, **34**, 327.

5. E. Carafoli, *Proc. Nat. Acad. Sci.*, 2002, **99**, 1115.
6. L. Aravind, M.Y. Galperin and E.V. Koonin, *Trends Biochem. Sci.*, 1998, **23**, 127.
7. C. Toyoshima, M. Nakasako, H. Nomura and H. Ogawa, *Nature*, 2000, **405**, 647.
8. C. Toyoshima and H. Nomura, *Nature*, 2002, **418**, 605.
9. T.L.-M. Sørensen, J.V. Møller and P. Nissen, *Science*, 2004, **304**, 1672.
10. C. Toyoshima and T. Mizutani, *Nature*, **430**, 529.
11. T.L.-M. Sørensen, J.D. Clausen, A.-M. L. Jensen, B. Vilsen, J.V. Møller, J.P. Andersen and P. Nissen, *J. Biol. Chem.*, 2004, **279**, 46355.
12. C. Toyoshima, H. Nomura and T. Tsuda, *Nature*, 2004, **432**, 361.
13. C. Olesen, T.L.-M. Sørensen, R.C. Nielsen, J.V. Møller and P. Nissen, *Science*, 2004, **306**, 2251.
14. D. Levy, M. Seigneuret, A. Bluzat, J.-L. Rigaud. *J. Biol. Chem.*, 1990, **265**, 19524.
15. F. Cornelius and J.V. Møller, *FEBS Letters.*, 1991, **284**, 46.
16. X. Yü, S. Carroll, J.-L. Rigaud and G. Inesi, *Biophys. J.*, 1993, **64**, 1232.
17. U. Gerdes and J.V. Møller, *Biochim. Biophys. Acta*, 1983, **734**, 191.
18. X Yu, L. Hao and G. Inesi, *J. Biol. Chem.*, 1994, **269**, 16656.
19. L. de Meis, A.P. Arruda, W.S. Da-Silva, M. Reis and D.P. Carvalho, *Ann. N. Y. Acad. Sci.*, 2003, **985**, 481.
20. C. Tanford, *J. Gen. Physiol.*, 1981, **77**, 223.
21. C.M. Pickart and W.P. Jencks, *J. Biol. Chem.*, 1984, **259**, 1629.
22. N.M. Green and D.L. Stokes, *Acta Physiol. Scand.*, 1992, **146**, 59.
23. P. Zhang, C. Toyoshima, K. Yonekura, N.M. Green and D.L. Stokes, *Nature*, 1998, **392**, 835.
24. D.L. Stokes and N.M. Green, *Biophys. J.*, 2000, **78**, 1765.
25. P. Champeil, T. Menguy, S. Soulié, B. Juul, A.G. de Gracia, F. Rusconi, R. Falson, L. Denoroy, F. Henao, M. le Maire, J.V. Møller, *J. Biol. Chem.*, 1998, **273**, 6619.
26. D.M. Clarke, T.W. Loo, G. Inesi and D.H. MacLennan, *Nature*, 1989, **339**, 476.
27. J.P. Andersen and B. Vilsen, *J. Biol. Chem.*, 1992, **267**, 19383.
28. J.P. Andersen, *Biosci. Rep.*, 1995, **15**, 243.
29. O. Thastrup, P.J. Cullen, B.K. Drobak, M.R. Hanley and A.P. Dawson, *Proc. Nat. Acad. Sci.*, 1990, **87**, 2466.
30. C. Patlak, *Bull. Math. Biol.*, 1957, **19**, 209.
31. W. Wilbrandt and T. Rosenberg, *Pharmacol. Rev.*, 1961, **13**, 109.
32. O. Jardetzky, *Nature*, 1966, **211**, 969.
33. C. Tanford, *Annu. Rev. Biochem.*, 1983, **52**, 379.
34. P. Läuger, *Electrogenic Ion Pumps*, Sinauer Associates, MA, 1991.
35. R. MacKinnon, *FEBS Lett.*, 2003, **555**, 62.
36. A. Accardi and C. Miller, *Nature*, 2004, **427**, 803.
37. D.C. Gadsby, *Nature*, 2004, **427**, 795.
38. L. de Meis, V.A. Suzano and G. Inesi, *J. Biol. Chem.*, 1990, **265**, 18848.
39. L. de Meis and G. Inesi, *FEBS Lett.*, 1992, **299**, 33.
40. Y. Dupont, *Eur. J. Biochem.*, 1980, **109**, 231.
41. H. Takisawa and M. Makinose, *J. Biol. Chem.*, 1983, **258**, 2986.

42. E.H. Serpersu, U. Kirch and W. Schoner, *Eur. J. Biochem.*, 1982, **122**, 347.

43. B. Vilsen and J.P. Andersen, *J. Biol. Chem.*, 1992, **267**, 3539.

44. C. Coan, J.-Y. Ji and J.A. Amaral, *Biochemistry*, 1994, **33**, 3722.

45. A. Trouillier, J.-L. Girardet and Y. Dupont, *J. Biol. Chem.*, 1992, **267**, 22821.

46. C.M. Pickart and W.P. Jencks, *J. Biol. Chem.*, 1982, **257**, 5319.

47. Y. Takakuwa and T. Kanazawa, *J. Biol. Chem.*, 1982, **257**, 426.

48. J.D. Clausen, D.B. McIntosh, B. Vilsen, D.G. Woolley and J.P. Andersen, *J. Biol. Chem.*, 2003, **278**, 20245.

49. S. Orlowski and P. Champeil, *Biochemistry*, 1991, **30**, 352.

50. M. Picard, C.Toyoshima and P. Champeil, *J. Biol. Chem.*, 2005, **280**, 18745.

51. S. Danko, T. Daiho, K. Yamasaki, M. Kamidochi, H. Suzuki and C. Toyoshima, *FEBS Lett.*, 2001, **489**, 277.

52. A.G. Lee and J.M. East, *Biochem. J.*, 2001, **356**, 665.

53. G. Inesi, H. Ma, D. Lewis and C. Xu, *J. Biol. Chem.*, 2004, **279**, 31629.

54. A.P. Einholm, B. Vilsen and J.P. Andersen, *J. Biol. Chem.*, 2004, **279**, 15888.

55. J.D. Clausen, B.Vilsen, D.B. McIntosh, A.P. Einholm and J.P. Andersen, *Proc. Nat. Acad. Sci.*, 2004, **101**, 2776.

56. S. Danko, K. Yamasaki, T. Daiho and H. Suzuki, *J. Biol. Chem.*, 2004, **279**, 14991.

57. Y. Sugita, N. Miyashita, M. Ikeguchi, A. Kidera and C. Toyoshima, *J. Am. Chem. Soc.*, 2005, **127**, 6150.

58. V. Forge, E. Mintz and F. Guillain, *J. Biol. Chem.*, 1993a, **268**, 10953.

59. V. Forge, E. Mintz and F. Guillain, *J. Biol. Chem.*, 1993b, **268**, 10961.

60. S. Wakabayashi and M. Shigekawa, *Biochemistry*, 1990, **29**, 7309.

61. S. Soulié, J.-M. Neumann, C. Berthomieu, J.V. Møller, M. le Maire and V. Forge, *Biochemistry*, 1999, **38**, 5813.

62. G. Nielsen, A. Malmendal, A. Meissner, J.V. Møller and N.C. Nielsen, *FEBS Lett.*, 2003, **544**, 50.

63. I.S. Skerjanc, T. Toyofuku, C. Richardson and D.H. MacLennan, *J. Biol. Chem.*, 1993, **268**, 15944.

64. C. Strock, M. Cagagna, W.E. Peiffer, C. Sumbilla, D. Lewis and G. Inesi, *J. Biol. Chem.*, 1998, **273**, 15104.

65. P. Falson, T. Menguy, F. Corre, L. Bouneau, A.G. de Gracia, S. Soulié, F. Centeno, J.V. Møller, P. Champeil and M. Le Maire, *J. Biol. Chem.*, 1997, **272**, 17258.

66. T. Menguy, F. Corre, B. Juul, L. Bouneau, D. Lafitte, P.J. Derrick, P.S. Sharma, P. Falson, B.A. Levine, J.V. Møller and M. le Maire, *J. Biol. Chem.*, 2002, **277**, 13016.

67. L de Meis and A.L.Vianna, *Annu. Rev. Biochem.*, 1979, **48**, 275.

68. J.V. Møller, G. Lenoir, C. Marchand, C. Montigny, M. le Maire, C. Toyoshima, B.S. Juul and P. Champeil, *J. Biol. Chem.*, 2002, **277**, 38647.

69. P.L. Jørgensen and J. Petersen, *Biochim. Biophys. Acta*, 1985, **821**, 319.

70. J.V. Møller, G. Lenoir, M. Le Maire, B.S. Juul and P. Champeil, *Ann. N. Y. Acad. Sci.*, 2003, **986**, 82.

71. G.Wang, K. Yamasaki, T. Daiho and H. Suzuki, *J. Biol. Chem.*, 2005, **280**, 26508.

72. J.P. Andersen, B. Vilsen, E. Leberer and D.H. MacLennan, *J. Biol. Chem.*, 1989, **264**, 21018.

73. W.P. Jencks, T. Yang, D. Peisach and J. Myung, *Biochemistry*, 1993, **32**, 7030.
74. J. Myung and W.P. Jencks, *Biochemistry*, 1994, **33**, 8775.
75. R.J. Webb, Y.M. Khan, J.M. East and A.G. Lee, *J. Biol. Chem.*, 2000, **275**, 977.
76. A. Vieyra, E. Mintz, J. Lowe and F Guillain, *Biochim. Biophys. Acta*, 2004, **1667**, 103.
77. C. Toyoshima and G. Inesi, *Annu. Rev. Biochem.*, 2004, **73**, 269.
78. C. Xu, W.J. Rice, W. He and D.L. Stokes, *J. Mol. Biol.*, 2002, **316**, 201.
79. Z. Zhang, D. Lewis, C. Sumbilla, G. Inesi and C. Toyoshima, *J. Biol. Chem.*, 2001, **276**, 15232.

CHAPTER 17

Comparison of the Multidrug Transporter EmrE Structures Determined by Electron Cryomicroscopy and X-ray Crystallography

C.G. TATE

MRC Laboratory of Molecular Biology, Hills Road, Cambridge CB2 2QH, UK

1   Introduction

Multidrug resistance in bacteria is becoming a serious worldwide threat to human health.[1,2] Simple bacterial infections acquired in hospitals can quickly escalate into life-threatening diseases because of the ability of bacteria to multiply in the presence of antibiotics. Different strategies have evolved in bacteria to confer a drug-resistant phenotype and include the destruction or chemical modification of the antibiotic, or the mutation of the target protein to which the antibiotic normally binds. The simplest strategy, which is the topic of this chapter, is to remove the drug by efflux of the antibiotic or antiseptic out of the cell. All the strategies are thought to have evolved to combat naturally occurring toxins in the environment or antibiotics secreted by fungi. It is unfortunate that the majority of antibiotics and antiseptics are based on natural products, which existing bacterial detoxification mechanisms can either cope with or require only minor modification by a few mutations to increase their efficacy towards the new compounds.[3,4]

Drug efflux from bacteria is catalysed by at least five different superfamilies of integral membrane proteins[5,6] (Figure 1). The energy for drug efflux is derived from either direct hydrolysis of ATP, as in the ABC superfamily of transporters, or the inward movement of protons down their concentration gradient, driven by the proton motive force. These drug antiporters include members of the major facilitator superfamily (MFS), the resistance/nodulation/cell division (RND) and the drug/metabolite transporter (DMT) superfamilies, and the multiantimicrobial extrusion (MATE) family.[6] A further characterisation of the transporters is made according to the mode of recognition of their substrates. Many transporters recognise only a single antibiotic, such as the tetracycline antiporter, a member of the MFS. In contrast, multidrug transporters recognise a multitude of substrates that share the same chemical characteristics. For example, EmrE transports compounds that are large, hydrophobic cations.[7] The most detailed structural knowledge we have for a multidrug transporter is for AcrB,[8] a member of the RND superfamily, that couples the proton motive force to the efflux of drugs across both the cytoplasmic and periplasmic membranes in Gram-negative bacteria. The structure of AcrB has also been determined with various different substrates bound;[9] this structure supports the hypothesis that the recognition of diverse substrates is accomplished by having a large, flexible hydrophobic binding site.[10] Three-dimensional (3D) structures for representatives of the MFS[11,12] and ABC[13–15] superfamilies have also been determined by X-ray crystallography to a resolution of 3.2–4.5 Å (Figure 1). These have given a valuable insight into the mechanism of transport, but specific multidrug transporter structures for these families have not yet been obtained. Structures for representatives of the MATE family also have not yet been obtained.

EmrE is the archetypal member of the small multidrug resistance (SMR) family[16] that is part of the larger DMT superfamily.[17] EmrE was originally characterised as the factor that gave *Escherichia coli* their resistance to ethidium and methyl viologen.[18,19] In addition, a homologue was isolated from a plasmid in clinical isolates of a multidrug resistant strain of *Staphylococcus aureus*, along with the MFS transporter, QacA.[20] EmrE was subsequently characterised as a multidrug transporter with a wide substrate specificity that included quaternary amine compounds often used as antiseptics and disinfectants.[7] Structural models for EmrE have been obtained by two different techniques, X-ray crystallography[21] and electron crystallography[22] to 3.8 and 7.5 Å resolution, respectively. The fundamental difference in data collection between X-ray and electron crystallography arises owing to the different nature of the crystals. In electron crystallography, the membrane protein is in a 2D crystal (*i.e.* it is one molecule thick), which is grown by reconstitution of the membrane protein into a lipid bilayer at very high protein–lipid ratios.[23] Thus, the membrane protein is reintroduced back into its native lipid environment. In contrast, X-ray crystallography requires 3D crystals that contain the membrane protein surrounded by detergent in an attempt to mimic the hydrophobic lipid environment immediately around the transmembrane regions of the membrane protein.[24] High-resolution X-ray structures have been determined for many membrane proteins;[25] the ability of the structures to explain the complex biochemical phenomena of transport, electron transfer and energy generation shows the importance of X-ray crystallography in membrane protein structure determination. However, in the case of EmrE, the two different environments for the molecule resulted in two different structures, a phenomenon that was previously considered extremely unlikely. These two structures will be discussed in the context of biochemical and biophysical data for EmrE.

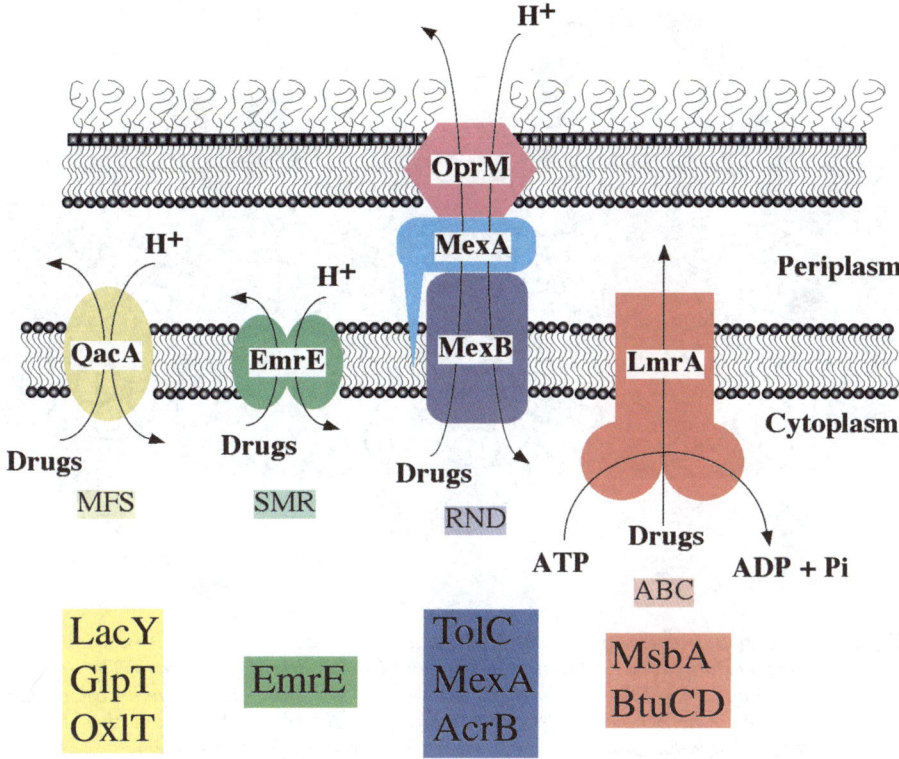

Figure 1 *The major multidrug efflux systems in bacteria. Superfamilies of transporters are represented by a single archetype and the mode of energy transduction to drug efflux is indicated. Underneath each structure is a list of proteins whose 3D structures have been determined to either high resolution by X-ray crystallography (AcrB,[8,9] TolC,[61] MexA,[62] MsbA,[13,14] BtuCD,[15] LacY,[11] GlpT,[12] EmrE[21]) or to medium resolution by electron crystallography (OxlT,[63] EmrE[22])*

2 The Oligomeric State of EmrE

EmrE is a remarkably small protein of 110 amino acid residues, which is typical of all members of the SMR family. Hydropathy analysis predicted the presence of four hydrophobic regions (Figure 2) that are sufficiently long to span a biological membrane.[7] Further support for this model was obtained by NMR analysis of EmrE purified in organic solvent (chloroform–methanol–water), which showed four regions of α-helix.[26] This correlated well with FTIR studies in both organic solvent and membrane vesicles, which indicated that EmrE was ~80% α-helical.[27] The use of organic solvent in these structural studies was apparently justified because EmrE has the unusual property of being able to spontaneously refold into a lipid bilayer when it is dried down from chloroform–methanol solutions containing *E. coli* lipids.[7] However, despite the presence of secondary structure, there appears to be no significant tertiary structure associated with EmrE in organic solvent and, in addition, it

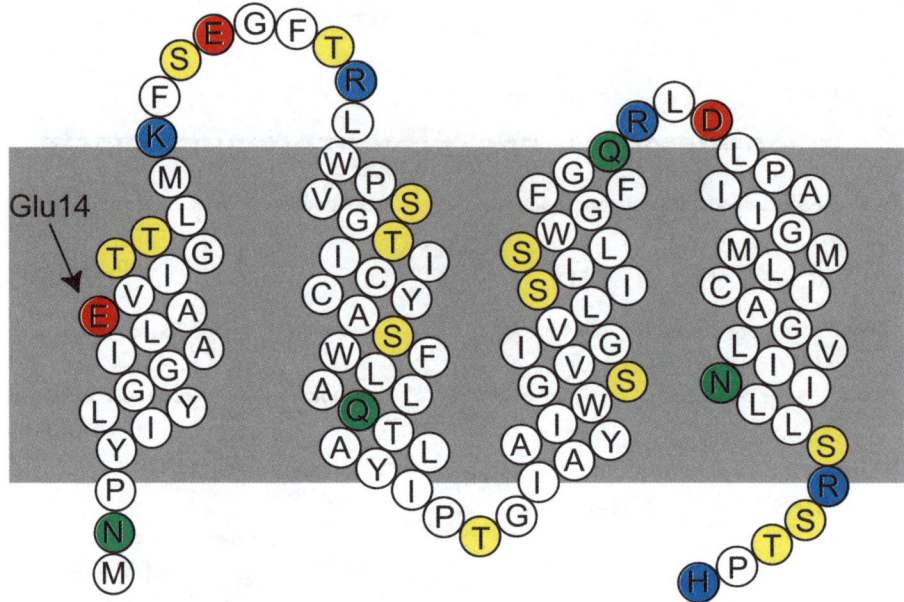

Figure 2 *The amino acid sequence of EmrE is arranged into four α-helices, with acidic residues in red, basic residues in blue and Glu14 shown by an arrow. The approximate position of the membrane is shown by a grey box. The first and last residues of each α-helix have not been determined exactly for EmrE in the native state, so the hydropathic profile of EmrE was used as a guide for this model*

appears to be monomeric.[28] This is in contrast to studies on EmrE in the native membrane environment. A negative dominance study performed on EmrE, purified in organic solvent and reconstituted into a lipid bilayer, clearly showed that EmrE functioned as an oligomer, as assessed by the uptake of ^{14}C-labelled methyl viologen into vesicles.[29] This is the only study to date that attempts to determine the oligomeric state in the biological context of a membrane, and it suggests that EmrE could be larger than a dimer. Further work is needed to substantiate this. It is not clear whether the primary assumption of negative dominance studies (*i.e.* that the mutants and wild-type molecules associate with the same affinity) holds true for EmrE, especially as we now know that there is no appreciable tertiary structure associated with EmrE in organic solvents. Unfortunately, it is notoriously difficult to determine the oligomeric state of a membrane protein in its native environment,[30] especially using biophysical techniques, so biochemical techniques such as cross-linking are often employed. Currently, cross-linking data indicate that there is no significant difference between the oligomeric state of EmrE in the membrane compared to EmrE solubilised in mild detergents,[31] although recent data purport to detect the presence of a tetramer.[32]

The oligomeric state of EmrE, purified in the detergent dodecylmaltoside (DDM), has been studied by both biochemical and biophysical techniques. It is important to note that EmrE can be purified in a functional state in DDM as ascertained by binding of the radiolabelled substrate tetraphenylphosphonium (TPP$^+$),[33] the affinity of binding being similar to the affinity in the membrane-bound state.[34] EmrE purified in DDM was also used to produce 2D crystals (see below) and to reconstitute EmrE transport activity in proteoliposomes.[32,35] The stoichiometry of TPP$^+$ binding to EmrE is one molecule of TPP$^+$ per two molecules of EmrE,[36] indicating that EmrE probably contains an even number of subunits, although whether this was a dimer, tetramer or higher oligomeric state could not be ascertained from these data. Cross-linking studies[31] and the formation of hetero-oligomers,[32,37] both *in vivo* and *in vitro*, also showed that EmrE was at least a dimer, but could not rule out higher oligomeric states.

Two biophysical studies showed unambiguously that the minimal functional unit of EmrE is a dimer. The first study arose from projection maps of EmrE derived from electron cryomicroscopy (cryo-EM) of 2D crystals.[38] EmrE forms an asymmetric homodimer in the membrane (Section 4) when crystals were grown in the absence of a drug substrate. If the substrate TPP$^+$ was included during crystallisation, the structure of EmrE looked very similar to the native form, except for the change of tilt in one helix and extra density that represented TPP$^+$.[34] The site of TPP$^+$ binding was determined by a careful comparison of the two projection maps, and was found to be in the centre of the asymmetric homodimer.[34] This strongly suggested that each homodimer was capable of binding TPP$^+$. EmrE in the 2D crystal was fully functional since TPP$^+$ bound with high affinity.[39] Moreover, it is likely that all the EmrE within the crystal bound TPP$^+$, because the small conformational change that occurred destroyed the order within the 2D crystal and altered the packing arrangement of EmrE from the planar space group c222 to p2.

EmrE purified in DDM was also subjected to analysis by analytical ultracentrifugation,[36] a biophysical method acknowledged as the most accurate means for determining the oligomeric state of a detergent-solubilised membrane protein.[40] Data from sedimentation equilibrium experiments were analysed using a density increment for the EmrE–lipid–detergent complex determined by direct measurement of the density of an EmrE solution. This showed that EmrE existed in solution in monomer–dimer equilibrium, with no evidence for higher oligomeric states.[36] Sedimentation velocity experiments showed that EmrE was predominantly a dimer and that there was no change in oligomeric state in the presence of substrate. During purification, dimeric EmrE migrates on a size exclusion column with an apparent molecular weight of 137 kDa.[36] A detergent and lipid analysis showed that for every dimer of EmrE (30.4 kDa, including purification tags) there was 209 molecules of DDM in the surrounding micelle (106.8 kDa) and 5.4 lipid molecules (4 kDa), giving a total molecular weight of 141.2 kDa for the EmrE–detergent–lipid complex. The dissociation of dimeric EmrE is a very slow process at 4 °C and full dissociation is only achieved by heating DDM-solubilised EmrE for 15 min at 80 °C.[37] Recently, the monomeric form of EmrE was purified and analysed by analytical ultracentrifugation by purifying EmrE in organic solvents and then resolubilising the dried-down protein in DDM.[41] There is, however, no evidence for monomeric EmrE *in vivo*, and the fact that EmrE exists predominantly as a dimer in DDM suggests that EmrE in the membrane is either a dimer or in a higher oligomeric state containing an even number of subunits.

3   Transport Activity of EmrE

Inspection of the amino sequence of EmrE shows that there is only one charged residue in the transmembrane region, Glu14, and seven other charged residues in loop regions (Figure 2). Glu14 is the only charged residue that, when mutated to Cys, completely abolishes uptake activity.[42,43] Glu14 is also absolutely conserved throughout the SMR family of multidrug transporters.[42,44] The only residue to which Glu14 can be changed and that maintains substrate binding is Asp; however, transport is severely impaired. The pK of Glu14 is very high, with the most recent determination[45] estimating the pK as ~8.5. In comparison, the pK of Asp14 is estimated to be ~6, which would explain the inability of this mutant to transport substrate despite being able to bind substrates with reasonably high affinity. From cross-linking studies[31] and spin-labelling studies,[46] the two Glu14 residues, one from each monomer in the asymmetric dimer, are predicted to be in close proximity to each other. This is consistent with the hypothesis that both Glu14 residues are required for binding a hydrophobic cation and that displacement of the cation into the periplasmic space requires the binding of two protons, one to each Glu14 in the dimer. The fact that transport of a substrate bearing a single positive charge is electrogenic, whereas transport of the divalent cation methyl viologen is electroneutral, further supports this hypothesis.[47] Thus, binding of the substrate and the protons is mutually exclusive and has been described as a "time-sharing" mechanism.[48,49] Elegant experiments have demonstrated that changing one of the Glu14 residues to an Asp residue reduces the affinity of hydrophobic cation binding, highlighting the importance of both Glu14 residues for transport.[37] In addition, TPP$^+$ protects both Glu14 residues from chemical modification with a hydrophobic carbodiimide.[50]

Glu14 is not the only important residue for substrate translocation. Residues clustered around Glu14 on the same face of helix 1 all affect transport or binding when mutated.[51] In addition, Trp63 in helix 3 is intimately involved in substrate recognition as its intrinsic fluorescence changes on substrate binding.[52] Systematic mutation of residues in transmembrane helices 2 and 3 identified other hydrophobic amino acid residues that are essential for substrate recognition.[53,54] These data suggested that substrate transport occurs through a hydrophobic channel within the EmrE homodimer, which is consistent with the proposed binding site for TPP$^+$ in the middle of a 6-helix bundle (see below). In contrast, no residues mutated in helix 4 altered substrate recognition or affected transport.

4   Structure of EmrE Determined by Electron Cryomicroscopy

EmrE purified in DDM was dialysed in the presence of the lipid dimyristoylphosphatidylcholine to remove the detergent and resulted in well-ordered 2D crystals.[38] For the subsequent discussions comparing the cryo-EM structure of EmrE with the X-ray structure, it is important to note that at all stages of the 2D crystallisation process, EmrE was functional. The particular construct of EmrE used for the structure determination had a Myc-His tag at the C-terminus and was shown to be functional *in vivo* for drug efflux[33] and *in vitro* after reconstitution into lipid vesicles.[35]

This particular EmrE construct was also used in the assays described above to determine the stoichiometry of TPP^+ binding and its affinity for TPP^+ in native membranes, after purification in DDM and after reconstitution into 2D crystals is, respectively, 1.9 nM (Tate, unpublished data), 2.6 nM[34] and 2.3 nM.[39] Four different crystal forms have been obtained so far. Two native EmrE crystal forms (c222 and p222$_1$) were grown in the absence of substrate.[34,38] The addition of substrate caused the c222 crystal form to become disordered, but there were low-resolution reflections that could be indexed to a p2 space group. When either of the substrates TPP^+ or ethidium was included during crystallisation, crystals with a p2 space group were obtained and the projection structures of EmrE were essentially identical, although the crystals of TPP^+-bound EmrE were much better ordered. The position in EmrE, where TPP^+ bound, was determined in projection by an analysis of the various crystal forms,[34] the site of binding being identical to the position of density assigned to TPP^+ in the 3D structure of EmrE determined from the same p2 crystals.[22]

EmrE is composed of a bundle of eight transmembrane helices that are arranged in an asymmetric manner, so that no four helices can be simply transposed onto another four.[22] There are two classes of helices, depicted in either red or yellow in Figure 3. The yellow helices form a 6-helix bundle around a cavity having a density assigned to TPP^+. All these six α-helices are sufficiently close to the substrate, such that amino acid residue side chains could make direct interactions with TPP^+ and determine the specificity of binding. In contrast, the two helices depicted in red are separated from the substrate-binding cavity by two highly tilted helices, and therefore they are unlikely to make any direct contribution to substrate specificity. However, these two red helices are unique in the structure as they are about 9 Å apart, centre to centre, along their whole lengths, and thus probably represent a very tightly packed interface that could be important in providing stability for the EmrE dimer during substrate transport. The substrate-binding cavity is centred in one leaflet of the lipid bilayer and provides potential access from both the aqueous phase and possibly from the hydrophobic phase of a single leaflet of the lipid bilayer. Access to the opposite side of the membrane from the open pocket is blocked by a single, highly tilted helix, helix H in Figure 3. Clearly, for transport to occur, the binding pocket must be reoriented to face the opposite side of the membrane. This could be achieved most simply by changing the tilt of part of helix H in a manner analogous to the conformational change in bacteriorhodopsin during proton pumping.[55] Additional structures of different conformational states of EmrE will be needed to clarify the mechanism of drug translocation.

The cryo-EM structure of EmrE was determined to an in-plane resolution of 7.5 Å, and at this resolution it is not possible to assign the amino acid sequence to the density directly, or to determine unambiguously the densities for each monomer in the dimer, although we have made tentative assignments based on a series of rationalisations and logical arguments.[22] The reason for the asymmetry in the EmrE homodimer is currently uncertain, but there are a number of possibilities for how the asymmetry arises. One interesting possibility suggested by the structural data is that the asymmetry arises because of the different membrane orientations of the two monomers in the dimer.[22] This is suggested from an in-plane pseudo two-fold axis that relates helices A–C and F–H in Figure 3, the asymmetry arising because

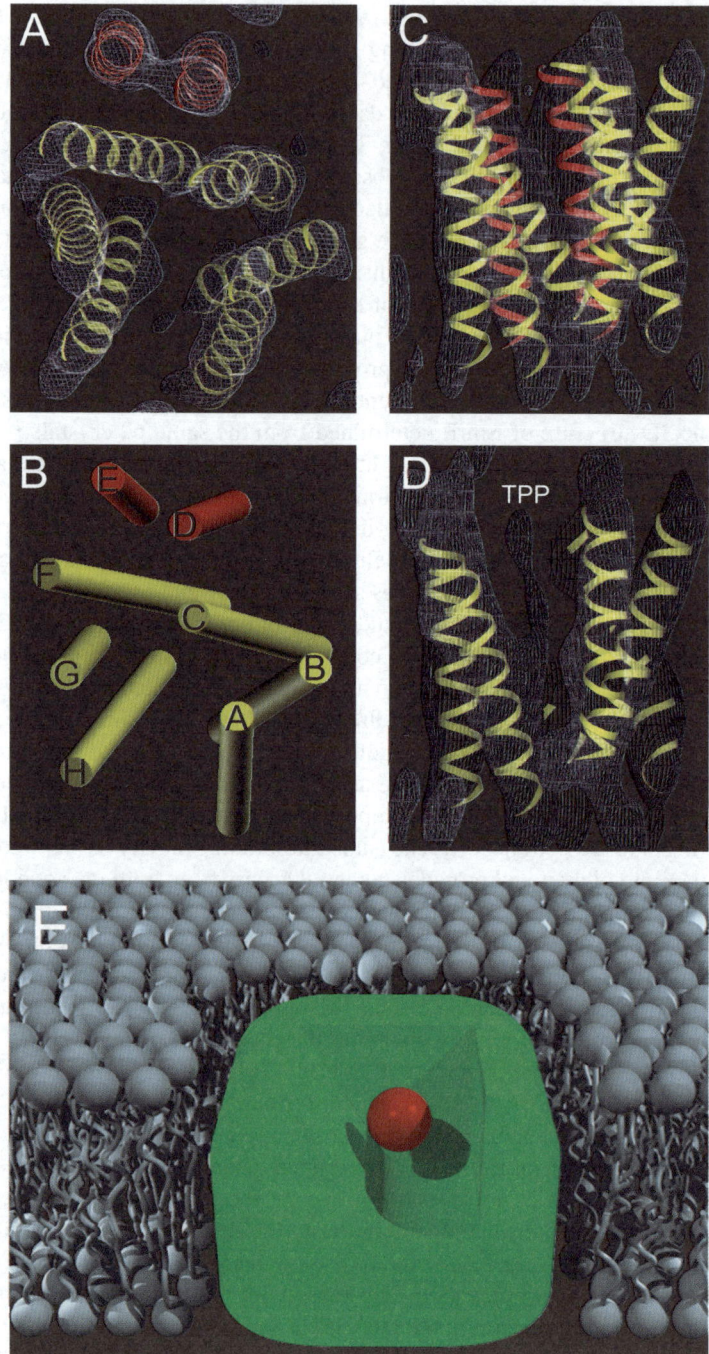

Figure 3 *Three-dimensional structure of EmrE determined by electron crystallography. (A) View perpendicular to the membrane plane and (B) the same view but with α-helices depicted as cylinders. (C) View parallel to the membrane plane and*

transposition of helix D does not result in its overlying helix E. The apparent close association of helices D and E in EmrE could explain how the dimer is formed and why it is asymmetric. However, this hypothesis is challenged by biochemical studies on the topology of EmrE in *E. coli* that apparently indicate that all the EmrE monomers have the same orientation in the membrane, with their N- and C-termini in the cytoplasm.[56] However, recently, a genome-wide study of the topologies of *E. coli* membrane proteins identified EmrE as possibly having dual topology in the membrane.[57] Another possibility for the asymmetry is that during the transport cycle, the conformation of the monomers alternates between two states A and B, so that reorientation of the binding pocket occurs when the conformation changes from A–B to B–A.[38] A structure at higher resolution will be required to resolve these issues and to determine the origin of the unusual asymmetry in the homodimer.

5 Comparison of the EmrE Structure Determined by Electron Crystallography with a 3.8 Å Resolution Structure Determined by X-ray Crystallography

After the cryo-EM structure of EmrE was determined, a structure of *E. coli* EmrE to 3.8 Å resolution by X-ray crystallography was published[21] (Figure 4). The modest resolution of the crystals necessitated a careful analysis of mercury binding sites in native and mutant EmrE to unambiguously trace each polypeptide chain in the oligomer. It is therefore probable that the EmrE structure determined is an accurate representation of the EmrE conformation within the 3D crystal. The striking feature of the X-ray structure is that EmrE is tetrameric and the arrangement of the α-helices does not in any respect resemble the arrangement of α-helices in the cryo-EM structure. This conclusion was drawn from direct inspection of the structures and by a comparison of the crossing angles between each pair of helices. There is insufficient similarity between the two structures to make any direct comparison by attempting to dock the X-ray structure into the EM density. The position of helix 4 in the X-ray structure is particularly peculiar, because, despite its hydrophobicity (Figure 2), it lies at an obtuse angle to the other helices and would therefore be outside the hydrophobic core of the membrane. Cross-linking data indicate that helix 4 from each monomer are adjacent to each other and can be cross-linked along their whole length,[31] whereas in the X-ray structure they are far apart. The fact that crystal contacts between adjacent tetramers are made by helix 4 could explain its strange position relative to the other helices. The quaternary structure of EmrE in the X-ray

---

(D) the same view but α-helices obscuring the density for TPP⁺ have been removed. The idealised α-helices in A, C and D were manually inserted into the density (mesh) without refinement. Yellow α-helices form the substrate binding pocket whereas the two red α-helices are not in contact with TPP⁺. (E) Schematic representation of EmrE (green) showing the access of the substrate (red) into the binding pocket either from the aqueous phase or possibly from one leaflet of the lipid bilayer. The density for EmrE is available from the EM Data Bank at www.ebi.ac.uk/msd/index.html under the accession code 1087. Reprinted by permission from EMBO Journal[22], copyright 2003 Macmillan Publishers Ltd.

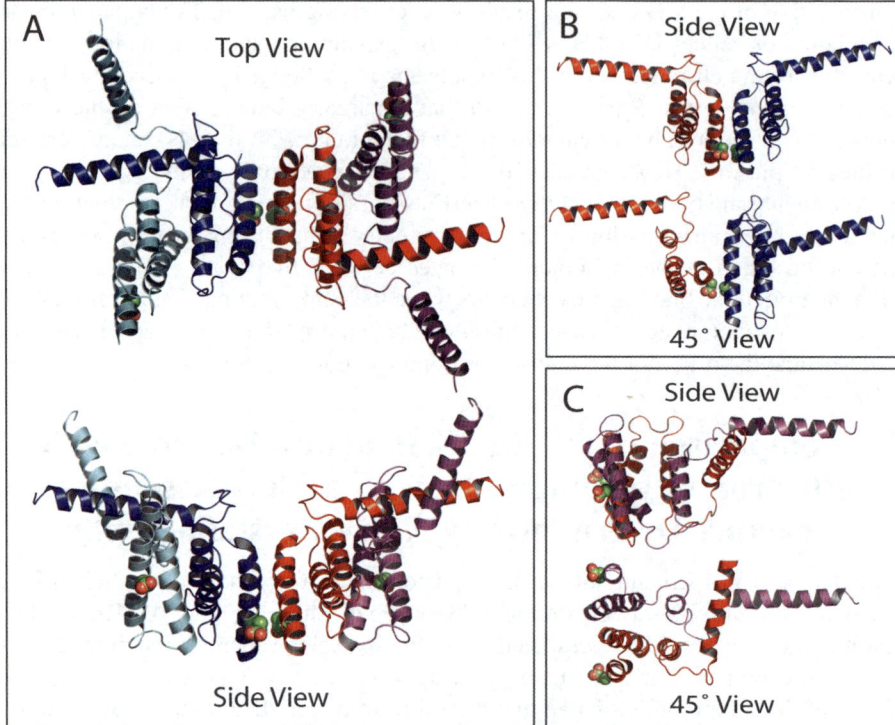

Figure 4 *Three-dimensional structure of EmrE determined by X-ray crystallography. Three different viewing perspectives of EmrE are shown: 'Top View' is the view perpendicular to the proposed membrance plane, 'Side view' is the view parallel to the membrane plane and '45° View' is viewed after the EmrE model has been rotated through 45° to allow visualisation parallel to the majority of the α-helices, as is seen in the cryo-EM structure (Figure 3A). (A) The EmrE tetramer. (B) Chain a and chain c of the EmrE tetramer. (C) Chain c and chain d of the EmrE tetramer. The side chain for Glu14 in each of the monomers is depicted as a space-filling model. Diagrams were made using Pymol from the pdb file 1S7B downloaded from www.rcsb.org/pdb.*

structure, a tetramer, must also be questioned since all the available data indicate that detergent-solubilised EmrE is a dimer (see above). In the X-ray study, EmrE was purified in nonylglucoside before 3D crystallisation and it is possible that the relative harshness of this detergent compared to DDM could be detrimental to the functional state of EmrE. The X-ray structure seems to have one binding site for a hydrophobic substrate where two Glu14 residues at the dimer–dimer interface are close to one another. This is directly opposed to the cryo-EM data that show the position of TPP$^+$ in the centre of the asymmetric dimer, implying that the binding site is at the monomer–monomer interface.[22,36] Another area where the X-ray structure does not correspond to the biochemical data is the prediction from the structure that there are two distinct types of binding sites, one composed of two adjacent Glu14 residues at the centre of the tetramer, and two other binding sites each with a single

Glu14 that are proposed to be where only protons bind. Cross-linking[31] and spin-labelling studies[46] support the idea that Glu14 from each monomer are juxtaposed as they appear in the centre of the structure. However, the available data do not support the presence of two distinct populations of Glu14 residues,[50] but instead point towards a mechanism where there is mutually exclusive binding of the transported hydrophobic cation and two protons.[32,37,45,47,49]

How did the conformation of EmrE in the 3D crystals come to be so different from the native conformation in the 2D crystals? From the published work available for the EmrE construct used in the EM study (Emr-His), it is clear that it is fully functional before, during and after crystallisation (see above). No corresponding data have been published for the N-terminal fusion construct that was used to produce EmrE for the 3D-crystallisation study (the tag was cleaved off before crystallisation), although passing mention was made to the activity of this construct in *E. coli*.[58] The effect of the low pH and nonylglucoside used for 3D crystallisation on the stability of EmrE has also not been characterised. In the absence of functional data, the conclusion drawn from the above arguments is that the X-ray structure represents a nonbiologically relevant form of EmrE.

6 Conclusions

This is the first time that there has been a significant discrepancy between the tertiary structures of the same integral membrane protein determined by two different techniques. Previous differences in structures arose in the quaternary structure and are readily explainable owing to different packing of domains in the 3D crystal as opposed to the native state. In contrast, it was always assumed that the tertiary structure of a protein represented the native fold of the protein. Clearly, for some membrane proteins, it is possible to crystallise them in alternative conformations, but rigorous assessment of the published biophysical and biochemical data is the key to defining the biological relevance of any claims arising from structural data. The ongoing debate[59] on the biological relevance of the structure of the voltage-gated potassium channel K_vAP determined by X-ray crystallography[60] is another recent example where the structure of a membrane protein may have been distorted during crystallisation.

References

1. S.B. Levy and B. Marshall, *Nat. Med.*, 2004, **10**, S122.
2. D.M. Morens, G.K. Folkers and A.S. Fauci, *Nature*, 2004, **430**, 242.
3. C. Walsh, *Nat. Rev. Microbiol.*, 2003, **1**, 65.
4. B.G. Hall, *Nat. Rev. Microbiol.*, 2004, **2**, 430.
5. M.I. Borges-Walmsley, K.S. McKeegan and A.R. Walmsley, *Biochem. J.*, 2003, **376**, 313.
6. M.H. Saier, www.tcdb.org/tcdb/
7. H. Yerushalmi, M. Lebendiker and S. Schuldiner, *J. Biol. Chem.*, 1995, **270**, 6856.

8. S. Murakami, R. Nakashima, E. Yamashita and A. Yamaguchi, *Nature*, 2002, **419**, 587.

9. E.W. Yu, G. McDermott, H.I. Zgurskaya, H. Nikaido and D.E. Koshland, Jr., *Science*, 2003, **300**, 976.

10. E.E. Zheleznova, P. Markham, R. Edgar, E. Bibi, A.A. Neyfakh and R.G. Brennan, *Trends Biochem. Sci.*, 2000, **25**, 39.

11. J. Abramson, I. Smirnova, V. Kasho, G. Verner, H.R. Kaback and S. Iwata, *Science*, 2003, **301**, 610.

12. Y. Huang, M.J. Lemieux, J. Song, M. Auer and D.N. Wang, *Science*, 2003, **301**, 616.

13. G. Chang and C.B. Roth, *Science*, 2001, **293**, 1793.

14. C.L. Reyes and G. Chang, *Science*, 2005, **308**, 1028.

15. K.P. Locher, A.T. Lee and D.C. Rees, *Science*, 2002, **296**, 1091.

16. I.T. Paulsen, R.A. Skurray, R. Tam, M.H. Saier, Jr., R.J. Turner, J.H. Weiner, E.B. Goldberg and L.L. Grinius, *Mol. Microbiol.*, 1996, **19**, 1167.

17. D.L. Jack, N.M. Yang and M.H. Saier, Jr., *Eur. J. Biochem.*, 2001, **268**, 3620.

18. M. Midgley, *J. Gen. Microbiol.*, 1986, **132**, 3187.

19. A.S. Purewal, *FEMS Microbiol. Lett.*, 1991, **66**, 229.

20. T.G. Littlejohn, I.T. Paulsen, M.T. Gillespie, J.M. Tennent, M. Midgley, I.G. Jones, A.S. Purewal and R.A. Skurray, *FEMS Microbiol. Lett.*, 1992, **74**, 259.

21. C. Ma and G. Chang, *Proc. Natl. Acad. Sci. USA*, 2004, **101**, 2852.

22. I. Ubarretxena-Belandia, J.M. Baldwin, S. Schuldiner and C.G. Tate, *EMBO J.*, 2003, **22**, 6175.

23. W. Kuhlbrandt, *Q. Rev. Biophys.*, 1992, **25**, 1.

24. W. Kuhlbrandt, *Q. Rev. Biophys.*, 1988, **21**, 429.

25. H. Michel, www.mpibp-frankfurt.mpg.de/michel/public/memprotstruct.html

26. M. Schwaiger, M. Lebendiker, H. Yerushalmi, M. Coles, A. Groger, C. Schwarz, S. Schuldiner and H. Kessler, *Eur. J. Biochem.*, 1998, **254**, 610.

27. I.T. Arkin, W.P. Russ, M. Lebendiker and S. Schuldiner, *Biochemistry*, 1996, **35**, 7233.

28. C. Klammt, F. Lohr, B. Schafer, W. Haase, V. Dotsch, H. Ruterjans, C. Glaubitz and F. Bernhard, *Eur. J. Biochem.*, 2004, **271**, 568.

29. H. Yerushalmi, M. Lebendiker and S. Schuldiner, *J. Biol. Chem.*, 1996, **271**, 31044.

30. L.M. Veenhoff, E.H. Heuberger and B. Poolman, *Trends Biochem. Sci.*, 2002, **27**, 242.

31. M. Soskine, S. Steiner-Mordoch and S. Schuldiner, *Proc. Natl. Acad. Sci. USA*, 2002, **99**, 12043.

32. Y. Elbaz, S. Steiner-Mordoch, T. Danieli and S. Schuldiner, *Proc. Natl. Acad. Sci. USA*, 2004, **101**, 1519.

33. T.R. Muth and S. Schuldiner, *EMBO J.*, 2000, **19**, 234.

34. C.G. Tate, I. Ubarretxena-Belandia and J.M. Baldwin, *J. Mol. Biol.*, 2003, **332**, 229.

35. P. Curnow, M. Lorch, K. Charalambous and P.J. Booth, *J. Mol. Biol.*, 2004, **343**, 213.

36. P.J. Butler, I. Ubarretxena-Belandia, T. Warne and C.G. Tate, *J. Mol. Biol.*, 2004, **340**, 797.

37. D. Rotem, N. Sal-man and S. Schuldiner, *J. Biol. Chem.*, 2001, **276**, 48243.

38. C.G. Tate, E.R. Kunji, M. Lebendiker and S. Schuldiner, *EMBO J.*, 2001, **20**, 77.

39. I. Ubarretxena-Belandia and C.G. Tate, *FEBS Lett.*, 2004, **564**, 234.

40. P.J.G. Butler and C.G. Tate, in *Modern Analytical Ultracentrifugation: Techniques and Methods*, D.J. Scott (ed), Royal Society of Chemistry, London, 2005, 133.

41. T.L. Winstone, M. Jidenko, M. le Maire, C. Ebel, K.A. Duncalf and R.J. Turner, *Biochem. Biophys. Res. Commun.*, 2005, **327**, 437.

42. H. Yerushalmi and S. Schuldiner, *J. Biol. Chem.*, 2000, **275**, 5264.

43. H. Yerushalmi, S.S. Mordoch and S. Schuldiner, *J. Biol. Chem.*, 2001, **276**, 12744.

44. S. Ninio, D. Rotem and S. Schuldiner, *J. Biol. Chem.*, 2001, **276**, 48250.

45. M. Soskine, Y. Adam and S. Schuldiner, *J. Biol. Chem.*, 2004, **279**, 9951.

46. H.A. Koteiche, M.D. Reeves and H.S. McHaourab, *Biochemistry*, 2003, **42**, 6099.

47. D. Rotem and S. Schuldiner, *J. Biol. Chem.*, 2004, **279**, 48787.

48. H. Yerushalmi and S. Schuldiner, *FEBS Lett.*, 2000, **476**, 93.

49. H. Yerushalmi and S. Schuldiner, *Biochemistry*, 2000, **39**, 14711.

50. A.B. Weinglass, M. Soskine, J.L. Vazquez-Ibar, J.P. Whitelegge, K.F. Faull, H.R. Kaback and S. Schuldiner, *J. Biol. Chem.*, 2005, **280**, 7487.

51. N. Gutman, S. Steiner-Mordoch and S. Schuldiner, *J. Biol. Chem.*, 2003, **278**, 16082.

52. Y. Elbaz, N. Tayer, E. Steinfels, S. Steiner-Mordoch and S. Schuldiner, *Biochemistry*, 2005, **44**, 7369.

53. M. Lebendiker and S. Schuldiner, *J. Biol. Chem.*, 1996, **271**, 21193.

54. S.S. Mordoch, D. Granot, M. Lebendiker and S. Schuldiner, *J. Biol. Chem.*, 1999, **274**, 19480.

55. S. Subramaniam and R. Henderson, *Nature*, 2000, **406**, 653.

56. S. Ninio, Y. Elbaz and S. Schuldiner, *FEBS Lett.*, 2004, **562**, 193.

57. D.O. Daley, M. Rapp, E. Granseth, K. Melen, D. Drew and G. von Heijne, *Science*, 2005, **308**, 1321.

58. C. Ma and G. Chang, *Acta Crystallogr. D Biol. Crystallogr.*, 2004, **60**, 2399.

59. F. Bezanilla, *Trends Biochem. Sci.*, 2005, **30**, 166.

60. Y. Jiang, A. Lee, J. Chen, V. Ruta, M. Cadene, B.T. Chait and R. MacKinnon, *Nature*, 2003, **423**, 33.

61. V. Koronakis, A. Sharff, E. Koronakis, B. Luisi and C. Hughes, *Nature*, 2000, **405**, 914.

62. M.K. Higgins, E. Bokma, E. Koronakis, C. Hughes and V. Koronakis, *Proc. Natl. Acad. Sci. USA*, 2004, **101**, 9994.

63. T. Hirai, J.A. Heymann, D. Shi, R. Sarker, P.C. Maloney and S. Subramaniam, *Nat. Struct. Biol.*, 2002, **9**, 597.

CHAPTER 18

Structure of Photosystems I and II

RAIMUND FROMME, INGO GROTJOHANN AND
PETRA FROMME

Department of Chemistry and Biochemistry, Arizona State University, Box
871604, Tempe, AZ 85287-1604, USA

1   Introduction to Oxygenic Photosynthesis

Photosynthesis is the main process on earth that converts light energy from the sun
into chemical energy. One and a half billion years ago, the invention of oxygenic
(oxygen evolving) photosynthesis by ancestors of cyanobacteria changed the early
reducing atmosphere of the earth to an oxygenic atmosphere. Three different classes
of organisms (plants, green algae, and cyanobacteria) are able to perform oxygenic
photosynthesis, thereby providing all higher life on earth with food and creating an
oxygen-rich atmosphere.

The primary steps of photosynthesis have been conserved over 1.5 billion years of
evolution, and are still essentially the same in plants, green algae, and cyanobacteria. The membrane protein complexes involved in the initial steps of photosynthesis,
i.e. the light reactions, are the main focus of this chapter.

In plants and green algae, photosynthesis takes place in chloroplasts, special
organelles that contain a highly differentiated membrane system, known as thylakoids. Chloroplasts share a common ancestor with cyanobacteria and it has been
established that they were acquired via endosymbiosis. Classically, the photosynthetic reactions are divided into the light reactions, which convert the light energy
into the high-energy substrates ATP and NADPH, and the dark reactions that con-
sume ATP and NADPH for the production of carbohydrates by CO_2 fixation. The
light reactions are catalyzed by four large membrane protein complexes: photosystem II (PSII), the cytochrome b_6f (cyt b_6f) complex, PSI and ATP synthase, all of
which are located in the thylakoid membrane (see Figure 1). Photosynthesis is the
only membrane-related process, where at least partial structural information is
known for all membrane protein complexes involved. This chapter will mainly focus

on the structure and function of PSI and PSII. However, we will take the opportunity to briefly describe the complete light reactions of the photosynthetic process, and the main structural and functional features of the b_6f complex and the ATP synthase. For details of these structures the reader is referred to the original publications on these enzymes.[1–8]

In oxygenic photosynthesis, two large membrane protein complexes catalyze the main step of energy conversion – light-induced charge separation. They both capture the light energy of the sun by an internal antenna system consisting of chlorophylls and carotenoids, and use the energy to perform charge separation across the thylakoid membrane. In addition, peripheral antenna complexes are associated with both photosystems, which increase the cross-section of light capturing. In plants, the light harvesting complex II (LHCII) is associated with PSII and the LHCI is associated with PSI. A high-resolution structure of LHCII was recently determined using X-ray crystallography.[9] Structural data on the supercomplex of PSII with LHCII is limited to low-resolution electron microscopy. More detailed information for the PSI supercomplex with LHCI from a recent X-ray structure at a resolution of 4.4 Å is available.[10]

The ETC works as follows: PSII catalyzes the light-driven ET from water to mobile plastoquinone (PQ), which, after double reduction, binds two protons and leaves the binding site as PQH_2. The hydrophobic PQH_2 is a mobile electron and proton carrier that exchanges with a PQ pool and carries both the electron and proton to the cyt b_6f complex. The b_6f complex releases two protons into the thylakoid lumen, subsequently reducing two molecules of plastocyanin (PC) (in plants) or cyt c_6 (in some cyanobacteria), and pumps an additional proton across the membrane in a process known as the Q-cycle. The structure of the b_6f complex has recently been determined both from green algae[4] and cyanobacteria[1]. It is a functional dimer that consists of 16 protein subunits. The electron transport chain (ETC) is shown in Figure 1B. When PQH_2 binds to the complex at the PQH_2 binding site (close to the low-potential heme b_L), two protons are released on the lumenal side of the membrane.

The iron–sulfur cluster (Fe_2S_2) takes one electron from PQH_2 and passes it to heme f in cyt f, where it is picked up by PC. The Fe_2S_2 cluster is attached to an iron–sulfur protein (Rieske protein), which may move on a hinge to bridge the gap between heme b_L and f. The second electron passes via heme b_L and heme b_H to a PQ bound at a site near the stromal membrane surface. Plastoquinone binds a proton, adding to the pH gradient across the membrane. The newly discovered heme c_i is located close to the stromal surface and may be involved in cyclic electron flow.

Plastocyanin and cyt c_6 are both soluble electron-carrier proteins that are located in the lumenal interior of the thylakoids. They serve as mobile electron carriers between the cyt b_6f complex and PSI.

Upon excitation with visible light, PSI catalyzes the light-driven ET from PC or cyt c_6 at the lumenal side to the soluble electron carrier ferredoxin at the stromal side of the membrane. Under Fe deficiency, flavodoxin replaces ferredoxin as the soluble electron acceptor of PSI. Ferredoxin finally brings the electron over to the soluble ferredoxin:NADPH reductase (FNR), which reduces $NADP^+$ to NADPH.

Photosynthetic ET leads to the establishment of an electrochemical potential across the membrane, which is the sum of the electrical potential $\Delta\Psi$ (the membrane

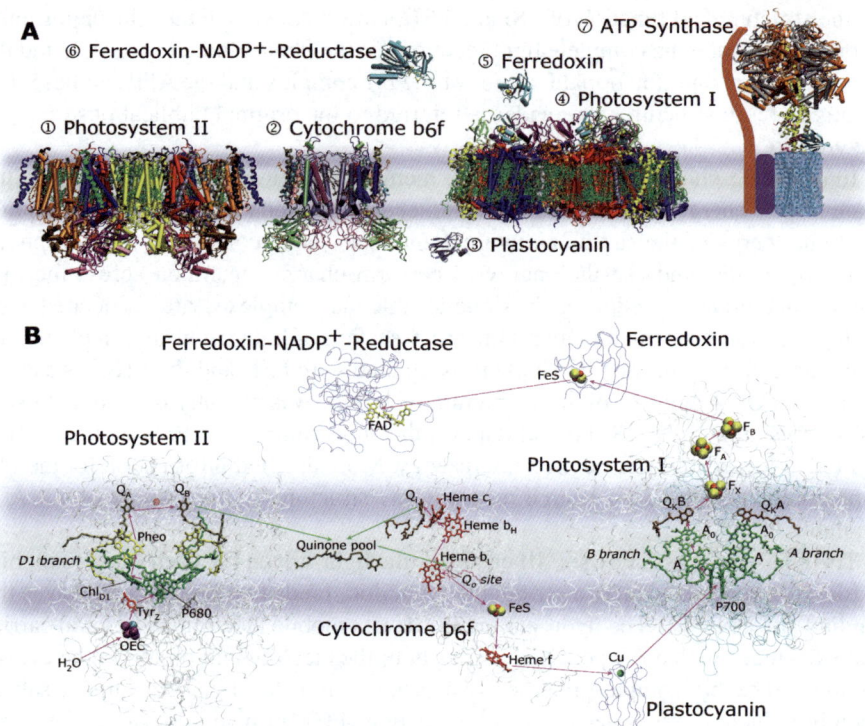

Figure 1 *The components of the light reactions of cyanobacterial oxygenic photosynthesis. (A) Structural models of the protein components of the electron transport chain (ETC) and the ATP synthase in the order how they are involved in ET. The proteins shown and the pdb files used are (a) a dimer of photosystem II (PSII) from Thermosynechococcus elongatus (1S5L),[19] (b) a dimer of cyt b_6f from Mastigocladus laminosus (1VF5),[1] (c) the reduced form of the soluble protein PC from Synechococcus sp. PCC7942 (1BXV),[96] (d) the trimeric complex of PSI from T. elongatus (1JB0),[22] (e) the soluble protein ferredoxin from Anabaena PCC7119 (1CZP),[97] and (f) the ferredoxin:NADP$^+$ reductase from Anabaena PCC7119 (1QUE).[98] The final protein, the ATP synthase, is not directly part of the ETC, but produces ATP from ADP and phosphate by using the electrochemical gradient that is generated during ET. Whereas all other models are available from cyanobacterial sources, we used the model of the (g) bovine mitochondrial F_1-ATPase (1H8E)[99] as placeholder for the membrane-extrinsic F_1 part of the protein. The membrane-intrinsic F_o part is even less well characterized, which is indicated by the structures. A glimpse at how the rotor ring composed of c subunits might look like is given by the structure derived from a similar enzyme, the F-type Na$^+$ ATPase from Ilyobacter tartaricus (1YCE).[100] (B) The functional components of the ETC. Here, the same structural files as in panel (A) were used. The multimeric complexes of the PSI and PSII are represented by one monomer each, whereas the dimeric cyt b_6f is reduced to one functional half. The path of the electrons through the system can be seen by following the arrows. The magenta arrows represent direct ET, whereas the green arrows indicate the transport of an electron together with a proton, i.e., in the form of a reduced PQ (plastohydroquinone). Oxidized PQs follow those arrows in the opposite direction. The blue molecule in the cytochrome marked "Q_o site" is a specific inhibitor sitting in the lumenal plastohydroquinone binding site of the*

is charged positively inside (lumen) and negatively outside (stroma)) and the chemical potential of the proton gradient ΔpH. The ΔpH is established by the release of protons into the lumen in the process of water oxidation by PSII, the pumping/release of protons by the cyt b_6f complex, and the depletion of protons at the stromal side by the reduction of $NADP^+ + H^+$ to NADPH.

The electrical gradient and the proton gradient (*i.e.*, the electrochemical gradient) drive the synthesis of ATP, which is catalyzed by ATP synthase. This enzyme can be found not only in the photosynthetic membrane but also in essentially all living cells. It is a molecular motor which catalyzes the synthesis of ATP from ADP and phosphate, driven by an electrochemical gradient of protons or even sodium ions.[11,12] The enzyme consists of a large membrane-extrinsic "head" called the F_1 part that consists of three α subunits and three β subunits and the γ subunit, which is located in the center and has been shown by Yoshida and co-workers to rotate during ATP hydrolysis.[13,14] Each β subunit contains one catalytic nucleotide binding site, and the catalytic mechanism follows the binding change mechanism proposed by Paul Boyer, 30 years ago.[15]

The catalytic F_1 part is connected to the membrane-intrinsic proton-translocating F_o-part by two "stalks." The central stalk is part of the rotor and harbors the subunits γ, δ, and ε. The proton-translocating part of F_o consists of a ring of 10–14 c subunits that rotate in the membrane, one a subunit which contains the proton channel, and two b subunits each containing one transmembrane helix and a long stromal domain that attaches to the alpha–beta head and forms the second stalk. Subunits a and b form the stator part of F_o. The first structure of the F_1 part was solved by John Walker and co-workers[5] from the beef heart enzyme. In the meantime, structural information is also available on the chloroplast CF_1 part. The structural information on the F_o part is still incomplete. The transmembrane part of b has been determined, and recently structures from the c subunit from vacuolar ATPases and a sodium pumping enzyme have been determined. However, no structural information is so far available for the intact complete ATP synthase. Because no structure of the proton-translocating

---

structure. For a detailed description of the processes in the ETC, see text. Photosystem II transports electrons from light-induced charge separation from the lumenal (lower) side of the membrane to the stromal (upper) side, where the electrons, together with protons from the stroma, are used to reduce PQ to plastohydroquinone. The plastohydroquinone contributes to the reduced species in the adjacent quinone pool. Photosystem II is re-reduced with electrons derived from water by the action of the oxygen-evolving complex (OEC). Plastohydroquinone binds to the lumenal binding pocket of cyt b_6f. One electron is transported to the stromal side of the cytochrome, where it is given to an (oxidized) PQ in the stromal binding pocket, which can take up protons from the stroma. The other electron is given via several mediators to the soluble carrier PC, and the protons are released into the lumen. In the meanwhile, PSI performs light-induced ET from P700 to the stromal side of the membrane, where it is used to reduce the soluble carrier ferredoxin. Photosystem I gets re-reduced from the lumenal side by the reduced PC coming from cyt b_6f. Ferredoxin is delivering electrons to the ferredoxin:$NADP^+$ reductase, where they are finally used to reduce $NADP^+$ to NADPH. All model images were done using the graphic program VMD[101]

subunit a has been determined so far, the unraveling of the mechanism of the proton-coupled ATP synthesis still awaits future discoveries.

2 Photosystem II

2.1 Overview

Photosystem II catalyzes the light-driven ET from water to PQ. The electrons are extracted from water during the process of water splitting:

$$H_2O \rightarrow O_2 + 4H^+ + 4e^-$$

Thereby, oxygen is evolved and provided for respiration, which is essential for all higher nonphotosynthetic organisms, including humans. With respect to photosynthesis, oxygen is only a by-product and, owing to its high oxidation potential, is even dangerous for cells.

In both photosystems, the light-induced charge separation is initiated by capturing light using an internal antenna system. The antenna system in PSII is smaller than that in PSI and consists of 35–40 chlorophylls. The excitation energy is transferred to the center of the complex where the ETC is located. The excitation of a chlorophyll (or chlorophylls) with a maximum absorption at 680 nm (P680) leads to a charge separation across the photosynthetic membrane, where the electron is transferred to a mobile quinone, Q_B, via a chain of electron carriers that include a pheophytin and the tightly bound quinone Q_A. After two subsequent ET steps, the doubly reduced Q_B^- binds two protons and leaves the binding pocket as PQH_2. The empty pocket is then refilled by a PQ molecule from the PQ pool.

In each round of the photocycle, P680$^+$ is re-reduced by extracting one electron from the Mn cluster, with a redox active tyrosine functioning as the intermediate electron carrier between P680$^+$ and the Mn cluster. After four subsequent ET steps, i.e., when four positive charges are accumulated at the Mn cluster, one molecule of oxygen is released.

Photosystem II experiences severe photodamage, which is caused by the cationic radical P680$^{+\bullet}$, which has the high redox potential of +1.1 V. Photodamage might occur by direct oxidation of the protein by P680$^{+\bullet}$ or by the formation of the ^{3}P680 triplet and highly reactive singlet oxygen, which would lead to irreversible damage of one core protein of PSII, D1, that binds most cofactors of the ETC including the Mn cluster. This protein has to be replaced every 30 min in plants in bright sunlight. The repair mechanism of PSII has been intensively studied.[16]

Photosystem II forms a dimer in the photosynthetic membrane of both plants and cyanobacteria. It consists of 17 protein subunits to which 35–40 chlorophylls and 8–12 carotenoids are noncovalently bound. The first crystals of PSII that were able to split water had been grown from the thermophilic cyanobacterium *Thermosynechococcus elongatus*,[17] leading to the first X-ray structural model of the intact PSII complex at 3.8 Å resolution.

This model provided the first insight into the structure of the water-splitting complex of PSII as the basis for the discussion of functional aspects with regard to the mechanism of water oxidation, ET, and the process of light capturing.

2.2 The Protein subunits in Photosystem II

PSII of *T. elongatus* consists of 17 protein subunits: PsbA to PsbO, PsbU, PsbV, and PsbX. Fourteen of these subunits are membrane intrinsic, whereas PsbO, PsbU, and PsbV do not contain transmembrane α-helices and are located at the lumenal side of the complex. In the structural model at 3.8 Å resolution,[18] the origin of 36 transmembrane helices was analyzed with respect to the localization of the individual subunits, taking into account the available data from biochemical investigations. In the meantime, several more structures from PSII have been published at 3.7–3.0 Å resolution, revealing more details of the structures, including assignments of most of the amino acid side chains and identification of the small membrane-intrinsic subunits. Figure 2 shows the structural model of PSII on the basis of the 3.5 Å structure.[19]

2.2.1 The Core Subunits D1 and D2 (PsbA and PsbD)

The central core of PSII, which harbors the ETC, is formed by a cluster of 2×5 transmembrane helices, assigned to the protein subunits D1 (PsbA) (blue) and D2 (PsbD) (red). The arrangement of these subunits resembles the structure of the L and M subunits of the reaction center of purple bacteria[20,21] and, to a lesser extent, the structure of the C-terminal domain of the large subunits PsaA and PsaB of PSI[22] (see Figure 6 for comparison between PSI and PSII). The similarity reveals that all actual existing photoreaction centers might have evolved from a common ancestor as previously proposed.[23,24]

2.2.2 The Antenna Proteins CP47 and CP43 (PsbB and PsbC)

The central core, consisting of D1and D2, is flanked by the antenna proteins CP47 (PsbB) and CP43 (PsbC). Each of these subunits consists of six transmembrane helices. CP47 coordinates 16 chlorophyll molecules and CP43 binds 14 chlorophylls. A comparison with the structure of PSI reveals that the arrangement of the transmembrane helices of these proteins is very similar to the arrangement of the transmembrane helices in the N-terminal part of the PsaA/PsaB in PSI (see Figure 6).

Both CP43 and CP47 have extended loops at the lumenal side, which are important for the function of PSII. The most important is the loop between helices 5 and 6 of CP43, as it may provide one of the ligands of the Mn cluster.

2.2.3 Cytochrome b_{559} (PsbE and PsbF)

Two helices, which are located in close vicinity to helix A of D2, were assigned to the two proteins PsbE and PsbF, constituting the membrane-bound cyt b_{559}. The cyt b_{559} is essential for the function of PSII as deletion of either PsaE or PsaF is lethal

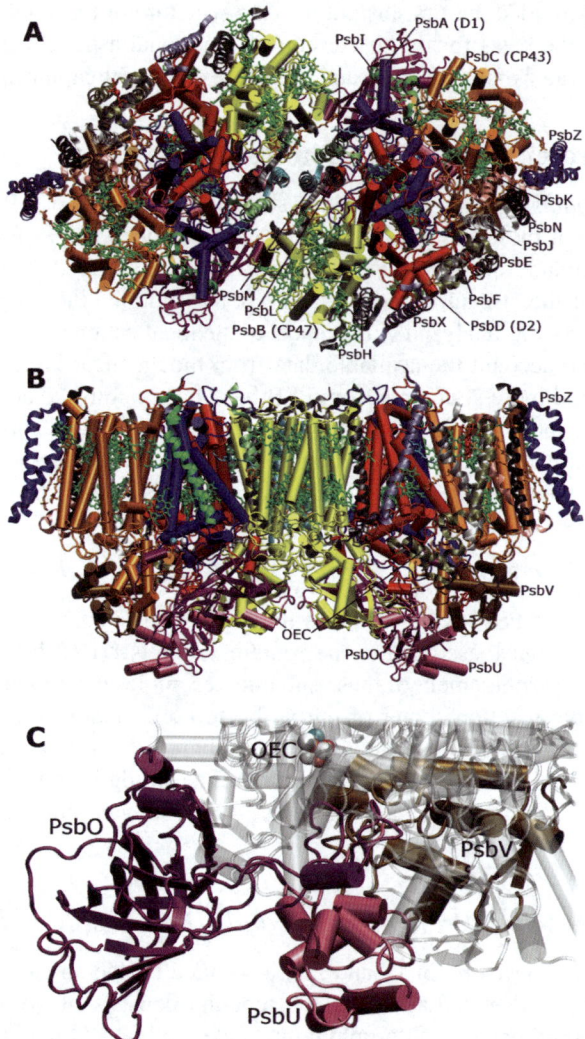

Figure 2 *The dimeric structure of PSII from cyanobacteria. (A) The view direction is from the stromal side onto the membrane plane. The subunits of the reaction center, the core antennas, and the lumenal-extrinsic subunits are shown in structural representations, whereas the smaller subunits are depicted as ribbons. The names are given for the right-hand PSII monomer, and the correspondent subunits in the other monomer have the same color-coding. Chlorophylls are shown in green and carotenoids in orange. (B) The same dimer, but this time with a view from parallel to the membrane plane. Here, the membrane-extrinsic subunits of the lumenal side and the flat stromal surface are visible. (C) The three membrane-extrinsic subunits, PsbO, PsbU and PsbV, which surround the site where the OEC is bound*

for the function of PSII. The functional role of this cytochrome and its heme cofactor is a hot topic of debate.[25–27,103] The structure implies that the heme might assist in the nonradiative charge recombination between the singly reduced Q_B and P680 to prevent excessive photodamage.

2.2.4 The Small Membrane-Intrinsic Subunits

All other small membrane-intrinsic proteins are located peripheral to the central core. The structural assignment of the small subunits is still under debate. The structural model at 3.5 Å[19] contains assignments of ten small membrane-intrinsic subunits, whereas the authors of the publication of the structural model at the nominally higher resolution of 3.2 Å[28] state that an unambiguous assignment is not possible at the given resolution and quality of their electron density map. One helix, present in the 3.8 Å model[29] and 3.2 Å model[28] of PSII, is missing from the 3.5 Å model. The loss may be caused by the higher detergent concentration used for the isolation of PSII in the work of Ferreira *et al.*[19]

The structure at 3.5 Å places subunits PsbL, PsbM, and PsbT in the dimerization domain and subunit PsbI close to helices A and B of the D1 protein. At the membrane exposed periphery of the dimeric complex, helices have been tentatively assigned to subunits PsbH and PsbX (in close vicinity to D2) and PsbJ, PsbN, PsbK, and PsbZ (in close vicinity to CP43). As the electron density is better defined in the dimerization domain than in the periphery, the assignments of PsbL, PsbM, and PsbT are clearer than the assignments of the remaining subunits. A detailed discussion of their functions must therefore await a structure of PSII at higher resolution.

2.2.5 The Lumenal Subunits PsbO, PsbV, and PsbU

In addition to the transmembrane subunits, PSII contains three extrinsic subunits, which are located at the lumenal side of the core complex: the 33 kDa protein (PsbO), the 12 kDa protein (PsbU), and the cyt c_{550} (PsbV) (see Figure 2C). The latter two subunits are unique to cyanobacteria, whereas PsbO is also present in the PSII complex of higher plants. In the structure at 3.8 Å resolution, the main body of PsbO was identified as a β-barrel structure, which was confirmed in the further structures. PsbO is involved in the stabilization of the Mn cluster, even if it does not directly provide a ligand. PsbV (cyt c_{550}) is located at the side of the lumenal hump. Its function is somewhat mysterious as the reduction of the heme has not been reported. PsbV has strong structural similarity to cyt c_6, the electron donor to PSI. One hypothesis describes PsbV as the old electron donor to the nonoxygenic ancestor of PSII that has been trapped and made an extrinsic subunit of PSII that now stabilized the oxygen-evolving complex.[30] PsbU is located at the periphery of the stromal hump and may further stabilize the complex.

2.3 The Electron Transport Chain of Photosystem II

The ETC in PSII is shown in Figure 3. It consists of four chlorophylls *a*, two pheophytins, two PQs, one redox active tyrosine Tyr_Z, and the Mn cluster.

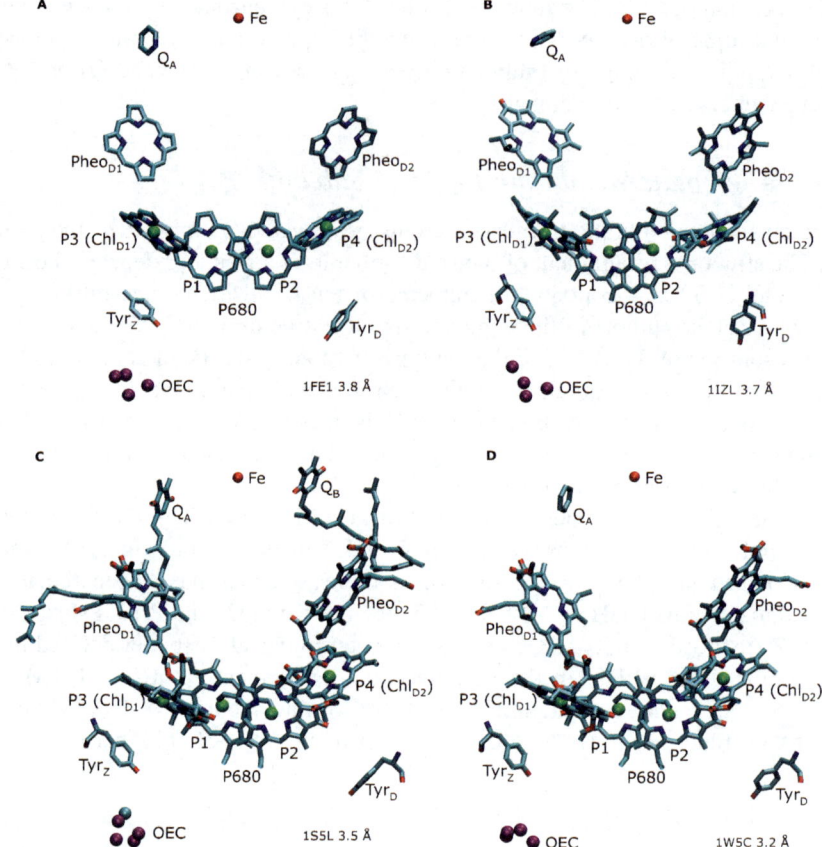

Figure 3 *Comparison of the structural models of the ETCs of PSII from different publications at (A) 3.8 Å [18], (B) 3.7 Å [29], (C) 3.5 Å [19] and (D) 3.2 Å resolution [28]. Though many elements are similar, only the 3.5 Å structure contains the plasto-quinone Q_B and a Ca^{2+} ion in the OEC. The orientation of Tyr_D varies widely. The most important point to notice, though, is that, although all structures agree on four manganese atoms in the OEC, the position of these manganese atoms is different in all of the structures. Here the question on the real position, depending on the different S states, waits still to be elucidated*

2.3.1   The Acceptor Site of the Electron Transport Chain in Photosystem II

The acceptor site and the position and orientation of the large organic cofactors of the ET chain (*i.e.*, the chlorophylls and the pheophytins) are essentially similar but not identical in all improved structures of PSII, with the exception of the Mn cluster. The acceptor site of PSII with the pheophytin, the quinone, and the nonheme iron resembles the structural arrangement of the ETC in reaction centers of purple bacteria. This result is not astonishing, as the physical-chemical properties of these cofactors are quite similar in PSII and PbRC. The electron is transferred along the D1 branch from

P1 (or P3) (see Section 2.3.2) to the pheophytin and from there to the tightly bound PQ, Q_A. The electron is then transferred to the mobile quinone Q_B. The nonheme iron that is located between Q_A and Q_B has only a structural function and is not directly involved in ET from Q_A to Q_B. A second charge-separation event then leads to the doubly reduced Q_B, which binds two protons and leaves the binding pocket as PQH_2. It is then replaced by a PQ from the PQ-pool. As PQ is mobile, the PQ in the Q_B site is only visible in the structure at 3.5 Å resolution. At this resolution and quality of the map, it is difficult to draw definitive conclusions regarding the Q_B binding site in PSII. This site is of special interest as it is the site of action for many herbicides.

2.3.2 The Donor Site of the Electron Transfer Chain of Photosystem II

The donor site of the ETC differs between the PbRC and PSII and is unique for PSII. It will now be discussed in more detail. It consists of P680$^{+\bullet}$, the redox active tyrosine Tyr_Z, and the Mn cluster, which catalyzes the water splitting. There is another tyrosine Tyr_D located at the D2 site of the ETC. This tyrosine Tyr_D is not directly involved in ET or water splitting.[31]

One electron is extracted from the Mn cluster in four subsequent charge-separation events before oxygen is evolved. The oxidation states of the oxygen-evolving center (OEC) are named S states and the "clock of water splitting" cycles between S_0 (no charge), S_1 (+1), S_2 (+2), S_3 (+3), and S_4 (+4). The water molecules are bound in the transition from S_4 to S_0. The mechanism of water splitting and the oxidation states of the Mn cluster are still under "hot" discussion and no final model exists which can explain all the experimental evidence.

2.3.2.1 The primary electron donor: P680. The most important differences between PSII and the primary donor of PSI or the PbRC is the redox potential of P680$^{+\bullet}$, the primary donor of PSII, which is above 1100 mV, and thereby provides a sufficient redox potential for the unique function of water oxidation. Which molecules form P680? Four chlorophyll molecules are located in the center of the D1/D2 core. They are arranged in two symmetrically related pairs, P1/P2, and P3 and P4. The chlorophylls named P1 and P2 are oriented perpendicular to the membrane plane. The center-to-center distance varies between the different structures: 10 Å in the 3.8 Å structure,[18] 9.56 Å in the 3.7 Å structure,[29] 8.6 Å in the 3.6 Å structure,[32] 8.2 Å in the 3.5 Å structure,[19] and 8.3 Å in the 3.2 Å structure.[28] The decrease in distance with increase in resolution may be caused by clearer assignment of the 5/6 ring of the chlorin system at higher resolution. Even with the shortest distance reported to be 8.2 Å, the chlorophylls are more separated from each other in PSII than the chlorophylls in P700 or in the special pair of the PbRC. The longer distance between the two chlorophylls in PSII indicates that the two chlorophylls are not strongly excitonically coupled and, therefore, may be regarded as chlorophyll monomers. The second pair of chlorophylls is located at a distance of ≈10 Å from the first pair of chlorophylls. The plane of the chlorin head group of these chlorophylls is tilted at an angle of 30° to the membrane plane.

The answer as to which of the four chlorophylls may represent P680 may have to be more clearly specified, with respect to whether we want to refer the P680 as the cationic radical P680$^{+\bullet}$, to the triplet state ^{3}P680, or to the excited state P680*. The cationic radical state may be located on P1 because the distance between P680$^{+\bullet}$ and Tyr$_z$ has been determined by EPR investigations to be in the range of 7.9 ± 0.2 Å,[33–35] which fits only for the P1 chlorophyll molecule. However, this leads to the question of why the neighboring chlorophylls P2 and P3 may not be oxidized instead of the Tyr$_Z$ (redox potential 1.0 V) by P1. Taking into account that Pheo$^{-\bullet}$ has a redox potential of 1.4 V, we can assume that all four chlorophylls must have a high redox potential between 1.0 and 1.3 V and may be able to perform the initial charge separation as discussed in Ref. 36. Therefore, the excited state P680* may be delocalized — at least at room temperature — among all four chlorophylls. The last question regards as to which chromophore the triplet state of P680 may be localized on, which may be responsible for the photodamage of D1. Optical spectroscopy investigations on partially oriented samples provided evidence that the triplet state ^{3}P680 is located on a chlorophyll molecule with an inclination of 30° to the membrane plane.[37] This requirement is fulfilled by P3 and P4, and the location of the triplet on one of the "accessory" chlorophylls has since then been proven by EPR measurements on single PSII crystals.[38] However, the triplet state must be primarily formed from the state P680$^{+\bullet}$ located on P1 by intersystem crossing, therefore P4 can be excluded as a possible candidate and P3 remains as the most probable molecule for the location of ^{3}P680.

The question remains as to what causes the very positive redox potential of P680$^{+\bullet}$. The binding pocket for P680 is very hydrophobic, with a histidine providing the fifth ligand to the central Mg^{2+} ion. However, PSII contains several potentially positively charged amino acids in the vicinity of P680, which may destabilize the cation and increase the redox potential as discussed in Ref. 30. Furthermore, the close proximity of Trp191, which may be π-stacked with D2-His197, which provides the ligand for P2, was discussed as being important for the redox potential shift in Refs. 28 and 39.

2.3.2.2   The water-oxidizing complex.   The most interesting feature of PSII is the ability to oxidize water to O$_2$ and 4 H$^+$. Photosystem II performs this reaction by a cluster of four Mn ions bound to the protein D1. The first structure of PSII at 3.8 Å resolution showed a papaya-shaped structure of the electron density for the Mn cluster, which gave implications for a 3 + 1 organization of the four Mn atoms of the cluster. The subsequent structural models at improved resolution confirmed this principal structural arrangement. However, the arrangements of the Mn atoms differ significantly in the recent structural models, as shown in Figure 3. Other cofactors that play an important role in the process of water splitting are Ca^{2+} and Cl$^-$. The location of the Ca^{2+} ion has been identified for the first time in the structure of PSII at a resolution of 3.5 Å[19] and is indicated as a green dot in Figure 3. Ca^{2+} is clustered together with the three Mn atoms, thereby possibly forming a distorted cubane arrangement consisting of three Mn and one Ca ion, with the fourth Mn being more distal to the distorted cubane. This model of the 4Mn–Ca cluster is in agreement with EPR and XAFS data (see Ref. 40 and references therein).

The ligands to the Mn cluster are still under discussion, because they differ in the different X-ray structures and also do not agree in all points with mutagenesis studies. Ferreira *et al.*[19] have proposed Asp170, Glu189, His332, Glu333, Asp342 of D1, and E354 of CP43 as ligands for the Mn cluster. However, the structure of PSII at 3.2 Å resolution shows a similar but not identical arrangement for the Mn atoms and the ligands. Both structures agree on Asp170, His332, and Glu333 being possible ligands to the Mn cluster. However, Biesiadka *et al.*[28] consider Glu189 as more distant to the Mn cluster, in a position that may bridge the Mn cluster and Tyr_Z. In this respect, it is remarkable that Glu189 can be replaced by many other amino acids without affecting oxygen evolution,[41] which leaves the question about the role of Glu189 completely open. The same is true for CP43-E354, which has been replaced by other amino acids with only a moderate influence on oxygen evolution, whereas the mutation of CP43-R357 (an amino acid which is close, but was not proposed to be a ligand) is lethal. Another controversial question deals with the CP43 C-terminus, Ala344. Mutagenesis studies and FTIR experiments propose that it should provide a ligand to the Mn cluster. Indeed, it was supposed to be a ligand in the 3.7 Å structure of PSII[29] but is too distant in both the 3.5 and 3.2 Å structures.[19,28]

The final answers to the multitude of questions and uncertainties must be provided by improved X-ray data in combination with spectroscopic and mutagenesis studies. It should also be emphasized that none of the present structural models has a sufficient resolution to identify individual Mn atoms, the bridging oxygens, or the substrate water molecules bound to the cluster, so that the elucidation of the structure of the Mn cluster and the mechanism of water splitting by PSII is still an open field for future discoveries.

2.4 The Antenna System of Photosystem II

A total of 42 cofactors have been identified at 3.8 Å resolution: 26 antenna chlorophyll molecules bound to CP47/43, 6 chlorophylls coordinated by D1/D2 as well as the Mn cluster (consisting of 4 Mn ions), 1 nonheme iron located at the D1/D2 interface on the stromal side of the membrane, and the 2 heme groups of cyt b_{559} and cyt c_{550}, respectively. In the structural model at 3.6 Å resolution,[18,32] three additional chlorophylls bound to CP43/CP47 were identified, so that the total amount of chlorophylls was close to the number of 30 chlorophylls assigned at 3.5 Å resolution.[19] All X-ray structural models show very similar locations for the chlorophylls. Figure 6 compares the chlorophylls assigned in the structural model at 3.5 Å with the chlorophyll arrangement in PSI. The most striking difference between the two is the lack of the central antenna domain in PSII. Whereas more than 50 chlorophylls surround the reaction center domain carrying the ETC in PSI, the central core in PSII contains only two chlorophylls (ChlZD1 and ChlZD2). This lack of the central antenna domain may be responsible for the lower efficiency of the excitation energy transfer in PSII compared to PSI. The absence of the central antenna domain may be the price PSII has to pay for its ability to use water as an unlimited electron source, creating the sensitivity to photodamage with the need for repair of the D1 protein. If PSII contained a central antenna domain, all central chlorophylls would have also to be replaced with the D1 protein, which would be an extremely resource-wasting process.

3   Photosystem I

3.1   Overview

Photosystem I catalyzes the light-driven ET from PC to ferredoxin. In some cyanobacteria, cyt c_6 is used as the alternate electron donor to PSI. The light energy is captured by a large antenna system, consisting of 90 chlorophylls and 22 carotenoids, and transferred into the center of the complex, where the charge separation takes place, when a pair of chlorophyll molecules (P700) is excited. The electron is transferred across the membrane by a chain of electron carriers A (chlorophyll), A_0 (chlorophyll), A_1, a phylloquinone, and three FeS clusters - F_X, F_A, and F_B. In cyanobacteria, the dominant oligomeric form of PSI *in vivo* is the trimer. The formation of intact trimers is essential for the growth of the cells at low light intensity.

The trimer, with a molecular weight of around 1,080,000 Da has been isolated and crystallized.[42] Photosystem I is the largest and most complex membrane protein for which a high-resolution structure has been determined.[22]

Figure 4 shows a simplified picture of the trimeric complex. This picture makes another remarkable feature of trimeric PSI visible: high cofactor content. Thirty percent of the total mass of PSI consists of cofactors, which are therefore not only important for the function of the protein but also play an essential role in the assembly and structural integrity of PSI.

One monomeric unit of the cyanobacterial PSI (see Figure 5) consists of 12 different proteins (PsaA, PsaB, PsaC, PsaD, PsaE, PsaF, PsaI, PsaJ, PsaK, PsaL, PsaM, and PsaX) to which 127 cofactors are noncovalently bound. In plants, PSI is monomeric. The plant PSI contains at least four additional subunits PsaG, PsaH, PsaO, and PsaN. The structure of the plant PSI in a complex with four subunits of its peripheral antenna (Lhca1–4) has been determined at 4.4 Å resolution,[10] with PsaG interacting with Lhc1 and PsaH hindering the trimerization of the plant PSI.

The structural comparison between plant and cyanobacterial PSI shows that the core proteins, including all cofactor-binding sites, are well conserved between plants and cyanobacteria, whereas PsaX and PsaM may be unique to cyanobacteria.

3.2   The Protein Subunits of Photosystem I

Figures 4 and 5 show the organization of protein subunits in PSI. The two most important proteins are the large subunits PsaA and PsaB. They harbor most of the antenna system, as well as most of the cofactors of the ETC from P700 to the first FeS cluster, F_X.

All of the small subunits are located at the periphery of PsaA and PsaB. The subunits PsaC, PsaD, and PsaE do not contain transmembrane helices. They form the stromal hump of PSI, which extends from the membrane by approximately 90 Å. PsaC carries the two terminal FeS clusters, F_A and F_B.

The other small subunits contain between one and three transmembrane helices and are located at the periphery of PsaA and PsaB.

These subunits exist in quite different environments. The subunits PsaL and PsaI are located in the center of the trimeric complex and form the "trimerization domain," while PsaM is located at the interface between the monomers.

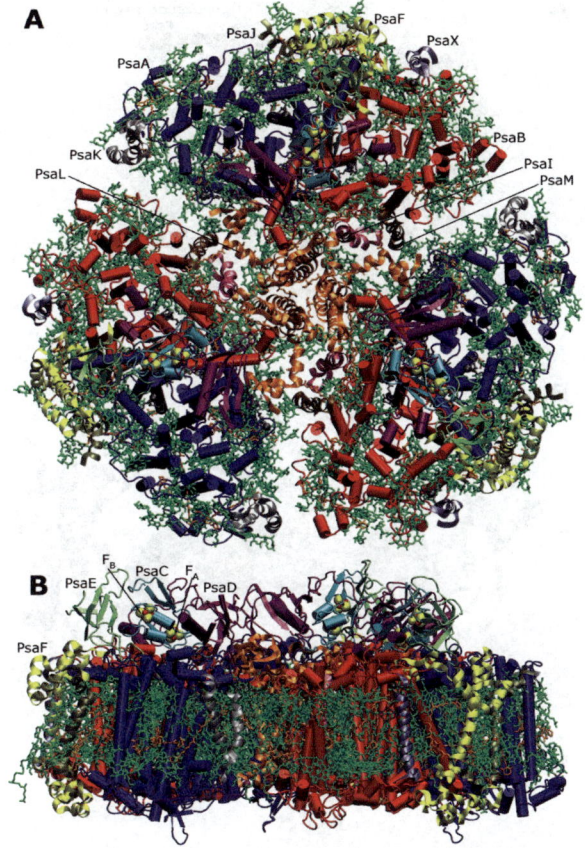

Figure 4 *The trimeric structure of PSI from cyanobacteria. (A) The view direction is from the stromal side onto the membrane plane. The proteins are shown in a backbone representation, with the helices of subunits PsaA and PsaB, as well as those of the stromal hump, shown as columns. Small transmembrane proteins are shown in ribbon representation. The subunits are color-coded, and the names are given for the topmost monomer. The chlorophylls (depicted in green) are represented by their chlorin head groups, their phytyl tails have been omitted for clarity; the carotenoids are depicted in orange and the lipids in cyan. (B) The same molecule, seen from within the membrane plane. Here, the extrinsic stromal subunits with their FeS clusters are seen best, as is the very flat lumenal surface of the trimer*

PsaJ and PsaF, as well as PsaK and PsaX, are located at the membrane exposed surface of the trimeric PSI complex. They stabilize the core antenna system of PSI and may also be involved in the interaction of PSI with its external antenna systems.

3.2.1 The Core of Photosystem I: The Large Subunits PsaA and PsaB

The major subunits of PSI, PsaA and PsaB, exhibit considerable homology to each other and are suggested to have evolved via gene duplication. Each of them

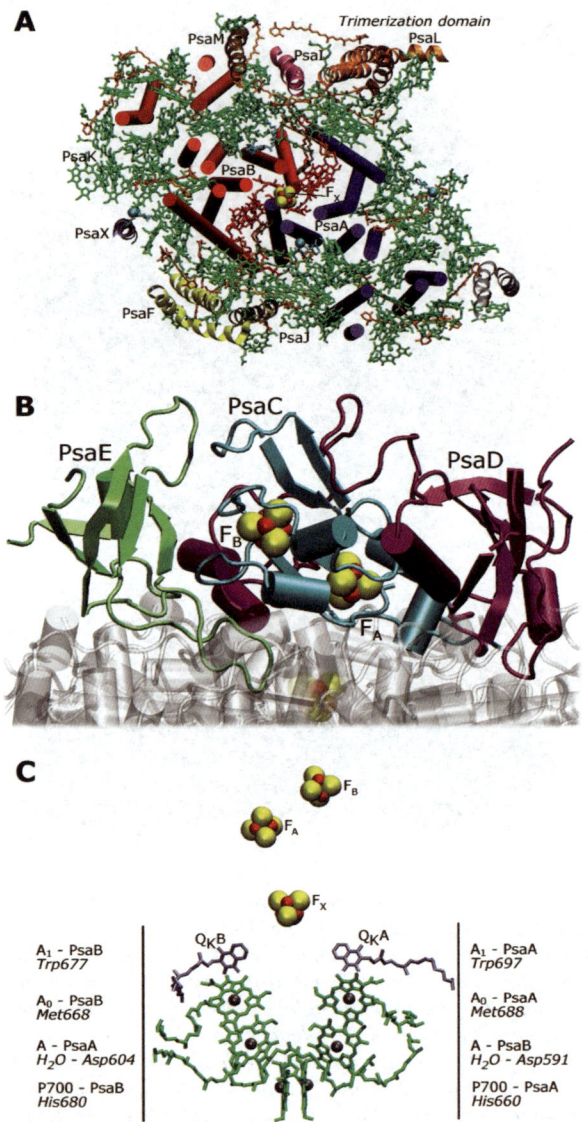

Figure 5 *Details from PSI. (A) Arrangement of the transmembrane helices of protein subunits and cofactors in one monomeric unit of PSI. The three stromal subunits and all loops have been omitted for clarity. Transmembrane helices of the membrane-intrinsic subunits are represented as cylinders. Color-coding is the same as in Figure 4. The two main subunits, PsaA and PsaB, together build the reaction center in the core, and harbor the core antenna. The smaller transmembrane subunits can be found on the outside of the molecule and serve functions like the connection to other PSI monomers in the trimerization region. (B) The stromal hump of PSI. The view is parallel to the membrane plane. The three stromal subunits PsaE, PsaC, and PsaD form a stromal hump on top of the transmembrane region and provide the docking site for ferredoxin. PsaC contains the two FeS clusters F_A and F_B. (C) The*

contains 11 transmembrane helices. The arrangement of transmembrane helices in PSI is depicted in Figures 4 and 5. PsaA is depicted in blue and PsaB in red. The area of transmembrane helices for PsaA/PsaB can be divided into two domains: the five C-terminal α-helices surround the ETC, whereas the six N-terminal helices flank the central region at both sides, forming an arrangement of "trimers of helix-pairs," where three pairs are formed by transmembrane helices a/b, c/d, and e/f. The C-terminal domain shows some similarity to the arrangement of the L and M proteins in bacterial reaction centers[43] and the D1 and D2 proteins in PSII, whereas the N-terminal domain of PsaA and PsaB shows a striking similarity to the arrangement of helices in the core antenna proteins of PSII (PsbB and PsbC) (see Figure 6). It is therefore very likely that all photoreaction centers have evolved from a common ancestor and the genes for PsaA and PsaB have evolved by a gene fusion of an ancient RC protein containing five transmembrane helices and an antenna protein consisting of six transmembrane helices, as suggested earlier.[23]

Whereas the transmembrane helices nearly perfectly match the twofold symmetry between PsaA and PsaB, some of the loops show striking differences in sequence, length, and secondary structural elements, making the system more asymmetric. The functional reason for this asymmetry is its ability to attach the stromal hump and the small membrane-intrinsic subunits at a proper position to the core of PSI. PsaA and PsaB coordinate the majority of the cofactors of the ETC (P700, A, A_0, A_1, and F_x) and 79 of the 90 antenna chlorophylls in PSI. In addition, most of the carotenoids hydrophobically interact with either PsaA or PsaB. The cofactors of the ETC from P700 to F_x are coordinated by the C-terminal region of PSI.

Most, but not all, of the chlorophylls are coordinated by histidines. It is very interesting that most of the coordination sites were conserved over millions of years during the evolution from cyanobacteria to plants.[44]

In addition, PsaA and PsaB also play an essential role in the docking of the soluble electron donors to PSI. Plastocyanin and cyt c_6 bind to PSI at an indentation at the lumenal side to re-reduce the primary electron donor, P700$^+$. The major interaction site is formed by two helices in the loop between helices i and j, located at the lumenal indentation close to P700$^+$. They provide a hydrophobic docking site for PC and cyt c_6. In contrast to cyanobacteria, where only PsaA and PsaB are involved in the docking of PC, in plants PsaF may also play an important role in the docking of PC (see Section 3.2.2).

---

components of the ETC of PSI. The view is parallel to the membrane plane. The organic cofactors of the ETC are arranged in two branches. The left branch is called the B-branch, whereas right branch is the A-branch. The three FeS clusters, F_X, F_A, and F_B are located at the stromal side of the membrane. The spectroscopic names of the chlorophylls and the phylloquinone in the ETC are given at the sides, together with the coordinating amino acids from PsaA and PsaB. This shows that the amino acids that coordinate a certain component of the ETC are conserved between the two protein subunits

PS I

PS II

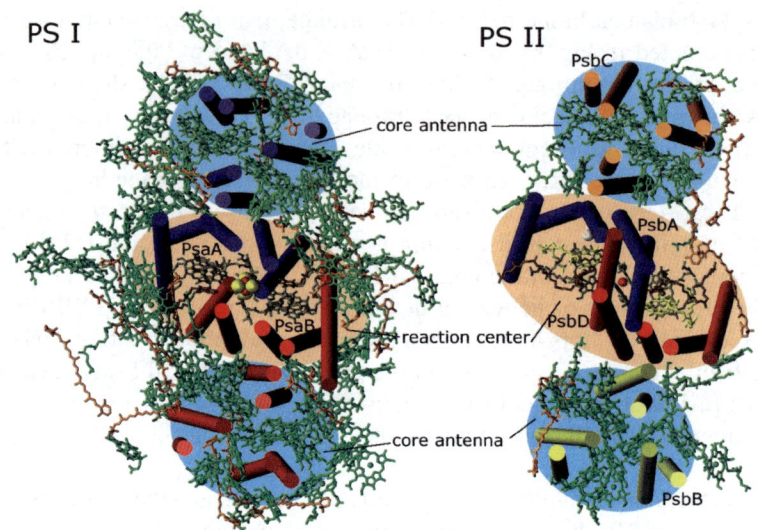

PsbC

core antenna

PsbA

PsaA

PsbD

reaction center

PsaB

core antenna

core antenna

PsbB

Figure 6 *Comparison of the core region and the antenna systems in PSI and PSII. Only the transmembrane helices of PsaA and PsaB for PSI or PsbA, PsbD, PsbB, and PsbC for PSII are shown as columns. The colored ovals show that the general building plan of both molecules is very similar, with a reaction center built from ten transmembrane helices from adjacent subunits in the center, flanked by two core antenna built from a trimer of dimers of transmembrane helices. This shows that both photosystems share a common ancestor. As a difference, PSII lacks the central domain of chlorophylls that surround the reaction center domain in PSI. The PSII antenna is optimized for rapid exchange of PsbA during repair processes that are necessary because of increased photodamage (see text). The PSI antenna is optimized for a maximum energy transfer efficiency and photoprotection*

3.2.2   The Small Transmembrane Subunits in Photosystem I

Photosystem I from cyanobacteria contains seven small membrane-intrinsic subunits. These subunits in PSI can be subdivided into two categories according to their location: the subunits PsaI, PsaL, and PsaM facing the monomer–monomer interface and the subunits PsaF, PsaJ, PsaK, and PsaX that are located at the membrane-exposed surface of the trimeric PSI.

3.2.2.1   The monomer–monomer interface: PsaL, PsaI, and PsaM.
PsaL, PsaI, and PsaM are located in the region where the adjacent monomers face each other in the trimeric PSI complex. PsaL and PsaI are located in the trimerization domain, whereas PsaM is located at the monomer–monomer interface.

PsaL forms the trimerization domain of PSI, making most of the contacts between the monomers. Furthermore, PsaL coordinates three antenna Chl*a* and forms hydrophobic contacts with carotenoids, and may therefore be important for the excitation energy transfer between the monomers.[45] PsaL contains three transmembrane helices, forming hydrophobic contact sites between the monomers within the trimer.

Most of the other contact sites between the monomers in the trimerization domain are provided by hydrogen bonds and electrostatic interactions within the loop regions. The trimer is further stabilized by Ca^{2+}, which is coordinated by amino acid side chains of PsaL in two adjacent PSI monomers and by PsaA. PsaL coordinates three chlorophylls and binds three carotenoids and may be essential for excitation energy transfer between the trimers.[29]

PsaI contains one transmembrane helix. It does not bind any Chl*a*, but forms hydrophobic interactions with carotenoid molecules while forming few contacts with the adjacent monomer. PsaI is located between PsaL and PsaM. Schluchter *et al.*[46] showed that the deletion of PsaI leads to a drastic decrease in trimer formation and influences the stability of PsaM and PsaL in the PSI complex of *Synechococcus PCC7002*.

The existence of close interactions of PsaI and PsaL in higher plants suggests that the arrangement of these small subunits is a motif that was conserved during evolution.[47,48] This is remarkable, taking into account the fact that plant PSI is a monomer and the region of PsaI and PsaL (and PsaH) may function in forming interactions with the LHCII.[47-49] This raises questions on the function of the trimer in cyanobacteria. Recent results suggest that the trimer is essential for optimal light capturing in cyanobacteria. In plants, an additional subunit (PsaH) is located in close vicinity to PsaI and PsaL.[10] This region may be the contact site between PSI and the LHCII complex.

PsaM is the smallest subunit of PSI with an MW of 3.4 kDa. This subunit contains only one transmembrane α-helix. This subunit is unique to cyanobacterial PSI. Although an open reading frame for this subunit was also found in the liverwort chloroplast genome,[50] this subunit is yet to be identified in any preparation of plant PSI and is also not present in the 4.4 Å structure of pea PSI.[10] PsaM is located close to the monomer–monomer interface, in the neighborhood of PsaI and PsaB. It coordinates one chlorophyll which is important for the excitation energy transfer between the monomers.[51]

3.2.2.2 The membrane-exposed subunits: PsaF, PsaJ, PsaK, and PsaX.

Four hydrophobic protein subunits - PsaF, PsaJ, PsaK, and PsaX - are located at the detergent-exposed surface of PSI. PsaF and PsaJ are located symmetrically to the trimerization domain and form various contacts with PsaA, PsaB, and PsaE, whereas PsaX is located at the periphery of PsaA and is only in contact with PsaB and PsaK. All four proteins are involved in the stabilization of the core antenna system of PSI and may play an additional important role in forming interactions with the membrane-intrinsic peripheral antenna system, the IsiA ring.[52-54]

PsaF is the most astonishing subunit in PSI. The structure of subunit PsaF consists of three domains. The N-terminal domain is located in the lumen, followed by a transmembrane domain with one transmembrane helix and two short helical pieces in a V-shaped arrangement. The C-terminus is located in the stroma and is sandwiched between PsaA and PsaE. The N-terminal domain is located at the lumenal side of the complex, with two α-helices (F-c and F-d) being the most prominent features. These hydrophilic α-helices are parallel to the membrane plane and are located approximately 15 Å from the putative docking site of cyt c_6. The function of the N-terminal domain of PsaF is different in plant and cyanobacterial PSI. In plants and green algae, subunit PsaF contains 25 amino acids inserted into its lumenal domain, which is responsible for formation of a tight complex between PC and PSI.[55] A possible role of the transmembrane part of PsaF could be a means by which the carotenoids and chlorophylls are shielded from the lipid phase. A further possible function of PsaF in cyanobacteria could be its structural and functional interaction with the external antenna systems. In plants, the direct contact of PsaF with the plant light harvesting systems has been suggested by experiments in which plant subunit PsaF was isolated as a Chl–protein complex with LHCI proteins.[56] Interaction sites between PsaF and the LHCI proteins were also identified in the structure of PSI from pea plants.[10]

PsaJ contains one transmembrane α-helix, which is located in close vicinity to PsaF. It binds three chlorophylls and is in hydrophobic contact with carotenoids. PsaJ may play an important role in the stabilization of PsaF and the pigment clusters located at the interface between PsaJ/PsaF and the PsaA/PsaB core. The three chlorophylls that are coordinated by this subunit are supposed to play an important role in the excitation energy transfer from the IsiA ring, which serves as a peripheral antenna under Fe deficiency, to the PSI core.

PsaK contains two transmembrane α-helices. This subunit is located at the periphery of the PSI complex, forming only protein contacts with PsaA. This subunit seems to be weakly attached to the core and is the least ordered subunit in the PSI complex. The two helices are connected in the stroma, so that both the C- and N-termini are located in the lumen. The protein coordinates two chlorophylls and forms contacts with carotenoids. It may also play an important role in the interaction with the IsiA antenna ring under iron deficiency. In plants, PsaK has been shown to interact with the LHCI proteins. Furthermore, a role of PsaK in state transitions has been suggested.[57] Unlike plant PSI, cyanobacterial PSI does not contain a subunit corresponding to PsaG, a plant-PSI subunit which exhibits sequence homology to PsaK. These proteins have the same genetic origin[58] and PsaG has evolved via gene duplication. The 4.4 Å structure of the pea PSI shows that this subunit is located at a position symmetrical to PsaK in the vicinity of PsaB. It is the only subunit of the plant PSI core that forms direct contacts via a transmembrane helix with the Lhca1.[10]

PsaX is the 12th subunit of PSI. It coordinates one chlorophyll and forms hydrophobic contacts with several carotenoid molecules and one of the lipids. PsaX is present at the membrane-exposed surface of PSI, therefore one can speculate that PsaX may also play a role in the interaction of PSI with the IsiA antenna ring formed under iron deficiency.

This protein is not present in plant PSI, as recently shown by the 4.4 Å structure of the PSI from the pea plant.[10]

3.2.3   The Stromal Hump of PSI: PsaC, PsaD, and PsaE

Three subunits (the subunits PsaC, PsaD, and PsaE) are located at the stromal side of PSI and are involved in the docking of ferredoxin. The structures of the three subunits in the stromal hump and the potential docking site of ferredoxin are shown in Figure 5B.

PsaC (8.9 kDa) carries the two terminal FeS clusters F_A and F_B. It is the best conserved subunit in PSI and has essentially identical structures in cyanobacterial and plant PSI.[10,44] The central part of PsaC consists of two short α-helices connecting the two FeS clusters. This part is very similar in PsaC and ferredoxin. F_B is the terminal FeS cluster that transfers the electron to ferredoxin. The C- and N-termini of PsaC are very important for the correct docking of PsaC to the PSI core (for more details, see Ref. 59).

PsaD is essential for the ET from PSI to ferredoxin.[60,61] It is located at the stromal hump, close to the "connecting domain." The main part of PsaD consists of a large antiparallel four stranded β-sheet. The fourth β-strand is connected to the only α-helix of PsaD by a short loop. This helix forms interactions both with PsaC and with PsaA. A short antiparallel β-sheet after the helix is followed by a very prominent and remarkable feature of PsaD: its stromal clamp consisting of the sequence region between D95 and D123. This part of PsaD wraps around PsaC, forming several contacts between PsaD and both PsaC and PsaE. It is a critical stabilization factor of the electron acceptor sites in PSI and plays an important role in keeping PsaC in its correct orientation.[62,63]

In addition to its stabilizing function, PsaD is actively involved in the docking of ferredoxin. Cocrystals between PSI and ferredoxin have been reported which may serve as a basis for a structure of the PSI ferredoxin complex.[64,102]

PsaE is directly involved in anchoring ferredoxin,[65–67] plays a role in cyclic ET,[68] and can be crosslinked in barley with FNR via its N-terminal extension.[69] The structure of PsaE in solution was first determined by ^{1}H and ^{15}N NMR.[70] It shows a compact structure of five antiparallel β-strands. The core structure of PsaE is essentially the same in solution and in PsaE attached to the PSI complex. The main difference between the free and bound PsaE are the conformations of the loops and the C- and N-termini. Recent studies show that PsaE can assemble into the PSI complex without the help of assembly factors and that it is driven by electrostatic interactions.[71] The fact that PsaE is involved in the docking of ferredoxin and flavodoxin[72,73] was questioned by the finding that PsaE deletion mutants are still able to grow photoautotrophically. This contradiction was solved by the discovery that PsaE deletion mutants increased the level of ferredoxin in the cells by orders of magnitude to compensate for deficits caused by the lack of PsaE.[74]

3.3 The Electron Transfer Chain of Photosystem I

The ETC is located at the center of the PSI complex, representing the heart of PSI. Figure 5C shows the structural organization of the cofactors of the ETC. It consists of six chlorophylls, two phylloquinones, and three clusters.[75] A detailed sequence comparison and alignment between PSI complexes from cyanobacteria, green algae, and higher plants reveals that all amino acids involved in axial coordination or hydrogen bonding to the electron carriers of the ETC are strictly conserved. The organic cofactors (*i.e.*, the chlorophylls and the phylloquinones) of the ETC are arranged in two branches. They are named the A and B branches because most — but not all — cofactors of the A and B branches are coordinated by PsaA and PsaB proteins, respectively. Most of the molecules involved in ET were first identified by spectroscopy. With respect to the mechanism, one of the most exciting questions is whether one or both branches are active in ET. First, we will discuss the structure and function of the individual cofactors and then will come back to the question of whether one or two branches are active.

3.3.1 P700: The Primary Electron Donor

The pair of chlorophylls assigned to P700 is located close to the lumenal surface of PSI. It consists of two chemically different chlorophyll molecules. The chlorophyll on the B-branch is the "common" chlorophyll *a (Chla)* molecule, chemically identical to all the other 95 chlorophylls in PSI, whereas the chlorophyll at the A-branch is Chla', the epimer at the C13 position of the chlorin ring system.

The existence of at least one Chla' molecule was first suggested by Watanabe and co-workers[76] on the basis of chlorophyll extraction experiments. The chlorophyll on the B-branch of P700 is axially coordinated to His B660. No hydrogen bonds are formed between the surrounding protein and the chlorin head group of this chlorophyll in contrast to the Chla', which forms three hydrogen bonds. The distance between the central Mg^{2+} ions of the two chlorophylls assigned to P700 is 6.3 Å, which is shorter than the corresponding distance between the bacteriochlorophylls in the special pair of PbRC and shorter than the distance of the two chlorophylls P1 and P2 in PSII. ENDOR studies in solution[77] and on single crystals of PSI[78] revealed that the spin density in $P700^{+\bullet}$ is asymmetric, with more than 85% of the spin density located on the B-branch Chla of P700. Molecular orbital studies of the electronic structure of P700, show that the two chlorophylls are tightly coupled and P700 is a supermolecule.[79] The asymmetry of the spin density can be explained by the interplay between the asymmetric hydrogen bonding, differences in the protein environment, and the chemically different natures of the two chlorophyll molecules. Whether the asymmetry is essential for the function of PSI still remains to be proven. However, the fact that Chla' is a constituent in cyanobacterial, algae, and plant PSI may suggest that Chla' plays a key role for the function of PSI. It may be speculated that the asymmetrical spin-density distribution could be responsible for the "gating" of the electron along the two cofactor branches, however there is thus far no experimental proof for this suggestion.

The question of how the Chla' is assembled into the PSI complex or if it may be isomerized by PSI in a photoactivated process also remains unanswered.

3.3.2 A: The Initial Electron Acceptor

Three pairs of chlorophylls are present in the ETC as found in the X-ray structural model of PSI at a resolution of 2.5 Å. From the electronically excited singlet state $^1P700^*$ (represented by the first pair of chorophylls) the electron is transferred via one of the chlorophylls from the second pair chlorophylls (A) to the first stable electron acceptor, A_0, which may be located on one of the chlorophylls in the middle of the membrane (see Figure 5C). These first steps of ET occur in less than 3 ps; as such the second pair of chlorophylls has not yet been detected by spectroscopy. Recent results suggest that the charge separation may start from the electronically excited singlet state of the accessory chlorophyll at the B-branch rather than from $^1P700^*$,[80] but this idea is still a matter of debate.

In both branches, a water molecule provides the fifth ligand to the central Mg^{2+} ion of the second pair of chlorophylls. Remarkable differences are observed in the orientation and coordination of the second pair of chlorophylls between PSI and PSII (and the PbRC) (compare Figure 3 with Figure 5C).

3.3.3 A_0: The First Stable Electron Acceptor

The second and third pairs of chlorophylls are located in close vicinity to each other in PSI (3.8 Å). Even if these chlorophylls (or one of them) are supposed to correspond to the spectroscopically identified electron acceptor A_0, it is very likely that the spectroscopic and redox properties of these chlorophylls may be influenced by the second pair of chlorophylls, as indicated by the finding that the difference spectrum $A_0/A_0^{-\bullet}$ contains contributions from more than one chlorophyll.[81] The chlorophyll molecules assigned to A_0 are located in the middle of the membrane at a position that exhibits positional similarities to the pheophytin of both PbRC[43] and PSII[19] (compare Figure 3 with Figure 5C). The axial ligands of eC-A3 and eC-B3 are unusual as the sulfur atoms of methionine residues A688 and B668 provide the fifth "ligands" of the Mg^{2+} ions. This structural result is remarkable because the concept of hard and soft acids and bases predicts only weak interactions between the hard acid Mg^{2+} and methionine sulfur as a soft base. The mutation of the methionine to His provides a strong ligand to Mg^{2+} and thereby blocks or slows down the ETC along the corresponding branches.[82,83] In cyanobacteria the mutation of the B-branch has more severe effects on photoautotrophic growth of green algae than the mutation of the A-branch, leading to the suggestion that the A-branch is less important than the B-branch at least in green algae. However, the question as to whether there are one or two active branches in PSI is still the most controversially discussed topic in PSI research. It may even be the case that the branching of the electron along the two chains could differ between different organisms.

3.3.4 A_1: The Phylloquinone

In the next step of the ETC the electron is further transferred from A_0 to one of the phylloquinones, one or both of them represent the electron acceptor A_1. The two phylloquinones, Q_KA and Q_KB, are located at the stromal side of the membrane. The binding pockets are identical on both sites but differ significantly from all other quinone-binding pockets found in proteins so far. Both quinones are π-stacked with a tryptophan residue and both show asymmetrical hydrogen bonding: only one of the two oxygen atoms forms an H-bond to an NH backbone group, whereas the other oxygen atom is not hydrogen bonded at all. This could lead to a protein-induced asymmetry in the distribution of the unpaired electron in the radical state $A^{1-\bullet}$. This may answer the question of why A_1 has the most negative redox potential (-770 mV) of all the quinones so far found in nature. The electron proceeds from A_1 to the FeS cluster, F_x. This is the rate limiting step of the ET in PSI. Still, a lively scientific discussion continues regarding the question of whether one or both branches are active.

There is experimental evidence that the ET can proceed along both branches, but at different rates. In the green algae *Chlamydomonas reinhardtii*, the ET is about a factor of 50 slower on the A- than on the B-branch.[84,85] This could be the result of a higher activation energy barrier on the A- compared to the B-branch. This finding raises the question of the structural reason for this functional difference. There is no significant difference in the protein environments of both branches, but there are two lipid molecules located close to the pathway from A_1 to F_x which could be responsible for the asymmetry. A negatively charged phospholipid is located on the slower A-branch, which probably hampers the ET, whereas on the faster B-branch a neutral galactolipid has replaced the phospholipid. This is the most reasonable explanation for the higher activation energy barrier of the ET between Q_KA and F_x compared to the ET between Q_KB and F_x.

3.3.5 F_X: The First FeS Cluster

F_x is a rare example of an interprotein FeS cluster. Two cysteine ligands are provided by PsaA and two by PsaB. The ligands are located in the loop connecting helices h and i. This loop is highly conserved and essentially identical in all PSI species from plants, algae, and cyanobacteria. F_x is functional in ET and also plays an important role in the stabilization and assembly of the PSI complex. The PSI complex cannot assemble without the help of a rubredoxin (RubA),[86] which is not a part of PSI, but is found to be attached to the monomeric PSI during assembly. The assembly of the whole stromal hump depends on the functional assembly of F_x.

3.3.6 F_A and F_B: The Terminal FeS Clusters

The two terminal FeS centers are bound to the extrinsic subunit PsaC (see Section 3.2.3). Mutagenesis in combination with EPR investigations[87,88] showed that the cluster in close proximity to F_x (center-to-center distance is 14.9 Å), represents F_A, whereas the distal cluster is F_B,[89,90] which donates electrons to ferredoxin.

A sequential ET from F_x to F_A to F_B is confirmed by a large number of spectroscopic and biochemical studies (see Refs. 91, 33, and references therein).

3.4   The Antenna System of Photosystem I

The core antenna system of PSI consists of 90 Chl*a* molecules and 22 carotenes. The arrangement of the pigments is shown in Figure 5A. The function of the antenna chlorophylls (shown in green) is to capture light and transfer the excitation energy to the center of the complex, where the ET takes place. After excitation of any of the antenna chlorophylls, the chance that the energy is successfully transferred to P700 and subsequent charge separation occurs is 99.98% at room temperature. The carotenoids (orange) have a dual function as they serve in light harvesting as well as in the photoprotection.

3.4.1 The Chlorophylls

The arrangement of the antenna chlorophylls in PSI is unique. Instead of forming a symmetric ring surrounding the reaction center core, as is the case in the light harvesting systems of purple bacteria, the chlorophylls form a clustered network.[92] Each of the chlorophylls has several neighbors at a center-to-center distance of less than 15 Å, so that energy can be efficiently transferred via multiple pathways to the center of the complex. The system can be compared to the nerve-network system in the brain where multiple connections are responsible for the high efficiency of information transfer. The antenna system in PSI is highly optimized for efficiency and robustness.[93] The core antenna system of PSI can be divided into a central domain, which surrounds the ETC, and two peripheral domains, flanking the core on both sides. In the peripheral domains, the antenna chlorophylls are arranged in two layers, one close to the stromal surface of the membrane and the other close to the lumenal surface of the membrane. When a peripheral antenna chlorophyll becomes electronically excited, this energy will first be transferred from this "two-dimensional" layer to the central domain. In the central domain, chlorophylls are distributed over the full depth of the membrane, *i.e.*, the excitation energy can be exchanged between the two layers. The excitation energy is then transferred from the chlorophylls of the central domain to the ETC. There are two chlorophylls (named connecting chlorophylls) that seem to be the structural link between the antenna system and the ETC. Recently, mutagenesis experiments were performed on the ligands of these connecting chlorophylls.[94] The mutants exhibit minor alterations in the trapping of the excitation energy, but the question as to whether they play a crucial role in excitation energy transfer still remains to be answered.

Photosystem I contains pigments that absorb light at wavelengths $\lambda > 700$ nm. They are called "red" or "long-wavelength" chlorophylls. The long-wavelength chlorophylls may function in increasing the spectral width of the light absorbed and are used by PSI. Another function of the red chlorophylls may be to "funnel" the excitation energy to the center of the complex. The latter function may be provided by chlorophylls that absorb wavelengths between 685 and 705 nm, which are present in all PSI complexes from plants, algae, and cyanobacteria. The exact location of the red chlorophylls is still a matter of further investigation.

On the basis of the structure of PSI, several theoretical studies investigated the excitation energy transfer and trapping in PSI. These studies show that the excitation energy transfer in PSI is probably trap limited and is highly optimized for robustness and efficiency.[95]

3.4.2 The Carotenoids

Twenty-two carotenoids have been identified in the cyanobacterial structure of PSI. The carotenoids fulfill three functions in PSI: they play a structural role, function as additional antenna pigments, and prevent the system from damage by overexcitation caused by excess light (photoinhibition). The latter function is absolutely critical to the whole system. Chlorophylls are, in principle, dangerous and reactive molecules.

Superfluous excitation can lead to the formation of chlorophyll triplets (^{3}Chl), which react with oxygen under formation of the singlet oxygen $^1\Delta_g$ O_2, acting as a very dangerous cell poison. Multiple interactions can be observed between the carotenoids and the chlorophylls of the antenna system. The carotenoids are distributed over the whole antenna system (see Figures 4 and 5) and prevent photodamage by the quenching of chlorophyll triplet states. The energy from the triplet chlorophylls is transferred to the carotenoids that form the carotenoid triplet state ^{3}Car.

3.4.3　The Lipids in Photosystem I

Four lipids have been identified in the structure of PSI at a resolution of 2.5 Å: three molecules of phosphatidylglycerol (PG) and one molecule of monogalactosyldiacylglycerol (MGDG). Two of these lipids are located close to the ETC. They are incorporated into PSI at a very early stage of the assembly process, because their head groups are not solvent accessible but are covered by the loops of PsaA and PsaB and the three stromal subunits PsaC, PsaD, and PsaE. The lipids are located close to the ETC and may even play an important role for the difference in the rates of ET between the two different branches.

4　Conclusion and Outlook

The current X-ray structure of PSI and the biochemical–biophysical data already provide a clear picture of the function of PSI. Nevertheless, there are still a lot of exciting questions open, such as the role of the lipids, the specific function of the *cis*-carotenoids, the interaction with peripheral antenna systems, and the still open discussion as to which of the ET branches is the active one, the A- or the B-branch. As far as PSII is concerned, the last few years have brought great improvements towards better insight into its structure and function. The central questions, such as the mechanism of water splitting, the structure of the oxygen-evolving complex, its ligandation, the role and position of the carotenoids, the function of the small subunits, and the repair of PSII, still await answers. These questions can only be answered by a combination of new discoveries in spectroscopy, molecular biology, model calculations, and finally the determination of a high-resolution structure of PSII (to at least 2.5 Å) in well-defined oxidation states of the Mn cluster.

References

1. G. Kurisu, H. Zhang, J.L. Smith and W.A. Cramer, *Science*, 2003, **302**, 1009–1014.
2. N. Dashdorj, H. Zhang, H. Kim, J. Yan, W.A. Cramer and S. Savikhin, *Biophys. J.*, 2005, **88**, 4178–4187.
3. H. Zhang, A. Primak, J. Cape, M.K. Bowman, D.M. Kramer and W.A. Cramer, *Biochemistry*, 2004, **43**, 16329–16336.
4. D. Stroebel, Y. Choquet, J.L. Popot and D. Picot, *Nature*, 2003, **426**, 413–418.

5. J.P. Abrahams, A.G. Leslie, R. Lutter and J.E. Walker, *Nature*, 1994, **370**, 621–628.

6. R.J. Carbajo, F.A. Kellas, M.J. Runswick, M.G. Montgomery, J.E. Walker and D. Neuhaus, *J. Mol. Biol.*, 2005, **351**, 824–838.

7. D.J. Gordon-Smith, R.J. Carbajo, J.C. Yang, H. Videler, M.J. Runswick, J.E. Walker and D. Neuhaus, *J. Mol. Biol.*, 2001, **308**, 325–339.

8. G. Groth and E. Pohl, *J. Biol. Chem.*, 2001, **276**, 1345–1352.

9. Z. Liu, H. Yan, K. Wang, T. Kuang, J. Zhang, L. Gui, X. An and W. Chang, *Nature*, 2004, **428**, 287–292.

10. A. Ben-Shem, F. Frolow and N. Nelson, *Nature*, 2003, **426**, 630–635.

11. J. Weber and A.E. Senior, *FEBS Lett.*, 2003, **545**, 61–70.

12. J. Vonck, T.K. von Nidda, T. Meier, U. Matthey, D.J. Mills, W. Kuhlbrandt and P. Dimroth, *J. Mol. Biol.*, 2002, **321**, 307–316.

13. M. Yoshida, E. Muneyuki and T. Hisabori, *Nat. Rev. Mol. Cell Biol.*, 2001, **2**, 669–677.

14. H. Noji, R. Yasuda, M. Yoshida and K. Kinosita Jr., *Nature*, 1997, **386**, 299–302.

15. P.D. Boyer, *FEBS Lett.*, 1975, **58**, 1–6.

16. E. Baena-Gonzalez and E.M. Aro, *Philos. Trans. R. Soc. Lond. B Biol. Sci.*, 2002, **357**, 1451–1459; discussion 1459–1460.

17. A. Zouni, C. Lueneberg, P. Fromme, W.D. Schubert, W. Saenger and H.T. Witt, in *Photosynthesis: Mechanisms and Effects*, G. Garab (ed), Kluwer, Dordrecht, 1998, 925–928.

18. A. Zouni, H.T. Witt, J. Kern, P. Fromme, N. Krauss, W. Saenger and P. Orth, *Nature*, 2001, **409**, 739–743.

19. K.N. Ferreira, T.M. Iverson, K. Maghlaoui, J. Barber and S. Iwata, *Science*, 2004, **303**, 1831–1838.

20. J. Deisenhofer, O. Epp, I. Sinning and H. Michel, *J. Mol. Biol.*, 1995, **246**, 429–457.

21. I. Muegge, J. Apostolakis, U. Ermler, G. Fritzsch, W. Lubitz and E.W. Knapp, *Biochemistry*, 1996, **35**, 8359–8370.

22. P. Jordan, P. Fromme, H.T. Witt, O. Klukas, W. Saenger and N. Krauss, *Nature*, 2001, **411**, 909–917.

23. W.D. Schubert, O. Klukas, W. Saenger, H.T. Witt, P. Fromme and N. Krauss, *J. Mol. Biol.*, 1998, **280**, 297–314.

24. R.E. Blankenship, *Photosynth. Res.*, 1992, **33**, 91–111.

25. N. Bondarava, L. De Pascalis, S. Al-Babili, C. Goussias, J.R. Golecki, P. Beyer, R. Bock and A. Krieger-Liszkay, *J. Biol. Chem.*, 2003, **278**, 13554–13560.

26. S. Vasil'ev, G.W. Brudvig and D. Bruce, *FEBS Lett.*, 2003, **543**, 159–163.

27. F. Morais, K. Kuhn, D.H. Stewart, J. Barber, G.W. Brudvig and P.J. Nixon, *J. Biol. Chem.*, 2001, **276**, 31986–31993.

28. J. Biesiadka, B. Loll, J. Kern, K.D. Irrgang and A. Zouni, *Phys. Chem. Chem. Phys.*, 2004, **6**, 4733–4736.

29. N. Kamiya and J.R. Shen, *Proc. Natl. Acad. Sci. USA*, 2003, **100**, 98–103.

30. I. Grotjohann, C. Jolley and P. Fromme, *Phys. Chem. Chem. Phys.*, 2004, **6**, 4743–4753.

31. M. Sugiura, F. Rappaport, K. Brettel, T. Noguchi, A.W. Rutherford and A. Boussac, *Biochemistry*, 2004, **43**, 13549–13563.
32. P. Fromme, J. Kern, B. Loll, J. Biesiadka, W. Saenger, H.T. Witt, N. Krauss and A. Zouni, *Philos. Trans. R. Soc. Lond. B Biol. Sci.*, 2002, **357**, 1337–1344.
33. K.V. Lakshmi, Y.S. Jung, J.H. Golbeck and G.W. Brudvig, *Biochemistry*, 1999, **38**, 13210–13215.
34. S.G. Zech, J. Kurreck, G. Renger, W. Lubitz and R. Bittl, *FEBS Lett.*, 1999, **442**, 79–82.
35. A.T. Gardiner, S.G. Zech, F. MacMillan, H. Kass, R. Bittl, E. Schlodder, F. Lendzian and W. Lubitz, *Biochemistry*, 1999, **38**, 11773–11787.
36. J. Barber, *Bioelectrochemistry*, 2002, **55**, 135–138.
37. F.J.E. Vanmieghem, K. Satoh and A.W. Rutherford, *Biochim. Biophys. Acta*, 1991, **1058**, 379–385.
38. M. Kammel, J. Kern, W. Lubitz and R. Bittl, *Biochim. Biophys. Acta Bioenerg.*, 2003, **1605**, 47–54.
39. A.T. Keilty, D.V. Vavilin and W.F.J. Vermaas, *Biochemistry*, 2001, **40**, 4131–4139.
40. K. Sauer and V.K. Yachandra, *Biochim. Biophys. Acta*, 2004, **165**, 140–148.
41. J. Clausen, S. Winkler, A.M. Hays, M. Hundelt, R.J. Debus and W. Junge, *Biochim. Biophys. Acta*, 2001, **1506**, 224–235.
42. P. Fromme, Habilitation in Max Volmer Institute, Department of Chemistry, Technical University Berlin, Germany, 1998.
43. J. Deisenhofer, O. Epp, I. Sinning and H. Michel, *J. Mol. Biol.*, 1995, **246**, 429–457.
44. C.C. Jolley, A. Ben-Shem, N. Nelson and P. Fromme, *J. Biol. Chem.*, 2005, **280**(39), 33627–33636.
45. M.K. Sener, C. Jolley, A. Ben-Shem, P. Fromme, N. Nelson, R. Croce and K. Schulten, *Biophys. J.*, 2005, **89**, 1630–1642.
46. W.M. Schluchter, G. Shen, J. Zhao and D.A. Bryant, *Photochem. Photobiol.*, 1996, **64**, 53–66.
47. B. Andersen and H.V. Scheller, in *Plastids: Synthesis and Assembly*, C. Sundquist (ed), Academic Press, San Diego, 1993, 383–418.
48. S. Janson, B. Andersen and H.V. Scheller, *Plant. Physiol.*, 1996, **112**, 409–420.
49. H.V. Scheller, P.E. Jensen, A. Haldrup, C. Lunde and J. Knoetzel, *Biochim. Biophys. Acta*, 2001, **1507**, 41–60.
50. K. Ohyama, H. Fukazawa, T. Kohchi, H. Shirai, S. Tohru, S. Sano, K. Umesono, Y. Shiki, M. Takeuchi, Z. Chang, S.I. Aota, H. Inokuchi and H. Ozeki, *Nature*, 1986, **322**, 572–574.
51. M.K. Sener, S. Park, D. Lu, A. Damjanovic, T. Ritz, P. Fromme and K. Schulten, *J. Chem. Phys.*, 2004, **120**, 11183–11195.
52. R. Kouril, N. Yeremenko, S. D'Haene, A.E. Yakushevska, W. Keegstra, H.C. Matthijs, J.P. Dekker and E.J. Boekema, *Biochim. Biophys. Acta*, 2003, **1607**, 1–4.
53. T.S. Bibby, J. Nield and J. Barber, *Nature*, 2001, **412**, 743–745.
54. E.J. Boekema, A. Hifney, A.E. Yakushevska, M. Piotrowski, W. Keegstra, S. Berry, K.P. Michel, E.K. Pistorius and J. Kruip, *Nature*, 2001, **412**, 745–748.

55. M. Hippler, F. Drepper, W. Haehnel and J.D. Rochaix, *Proc. Natl. Acad. Sci. USA*, 1998, **95**, 7339–7344.
56. S. Anandan, A. Vainstein and J.P. Thornber, *FEBS Lett.*, 1989, **256**, 150–154.
57. C. Varotto, P. Pesaresi, P. Jahns, A. Lessnick, M. Tizzano, F. Schiavon, F. Salamini and D. Leister, *Plant Physiol.*, 2002, **129**, 616–624.
58. S. Kjaerulff, B. Andersen, V.S. Nielsen, B.L. Moller and J.S. Okkels, *J. Biol. Chem.*, 1993, **268**, 18912–18916.
59. M.L. Antonkine, P. Jordan, P. Fromme, N. Krauss, J.H. Golbeck and D. Stehlik, *J. Mol. Biol.*, 2003, **327**, 671–697.
60. P. Setif, *Biochim. Biophys. Acta*, 2001, **1507**, 161–179.
61. P. Setif, N. Fischer, B. Lagoutte, H. Bottin and J.D. Rochaix, *Biochim. Biophys. Acta*, 2002, **1555**, 204–209.
62. N. Li, P.V. Warren, J.H. Golbeck, G. Frank, H. Zuber and D.A. Bryant, *Biochim. Biophys. Acta*, 1991, **1059**, 215–225.
63. B. Lagoutte, J. Hanley and H. Bottin, *Plant Physiol.*, 2001, **126**, 307–316.
64. P. Fromme, H. Bottin, N. Krauss and P. Setif, *Biophys. J.*, 2002, **83**, 1760–1773.
65. K. Sonoike, H. Hatanaka and S. Katoh, *Biochim. Biophys. Acta*, 1993, **1141**, 52–57.
66. H. Strotmann and N. Weber, *Biochim. Biophys. Acta*, 1993, **1143**, 204–210.
67. F. Rousseau, P. Setif and B. Lagoutte, *EMBO J.*, 1993, **12**, 1755–1765.
68. L. Yu, J. Zhao, U. Muhlenhoff, D.A. Bryant and J.H. Golbeck, *Plant Physiol.*, 1993, **103**, 171–180.
69. B. Andersen, H.V. Scheller and B.L. Moller, *FEBS Lett.*, 1992, **311**, 169–173.
70. C.J. Falzone, Y.H. Kao, J. Zhao, D.A. Bryant and J.T. Lecomte, *Biochemistry*, 1994, **33**, 6052–6062.
71. A. Lushy, L. Verchovsky and R. Nechushtai, *Biochemistry*, 2002, **41**, 11192–11199.
72. K. Meimberg, B. Lagoutte, H. Bottin and U. Mühlenhoff, *Biochemistry*, 1998, **37**, 9759–9767.
73. U. Muhlenhoff, J. Zhao and D.A. Bryant, *Eur. J. Biochem.*, 1996, **235**, 324–331.
74. J.J. van Thor, T.H. Geerlings, H.C. Matthijs and K.J. Hellingwerf, *Biochemistry*, 1999, **38**, 12735–12746.
75. P. Fromme, P. Jordan and N. Krauss, *Biochim. Biophys. Acta*, 2001, **1507**, 5–31.
76. T. Watanabe, M. Kobayashi, A. Hongu, M. Nakazato and T. Hiyama, *FEBS Lett.*, 1985, **235**, 252–256.
77. H. Kass, E. Bittersmannweidlich, L.E. Andreasson, B. Bonigk and W. Lubitz, *Chem.l Phys.*, 1995, **194**, 419–432.
78. H. Kass, P. Fromme, H.T. Witt and W. Lubitz, *J. Phys. Chem. B*, 2001, **105**, 1225–1239.
79. M. Plato, N. Krauss, P. Fromme and W. Lubitz, *Chem. Phys.*, 2003, **294**, 483–499.
80. M.G. Muller, J. Niklas, W. Lubitz and A.R. Holzwarth, *Biophys. J.*, 2003, **85**, 3899–3922.
81. G. Hastings, S. Hoshina, A.N. Webber and R.E. Blankenship, *Biochemistry*, 1995, **34**, 15512–15522.

82. W.V. Fairclough, A. Forsyth, M.C. Evans, S.E. Rigby, S. Purton and P. Heathcote, *Biochim. Biophys. Acta*, 2003, **1606**, 43–55.

83. V.M. Ramesh, K. Gibasiewicz, S. Lin, S.E. Bingham and A.N. Webber, *Biochemistry*, 2004, **43**, 1369–13675.

84. M. Guergova-Kuras, B. Boudreaux, A. Joliot, P. Joliot and K. Redding, *Proc. Natl. Acad. Sci. USA*, 2001, **98**, 4437–4442.

85. B. Boudreaux, F. MacMillan, C. Teutloff, R. Agalarov, F. Gu, S. Grimaldi, R. Bittl, K. Brettel and K. Redding, *J. Biol. Chem.*, 2001, **276**, 37299–37306.

86. G. Shen, M.L. Antonkine, A. van der Est, I.R. Vassiliev, K. Brettel, R. Bittl, S.G. Zech, J. Zhao, D. Stehlik, D.A. Bryant and J.H. Golbeck, *J. Biol. Chem.*, 2002, **277**, 20355–20366.

87. J. Zhao, N. Li, P.V. Warren, J.H. Golbeck and D.A. Bryant, *Biochemistry*, 1992, **31**, 5093–5099.

88. T. Mehari, F. Qiao, M.P. Scott, D.F. Nellis, J. Zhao, D.A. Bryant and J.H. Golbeck, *J. Biol. Chem.*, 1995, **270**, 28108–28117.

89. L. Yu, D.A. Bryant and J.H. Golbeck, *Biochemistry*, 1995, **34**, 7861–7868.

90. L. Yu, I.R. Vassiliev, Y.S. Jung, D.A. Bryant and J.H. Golbeck, *J. Biol. Chem.*, 1995, **270**, 28118–28125.

91. N. Fischer, P. Setif and J.D. Rochaix, *J. Biol. Chem.*, 1999, **274**, 23333–23340.

92. G. Mcdermott, S.M. Prince, A.A. Freer, A.M. Hawthornthwaitelawless, M.Z. Papiz, R.J. Cogdell and N.W. Isaacs, *Nature*, 1995, **374**, 517–521.

93. M. Sener, S. Park, D.Y. Lu, A. Damjanovic, T. Ritz, P. Fromme and K. Schulten, *Biophys. J.*, 2003, **84**, 274a.

94. K. Gibasiewicz, V.M. Ramesh, S. Lin, K. Redding, N.W. Woodbury and A.N. Webber, *Biophys. J.*, 2003, **85**, 2547–2559.

95. B. Gobets, L. Valkunas and R. van Grondelle, *Biophys. J.*, 2003, **85**, 3872–3882.

96. T. Inoue, H. Sugawara, S. Hamanaka, H. Tsukui, E. Suzuki, T. Kohzuma and Y. Kai, *Biochemistry*, 1999, **38**, 6063–6069.

97. R. Morales, M.H. Charon, G. Hudry-Clergeon, Y. Petillot, S. Norager, M. Medina and M. Frey, *Biochemistry*, 1999, **38**, 15764–15773.

98. L. Serre, F.M. Vellieux, M. Medina, C. Gomez-Moreno, J.C. Fontecilla-Camps and M. Frey, *J. Mol. Biol.*, 1996, **263**, 20–39.

99. R.I. Menz, J.E. Walker and A.G. Leslie, *Cell*, 2001, **106**, 331–341.

100. T. Meier, P. Polzer, K. Diederichs, W. Welte and P. Dimroth, *Science*, 2005, **308**, 659–662.

101. W. Humphrey, A. Dalke and K. Schulten, *J. Mol. Graph.*, 1996, **14**, 33.

102. R. Fromme, H. Yu, J. Grotjohann and P. Fromme, *FEBS J.*, **272**, 449.

103. B. Loll, J. Kern, W. Saenger, A. Zouni, J. Biesiadka, *Nature*, 2005, **438**, 1040–1044.

Glutamate Receptor Ion Channels: Structural Insights into Molecular Mechanisms

AVINASH GILL AND DEAN R. MADDEN

Department of Biochemistry, Dartmouth Medical School, Hanover, NH, U.S.A.

1 Introduction

L-Glutamate was first proposed as a candidate neurotransmitter in the late 1950s, but it was not until the 1980s that a growing body of pharmacological, physiological, and molecular biological evidence had unambiguously established the essential role of glutamergic signaling in the central nervous system (CNS) and identified several of the receptors involved.[1,2] Glutamate is now recognized as the major excitatory neurotransmitter of the CNS, where it activates three related families of ligand-gated ion channels and several classes of G-protein-coupled receptors.[3] This review focuses on the ionotropic glutamate receptors (iGluRs).

Even when genes for the receptor subunits were in hand, the iGluRs retained several surprises in terms of their molecular organization and function. Although iGluRs were initially classified as members of the nicotinic acetylcholine receptor family on the basis of hydropathy plots and mutagenesis studies, it emerged in the mid-1990s that they in fact constitute a structurally and evolutionarily independent ion-channel superfamily.[3] The subunits exhibit functionally suggestive domain homologies to the amino acid binding proteins on the one hand, and to the potassium channel pore domains on the other.[4,5,6] These homologies provided an attractive basis for structure—function analyses, and over the last 10 years, biophysical and biochemical characterization of iGluRs and their domains has continued to refine our knowledge of how these channels work. They now provide one of the best-understood examples of how ligand binding produces conformational changes that drive channel function.[1,7–10] The increasingly detailed understanding of the molecular nature of these channels reveals not only a number of functional similarities to the homologous domains but also several novel adaptations that presumably reflect the specific physiological requirements of iGluR signaling. This review will survey the current structural biological understanding of iGluR function and will highlight a number of issues that remain to be resolved.

1.1 Physiological and Pathophysiological Roles of the Ionotropic Glutamate Receptors

The iGluRs are classified according to their affinities for the synthetic agonists AMPA (α-amino-5-methyl-3-hydroxy-4-isoxazole propionic acid), NMDA (*N*-methyl-D-aspartate), and kainate, the differences being brought about by the subunits forming the homo- or heterotetrameric channels.[3,4] AMPA receptors are assembled from the subunits GluR1-GluR4, and kainate receptors from the subunits GluR5-GluR7 and KA1 and KA2. Functional NMDA receptors are formed as obligate heterotetramers of NR1 subunits with NR2A-D and NR3A-B subunits. They generally require coactivation by glycine or D-serine and glutamate,[11,12] although NR1/NR3 receptors may form excitatory receptors activated by glycine alone.[13] Within each subfamily, different combinations of subunits yield receptors with different electrophysiological characteristics, providing significant adaptability to physiological signaling requirements. Additional variability is provided by alternative splicing of subunits, RNA editing, posttranslational modifications, and regulation of membrane localization.[4,14]

In general, the iGluRs are permeable to the monovalent cations Na^+ and K^+, while permeability to the divalent cation Ca^{2+} is subtype dependent. The AMPA receptors, which mediate most fast synaptic transmission in the brain, are typically impermeable to Ca^{2+}, because of the presence of RNA-edited GluR2 subunit(s) in heteromeric channels.[15] Some specialized AMPA receptors lack GluR2 and are Ca^{2+} permeable. On the other hand, classical NMDA receptors, which exhibit slow kinetics, generally have high permeability to Ca^{2+} but are also subject to voltage-dependent block by Mg^{2+} ions.[4]

The iGluRs play crucial but diverse roles in several physiological and pathological processes.[4,16,17] In addition to neuronal signaling, these include mechanisms of long-term potentiation and long-term depression,[18] which are key components of the synaptic plasticity that is thought to underpin learning and memory. Conversely, the physiological significance of the iGluRs is mirrored in the number of pathological processes that appear to involve them.[4,16] Glutamate receptor activation can mediate neuronal cell death in neurodegenerative disorders such as amyotrophic lateral sclerosis (ALS), Huntington's disease, Parkinson's disease, and Alzheimer's disease.[19,20] Glutamate-induced excitotoxicity can also contribute to the pathology of cerebral ischemia, epilepsy, and traumatic brain injury.[4,20–23] The link between glutamate receptors and human cancers such as glioblastoma has been also a subject of study,[24,25] and it has been suggested that malignant gliomas "hijack" glutamate receptors to support their growth and migration.[26]

1.2 Medicinal Chemistry

Correspondingly, significant pharmaceutical efforts have been devoted to the formulation of compounds selectively targeting iGluR populations, which could help to treat the numerous chronic and acute pathological conditions in which these receptors are involved.[4,27] The search for agents that can be used to activate, block, or modulate glutamate receptors has resulted in candidate therapeutics, one of which has been approved: Namenda® (memantine HCl) is an uncompetitive NMDA receptor inhibitor that is prescribed to treat moderate to severe Alzheimer's disease.[28–31] Other examples include the noncompetitive AMPA receptor antagonists GYKI 53773/LY300164, which have entered clinical trials for conditions including epilepsy.[32] The AMPA receptor competitive antagonist YM872[33] and the antagonist BIIR 561, which is directed against both voltage-gated Na^+ channels and AMPA receptors,[34] have been shown to be useful for the treatment of acute ischemic stroke. Clinical trials for YM872 have been completed (http://clinicaltrials.gov/show/NCT00044057),[32] although further information is not yet available.

1.3 Ionotropic Glutamate Receptor Subunits Are Modular

The glutamate receptor ion channels from eukaryotes share a modular architecture, which, together with the domain homologies mentioned above, has made them excellent candidates for reductionist analysis. The typical receptor subunit consists of an N-terminal domain (NTD), a ligand-binding domain (LBD; also known as "S1S2" from the sequences that comprise it), three transmembrane (TM) segments, and a reentrant pore loop, followed by a cytoplasmic carboxy-terminal domain[35] (Figure 1). Bacterial iGluRs share a core structure consisting of an LBD and a helix/pore loop/helix TM topology,[36–38] but lack the NTD, the third TM domain, and the cytoplasmic C-terminus, which provide additional functionalities in the eukaryotic channels. The bacterial iGluR may represent an evolutionary intermediate, in which amino acid binding and ion-conducting moieties were fused to generate a primitive ligand-gated ion channel.[1,36,37]

2 Studies of the Ligand-Binding Domain

As suggested by this apparent modularity, ligand-binding domain constructs can be excised recombinantly from the subunits, yielding water-soluble, pharmacologically active proteins in quantities suitable for biochemical and crystallographic studies. The efforts of the Keinänen group[39] resulted in the first soluble version of an iGluR LBD, expressed and purified from insect cells. This approach has been used for much of the biophysical analysis of the conformational behavior of LBD in solution,[40–47] as well as the analysis of the GluR6 LBD structure in complex with domoate.[48] Subsequent development of a bacterial expression system for the periplasmic-binding protein (PBP)-homologous core of the GluR2 LBD[49,50] led to the first crystals diffracting to high resolution[51] and provided a foundation for most high-resolution structural studies.[38,48,52–59] A modified bacterial expression system

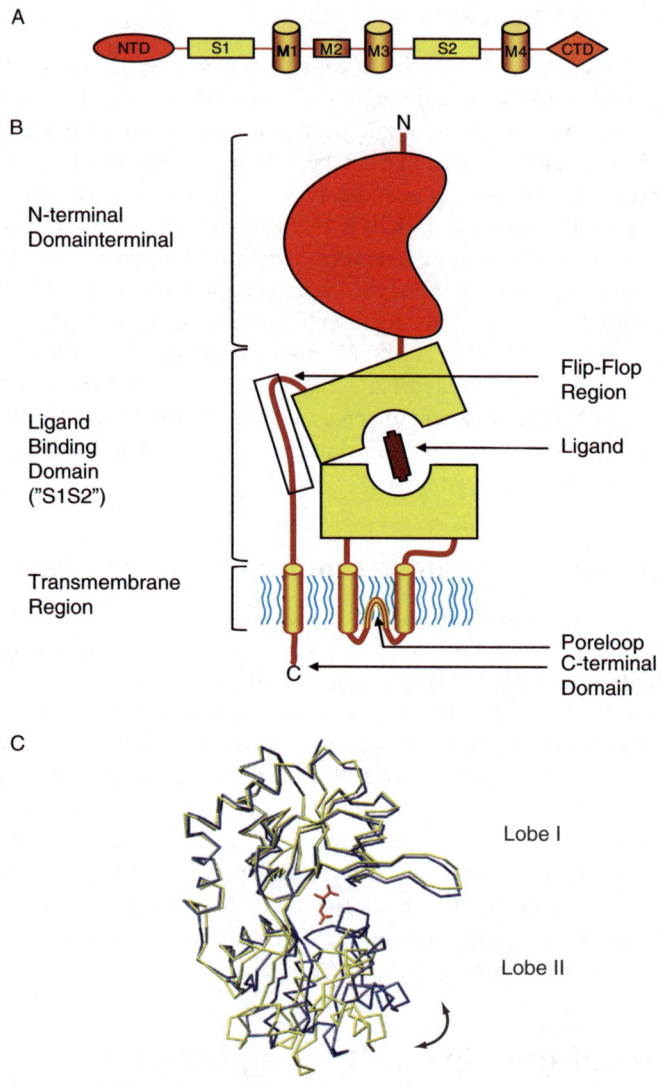

Figure 1 *Modular architecture of the glutamate receptor. (A) Linear organization of the domains in the ionotropic glutamate receptor (iGluR). (B) The extracellular moiety of a eukaryotic iGluR consists of an N-terminal domain (NTD) and a bilobate ligand-binding domain (LBD or "S1S2"). The transmembrane region consists of the three transmembrane domains (M1, M3, and M4) and a reentrant pore loop (M2). The carboxy-terminal region is located inside the cell. The LBD can be expressed as a soluble domain, and its structure in combination with different ligands has been determined crystallographically. Note that S1 and S2 each contribute to both of the lobes, although lobe 1 is formed primarily by S1 residues and lobe 2 primarily by S2 residues. (C) Crystal structures of the GluR2 LBD in the ligand-free (yellow) and glutamate-bound (dark blue) states (PDB entries 1FTO and 1FTJ, respectively) have been superimposed by least-squares fitting of the Cα atoms in lobe I only. The cleft closure is seen as a relative motion of lobe II. (Figure 1C was produced using the CHIMERA package)[153]*

that avoids the need for refolding has also been exploited to generate corresponding LBD core domains of the NR1 NMDA receptor subunit[60] and of the GluR5 and GluR6 kainate receptor subunits,[57,59] all of which have been crystallized. The crystallographic data from these LBD constructs have provided the basis for a tremendous increase in our understanding of the molecular mechanisms of iGluR functional transitions.

2.1   Overall Structure

The first crystallographic structure of the GluR2 LBD core, which was determined in the kainate-bound state, revealed a bilobate structure, with the attachment points for the NTD and the transmembrane domains located at opposite ends of the molecule (Figure 1C). The lobes are connected by a "hinge".[51] Kainate was observed to bind in the cleft formed between the two lobes, where it interacts with a number of stereochemically compatible functional groups. On the basis of the kainate cocrystal structure, a modified core LBD construct was designed, which has crystallized in the presence of a wide variety of ligands that also bind in the interlobe cleft. This permitted the determination of the structure of a complex with the physiological ligand glutamate, as well as a series of synthetic agonists and antagonists.[52–56,61–65] Structures have now also been determined from the GluR5 and GluR6 LBD and from the NR1 glycine-specific LBD[48,57,59,60] These reveal similar overall folds, with a slightly closer structural resemblance between the AMPA and kainate receptor LBD than between these and the NR1 domain.[59] Differences are observed, however, particularly in surface loops, some of which are quite prominent in the NR1 LBD.[60]

2.2   Pharmacological Specificity

Comparison of the different LBD complexes has revealed that the α-NH$_3^+$ and α-COO$^-$ functional groups (or their homologues) bind almost identically to a conserved site on the receptor. This applies not only to the AMPA receptor family but also to the NMDA receptor NR1 subunit[60] and to the GluR5 and GluR6 kainate receptor subunits.[48,57,59] In contrast, the agonist side-chain (or equivalent) occupies a more variable set of surfaces in the ligand-binding pocket, an interaction associated with cleft closure, as described below. It has been long known that the surface of the ligand-binding pocket exhibits a local excess of acidic functional groups and thus has an apparent electrostatic mismatch with the net negative charge of its physiological ligand glutamate.[51] However, detailed electrostatic calculations have recently shown that the overall potential of the GluR2 LBD at the surface of the binding pocket is indeed complementary to bound glutamate, and that the overall attractive (positive) potential of the apo-LBD extends quite far into solution, which may help to promote the initial binding interaction.[66]

One of the remarkable features of the iGluRs is the wide variety of chemical ligands that can bind to each subtype, with specificities that differ among receptor subtypes.[4,27] The structural analysis of LBD provided a clear explanation for the stereochemical promiscuity of the binding site: even in the closed conformation, the site itself is significantly larger than would be required to bind glutamate, with a number of subsites that are not engaged by the physiological ligand but are instead occupied by bound water molecules. Nonphysiological agonist functional moieties can be accommodated by different sets of subsites, with a corresponding rearrangement or displacement of bound water molecules.[67]

Furthermore, differences in the distribution and stereochemistry of the subsites found in the GluR2, NR1, and GluR5/6 LBDs satisfyingly explain the different selectivities of these domains. In particular, the substitution of a Thr by a Val and of a Leu by a Trp removes key binding partners and sterically inhibit the binding of glutamate, but not glycine, to the NR1 subunit, yielding a significantly smaller binding pocket.[60] In contrast, the GluR5 ligand-binding cleft accommodates water molecules in a volume 40% larger than the corresponding region in GluR2, but even in this case, steric restrictions discriminate among chemically similar compounds, as reflected in differential selectivity for quisqualate over AMPA.[57]

An interesting question concerns the evolutionary basis for the stereochemical promiscuity of the iGluR LBDs. Since synthetic agonists and antagonists play no role in natural selection, it is not immediately obvious why the iGluR binding sites contain the additional subsites that are occupied by water molecules in the presence of glutamate. It is also not clear why the kainate receptors, for example, have evolved even more capacious binding sites than GluR2, given that the physiological ligand is the same. One suggestion is that the observed stereochemical differences may reflect attempts to tune the kinetics and desensitization levels of the responses to glutamate among different subtypes.[10] Tuning of the stability of the glutamate/LBD complex has also been suggested to be important for the kinetics of fast synaptic transmission in the case of AMPA receptors.[68] Both the presence of bound water molecules and the formation of interlobe hydrogen bonds can contribute to these effects.

2.3 Ligand-Binding Domain Conformational Changes

The availability of crystal structures of the LBD in a wide variety of against- and antagonist-bound states also provides an excellent opportunity to study the conformational changes associated with ligand binding. The common theme underlying agonist interaction with the LBD is a "Venus-flytrap" mechanism, in which a "closed cleft" is stabilized in the presence of agonist, whereas the "open cleft" state predominates in its absence and is stabilized by the binding of competitive antagonists[51,52] (Figure 1C). Depending on the agonist bound, a spectrum of closures is observed, ranging from ~9° to >26°. In general, the maximal cleft closure observed for the NR1 and GluR5/6 LBD is noticeably greater (~26°) than that for the AMPA receptor LBD (~21°). However, in every case, the cleft closure remains significantly smaller than that observed for the homologous bacterial glutamine-binding protein QBP.[69] This may reflect the influence of conformational constraints imposed by interdomain and intersubunit packing interactions in the iGluR.[7] It also corresponds to the reorientation of the ligand to lie across the binding-site hinge, rather than along it, as observed for the bacterial glutamine- and glutamate-binding proteins.[38,41,69,70]

Corresponding conformational changes have now also been detected by small-angle X-ray scattering measurements of purified LBD in solution.[43] Solution scattering measurements also suggest that the GluR2 LBD exhibits a conformational equilibrium in the absence of ligand that may include both open and closed clefts.[43] Crystal structures of iGluR LBD in the apo state have revealed a range of conformations, up to and including a fully closed empty state.[38,52] Mutagenesis and stopped-flow kinetic analysis of agonist binding to the GluR4 LBD had previously revealed that the first step of the association involves the relatively rapid "docking" of the α-substituents into their conserved binding sites, which is then followed by a slower conformational change in the domain as a whole.[41] These observations, taken together, are consistent with a model in which agonist binding preferentially stabilizes one component of a preexisting equilibrium in LBD. However, given the variability observed in the degree of cleft closure associated with the binding of different agonists, it appears that the binding of agonist to the iGluR LBD may represent a hybrid of the Monod–Wyman–Changeaux (MWC) and induced-fit models.[71,72]

Evidence that an analogous cleft closure takes place in membrane-bound ion channels is provided by fluorescence resonance energy transfer (FRET) studies carried out on GFP-tagged GluR4 mutants lacking the NTD.[73] These subunits form functional channels, in which the FRET signature of kainate is intermediate between those of the apo state and glutamate-bound state. Furthermore, the calculated magnitudes of the donor—acceptor displacements correspond to the relative cleft closures observed crystallographically for the isolated LBD binding to kainate (partial closure) and glutamate (full closure).

Solution NMR studies of the GluR2 LBD indicate that not only the relative disposition of the lobes but also their dynamic flexibility varies in response to agonist binding. Of the two lobes, chemical exchange differences suggest that while lobe I is relatively rigid, lobe II is more flexible.[58] In addition, the binding of different full agonists causes distinct changes in the chemical shift behavior of the domain, indicating that they may differentially stabilize it.[74] Since lobe II forms the attachment site of the transmembrane domains, these changes in protein flexibility may be important for channel gating.[1] They may also have important implications for possible compliance in the coupling between the LBD and the channel pore (see Section 4).

2.4 Correlation with Channel Activation

Among the numerous synthetic agonists acting on the iGluR are several that induce submaximal currents even at saturating concentrations (corresponding to a reduced "intrinsic activity," also referred to as "efficacy").[75–77] A number of these have been cocrystallized with the corresponding LBD, and for the non-NMDA receptors, the extent of cleft closure associated with binding of a given agonist correlates with the magnitude of the current response it can induce at saturation. This correlation is particularly striking for the GluR2 LBD, for which the most cocrystal structures have been determined, and has been the subject of extensive review.[1,7,8,10,67,78] It strongly suggests that the observed cleft closure is a fundamental component of ion-channel activation. A similar correlation has also been observed in the case of the kainate receptor LBD, for which kainate is, in fact, a partial agonist[79] and induces a slightly smaller cleft closure in the GluR6 LBD

(23° relative to the GluR2 apo-LBD) than does glutamate (26°).[57] Domoate, which is additionally thought to be a partial agonist, also induces a less-than-complete cleft closure (15°).[48]

However, in the case of the NR1 LBD, the partial agonists D-cycloserine and 1-amino-cyclopropane- and 1-amino-cyclobutane-carboxylic acids induce the same degree of cleft closure as do the full agonists glycine and D-serine.[10,60] It thus appears that the reduced intrinsic activity of these agonists reflects their reduced selectivity for the closed-cleft *vs.* the open-cleft state, rather than stabilization of a more limited cleft closure. An analogous mechanism may explain the partial agonism of AMPA acting at the GluR2-L650T mutant channel, which can induce complete cleft closure in the corresponding LBD.[80] As a result, it appears that multiple mechanisms of partial agonism may be at work within the iGluR family (see Section 5).

2.5 Dimerization

An unexpected feature of the LBD crystal structures was the observation that they form back-to-back dimers in a large number of cases. Dimers are formed by the membrane distal lobes only, so that cleft closure in both LBDs acts to pull apart the membrane-proximal lobes[1,7,67] (Figure 2A). Envisaged in the context of a schematic half-receptor (Figure 2A), this motion could apply force to the transmembrane domains, opening the channel pore by a mechanism analogous to that proposed for the Ca^{++}-gated K^+-channel MthK.[81] LBD dimerization is also consistent with dimer-of-dimer models for the assembly of iGluR channels (see Section 3).

The dimer was first observed in crystals of the GluR2 LBD,[51,52] and has since been seen in numerous GluR2 LBD cocrystal structures formed in varying spacegroups. Furthermore, the dimerization interaction could be shown in solution for GluR0[38] and GluR2.[82] Thus, it does not appear to be an artifact of crystallization. However, the dimerization affinity of the eukaryotic LBD is extremely weak (~ 6 mM) [82], and is presumably only efficiently stabilized during crystallization by the extremely high protein concentrations found in the crystal (~15 to 20 mM for a 30 kDa protein crystallized with 50% solvent). *In vivo*, the dimerization interaction may be stabilized by the simultaneous interaction of multiple subunits ("chelate effect"), as well as by the entropically favorable restriction of membrane protein diffusion to essentially two dimensions. A similar dimer interface is observed in various GluR5 and GluR6 LBD crystal structures,[48,57] but not in the NR1 LBD.[60] A distinct, but partially overlapping, dimerization interface has been observed in an independent crystal structure of the GluR5/glutamate complex,[59] but its physiological significance is currently unclear.

2.6 Desensitization and the Stability of the Dimer Interface

The weakness of the GluR2 LBD dimerization turned out to be physiologically significant in the context of AMPA receptor desensitization. A feature exhibited by most ligand-gated ion channels, especially those involved in fast synaptic transmission, is that activation is followed by desensitization, a phenomenon in which the receptor closes even in the continued presence of bound ligand.[83] In the case of glutamatergic synapses, removal of glutamate from the synaptic cleft is relatively slow,

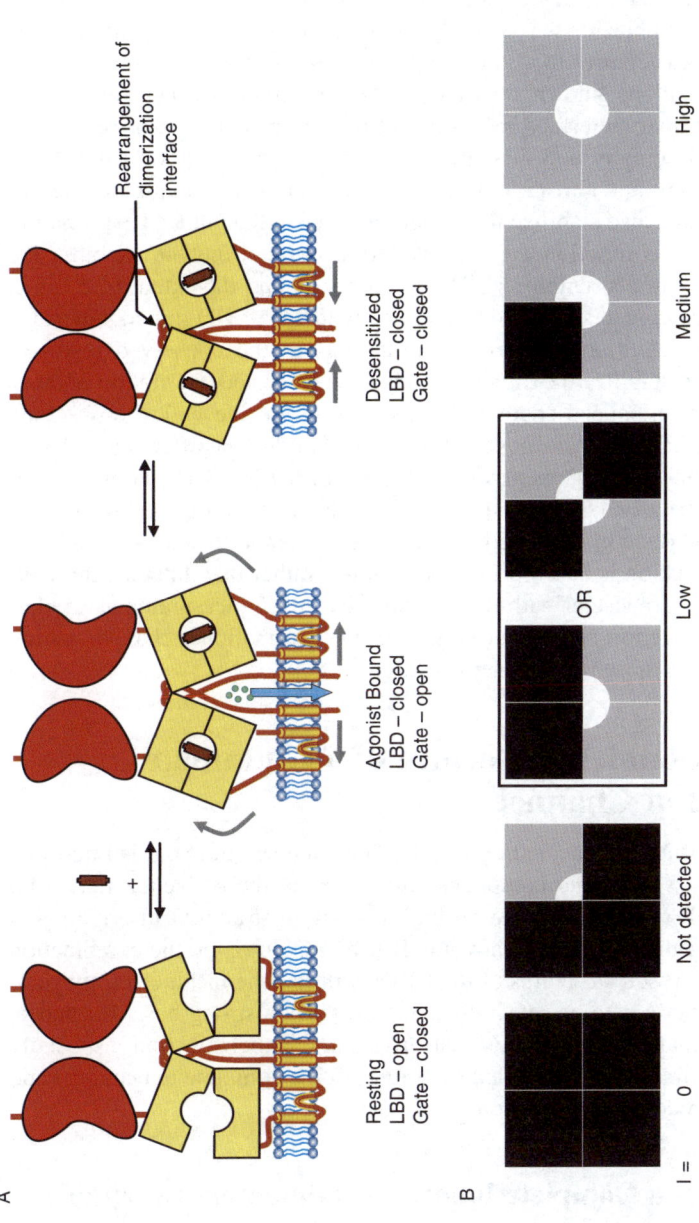

Figure 2 *Model of conformational transitions in glutamate receptor activation and desensitization. (A) As shown schematically for two of the four subunits comprising an iGluR, the transition from the resting to the activated state involves closure of the LBD (curved gray arrows, middle panel), which pulls apart the TM domains (straight gray arrows, middle panel), opening the ion-channel gate and permitting ion flow. Thereafter, the channel enters a desensitized state, in which the gate closes even though the LBD remains in the closed, agonist-bound state. This transition is thought to require the disruption of contacts between the dimer interfaces and subunit rearrangements (gray arrows, right panel). The light blue arrow indicates direction of ion flow. (B) The number of open subunit gating elements in a receptor controls current output. Black square: Subunit with closed gating element. Gray notched square: Subunit with open gating element. Associated current levels are indicated qualitatively below each combination*

but AMPA receptor desensitization occurs on a millisecond timescale. As a result, receptor desensitization can act to shape the frequency and magnitude of glutamatergic signals, which may be important for avoiding overstimulation of the neuronal circuitry. Consistent with this view, blockade of AMPA receptor desensitization by cyclothiazide was recently shown to increase the severity of pharmacologically induced seizures in a status epilepticus model.[84]

A number of mutations known to affect AMPA receptor desensitization map directly to LBD dimer interface,[82,85] supporting its physiological relevance. Furthermore, crystallography and analytical ultracentrifugation showed that LBD mutations (*e.g.*, L483Y) and allosteric modulators (cyclothiazide) that block desensitization in the channel also stabilize the dimer interface in the GluR2 LBD core; in general, a clear inverse correlation was observed for a series of mutant/cyclothiazide combinations between the strength of LBD dimerization and the extent of channel desensitization.[82] The contacts observed in GluR2 wild-type LBD dimers have also been used to design additional systematic mutations in the conserved residues, which are expected to stabilize the dimer assembly. As predicted, disruption of these dimerization contacts enhanced channel desensitization, in line with the observed correlation.[86] On the basis of these observations, it is thought that after channel activation, conformational rearrangement and disruption of the interfaces between subunits lead to desensitization (shown schematically at the right in Figure 2A). On the other hand, if the interface is stabilized, desensitization cannot occur. According to this model, desensitization is based primarily on inter- rather than intrasubunit conformational changes. Consistent with this hypothesis, FRET measurements of cleft closure following activation showed no large-scale conformational changes within LBD in the presence or absence of desensitization.[73]

3   The Functional Architecture of a Glutamate Receptor Ion Channel

Despite the wealth of insights available from LBD crystal structures, it is important to remember that even the dimer comprises only ~15% of the molecular mass of a fully assembly iGluR channel and that no high-resolution structural information is yet available for the other domains of any iGluR subunit. However, the combination of insights from the crystal structures of the LBDs with biochemical, electrophysiological, and mutagenetic data on intact channels provides a strong basis for understanding the conformational transitions that drive ion-channel function, especially within the emerging framework provided by studies of the complete ion channel and of additional domain:domain interactions.

3.1   Structure of a Complete Ionotropic Glutamate Receptor

The first step toward visualizing an intact glutamate receptor was the development of a system for expressing and purifying the full length, recombinant GluR2 receptor.[39, 87–89] The availability of pure protein with high specific ligand-binding activity paved the way for the first three-dimensional structure of a tetrameric AMPA receptor ion

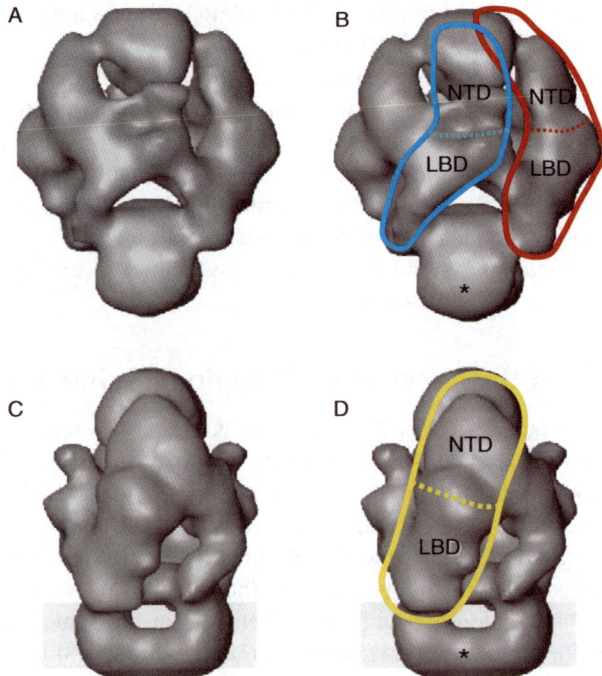

Figure 3 *The EM structure of a fully assembled iGluR viewed in the plane of the postsynaptic membrane. (A and C) Two views of the molecular envelope of GluR2 determine d by image reconstruction from negatively stained single particles.[90] (B and D) An outline of possible subunit (solid) and domain (dashed) boundaries within the ectodomain.[90] The proposed position of the transmembrane domain is indicated by the gray rectangles and asterisks (*), corresponding to the region where TARP subunits may bind.[95] In (C) and (D), the model is rotated by 90° around a vertical axis relative to its orientation in (A) and (B). The overall twofold symmetry of the tetrameric channel is reflected in the dimensions of the two perspectives*

channel by electron microscopic reconstructions of negatively stained single-particle images.[90] Analysis of the symmetry of the receptor provided direct evidence for a dimer-of-dimers assembly[90] (Figures 3A and 3C), a result consistent with biochemical studies on chimeric and tandem iGluR subunits.[91–94] On the basis of biochemical and functional characteristics of the channel, a hypothetical model for the approximate subunit and domain boundaries in the channel has been proposed (Figures 3B and 3D). GluR4 homomeric channels yield EM images similar to those of GluR2 (Tichelaar and Madden, unpublished data), suggesting that the overall architecture of the AMPA receptors is relatively well conserved.

Recently, heterotetrameric AMPA receptors purified from rat brains have also been analyzed by electron microscopy.[95] When solubilized in CHAPS, the purified receptors exhibit expanded transmembrane domains and can be labeled by antibodies specific for transmembrane AMPA-R regulatory proteins (TARPs) suggesting that they form a stable complex *in vivo* (see Section 5). On the basis of alignment of gross structural features and our proposed assignment of the transmembrane domains within the intact GluR2 structure, the TARP binding site(s) would correspond to the vicinity of the asterisks in Figures 3B and 3D.

In contrast to the homogeneous structure of the GluR2 homomers, it was also observed that the NTDs of the brain receptors exhibit striking conformational heterogeneity and that there were different sets of conformations in the presence or absence of agonist. The physiological relevance of this variability is not immediately apparent. It may reflect the role of NTD in functional transitions. Alternatively, it may be a result of biochemical destabilization of subunit interfaces: during purification the receptor-containing membranes are washed with 4 M urea to remove associated proteins prior to solubilization.[95]

3.2   The Role of the N-Terminal Domain in Subunit Assembly

The glutamate receptor NTD (or "X" domain) is about 400 residues long and is related to the bacterial leucine/isoleucine/valine-binding protein (LIVBP) and to the mGluRs. This domain is absent in the kainate binding proteins (KBP) of lower vertebrates (chick, frog, fish) and in bacterial iGluRs.[1] The NTD plays an important role in the oligomerization of eukaryotic iGluR subunits.[96,97] It also contributes to the subtype specificity of that assembly, ensuring that subunits of each family (AMPA, kainate, NMDA) associate only with other members of the same family.[91,96,97] An attractive model for iGluR assembly proposes that the dimer of dimers is formed in two stages (Figure 4). The initial step of dimerization is thought to be directly mediated by NTD interactions between compatible subunits (black lines in Figure 4), while the secondary dimerization (orange lines) requires LBD and TM domain compatibility as well.[7,91,98] In this view, the LBD dimer observed crystallographically would represent a component of the secondary dimerization interface.

Although there is still no direct structural information available on the NTD, sequence homology has been used to guide further experiments on the nature of its role in subunit–subunit interaction. The domain was subdivided into four regions, and AMPA/kainate chimeras were constructed.[99] The results suggest that there are two NTD surfaces that are required for the formation of functional tetramers, one of

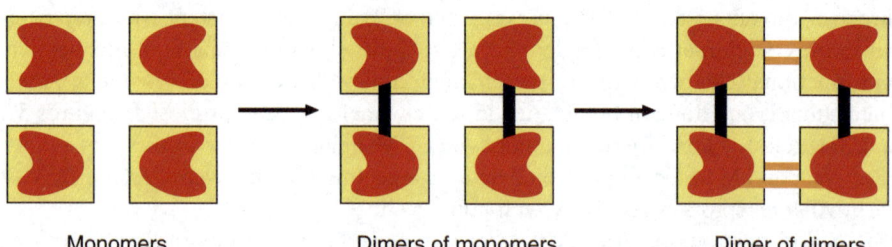

Monomers   Dimers of monomers   Dimer of dimers

Figure 4 *Stepwise assembly model for iGluR dimer of dimers. Initial dimerization (indicated by black lines) is thought to be mediated by interaction between the NTDs of compatible subunits. Secondary dimerization (indicated by orange lines) is brought about by interaction between a second surface on the NTD and by compatibility between LBD and TM domains, to result in a functional tetramer*

which contributes to the formation of the initial dimer, while the other may also participate in the secondary dimerization step (Figure 4). The requirement for a tetrameric network of NTD is also indicated by studies in which the domain was replaced by leucine-zipper peptides.[100] Only a tetrameric assembly of the synthetic peptide resulted in the formation of functional channels. Curiously, at least within the context of a homomeric AMPA receptor, complete removal of NTD yields functional channels that can bind ligand, conduct ion currents, and desensitize.[101] One possible explanation for this apparently contradictory result is that while subunit assembly in the absence of an NTD is possible, assembly in its presence requires complementarity. Indeed, the observation that KBP and bacterial iGluR lack an NTD but nevertheless form functional channels confirming that NTD is not absolutely required for assembly. In addition to ensuring subtype-specific oligomerization, the presence of the domain may also confer additional regulatory possibilities on the eukaryotic channels. A number of endogenous and synthetic modulatory substances have been found to bind in NTD.[102-110] Furthermore, the domain is known to influence receptor desensitization characteristics.[111,112]

3.3 The Organization of the Transmembrane Domains

As for the NTD, there is no direct structural information on the transmembrane domains of an iGluR. However, it is widely accepted today that each subunit of the glutamate receptor ion channel contributes three transmembrane segments and a reentrant pore loop (M2) (Figure 1). This topology shares essential features with the helix/pore loop/helix arrangement of the K^+ channels and is supported by sequence homology and extensive mutagenetic analysis.[6,35,113-115] As suggested by the assembly behavior of chimeric subunits,[91] the transmembrane helices, in particular M3, contribute to the stability of receptor assembly.[93,98] As would be expected on the basis of the tight coupling of binding and gating properties,[116] TM domain mutations also have clear effects on the gating and desensitization properties of the receptors.[117-125]

However, there are some important functional differences between the K^+-channels and the iGluR. The most obvious concerns ion selectivity: glutamate receptor ion channels are nonselective for monovalent cations and are also permeable to Ca^{2+} in some cases, whereas the classical K^+-channels are quite strictly selective. Permeation experiments with organic cations also confirm that the iGluR pore domain is significantly larger and most likely allows the passage of at least partially hydrated ions.[126,127] However, only modest conformational rearrangements may be required to expand the ion selectivity filter appropriately, since Ca^{2+}-channels and K^+-channel mutants are known to have larger and less selective ion pore than the strict K^+ channels, despite clear homologies.[128-130]

There are also likely to be some adjustments in the gating mechanism of the iGluR, relative to those of the ligand-gated K^+-channels. Shortly after the crystallographic analysis of the MthK gating helix, it was recognized that the hinge glycine, conserved in all K^+-channels and required for the kinking of the inner gating helix, was absent in the iGluR.[8] Subsequent mutagenesis experiments demonstrated that none of the other glycine residues in the helix was required for iGluR gating[120] and that the helices might be arranged with overall twofold, rather than fourfold, symmetry,[94] consistent with the dimer-of-dimers model of receptor assembly. Despite these differences, it seems likely that the iGluR transmembrane domains will preserve the gross features of the K^+-channel homologues, with functional and structural adaptations to accommodate their distinct physiological requirements. An example of such adaptations is provided by the calcium-release channel, whose very distinct ion permeation properties can be accounted for by a pore loop based homology model.[131]

4 A Working Model of AMPA Receptor Function

Within this framework, how do the observed changes in the LBD drive conformational changes in the TM domain and prime the receptor for subsequent desensitization? Combining the insights described above provides the basis for a strong working model of the molecular mechanisms involved, shown in simplified form in Figure 2A. First, in a fully assembled receptor, LBD is dimerized back-to-back, as seen in the GluR2 LBD crystal structures, and assembled into a dimer of dimers by subunit–subunit interactions that involve the NTD and the TM domains as well (Figure 4). Agonist docks in the open clefts of the LBD, where its presence stabilizes the closed-cleft conformation (curved gray arrows, Figure 2A, middle). Because of the relative orientation of the LBD, the closure increases the separation between the membrane-proximal lobes, moving them away from each other and from the pore axis of the channel and opening the gate (straight gray arrows, Figure 2A, middle). Since the gating elements are spontaneously closed, free energy of ligand binding must be expended to drive its opening. At the same time, cleft closure must also destabilize intersubunit interactions, priming the system for relaxation into the desensitized state. By allowing the subunits to shift position (gray arrows, Figure 2A, right) the ion-channel gate can close even while the LBD remains in its locked conformation, a transition driven by recovery of the free-energy cost associated with opening the gate.

While such a simple three-state model does not account for the complexity detected by kinetic investigations of channel gating,[125] it nevertheless exhibits considerable explanatory power. In particular, it provides a unifying framework that explains a host of mutagenetic and pharmacological observations, as detailed above. Furthermore, the structural models have demonstrated excellent predictive power. For example, mutagenetic probing of the LBD dimerization interface has confirmed the correlation between its stability and the ability of the receptor to desensitize.[86] In addition, mutagenesis has been used to study the contribution of individual stereochemical interactions within the ligand-binding cleft to the affinity and magnitude of the electrophysiological response. Analysis of the GluR2/kainate complex identified a leucine residue that appears to interact sterically with the kainate isopropenyl group and thus prevent full cleft closure. Consistent with the model, replacement of the leucine with smaller threonine or valine side chains yielded LBD conformational changes more similar to those seen with full agonists and increased kainate efficacy on the channels.[44,80] At the same time, as described below, the study of gating responses to partial and full agonists suggests that a more complex series of conformational transitions may take place within an iGluR subunit, requiring an extension of the model shown in Figure 2A.

4.1 LBD Mutations Affecting Binding and Gating

The structural understanding that emerges from the crystal structures of the iGluR LBD also permits a more quantitative dissection of the relative contributions of the processes of agonist binding and channel gating. A simple two-step binding/gating process is governed thermodynamically by two parameters: (1) the affinity of the initial interaction and (2) the efficacy with which the bound agonist stabilizes the closed-cleft/open-gate conformation. The affinity and the efficacy together determine the shape of the dose–response curve and thus establish the macroscopic EC_{50} (agonist concentration yielding half-maximal response) and intrinsic activity (response at saturating concentration of agonist) of the interaction.[75–77,116,132] Once the efficacy is sufficiently high to ensure that essentially all receptor molecules are activated at saturation, further increases in efficacy do not increase the observed intrinsic activity but instead are reflected only in changes to the EC_{50} (Ref. 132); such "cryptic" efficacy differences can be difficult to detect. Against this background, the mutagenetic investigation of agonist efficacy and affinity provides new opportunities to probe the influence of stereochemical interactions identified crystallographically.

Mutations of NMDA receptor subunits clearly showed that residues known or predicted to interact with agonist influenced both the affinity and the efficacy of the response, whereas residues involved in forming lobe-to-lobe contacts in the closed-cleft conformation affected only the efficacy.[132] In AMPA receptors, targeted mutagenesis of a threonine side chain involved in a *trans*-lobe hydrogen bond was observed to affect both the affinity and the efficacy of the response, as well as to speed recovery from desensitization.[68] The somewhat different interpretations of the role of *trans*-lobe interactions on agonist affinity may reflect differences in the models used, in addition to stereochemical differences between the different iGluR subfamilies. The overall reduction in efficacy associated with the T686 mutations also unmasked cryptic differences in the efficacies of the full agonists glutamate and quisqualate.[68] In any case, the observations confirm the importance of structurally identified interactions in determining binding and gating phenomena for the receptors. Consistent with this view, the crystal structure of the GluR5 LBD revealed that S674 is also involved in a *trans*-lobe hydrogen bond.[57] The S674A mutation had previously been shown to increase channel deactivation,[133] an effect most likely attributable to destabilization of the closed cleft.[57] Overall, there are more *trans*-lobe contacts in kainate receptors than in AMPA receptors. Since these contacts influence the kinetics of conformational transitions within the receptor,[68,132] they may be important in tuning differential responses in these two classes of iGluR.[57,59]

4.2 Subunit Gating Behavior

A major question that cannot be resolved entirely by studies of soluble domain constructs concerns the mechanism by which LBD cleft closure actually mediates channel gating. In this case, key insights have been provided by electrophysiological studies of single-channel gating events. Analysis of chimeric iGluR preloaded with high-affinity competitive antagonists permitted the resolution of subconductance levels that correspond to the opening of individual components of the overall channel gate contributed by each subunit.[134] Surprisingly, it appears that the iGluR subunits "gate" independently of each other. A spectrum of zero to four open subunit gates yields a corresponding distribution of subconductance levels, as indicated in Figure 2B. Jin and co-workers extended this analysis, measuring the subconductance levels activated by partial and full agonists.[64] Since partial agonists stabilize partial cleft closure in the LBD, it might have been expected that they would also stabilize partially open channels. Instead, these studies showed that both partial agonists and full agonists activate the same set of subconductance levels, albeit with different probabilities: at saturating concentrations, the full agonist preferentially activates higher subconductance levels and the partial agonist preferentially activates the lower subconductance states.[63] This implies that the conformational behavior of the subunit gate is essentially binary (open or closed). To reconcile this observation with the graded response of the LBD, the authors postulated a coupling efficiency, by which the extent of cleft closure is converted into the probability of gating in each subunit.[63] A binomial distribution of subconductance levels results.

One implication of a probabilistic relationship between LBD closure and subunit gating is that the two processes are not *strictly* coupled to each other or to ligand binding. As a result, there are possible intermediate "docked" and "locked" steps, in which the ligand has bound but the cleft is not yet closed and the subunit gating element has not opened (Figure 5A). These two states would correspond heuristically to the docking and locking steps of agonist binding to LBD.[41,68] In the model shown in Figure 5A for a single subunit, agonist binding stabilizes LBD cleft closure ("locked" state), which in turn favors the "activated" state of the associated gating helix. In principle, both of these transitions should influence the overall intrinsic activity associated with a given agonist (see Section 5).

5   Open Questions

While the working model described above provides an excellent description of iGluR function, there are several essential features that remain to be elucidated. In particular, these include the mechanism(s) of partial agonism and the structural correlates of multistate models of channel activation. They also include the modulatory influence of allosteric regulators and of auxiliary subunits. Presumably, there remains much to learn from high-resolution structures of other iGluR domains and domain assemblies, up to and including a fully assembled receptor.

5.1   Different Models of Partial Agonism

The lack of high-resolution multidomain structural information clearly limits our understanding of one of the most fundamental observations to emerge from the AMPA receptor LBD crystal structures: namely, the correlation of the intrinsic activity of an agonist with the degree of LBD cleft closure. The classical allosteric model predicts that a partial agonist is simply less efficacious at shifting the balance of a preexisting equilibrium between inactive and active receptors toward the active state.[135] This model seems to apply to the NMDA receptors, whose LBDs do not exhibit a graded degree of cleft closure in response to the binding of partial agonists

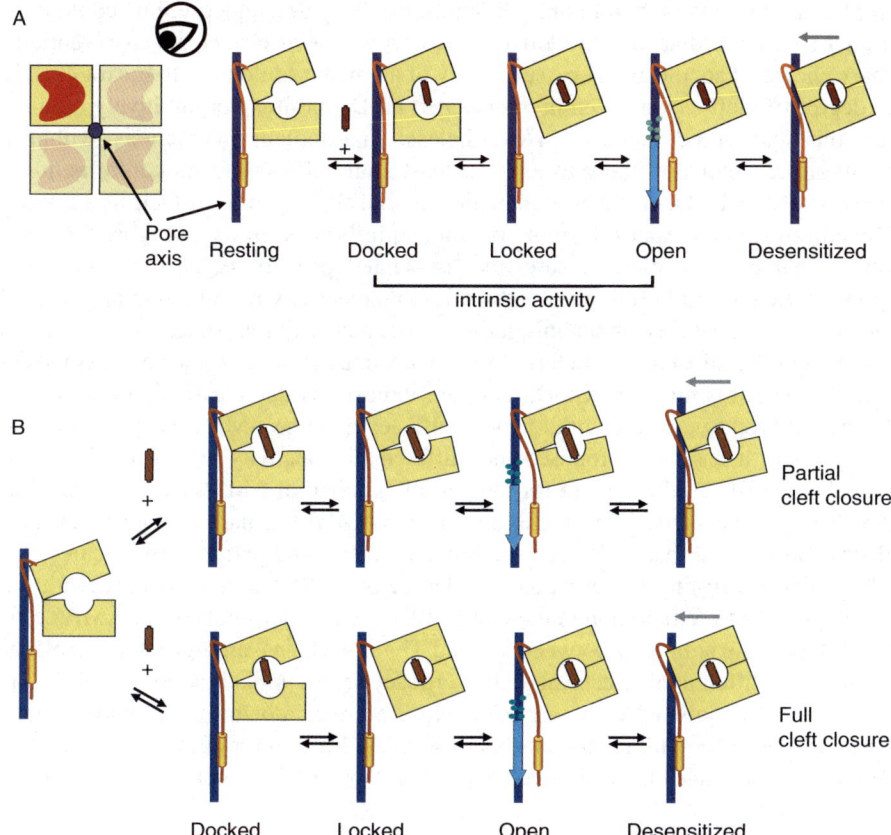

Figure 5 *Possible conformational transitions in the LBD and subunit gating elements. (A) Side view showing possible conformational states for the LBD and subunit gating element (represented as a cylinder) for a single subunit. The perspective is indicated at left. The LBD cleft can be open or closed, and the gating element can be in either a closed (touching the pore axis) or open (displaced away from the pore axis) conformation. The influence of agonist on both the "docked" ↔ "locked" and "locked" ↔ "open" equilibria can contribute to its intrinsic activity. (B) Possible parallel equilibria as a function of agonist identity. Following docking, agonists can stabilize either partial (top) or full (bottom) cleft closure in the "locked" state. Nevertheless, the subunit gating element exhibits a binary conformation regardless of the degree of cleft closure. At saturation, partial cleft closure may less efficiently stabilize the "open" state of each subunit gate,[63] leading to reduced macroscopic currents in the tetrameric receptor. Note that multiple partially closed equilibria could exist, in principle, each corresponding to a different degree of cleft closure*

but instead all show fully closed structures.[60,136] In the model shown in Figure 5A, this form of partial agonism can result primarily from an incomplete selectivity for the "locked" *vs.* the "docked" states.

However, such a model would not be consistent with the graded conformational response of the non-NMDA receptor LBD cores. One formal possibility is that interdomain and intersubunit contacts within the fully assembled iGluR could constrain LBD, such that *in situ* cleft closures would be restricted to an essentially binary choice. If so, the allosteric selectivity of agonists would have to mirror closely the degree of cleft closure seen in the isolated LBD, and it is not intuitively obvious why this should be the case. Nevertheless, interdomain interactions within a tetrameric receptor are likely to exert at least some effects on the conformational behavior of the LBD. Another explanation is that there could, in fact, be multiple "locked" and "open" states, forming parallel equilibria as shown in Figure 5B, each with a characteristic cleft closure and associated probability of subunit gating. Agonist efficacy would reflect a combination of selectivity for a particular "locked" state and the probability of subunit gating associated with that state.

Although the allosteric and cleft closure mechanisms have so far been associated with distinct subfamilies of iGluR, there are indications that classical allostery may also be required to explain partial agonist efficacy at some AMPA receptor mutants. For example, it has been proposed that differential selectivity among partially and fully closed LBD could account for the partial agonism of AMPA acting on GluR2-L650T mutant channels.[67] This mutation converts AMPA to a partial agonist, but crystal structures reveal that AMPA can stabilize both full and partial cleft closures.[80] It will also be interesting to determine the degree of cleft closure associated with the homologous L651V mutation in the GluR4 LBD, which does not reduce AMPA efficacy.[44] Given the similarity between Val and Thr side chains, a purely steric explanation for their differential effects on AMPA efficacy seems unlikely. Cases in which the correlation of agonist efficacy with cleft closure breaks down may provide experimental evidence for additional contributions to AMPA receptor efficacy. Such contributions could include classical selectivity in the "docked" ↔ "locked" equilibrium.

5.2 Multistate Kinetic Models

In general, resolution of the conformational equilibria within and between iGluR subunits may be aided by the analysis of functional transitions using high-resolution kinetic techniques[137–140] in combination with various multistate models of iGluR activation and desensitization.[68,125,132,141–143] It will also be interesting to extend the kinetic investigations of the isolated LBD. Stopped-flow analysis of agonist binding to the GluR4 LBD revealed significantly slower kinetics of "unlocking"[41] than were detected in an intact channel,[68] which may well reflect the destabilizing influence of conformational coupling to the transmembrane domains. As a result, systematic comparison of LBD and iGluR kinetics may provide insights into the nature of this coupling.

5.3 Structural Prospects

Ultimately, a detailed understanding of the iGluR activation mechanisms is likely to require structural information on the interaction of the LBD with the other domains and subunits of a fully assembled channel. Structural information on the transmembrane domains will also be required to understand the emerging functional differences that delineate the ion pore of the iGluR from those of the homologous potassium channels, both in terms of its functional stoichiometry[90,94] and the conformational mechanisms associated with gating.[8,120] Finally, the clear modulatory influence of small molecules that target the NTD reflects the functional importance of this domain and its contribution to the functional scaffolding of the receptor.

High-level bacterial expression of the NR2B NTD has been reported, although no structural data are yet available.[109,110] This offers an excellent prospect for the determination of the domain structure, and the resolution of the binding sites of modulatory substances. Given the likely dimeric assembly of the soluble domain,[97] it is possible that it will also provide novel insights into the assembly of the receptor, and to interfaces that may modulate channel activity.

Crystallographic analysis of intact iGluR or their transmembrane domains is likely to be complicated by the difficulties inherent in working with integral membrane proteins. The recent identification[37] of a series of prokaryotic homologues[37] offers the possibility of expressing and purifying large quantities of a primitive iGluR, perhaps in combination with the membrane-integrating protein Mistic.[144] An alternative would involve a hybrid approach, coupling single particle EM analysis[90,95] with high-resolution studies of the component domains.

5.4 Auxiliary Proteins

Finally, a particularly exciting development is the discovery that the glutamate receptor ion channels may function as part of larger complexes involving auxiliary proteins. In particular, recent evidence suggests that stargazin, a four-pass transmembrane protein, not only contributes to trafficking of AMPA-Rs[145–147] but apparently remains associated with the channel and can modulate its electrophysiological properties.[148,149] Consistent with this idea, stargazin/TARP family members appear to copurify with brain AMPA-R solubilized in CHAPS and contribute to the transmembrane density seen in the electron microscopic images.[95] In general, recent evidence indicates extensive interplay between channel function and trafficking[150–152] and a TARP/iGluR complex structure could provide insight into a whole new level of regulatory interactions for these physiologically versatile channels.

References

1. R.E. Oswald, *Adv. Protein Chem.*, 2004, **68**, 313.
2. J.C. Watkins, *Biochem. Soc. Trans.*, 2000, **28**, 297.
3. M. Hollmann and S. Heinemann, *Annu. Rev. Neurosci.*, 1994, **17**, 31.
4. R. Dingledine, K. Borges, D. Bowie and S.F. Traynelis, *Pharmacol. Rev.*, 1999, **51**, 7.
5. Z.G. Wo and R.E. Oswald, *Trends Neurosci.*, 1995, **18**, 161.
6. R. MacKinnon, *Neuron*, 1995, **14**, 889.
7. D.R. Madden, *Nat. Rev. Neurosci.*, 2002, **3**, 91.
8. D.R. Madden, *Curr. Opin. Drug Discov. Dev.*, 2002, **5**, 741.
9. M. Mayer, *Ann. N.Y. Acad. Sci.*, 2004, **1038**, 125.
10. M.L. Mayer, *Curr. Opin. Neurobiol.*, 2005, **15**, 282.
11. J.W. Johnson and P. Ascher, *Nature*, 1987, **325**, 529.

12. J.-P. Mothet, A.T. Parent, H. Wolosker, R.O. Brady Jr., D.J. Linden, C.D. Ferris, M.A. Rogawski and S.H. Snyder, *Proc. Natl. Acad. Sci. U.S.A.*, 2000, **97**, 4926.

13. J.E. Chatterton, M. Awobuluyi, L.S. Premkumar, H. Takahashi, M. Talantova, Y. Shin, J. Cui, S. Tu, K.A. Sevarino, N. Nakanishi, G. Tong, S.A. Lipton and D. Zhang, *Nature*, 2002, **415**, 793.

14. M. Sheng, *Proc. Natl. Acad. Sci. U.S.A.*, 2001, **98**, 7058.

15. B. Sommer, M. Kohler, R. Sprengel and P.H. Seeburg, *Cell*, 1991, **67**, 11.

16. T. Gillessen, S.L. Budd and S.A. Lipton, *Adv. Exp. Med. Biol.*, 2002, **513**, 3.

17. G. Riedel, B. Platt and J. Micheau, *Behav. Brain Res.*, 2003, **140**, 1.

18. R.C. Malenka and M.F. Bear, *Neuron*, 2004, **44**, 5.

19. S. Bleich, K. Romer, J. Wiltfang and J. Kornhuber, *Int. J. Geriatr. Psychiatry.*, 2003, **18**, S33.

20. B.S. Meldrum, *J. Nutr.*, 2000, **130**, 1007.

21. G.N. Barnes and J.T. Slevin, *Curr. Med. Chem.*, 2003, **10**, 2059.

22. S. Franciosi, *Cell. Mol. Life Sci.*, 2001, **58**, 921.

23. D.E. Pellegrini-Giampietro, J.A. Gorter, M.V.L. Bennett and R.S. Zukin, *Trends Neurosci.*, 1997, **20**, 464.

24. W. Rzeski, C. Ikonomidou and L. Turski, *Biochem. Pharmacol.*, 2002, **64**, 1195.

25. S. Ishiuchi, K. Tsuzuki, Y. Yoshida, N. Yamada, N. Hagimura, H. Okado, A. Miwa, H. Kurihara, Y. Nakazato, M. Tamura, T. Sasaki and S. Ozawa, *Nat. Med.*, 2002, **8**, 971.

26. H. Sontheimer, *Trends Neurosci.*, 2003, **26**, 543.

27. H. Brauner-Osborne, J. Egebjerg, E.O. Nielsen, U. Madsen and P. Krogsgaard-Larsen, *J. Med. Chem.*, 2000, **43**, 2609.

28. B. Reisberg, R. Doody, A. Stoffler, F. Schmitt, S. Ferris, H.J. Mobius and the Memantine Study Group, *N. Engl. J. Med.*, 2003, **348**, 1333.

29. R. Vandenberghe and J. Tournoy, *Postgrad. Med. J.*, 2005, **81**, 343.

30. J.L. Molinuevo, A. Llado and L. Rami, *Am. J. Alzheimers Dis. Other Demen.*, 2005, **20**, 77.

31. S.A. Lipton, *Curr. Alzheimer Res.*, 2005, **2**, 155.

32. J.N.C. Kew and J.A. Kemp, *Psychopharmacology*, 2005, **179**, 4.

33. K. Terai, M. Suzuki, M. Sasamata, S.-i. Yatsugi, T. Yamaguchi and K. Miyata, *Eur. J. Pharmacol.*, 2003, **467**, 95.

34. T. Weiser, *Curr. Pharm. Des.*, 2002, **8**, 941.

35. Z.G. Wo and R.E. Oswald, *Proc. Natl. Acad. Sci. U.S.A.*, 1994, **91**, 7154.

36. G.-Q. Chen, C. Cui, M.L. Mayer and E. Gouaux, *Nature*, 1999, **402**, 817.

37. T. Kuner, P.H. Seeburg and H.R. Guy, *Trends Neurosci.*, 2003, **26**, 27.

38. M.L. Mayer, R. Olson and E. Gouaux, *J. Mol. Biol.*, 2001, **311**, 815.

39. A. Kuusinen, M. Arvola and K. Keinänen, *EMBO J.*, 1995, **14**, 6327.

40. R. Abele, S. Dmitri, K. Keinänen, M.H.J. Koch and D.R. Madden, *Biochemistry*, 1999, **38**, 10949.

41. R. Abele, K. Keinänen and D.R. Madden, *J. Biol. Chem.*, 2000, **275**, 21355.

42. D.R. Madden, R. Abele, A. Andersson and K. Keinänen, *Eur. J. Biochemistry*, 2000, **267**, 4281.

43. D.R. Madden, N. Armstrong, D. Svergun, J. Perez and P. Vachette, *J. Biol. Chem.*, 2005, **280**, 23637.
44. D.R. Madden, Q. Cheng, S. Thiran, S. Rajan, F. Rigo, K. Keinänen, S. Reinelt, H. Zimmermann and V. Jayaraman, *Biochemistry*, 2004, **43**, 15838.
45. D.R. Madden, S. Thiran, H. Zimmermann, J. Romm and V. Jayaraman, *J. Biol. Chem.*, 2001, **276**, 37821.
46. D. Deming, Q. Cheng and V. Jayaraman, *J. Biol. Chem.*, 2003, **278**, 17589.
47. V. Jayaraman, S. Thiran and D.R. Madden, *FEBS Lett.*, 2000, **475**, 278.
48. M.H. Nanao, T. Green, Y. Stern-Bach, S.F. Heinemann and S. Choe, *Proc. Natl. Acad. Sci. U.S.A.*, 2005, **102**, 1708.
49. G.-Q. Chen and E. Gouaux, *Proc. Natl. Acad. Sci. U.S.A.*, 1997, **94**, 13431.
50. G.-Q. Chen, Y. Sun, R. Jin and E. Gouaux, *Protein Sci.*, 1998, **7**, 2623.
51. N. Armstrong, Y. Sun, G.-Q. Chen and E. Gouaux, *Nature*, 1998, **395**, 913.
52. N. Armstrong and E. Gouaux, *Neuron*, 2000, **28**, 165.
53. A. Frandsen, D.S. Pickering, B. Vestergaard, C. Kasper, B.B. Nielsen, J.R. Greenwood, G. Campiani, C. Fattorusso, M. Gajhede, A. Schousboe and J.S. Kastrup, *Mol. Pharmacol.*, 2005, **67**, 703.
54. A. Hogner, J.R. Greenwood, T. Liljefors, M.-L. Lunn, J. Egebjerg, I.K. Larsen, E. Gouaux and J.S. Kastrup, *J. Med. Chem.*, 2003, **46**, 214.
55. C. Kasper, M.-L. Lunn, T. Liljefors, E. Gouaux, J. Egebjerg and J.S. Kastrup, *FEBS Lett.*, 2002, **531**, 173.
56. M.-L. Lunn, A. Hogner, T.B. Stensbol, E. Gouaux, J. Egebjerg and J.S. Kastrup, *J. Med. Chem.*, 2003, **46**, 872.
57. M.L. Mayer, *Neuron*, 2005, **45**, 539.
58. R.L. McFeeters and R.E. Oswald, *Biochemistry*, 2002, **41**, 10472.
59. P. Naur, B. Vestergaard, L.K. Skov, J. Egebjerg, M. Gajhede and J.S. Kastrup, *FEBS Lett.*, 2005, **579**, 1154.
60. H. Furukawa and G. Eric, *EMBO J.*, 2003, **22**, 2873.
61. A. Hogner, J.S. Kastrup, R. Jin, T. Liljefors, M.L. Mayer, J. Egebjerg, I.K. Larsen and E. Gouaux, *J. Mol. Biol.*, 2002, **322**, 93.
62. B.B. Nielsen, D.S. Pickering, J.R. Greenwood, L. Brehm, M. Gajhede, A. Schousboe and J.S. Kastrup, *FEBS J.*, 2005, **272**, 1639.
63. R. Jin, T.G. Banke, M.L. Mayer, S.F. Traynelis and E. Gouaux, *Nat. Neurosci.*, 2003, **6**, 803.
64. R. Jin and E. Gouaux, *Biochemistry*, 2003, **42**.
65. R. Jin, M. Horning, M.L. Mayer and E. Gouaux, *Biochemistry*, 2002, **41**, 15635.
66. K. Speranskiy and M. Kurnikova, *Biochemistry*, 2005, **44**, 11508.
67. M.L. Mayer and N. Armstrong, *Annu. Rev. Physiol.*, 2004, **66**, 161.
68. A. Robert, N. Armstrong, J.E. Gouaux and J.R. Howe, *J. Neurosci.*, 2005, **25**, 3752.
69. Y.-J. Sun, J. Rose, B.-C. Wang and C.-D. Hsiao, *J. Mol. Biol.*, 1998, **278**, 219.
70. H. Takahashi, E. Inagaki, C. Kuroishi and H.T. Tahirov, *Acta Crystallogr. D Biol. Crystallogr.*, 2004, **60**, 1846.
71. D.E. Koshland, G. Nemethy and D. Filmer, *Biochemistry*, 1966, **5**, 365.
72. J. Monod, J. Wyman and J.P. Changeux, *J. Mol. Biol.*, 1965, **12**, 88.
73. M. Du, S.A. Reid and V. Jayaraman, *J. Biol. Chem.*, 2005, **280**, 8633.

74. E.R. Valentine and A.G. Palmer III, *Biochemistry*, 2005, **44**, 3410.
75. E.J. Ariëns, *Arch. Int. Pharmacodyn. Ther.*, 1954, **99**, 32.
76. R.F. Furchgott, in *Advances in Drug Research*, Vol 3, N.J. Harper and A.B. Simmonds (eds), Academic Press, San Diego, 1966, 21.
77. R.P. Stephenson, *Br. J. Pharmacol.*, 1956, **11**, 379.
78. E. Gouaux, *J. Physiol. (London)*, 2004, **554**, 249.
79. M.W. Fleck, E. Cornell and S.J. Mah, *J. Neurosci.*, 2003, **23**, 1219.
80. N. Armstrong, M.L. Mayer and E. Gouaux, *Proc. Natl. Acad. Sci. U.S.A.*, 2003, **100**, 5736.
81. Y. Jiang, A. Lee, J. Chen, M. Cadene, B.T. Chait and R. MacKinnon, *Nature*, 2002, **417**, 515.
82. Y. Sun, R. Olson, M. Horning, N. Armstrong, M. Mayer and E. Gouaux, *Nature*, 2002, **417**, 245.
83. M.V. Jones and G.L. Westbrook, *Trends Neurosci.*, 1996, **19**, 96.
84. F. Fornai, C.L. Busceti, A. Kondratyev and K. Gale, *Eur. J. Neurosci.*, 2005, **21**, 455.
85. J.C. Quirk, E.R. Siuda and E.S. Nisenbaum, *J. Neurosci.*, 2004, **24**, 11416.
86. M.S. Horning and M.L. Mayer, *Neuron*, 2004, **41**, 379.
87. K. Keinänen, G. Kohr, P.H. Seeburg, M.L. Laukkanen and C. Oker-Blom, *Biotechnology (NY)*, 1994, **12**, 802.
88. M. Safferling, W. Tichelaar, G. Kummerle, A. Jouppila, A. Kuusinen, K. Keinänen and D.R. Madden, *Biochemistry*, 2001, **40**, 13948.
89. D.R. Madden and M. Safferling, *Methods Mol. Biol.*, 2006, in press.
90. W. Tichelaar, M. Safferling, K. Keinänen, H. Stark and D.R. Madden, *J. Mol. Biol.*, 2004, **344**, 435.
91. G. Ayalon and Y. Stern-Bach, *Neuron*, 2001, **31**, 103.
92. S. Qiu, Y.-l. Hua, F. Yang, Y.-z. Chen and J.-h. Luo, *J. Biol. Chem.*, 2005, **280**, 24923.
93. S. Schorge and D. Colquhoun, *J. Neurosci.*, 2003, **23**, 1151.
94. A.I. Sobolevsky, M.V. Yelshansky and L.P. Wollmuth, *Neuron*, 2004, **41**, 367.
95. T. Nakagawa, Y. Cheng, E. Ramm, M. Sheng and T. Walz, *Nature*, 2005, **433**, 545.
96. W.D. Leuschner and W. Hoch, *J. Biol. Chem.*, 1999, **274**, 16907.
97. A. Kuusinen, R. Abele, D.R. Madden and K. Keinänen, *J. Biol. Chem.*, 1999, **274**, 28937.
98. E. Meddows, B. Le Bourdelles, S. Grimwood, K. Wafford, S. Sandhu, P. Whiting and R.A.J. McIlhinney, *J. Biol. Chem.*, 2001, **276**, 18795.
99. G. Ayalon, E. Segev, S. Elgavish and Y. Stern-Bach, *J. Biol. Chem.*, 2005, **280**, 15053.
100. S. Matsuda, Y. Kamiya and M. Yuzaki, *J. Biol. Chem.*, 2005, **280**, 20021.
101. A. Pasternack, S.K. Coleman, A. Jouppila, D.G. Mottershead, M. Lindfors, M. Pasternack and K. Keinänen, *J. Biol. Chem.*, 2002, **277**, 49662.
102. Y.-B. Choi and S.A. Lipton, *Neuron*, 1999, **23**, 171.
103. A. Fayyazuddin, A. Villarroel, A. Le Goff, J. Lerma and J. Neyton, *Neuron*, 2000, **25**, 683.
104. C.J. Hatton and P. Paoletti, *Neuron*, 2005, **46**, 261.

105. G.A. Herin and E. Aizenman, *Eur. J. Pharmacol.*, 2004, **500**, 101.

106. C.-M. Low, F. Zheng, P. Lyuboslavsky and S.F. Traynelis, *Proc. Natl. Acad. Sci. U.S.A.*, 2000, **97**, 11062.

107. P. Malherbe, V. Mutel, C. Broger, F. Perin-Dureau, J.A. Kemp, J. Neyton, P. Paoletti and J.N.C. Kew, *J. Pharmacol. Exp. Ther.*, 2003, **307**, 897.

108. P. Paoletti, F. Perin-Dureau, A. Fayyazuddin, A. Le Goff, I. Callebaut and J. Neyton, *Neuron*, 2000, **28**, 911.

109. F. Perin-Dureau, J. Rachline, J. Neyton and P. Paoletti, *J. Neurosci.*, 2002, **22**, 5955.

110. J. Rachline, F. Perin-Dureau, A. Le Goff, J. Neyton and P. Paoletti, *J. Neurosci.*, 2005, **25**, 308.

111. J.J. Krupp, B. Vissel, S.F. Heinemann and G.L. Westbrook, *Neuron*, 1998, **20**, 317.

112. F. Zheng, K. Erreger, C.M. Low, T. Banke, C.J. Lee, P.J. Conn and S.F. Traynelis, *Nat. Neurosci.*, 2001, **4**, 894.

113. T. Kuner, L.P. Wollmuth, A. Karlin, P.H. Seeburg and B. Sakmann, *Neuron*, 1996, **17**, 343.

114. V.A. Panchenko, C.R. Glasser and M.L. Mayer, *J. Gen. Physiol.*, 2001, **117**, 345.

115. L.P. Wollmuth and A.I. Sobolevsky, *Trends Neurosci.*, 2004, **27**, 321.

116. D. Colquhoun, *Br. J. Pharmacol.*, 1998, **125**, 924.

117. K.S. Jones, H.M.A. VanDongen and A.M.J. VanDongen, *J. Neurosci.*, 2002, **22**, 2044.

118. K. Kohda, Y. Wang and M. Yuzaki, *Nat. Neurosci.*, 2000, **3**, 315.

119. H. Ren, Y. Honse, B.J. Karp, R.H. Lipsky and R.W. Peoples, *J. Biol. Chem.*, 2003, **278**, 276.

120. A.I. Sobolevsky, M.V. Yelshansky and L.P. Wollmuth, *J. Neurosci.*, 2003, **23**, 7559.

121. M.V. Yelshansky, A.I. Sobolevsky, C. Jatzke and L.P. Wollmuth, *J. Neurosci.*, 2004, **24**, 4728.

122. H. Yuan, K. Erreger, S.M. Dravid and S.F. Traynelis, *J. Biol. Chem.*, 2005, **280**, 29708.

123. J. Zuo, P.L. De Jager, K.A. Takahashi, W. Jiang, D.J. Linden and N. Heintz, *Nature*, 1997, **388**, 769.

124. A. Villarroel, M.P. Regalado and J. Lerma, *Neuron*, 1998, **20**, 329.

125. K. Erreger, P.E. Chen, D.J. Wyllie and S.F. Traynelis, *Crit. Rev. Neurobiol.*, 2004, **16**, 187.

126. N. Burnashev, A. Villarroel and B. Sakmann, *J. Physiol. (London)*, 1996, **496**, 165.

127. A. Villarroel, N. Burnashev and B. Sakmann, *Biophys. J.*, 1995, **68**, 866.

128. L. Heginbotham, T. Abramson and R. MacKinnon, *Science*, 1992, **258**, 1152.

129. L. Heginbotham, Z. Lu, T. Abramson and R. MacKinnon, *Biophys. J.*, 1994, **66**, 1061.

130. E.W. McCleskey and W. Almers, *Proc. Natl. Acad. Sci. U.S.A.*, 1985, **82**, 7149.

131. W. Welch, S. Rheault, D.J. West and A.J. Williams, *Biophys. J.*, 2004, **87**, 2335.

132. T.L. Kalbaugh, H.M.A. VanDongen and A.M.J. VanDongen, *Mol. Pharmacol.*, 2004, **66**, 209.

133. G.T. Swanson, S.K. Kamboj and S.G. Cull-Candy, *J. Neurosci.*, 1997, **17**, 58.

134. C. Rosenmund, Y. Stern-Bach and C.F. Stevens, *Science*, 1998, **280**, 1596.

135. J. Li, W. Zagotta and H.A. Lester, *Q Rev. Biophys.*, 1997, **30**, 177.

136. A. Inanobe, H. Furukawa and E. Gouaux, *Neuron*, 2005, **47**, 71.

137. V. Jayaraman, *Biochemistry*, 1998, **37**, 16375.

138. G. Li, W. Pei and L. Niu, *Biochemistry*, 2003, **42**, 12358.

139. G. Li and L. Niu, *J. Biol. Chem.*, 2004, **279**, 3990.

140. G. Li, Z. Sheng, Z. Huang and L. Niu, *Biochemistry*, 2005, **44**, 5835.

141. D. Bowie and G.D. Lange, *J. Neurosci.*, 2002, **22**, 3392.

142. G. Popescu and A. Auerbach, *Nat. Neurosci.*, 2003, **6**, 476.

143. I.M. Raman and L.O. Trussell, *Biophys. J.*, 1995, **68**, 137.

144. T.P. Roosild, J. Greenwald, M. Vega, S. Castronovo, R. Riek and S. Choe, *Science*, 2005, **307**, 1317.

145. L. Chen, D.M. Chetkovich, R.S. Petralia, N.T. Sweeney, Y. Kawasaki, R.J. Wenthold, D.S. Bredt and R.A. Nicoll, *Nature*, 2000, **408**, 936.

146. L. Chen, A. El-Husseini, S. Tomita, D.S. Bredt and R.A. Nicoll, *Mol. Pharmacol.*, 2003, **64**, 703.

147. S. Tomita, V. Stein, T.J. Stocker, R.A. Nicoll and D.S. Bredt, *Neuron*, 2005, **45**, 269.

148. S. Tomita, H. Adesnik, M. Sekiguchi, W. Zhang, K. Wada, J.R. Howe, R.A. Nicoll and D.S. Bredt, *Nature*, 2005, **435**, 1052.

149. A. Priel, A. Kolleker, G. Ayalon, M. Gillor, P. Osten and Y. Stern-Bach, *J. Neurosci.*, 2005, **25**, 2682.

150. H.C.-H. Chang and C. Rongo, *J. Cell. Sci.*, 2005, **118**, 1945.

151. I.H. Greger, L. Khatri, X. Kong and E.B. Ziff, *Neuron*, 2003, **40**, 763.

152. S.J. Mah, E. Cornell, N.A. Mitchell and M.W. Fleck, *J. Neurosci.*, 2005, **25**, 2215.

153. E.F. Pettersen, T.D. Goddard, C.C. Huang, G.S. Couch, D.M. Greenblatt, E.C. Meng and T.E. Ferrin, *J. Comput. Chem.*, 2004, **25**, 1605.

CHAPTER 20

The Mitochondrial ADP/ATP Carrier

EVA PEBAY-PEYROULA

Institut de Biologie Structurale Jean-Pierre Ebel, UMR 5075 CEA-CNRS-Université Joseph Fourier, 41 rue Jules Horowitz, F-38027 Grenoble cedex 1

1 Introduction

Deciphering protein structures is a very powerful approach to understand biological functions at a molecular level. In addition, dynamic properties and structures of different conformations, correlated with biochemical and functional data, should shed light on the complete movie of a protein in action. In the last decade a tremendous progress was made in experimental approaches for the routine production of recombinant proteins opening possibilities for getting large and pure protein quantities and also modified proteins (mutations, incorporation of Seleno-methionines, ^{15}N or ^{13}C labeling). Crystallization robots now allow a large scale sampling of conditions. Micro-crystals can be analyzed on powerful X-ray synchrotron radiation sources with narrow or well-collimated beams, and sophisticated programs lead automatically from the crystallographic data to the final model. Determination of NMR structures also progresses, and several places in the world have implemented facilities of high magnetic field spectrometers (800–900 MHz). The results of all the efforts done so far are visible in the Protein Data Bank, where 30,000 structures are deposited (http://www.rcsb.org/pdb/). In contrast to soluble proteins, membrane proteins are far behind with about a hundred structures present in the PDB.[1] The low percentage is not all representative of the genomes where 20–30% of the predicted proteins are estimated to be located in the membranes.[2] In addition, they accomplish very important functions as they are implicated in all the communications and signaling between cells or cellular compartments, and are therefore major drug targets (70% of drug targets are estimated to be membrane proteins). Molecular biology and biochemistry of membrane proteins aiming at structural studies are rather difficult because of their dual hydrophilic and hydrophobic properties. In particular, expressing large quantities of proteins in a membrane or extracting the proteins from their natural environment, a lipid bilayer, are often severe bottlenecks. Even though protocols were successful in some cases, there are still no general rules for the production of membrane proteins for structural studies.[3] Crystallization is an additional limitation that despite the first bottleneck of production, should not be underestimated.[4,5] Indeed, even for membrane proteins that are naturally abundant, in particular proteins from the respiratory chain, crystallization was not straightforward and took several years. Among membrane proteins, transporters are of particular interest and structural biology will contribute to understand the specificity of the transport, by finding how the transported molecules are recognized by the protein, and the mechanisms that allow conformational changes to occur during the process by studying at least two different conformations. The difficulty lies in the fact that in order to be crystallized, transporters have to be stabilized in a single conformation in the solution as has been shown in a few cases. The Ca-ATPase of skeletal muscle sarcoplasmic reticulum was crystallized in the presence of various ligands or ligand analogs, therefore leading to the structures of different conformations (most recent structures[6,7]). Mutants might also stabilize one conformation by enhancing ligand binding, as exemplified by the structure of lactose permease.[8] Blocking a conformation with an inhibitor, as will be described herein is another possibility. This chapter focuses on the ADP/ATP carrier (AAC), a mitochondrial carrier for which the structure was recently solved.[9] It describes the experimental approach that led to the structure, from crystallization to model building, and discusses a possible transport mechanism in light of biochemical and functional data.

2   Mitochondrial Carriers and ADP/ATP Carrier

Mitochondrial carriers (MCF) form a family of nuclear encoded transporters that are located in the inner mitochondrial membrane.[10,11] They import and/or export many different small molecules implicated in various metabolic pathways occurring in mitochondria, such as energy-generating pathways (citric acid cycle or fatty acid β-oxidation), amino acid synthesis and degradation (urea cycle), replication, transcription and translation of mitochondrial DNA, and heat generation by dissipating the proton gradient. All these proteins are made of about 300 amino acids and share a maximum of 20% sequence identity.[12,13] The presence of a triplicate motif in their sequences indicates that MCF genes probably derived from an ancestral gene encoding for about a hundred amino acids that has been triplicated. All the MCF carriers contain a characteristic motif P-X-D/E-X-X-K/R present three times in the sequence.[14] Because of its natural abundance AAC is the best characterized member of the family and has been extensively studied since the seventies.[15–17] AAC is linked to ATP synthesis, it imports ADP to mitochondria and exports ATP toward the cytosol where it will be utilized. The carrier is specific for adenine nucleotides and transport measurements demonstrate that AAC is able to translocate ADP^{3-} against ATP^{4-} in an electrogenic way.[16,18] AAC contains a specific signature, RRRMMM predicted at the C-terminal end of the fifth transmembrane helix; this sequence is rather unique and therefore has certainly a functional implication. It is also generally acknowledged that the carrier functions as a dimer.[17,19,20,21] Two types of natural inhibitors, atractyloside (ATR) or carboxyatractyloside (CATR) from a Mediterranean thistle, and bongkrekic acid (BA) from a bacteria, both strong poisons, were discovered and characterized.[22] ATR or CATR are known to inhibit the transport by interacting with the protein from the intermembrane space (IMS), while BA blocks the transport from inside. Both inhibitors were described to interact with two different conformations that are likely to represent the two extreme conformations adopted by AAC during the transport of nucleotides. Indeed, AAC should be able to oscillate between two conformations, one open toward IMS ready to release ATP to IMS and then to recognize ADP, the second open toward the matrix where ATP is captured after releasing ADP. In the presence of ATR or CATR, AAC is blocked in the conformation open toward IMS, while in the presence of BA it is blocked in the conformation that binds ATP from the matrix. These molecules are therefore ideal tools and all the biochemical studies done in the presence of nucleotides or inhibitors provide an interesting background for structural studies.

3   Crystallization

Crystallization of membrane proteins is often the second bottleneck, after their production and purification. The challenge for membrane proteins is to obtain sufficient quantities of proteins that are stable and homogeneous also in terms of conformations. Therefore, it is important to understand the fundamental physical chemistry of crystallization in the presence of amphiphiles, detergents, and/or lipids. Membrane proteins are naturally embedded in a lidipic environment. In order to determine the structure at high resolution, three-dimensional crystals have to be grown for which the protein has to be extracted and purified from the natural medium. Because of the dual properties of their surfaces, partly hydrophobic and partly hydrophilic, they are difficult to handle. First the membrane has to be disrupted using detergents and then the protein purified. Detergents are amphiphilic molecules, which are soluble in water at low concentration. Above the critical micellar concentration (cmc) they will form micelles in which typically a few hundreds of molecules are involved and expose their hydrophilic moiety toward the water. Above the cmc, the detergent molecules form a belt around the membrane protein's hydrophobic surface as seen from neutron scattering experiments.[23–26] These mixed micelles, protein and detergent, can then be purified by affinity or size exclusion chromatography. In a very rough approximation, the solution of membrane proteins in detergent can be compared to a soluble protein solution and therefore 3D crystallization will be approached in a similar manner. However, there are two major differences with soluble proteins: first the homogeneity of the solution is intrinsically different, second the crystallization agent will not only modify the solubility of the protein but will also affect the phase behavior of the detergent.[27] In addition, proteins in detergent micelles are not very stable and are difficult to maintain in a native structure,[28,29] and new designed amphiphiles might be necessary to solve this problem.[30] Alternative solutions such as lipidic cubic phases are presently being developed,[31,32] which were successful in several cases.[33–36]

AAC was extracted from bovine heart mitochondria using 3-laurylamido-N, N'dimethylpropylaminoxide (LAPAO), an aminoxide detergent.[37] The choice of the detergent resulted from long-term experience of biochemists studying the protein. In particular, measuring the intrinsic fluorescence of AAC in the presence of nucleotides or inhibitors made it possible to compare the protein stability in various detergents.[38] The inhibitor CATR is present from the beginning, prior to membrane solubilization, in order to stabilize the protein and prevent denaturation. The major difficulty is the lack of affinity of AAC for any chromatography column. The protein is indeed purified from other proteins of the inner mitochondrial membrane by collecting the flow-through of a hydroxylapatite column that retains all proteins except AAC.[39] The purification method has a major impact on the following steps. Indeed, many lipids are copurified with the protein. They were identified and quantified by phosphate titration and thin layer chromatography. Roughly 170 phospholipids per protein monomer are co-purified. Although it was already described that three cardiolipins, abundant in the inner mitochondrial membrane, are bound to AAC,[40] all of the lipids are not necessarily bound to the protein. Most of them might float in the solution as small patches mixed with the detergent. The initial concentration of the detergent itself during the solubilization is at 2%.[41] It increases drastically during the protein concentration because the micelle sizes, especially if they contain lipids, are similar to that of mixed protein–detergent micelles. This is very often the case, even when the amount of lipids is not so high. The starting solution was therefore very inhomogeneous and composed of mixed protein–detergent-lipid micelles, pure detergent micelles, detergent-lipid micelles and monomeric detergent molecules. Preliminary crystallization attempts under these conditions systematically failed. In principle, excess detergent could be eliminated either on binding the protein on an affinity column and changing the buffer (not possible for AAC), or by dialyzing the protein against a solution containing low detergent concentrations. LAPAO has a very low cmc of 1.3 mM,[41] and efficient dialysis would have taken too long and been incompatible with protein stability. An alternative method was developed for 2D crystallization and consists in adsorbing the detergent on polystyrene beads.[42] A preliminary test was done by incubating the protein solution for two hours, in the presence of Bio-Beads (Bio-Rad). Crystallization tests led to small needles in the presence of 2-methyl-2,4-pentanediol (MPD). This condition was then used in order to improve the quality of the protein solution and follow its ability to crystallize. Using ^{14}C labeled detergent, the protein-to-detergent ratio was followed carefully as a function of time during the Bio-bead treatment (Figure 1A). At different ratios, crystallization set-ups were performed around the MPD condition previously determined. Figure 1B shows that crystal habit could be improved from a crystalline precipitate to micro needles and finally thin plates, and was optimal after 4 hours of treatment.

The optimal Bio-bead treatment reduced the detergent concentration from 20 to 10 g per protein g, and was used again to explore large screens of crystallization conditions. They revealed several precipitants, such as MPD, ethanol, tert-butanol and Jeffamine M600® (Hampton Research), with which micro-crystals (needles or plates) were obtained. By refining the conditions, highly diffracting crystals were obtained with the latter one. Nevertheless, although the Bio-bead procedure was relatively well calibrated, the crystallization results were not systematically reproduced and the crystallizability or the quality of the crystals was not the same from one protein preparation to another. Small differences in the initial ratios of lipid or detergent

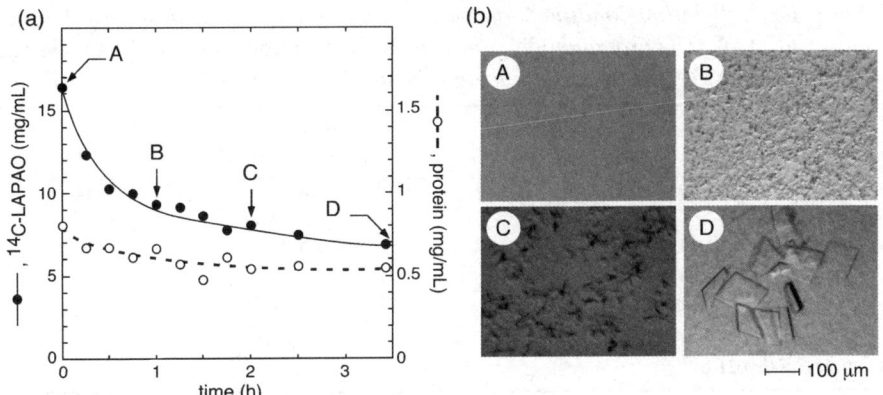

Figure 1 *Effect of the LAPAO-to-protein ratio on the ADP/ATP carrier crystallization. (a) Time course of ^{14}C-LAPAO removal. Purified bovine carrier (0.8 mg/mL) was incubated in the presence of 100 mg of SM2 Bio-Beads per mg of protein. Radioactivity associated to aliquots of the supernatant was quantified by liquid scintillation counting (solid line) and the protein amount was quantified by the bicinchoninic reagent kit (dotted line). (b) Crystallization experiments in the presence of 2-methyl-2,4-pentanediol. Protein was sampled during the Bio-Bead treatment at steps A, B, C, D shown in panel (a) with detergent-to-protein ratios of 20, 16, 13 and 10 g/g, respectively*
From Dahout-Gonzalez *et al.*, *Acta Cryst.*, 2003, **D59**, 2353–2355.

to protein might be a reason for instability. Dynamic light scattering was performed on several protein solutions prior to crystallization, but the polydispersity of the samples could not be related to crystallizability. This example shows how it is important to identify and quantify as much as possible all the molecules present in the protein solution prior to crystallization, in particular detergents and lipids. Recently, IR spectroscopy was shown to be a promising method to measure the amount of lipid and detergent in protein solutions.[43] The small volumes of protein (10–15 µL) needed for the measurement makes it attractive and deserves to be considered as a standard tool in laboratories involved in structural biology of membrane proteins.

Although the first crystal form diffracted well (native data to 2.2 Å resolution), the search for heavy atom derivatives was hindered by a crystal defect that existed in more than 90% of the crystals (see next paragraph). Another crystal form was found by lowering the salt content during the last step of purification (5 or even 0 mM NaCl instead of 100 mM), with the same precipitant but different additives. Table 1 indicates both crystallization conditions. Several aspects can be learned from the AAC crystallization. A good experience with the biochemistry of the protein is essential to start crystallization. In this case, knowing a favorable detergent and an inhibitor that stabilizes the protein in one conformation were both important parameters that helped in guiding the crystallization experiments. Tools for analyzing the detergent, lipid, and protein contents during the various steps of purification to the crystallization stage are indispensable, and different purification buffers might lead to different crystal forms.

Table 1 *Crystallization conditions and unit cells of the two crystal forms obtained for AAC. The protein buffers prior to crystallization contained 100 mM and 5 mM NaCl for the primitive and the centered crystal forms, respectively*

Reservoir content	Space group	Unit cell parameters (Å)	Resolution (Å)
30% Jeffamine M600® 20 mM NiSO$_4$ 100 mM HEPES pH 7.0	C222$_1$	$a = 79.874$ $b = 109.163$ $c = 89.315$	25–2.8
30% Jeffamine M600® 5 mM NaCitrate 100 mM TRIS pH 8.5	P2$_1$2$_1$2	$a = 85.437$ $b = 83.463$ $c = 49.922$	25–2.2

4   Diffraction, Phasing and Model Building

All the crystals obtained were usually small plates with typical dimensions of 100 × 50 × a few microns. These micro-crystals were rather unstable, appearing after 24 hours and sometimes disappearing within a week. This instability was not systematic and no rationale was found behind it. Because of the small size of the crystals, they were never tested at room temperature in capillaries but were directly frozen in liquid nitrogen from the mother liquor. The presence of 30% Jeff-M600 in the crystallization medium was enough to prevent the formation of crystalline ice during flash-freezing. Transfer to cryoprotecting solutions containing glycerol was also tested but turned out to be unsuccessful. In fact, it was never possible to find a solution stabilizing the crystals; they had to stay in the mother liquor in which the precipitant, lipid, detergent, and protein concentrations probably play a critical role in crystal stability. All the crystals were tested mainly on beam-line BM30A at ESRF-Grenoble (European Radiation Synchrotron Facility), except for a few larger ones that were tested on an in-house rotating anode equipped with Osmic mirrors. Most of the crystals did not diffract very well, and data sets were collected for about 10% of the crystals tested (at ESRF beam-lines BM30A, ID29, and ID14-2). Various annealing procedures were performed, consisting in blocking the nitrogen stream *in situ* either for 5 s at one time, or for 2 s repeated 5–10 times. Significant improvements in the diffraction could be observed but not enough to get good data sets. Small compact crystals diffracted better than larger plates. Therefore most of the diffraction was rather weak due to the small crystal sizes and signal-to-noise ratios had to be improved by adapting the beam size to that of the crystal. The crystals belonged to the *P*2$_1$2$_1$2 space group with typical unit cells of 85, 83, 50 Å. A good data set to 2.2 Å was obtained with crystals grown from high salt buffer, and diffraction intensities could be integrated with an R_{sym} of 8.3% (Table 2). Data were integrated with Denzo[44] and treated with the CCP4 package,[45] data collection statistics are summarized in Table 2. In order to determine the phases by multiple isomorphous replacement (MIR), heavy atom derivatives were searched by soaking crystals in solutions containing different compounds, varying concentrations, and

Table 2 *Data collection, phasing, and refinement statistics*

Data set	Native (1)	Native (2)	EthHgCl	PhHgAc	Thimerosal
Space group	P2$_1$2$_1$2	C222$_1$	C222$_1$	C222$_1$	C222$_1$
and unit cell	85.4 83.5	79.9 109.2	80.1 108.4	80.3 109.1	80.2 107.5
(Å)	49.9	89.3	89.5	89.2	89.3
Resolution	25–2.2	25–2.8	25–2.9	25–2.9	25–3.2
(Å)					
R_{merge} (%)*	8.3 (51.7)	5.8 (33.7)	7.0 (62.5)	6.3 (42.5)	8.8 (54.2)
Completeness	99.6 (99.6)	99.2 (99.2)	99.8 (99.8)	99.8 (99.8)	99.4 (99.4)
(%)					
<I/σ>	7.6 (1.4)	10.3 (2.1)	8.5 (1.1)	6.0 (1.3)	5.9 (1.4)
Phasing statistics					
Heavy atom	—	—	C128	C56, C128,	C56, C128,
binding sites				C256	C159, C256
R_{Cullis} (%)[†,‡]			0.69 (0.71)	0.73 (0.71)	0.70 (0.78)
Phasing			1.80 (1.18)	1.58 (1.12)	1.70 (1.02)
power[‡]					
FOM[§]		0.55 (0.69)			

Refinement statistics (native 1)		
Resolution (Å)	15–2.2	
Reflections working set (test set)	17,650 (894)	
R$_{cryst}$ (R$_{free}$) (%)[#]	22.6 (26.8)	
Number of non-hydrogen atoms	2761	

Note: Numbers in parentheses correspond to the highest resolution shell.

*$R_{merge} = \Sigma_h \Sigma_i |I_i(h) - <I(h)>| / \Sigma_h \Sigma_i I_i(h)$

[†]$R_{Cullis} = \Sigma_h (|F_{PH}(h) - F_P(h)| - F_{Hcalc}(h)) / \Sigma_h F_{PH}(h)$

[‡]Numbers are given for acentric reflections and for centric reflections in parentheses.

[§]FOM (figure of merit) calculated by the program MLPHARE, in parentheses the FOM from 25 to 2.5 Å after solvent flattening considering 45% of solvent and calculated by the program DM.

[#]$R_{cryst} = \Sigma_h |F_{obs}(h) - F_{calc}(h)| / \Sigma_h F_{obs}(h)$

soaking times, or alternatively adding heavy atoms directly to the crystallization solutions. Several hundred crystals were tested, most of them did not diffract but among the diffracting ones the majority had a double cell parameter along the c-axis (100 Å, instead of 50 Å for the native data set). A careful analysis of the diffraction data highlighted a crystal defect and later on, once the structure was known, the defect could be characterized. The crystal contains layers of proteins along the c-axis, some of the layers randomly distributed in the crystal are displaced by 2Å along the b-axis.[41] The combination of crystal defect, non-diffracting or non-isomorphous crystals made the search for heavy atom derivatives almost impossible with the primitive crystal form.

The second crystal form grown from low salt buffers belonged to the $C222_1$ space group with typical unit cells of 80, 110, 89 Å. Although the centered crystal form diffracted with strong anisotropy only to 2.8 Å maximum, it was easier to explore various heavy atoms. Non-isomorphism and crystal damage were overcome by adding mercuric compounds directly to the crystallization drop where crystals were present. A small volume of concentrated heavy atom solution was added to the drop, soaking times were short and crystals were backsoaked in another crystallization drop of the same composition. Therefore, crystals stayed in a solution that was very close to the mother liquor. Usual soaking procedures where crystals are transferred to a solution containing heavy atoms failed for non-isomorphism (large cell parameter variations were observed) and also for crystal damage. Indeed, anisotropy was always higher after soaking as judged from the analysis of mean F values along the three crystal axes given by the program Truncate. Three different derivatives were obtained with thimerosal (24 hours at 2.5mM), ethylmercury chloride (8 hours at 0.3 mM), and phenylmercury acetate (8 hours at 3 mM). Phases were calculated with MLPHARE[45] and phasing statistics are summarized in Table 2. Retrospectively, the heavy atom positions could be analyzed once the model was built. They are located on the four cysteines of AAC with different occupation rates. Three of the cysteines, C56, C159 and C256, are related by the gene triplication and follow the pseudo threefold symmetry of the backbone. The fourth, C128, is located within a trans-membrane helix and is the easiest to access as shown by high occupancies for all the derivatives. The pseudo threefold symmetry of three of them was obvious from the electron density maps (difference Fourier map $F_{derivative}$-F_{native} superimposed on F_{native} with MIR phases improved by solvent flattening) and it was intuited that the three positions could correspond to the three cysteines C56, C159, and C256, which guided the model building. Because of anisotropic diffraction, the electron density maps were rather poor. Six transmembrane helices were easily identified but side chains were difficult to assign and therefore the N- and C-termini of the helices were not straightforward. Also, loops could not be traced from the first experimental maps. After solvent flattening and a few cycles of refinement using CNS,[46] a partial model could be traced using the program O.[47] The model was placed by molecular replacement into the $P2_12_12$ space group crystal. In order to exploit at best the experimental phases, the MIR density map calculated in the $C222_1$ space group around the model was rotated and translated to the $P2_12_12$ crystal. This map combined to the $2F_{obs}$-F_{calc} map allowed the almost complete chain tracing. Missing side chains were progressively added by altering manual model building and energy minimization. In the final refinement steps, simulated annealing and individual B-factor refinements were also performed.

5   Structure Analysis

Both crystal forms, $C222_1$ and $P2_12_12$, consist of stacked protein layers in which the proteins are oriented as if they were sitting in a membrane. Molecular contacts in between layers are weak explaining the anisotropy of $C222_1$ crystals and the defects in $P2_12_12$ crystals. The weakness of protein–protein interactions along one direction is probably the reason why crystals grow as thin plates. Both crystal forms contain one monomer in their asymmetric unit. The following structure analysis is based on the model refined to 2.2 Å in the $P2_12_12$ space group and deposited in the PDB under code 1okc.

The whole protein is shaped as a basket opening a wide cavity toward the inter-membrane space (IMS) and closed toward the matrix (Figure 2A–C). IMS and

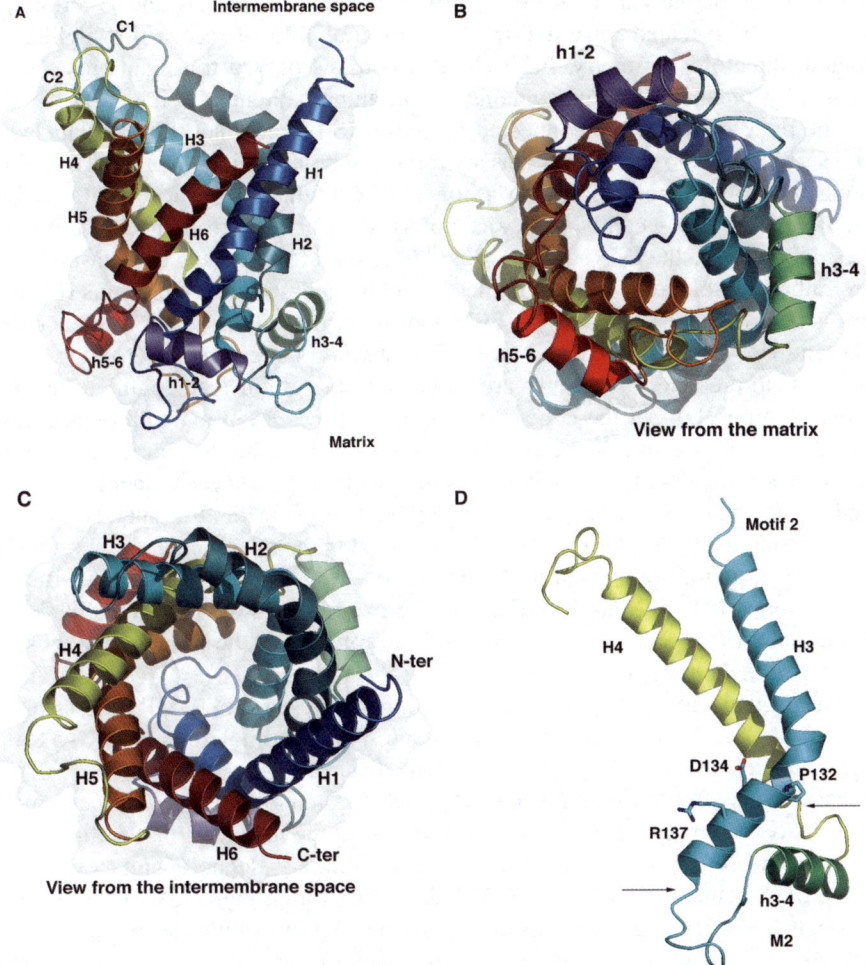

Figure 2 *Overall structure of the ADP/ATP carrier. The ribbon diagram is colored according to the sequence, blue (N-terminus) to red (C-terminus). (A) View from the side. Helices comprise following residues: 4–37 (H1), 53–64 (h1-2), 73–99 (H2), 108–142 (H3), 156–167 (h3-4), 176–199 (H4), 209–238 (H5), 253–264 (h5-6), 273–290 (H6). (B) View from the matrix. This view highlights the pseudo threefold symmetry induces by the gene triplication. It depicts also the closed conformation from the matrix induced by kinks in odd-numbered helices due to the presence of prolines (P27, P132, P229). (C) View from the intermembrane space. Transmembrane helices form a cavity with a large entrance accessible from the intermembrane space. (D) One single repeat. Residues 105–209 represent the second repeat present in AAC. The structure of each single motif comprises a kinked odd-numbered helix, a short amphipatic helix followed by a tilted even-numbered helix. Residues P132, D134 and R137 belong to the MCF motif PXD/EXXK/R, helix H3 is kinked at the level of P132*

Figure 2 and subsequent figures were drawn with the program PYMOL (http://www.pymol.org).

matrix sides of the carrier were identified by the fact that N- and C-termini are known to be oriented toward IMS.[48] The overall fold of the carrier highlights a pseudo threefold symmetry. Each element consists of two transmembrane helices related by a matrix loop that contains a short amphipatic helix (Figure 2D).

The three elements are connected by two IMS oriented short loops, labeled C1 and C2. In total the carrier contains six transmembrane helices labeled H1 to H6, as predicted from the sequence analyses of AAC and other MCF carriers, although the prediction of the second helix was not straightforward from hydropathy plots. Residues close to the IMS surface, N- and C-termini residues and cytoplasmic loops, are rather flexible as seen from higher B-factors and missing side chains (C2 loop) or residues (residues 1 and 294 to 297). Odd-numbered transmembrane helices, H1, H3, and H5, are sharply kinked (with kink angle ranging from 20° to 35°) and are longer than predicted. The kinks are induced by the prolines belonging to the MCF motif (Figure 2D). All the six transmembrane helices are tilted with respect to the membrane. As a result of the tilts and the kinks, these helices form a bundle that resembles a diaphragm in which helices interact tightly with each other at the matrix end while they form a circular arrangement toward IMS, the inner sides of the helices facing a hydrophilic cavity (Figure 2B). On the matrix side, the short helices h1-2, h3-4, and h5-6, located in the middle of each of the three repeats within the matrix loops M1, M2 and M3, encircle the extremity of the helix bundle. Toward IMS, the cavity opens as a wide cone that deeply enters the protein and ends by a narrow funnel closed at 10 Å from the matrix (Figure 3A).

The protein was purified and crystallized in the presence of CATR, known to bind strongly to AAC with a dissociation constant in the nanomolar range. A CATR molecule was clearly located within the cavity from $2F_{obs}-F_{calc}$ and $F_{obs}-F_{calc}$ electron density maps after the final refinements of the protein alone. The diterpene moiety with its two carboxylate groups is inserted toward the matrix while the glucose ring carrying two sulfates, an isovaleric group and a primary alcohol, points toward IMS. Most of the chemical groups of CATR interact strongly with AAC residues either through hydrogen bonds or van der Waals interactions, with the exception of the primary alcohol located on the sugar ring. These observations are in line with the previous studies in which various compounds derived from atractylosides were tested, only the modification of the primary alcohol had no effects on the inhibitory capacities.[15] Figure 4A highlights the cavity in which CATR is located.

Cardiolipins, which are abundant in the inner mitochondrial membrane, have been shown to be important for the function of AAC.[49] A stochoimetry of three cardiolipins per AAC monomer was also evidenced.[40] Cardiolipins are structurally unique lipids that carry four acyl chains and two negative charges. Indeed several lipid molecules were localized in the electron density maps. Two are clearly cardiolipins, as seen from the four characteristic acyl chains, a third although not complete is assumed to be a cardiolipin because it is related by the pseudo three-fold symmetry to the two others. Indeed, in a second crystal form, all three lipids are identified as cardiolipins.[50] In addition two phosphatidylcholine molecules as well as three LAPAO molecules were located. Cardiolipins are located on the inner leaflet of the inner mitochondrial membrane. The phosphatidyl groups and their glycerol linker interact with the short helices while the acyl chains interact with

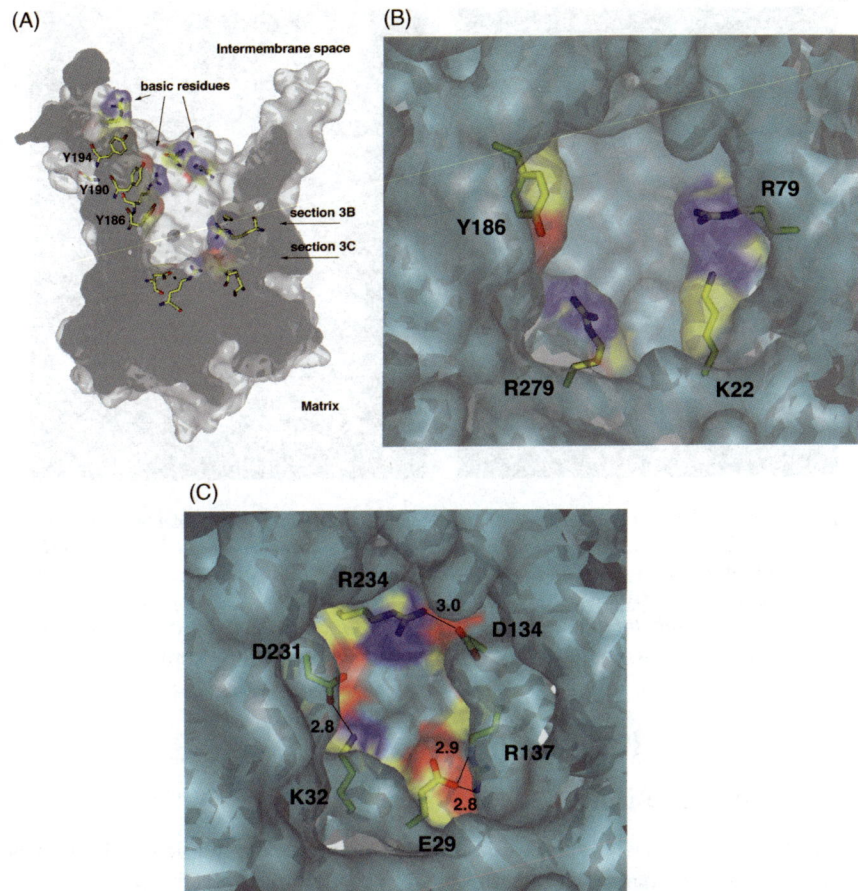

Figure 3 *Surface representation of the cavity. Highlighted residues are represented in ball-and-stick, their surfaces are colored in blue, red and yellow for nitrogen, oxygen and carbon, respectively. (A) Longitudinal section through the cavity. Tyrosines and basic residues present in the entrance of the cavity are shown. Sections along which Figure 3B and C are drawn, are indicated by arrows. (B) Selectivity filter. Y186 and basic residues K22, R79 and R279, located at about 6 Å from the bottom of the cavity restrict the access to the bottom. (C) Bottom of the cavity. Acidic and basic residues from the MCF motif close the cavity. Distances are given in Å. In B and C, the cavity is viewed from the intermembrane space*
From Pebay-Peyroula *et al.*, *Curr. Opin. Struct. Biol.*, 2004, **14**, 420–425.

the transmembrane helices in particular with aromatic residues: W70, Y173, F270, W274. Globally, lipids interact both through hydrogen bonds and polar interactions involving the phosphates or choline groups and also through van der Waals contacts between acyl chains and hydrophobic side chains (Val, Leu, Ile). Aromatic residues also interact through stacking interactions between the ring and the acyl chains.

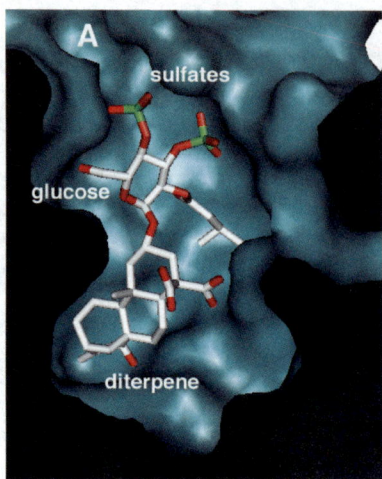 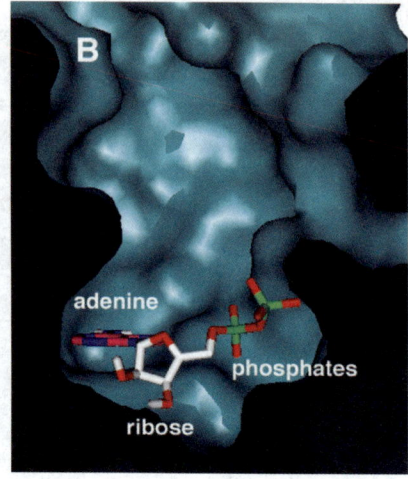

Figure 4 *Section through the carrier at the level of the cavity. (A) The CATR binding site determined from the crystallographic data. The diterpene moiety is facing the bottom of the cavity, while the two sulphate groups are oriented toward the intermembrane space. (B) A possible ADP site resulting from docking approaches. The figure highlights the differences in size between CATR and ADP. Most of the docking calculations located ADP at the very bottom of the cavity as shown here*

6   Functional Implications

6.1   Nucleotide Binding

AAC contains many basic residues that could potentially be implicated in the transport of negatively charged molecules such as ADP^{3-} or ATP^{4-}. Many of these residues were mutated in the yeast carrier in order to check their functional implication.[51,52] Most of the arginine or lysine residues affecting nucleotide binding and/or transport activity are located within the cavity and more specifically at its bottom, where CATR binds. Although CATR and ADP are different molecules (different chemical groups, different in size), it can be postulated from the location of the mutated residues that the nucleotide binds to the bottom of the cavity, which is not really a surprise. A careful analysis of the cavity highlights various interesting aspects. First, the acidic and basic residues belonging to the MCF motif cover the surface of the bottom of the cavity (Figure 3C) and form salt-bridges as previously described.[14] Second, the cavity presents a constriction at the near bottom, which is induced by four residues that are strictly conserved among ADP/ATP carriers and different for all other MCF carriers (except one of them, which is implicated in Grave's disease and whose function is still unknown). Three of the residues K22, R79, and R279 are basic and the fourth, Y186, is aromatic (Figure 3B). The size of the cavity at this level is restricted to less than 10 Å. The positive charges could attract the phosphates of ADP while the adenine moiety could be guided along the tyrosine. In addition, Y186 is aligned with two other tyrosines, Y190 and Y194, forming a ladder from the entrance of the cavity along which the adenine could glide to enter the cavity. Next to the tyrosine ladder, a basic patch might interact with the negatively charged phosphate groups. Docking ADP into the cavity gave several possible solutions at the very bottom of the cavity that could be classified in a few families. One of the docking results is shown in Figure 4B. However, the docking has to be considered carefully because of structural modifications that could have been induced by CATR. From these calculations, it is not possible to completely decide which is the exact binding site of ADP in the cavity. However, the different solutions indicate that ADP interacts with residues located at the very bottom of the cavity.

6.2 Conformational Changes

A striking feature of the AAC structure is the basket shape closed toward the matrix essentially through the odd-numbered helices that fold back toward the pseudo threefold axis of the protein (Figure 2B). Proline residues interrupt the hydrogen bond network that stabilizes α-helices. These residues are known to induce kinks in helices. In fact, proline residues adopt a broad range of psi angle values that are energetically equivalent.[53] Therefore local kinks in helices containing prolines can be driven by additional interactions occurring with surrounding residues. Indeed, prolines have been evidenced in helices of various protein structures with kink angles ranging from 0 to over 30°.[54] Modifying the interactions of the helix with surrounding residues can impose a different conformation on the helix and change the kink. Such a mechanism was suggested in bacteriorhodopsin, in which after light absorption conformational modifications induced by the retinal isomerization have to occur in order to expel a proton to the outside. During the photocycle, helix F, which is straight in the ground state, undergoes a structural modification from straight to kinked that might be induced by P186.[55] The bacteriorhodopsin example illustrates the fact that it is possible to modify a helix kink at the level of a proline by modulating the interactions of this helix with the neighboring residues. Therefore, proline residues contribute to the existence of different stable conformations, changing from one to the other necessitates to crossover an energy barrier. Modifying the tilt and/or the kink angles of the transmembrane helices will also induce conformational changes in the loops that are located on the matrix side. MCF carriers also have well-conserved glycines at the matrix extremity of even-numbered helices that could act as hinges. Indeed, glycines located at helix extremities in ion channels were described to be pivot elements that favor conformational changes.[56] Interestingly, there are major electrostatic interactions between the odd-numbered helices downstream from the kinks, which implicate the acidic and basic residues of the MCF motif. Figure 3C shows the three pairs of charged interactions formed by these residues. Because the MCF motif is generally conserved among MCF carriers, it can be postulated that the basket-shaped structure with the salt bridges at the bottom of the cavity is similar among all members of the family. The structure also suggests that in order to allow a nucleotide entering the cavity from the IMS to cross over to the matrix, a structural rearrangement of the helices has to be induced. The nucleotide itself could interfere with the salt bridges and modify the interactions between the 3 helices and possibly favor the conformational changes of these helices that would open the pathway toward the matrix. If this hypothesis is correct, it would mean that the mechanism is similar for all MCF carriers but that the access to the bottom of the cavity has to be selective for the transported molecule. Therefore, the idea of a possible selectivity filter formed by K22, R79, Y186, and R279 is attractive. The structure shows the necessity of large conformational changes on the matrix side for the transport to occur. Prolines of the MCF motif and conserved glycines are favorable residues for that as discussed above. In addition, amplitudes of large movements have to be controlled. This could be the role of the three cardiolipins located in the matrix-oriented leaflet.

6.3 Transport Regulation

All the transport experiments demonstrated the selectivity of AAC toward ADP or ATP. Any other nucleotides, such as GDP, GTP, or even AMP or deoxy-nucleotides are not transported by AAC. Transported nucleotides are not complexed to magnesium ions. In reconstituted liposomes, AAC is able to transport ADP or ATP against one or the other of these two nucleotides (ADP against ATP, ADP against itself, etc…). In the cell, the electrostatic potential due to the proton gradient produced by the respiratory chain drives the transport of ADP^{3-} into the matrix and ATP^{4-} toward the IMS. It was estimated that one-third of the energy gained by the respiratory chain is utilized for the nucleotide transport while two-third are needed for ATP synthesis.[16] Several biochemical data indicated that the transport unit should be a dimer transporting at the same time one ADP against one ATP.[57,58] It was even suggested that both nucleotides could cross through a single pathway formed by two monomers. The structure reveals that a single monomer forms a possible pathway at least for ADP coming from IMS. This structure was obtained in the presence of CATR, known to interact with AAC from the IMS possibly with the conformation that recognizes ADP. Combining all the biochemical and biophysical data with the structure leads to the hypothesis that two monomers forming a dimer could adopt opposite conformations. While one monomer is able to recognize ADP from IMS, the second binds ATP from the matrix. As described in the previous paragraph, ADP binding could favor a conformational change that could trigger the transport. The situation could be similar for ATP. Conformational changes in both monomers could produce a coherent effect that allows the translocation of both nucleotides. However, many experimental details are still missing. In particular, the structure of the dimer is one of the key elements. Two-dimensional crystals of yeast AAC showed a possible dimer.[59,60] However, the resolution of the projection maps is still low and has to be improved to shed more light on the protein-protein interactions present in this dimer.

7 Future Developments and Conclusions

There are still many open questions to understand the transport mechanism. The answers will come from the combination of several approaches. X-ray crystallography on three-dimensional crystals of AAC in complex with BA or nucleotide analogs will provide insights into other conformations. Electron diffraction on two-dimensional crystals might provide a better view of AAC in its natural environment, the lipid bilayer. Atomic force microscopy might provide interesting aspects of conformational changes. In general, the combination of various approaches is very promising in the near future and will help to bridge the gap between the molecular level and the situation in the cell where proteins are involved in larger complexes (protein complexes, protein in membranes, proteins with nucleic acids, etc.). The roles of all the residues assumed to be functionally important for nucleotide transport have to be tested. For these approaches the carrier must be overexpressed, purified, and incorporated in liposomes in order to measure transport activities of native and mutant carriers. At the present stage, the overexpression of human or bovine ADP/ATP carrier is still difficult. It would also be interesting to solve the structures of other MCF carriers and highlight in the structures the conserved and non-conserved residues, in order to identify which residues are important for the structure and which are necessary for the function.

Acknowledgments

The complementarity of three laboratories, *Laboratoire de Biochimie et de Biophysique des Systèmes Intégrés (CEA-CNRS-University)* in Grenoble and *Laboratoire de Physiologie Moléculaire et Cellulaire (CNRS-University)* in Bordeaux made it possible to solve the structure. All the coauthors of the paper[9] describing the structure of AAC are acknowledged for their contribution, in particular G. Brandolin and G. Lauquin. I thank also H. Nury for his present contribution aiming to determine the structure of another conformation, and P. Amara for her contribution on the docking of ADP to the cavity.

References

1. S.H. White, *Protein Sci.*, 2004, **13**, 1948.
2. E. Wallin and G. von Heijne, *Protein Sci.*, 1998, **7**, 1029.
3. C.G. Tate, *FEBS Lett.*, 2001, **504**, 94.
4. F. Reiss-Husson and D. Picot, in *Crystallization of Membrane Proteins*, A. Ducruix and R. Giégé (eds), Oxford University Press, Oxford, 1999.
5. C. Hunte and H. Michel, in *Membrane Protein Crystallization*, C. Hunte, H. Schaegger and G. von Jagow (eds), Elsevier, San Diego, CA, USA, 2003.
6. C. Toyoshima, H. Nomura and T. Tsuda, *Nature*, 2004, **432**, 361.
7. C. Olesen, T.L. Sorensen, R.C. Nielsen, J.V. Moller and P. Nissen, *Science*, 2004, **306**, 2251.
8. J. Abramson, I. Smirnova, V. Kasho, G. Verner, H.R. Kaback and S. Iwata, *Science*, 2003, **301**, 610.
9. E. Pebay-Peyroula, C. Dahout-Gonzalez, R. Kahn, V. Trezeguet, G.J. Lauquin and G. Brandolin, *Nature*, 2003, **426**, 39.
10. F. Palmieri, *Pflugers Arch.*, 2004, **447**, 689.
11. J.E. Walker and M.J. Runswick, *J. Bioenerg. Biomembr.*, 1993, **25**, 435.
12. M. Saraste and J.E. Walker, *FEBS Lett.*, 1982, **144**, 250.
13. J.E. Walker, *Curr. Opin. Struct. Biol.*, 1992, **2**, 519.
14. D.R. Nelson, C.M. Felix and J.M. Swanson, *J. Mol. Biol.*, 1998, **277**, 285.

15. P.V. Vignais, *Biochim. Biophys. Acta*, 1976, **456**, 1.
16. P.V. Vignais, M.R. Block, F. Boulay, G. Brandolin and G.J.-M. Lauquin, in *Molecular Aspects of Structure-Function Relationships in Mitochondrial Adenine Nucleotide Carrier*, CRC Press, Boca Raton, FL, 1985.
17. M. Klingenberg, *Arch. Biochem. Biophys.*, 1989, **270**, 1.
18. T. Gropp, N. Brustovetsky, M. Klingenberg, V. Muller, K. Fendler and E. Bamberg, *Biophys. J.*, 1999, **77**, 714.
19. M.R. Block, G. Zaccai, G.J. Lauquin and P.V. Vignais, *Biochem. Biophys. Res. Commun.*, 1982, **109**, 471.
20. H. Hackenberg and M. Klingenberg, *Biochemistry*, 1980, **19**, 548.
21. M. Hashimoto, E. Majima, S. Goto, Y. Shinohara and H. Terada, *Biochemistry*, 1999, **38**, 1050.
22. C. Fiore, V. Trezeguet, A. Le Saux, P. Roux, C. Schwimmer, A.C. Dianoux, F. Noel, G.J. Lauquin, G. Brandolin and P.V. Vignais, *Biochimie*, 1998, **80**, 137.
23. M. Roth, A. Lewit-Bentley, H. Michel, J. Deisenhofer, R. Huber and D. Oesterhelt, *Nature*, 1989, **340**, 659.
24. P.A. Timmins, E. Pebay-Peyroula and W. Welte, *Biophys. Chem.*, 1994, **53**, 27.
25. E. Pebay-Peyroula, R.M. Garavito, J.P. Rosenbusch, M. Zulauf and P.A. Timmins, *Structure*, 1995, **3**, 1051.
26. S. Penel, E. Pebay-Peyroula, J. Rosenbusch, G. Rummel, T. Schirmer and P.A. Timmins, *Biochimie*, 1998, **80**, 543.
27. R.M. Garavito and S. Ferguson-Miller, *J. Biol. Chem.*, 2001, **276**, 32403.
28. J.U. Bowie, *Curr. Opin. Struct. Biol.*, 2001, **11**, 397.
29. E. Pebay-Peyroula and J.P. Rosenbusch, *Curr. Opin. Struct. Biol.*, 2001, **11**, 427.
30. J.L. Popot, E.A. Berry, D. Charvolin, C. Creuzenet, C. Ebel, D.M. Engelman, M. Flotenmeyer, F. Giusti, Y. Gohon, Q. Hong, J.H. Lakey, K. Leonard, H.A. Shuman, P. Timmins, D.E. Warschawski, F. Zito, M. Zoonens, B. Pucci and C. Tribet, *Cell Mol. Life Sci.*, 2003, **60**, 1559.
31. E.M. Landau and J.P. Rosenbusch, *Proc. Natl. Acad. Sci. USA*, 1996, **93**, 14532.
32. P. Nollert, H. Qiu, M. Caffrey, J.P. Rosenbusch and E.M. Landau, *FEBS Lett.*, 2001, **504**, 179.
33. E. Pebay-Peyroula, G. Rummel, J.P. Rosenbusch and E.M. Landau, *Science*, 1997, **277**, 1676.
34. M.L. Chiu, P. Nollert, M.C. Loewen, H. Belrhali, E. Pebay-Peyroula, J.P. Rosenbusch and E.M. Landau, *Acta Crystallogr. D Biol. Crystallogr.*, 2000, **56** (Pt 6), 781.
35. A. Royant, P. Nollert, K. Edman, R. Neutze, E.M. Landau, E. Pebay-Peyroula and J. Navarro, *Proc. Natl. Acad. Sci. USA*, 2001, **98**, 10131.
36. G. Katona, U. Andreasson, E.M. Landau, L.E. Andreasson and R. Neutze, *J. Mol. Biol.*, 2003, **331**, 681.
37. G. Brandolin, J. Doussiere, A. Gulik, T. Gulik-Krzywicki, G.J. Lauquin and P.V. Vignais, *Biochim. Biophys. Acta*, 1980, **592**, 592.
38. G. Brandolin, Y. Dupont and P.V. Vignais, *Biochemistry*, 1985, **24**, 1991.
39. R. Kramer, H. Aquila and M. Klingenberg, *Biochemistry*, 1977, **16**, 4949.
40. K. Beyer and M. Klingenberg, *Biochemistry*, 1985, **24**, 3821.

41. C. Dahout-Gonzalez, G. Brandolin and E. Pebay-Peyroula, *Acta Crystallogr. D Biol. Crystallogr.*, 2003, **59**, 2353.

42. J.J. Lacapere, D.L. Stokes, A. Olofsson and J.L. Rigaud, *Biophys. J.*, 1998, **75**, 1319.

43. C.J. daCosta and J.E. Baenziger, *Acta Crystallogr. D Biol. Crystallogr.*, 2003, **59**, 77.

44. Z. Otwinowski and W. Minor, *Methods Enzymol.*, 1997, **276**, 307.

45. Collaborative Computational Project, number 4, *Acta Crystallogr.*, 1994, **D50**, 760.

46. A.T. Brunger, P.D. Adams, G.M. Clore, W.L. DeLano, P. Gros, R.W. Grosse-Kunstleve, J.S. Jiang, J. Kuszewski, M. Nilges, N.S. Pannu, R.J. Read, L.M. Rice, T. Simonson and G.L. Warren, *Acta Crystallogr. D Biol. Crystallogr.*, 1998, **54**(Pt 5), 905.

47. T.A. Jones, J.Y. Zou, S.W. Cowan and Kjeldgaard, *Acta Crystallogr. A*, 1991, **47**(Pt 2), 110.

48. G. Brandolin, F. Boulay, P. Dalbon and P.V. Vignais, *Biochemistry*, 1989, **28**, 1093.

49. F. Jiang, M.T. Ryan, M. Schlame, M. Zhao, Z. Gu, M. Klingenberg, N. Pfanner and M.L. Greenberg, *J. Biol. Chem.*, 2000, **275**, 22387.

50. H. Nury, C. Dahout-Gonzalez, V. Trézéguet, G. Lauquin, G. Brandolin and E. Pebay-Peyroula, *FEBS Lett.*, **759**, 6031.

51. D.R. Nelson, J.E. Lawson, M. Klingenberg and M.G. Douglas, *J. Mol. Biol.*, 1993, **230**, 1159.

52. V. Muller, D. Heidkamper, D.R. Nelson and M. Klingenberg, *Biochemistry*, 1997, **36**, 16008.

53. K.A. Williams and C.M. Deber, *Biochemistry*, 1991, **30**, 8919.

54. F.S. Cordes, J.N. Bright and M.S. Sansom, *J. Mol. Biol.*, 2002, **323**, 951.

55. S. Subramaniam and R. Henderson, *Nature*, 2000, **406**, 653.

56. I.R. Booth, M.D. Edwards and S. Miller, *Biochemistry*, 2003, **42**, 10045.

57. C. Duyckaerts, C.M. Sluse-Goffart, J.P. Fux, F.E. Sluse and C. Liebecq, *Eur. J. Biochem.*, 1980, **106**, 1.

58. V. Trézéguet, A. Le Saux, C. David, C. Gourdet, C. Fiore, A. Dianoux, G. Brandolin and G.J. Lauquin, *Biochim. Biophys. Acta*, 2000, **1457**, 81.

59. E.R. Kunji and M. Harding, *J. Biol. Chem.*, 2003, **278**, 36985.

60. E.R. Kunji, *FEBS Lett.*, 2004, **564**, 239.

Subject Index